POLYMER DEVOLATILIZATION

PLASTICS ENGINEERING

Founding Editor

Donald E. Hudgin

Professor
Clemson University
Clemson, South Carolina

1. Plastics Waste: Recovery of Economic Value, *Jacob Leidner*
2. Polyester Molding Compounds, *Robert Burns*
3. Carbon Black-Polymer Composites: The Physics of Electrically Conducting Composites, *edited by Enid Keil Sichel*
4. The Strength and Stiffness of Polymers, *edited by Anagnostis E. Zachariades and Roger S. Porter*
5. Selecting Thermoplastics for Engineering Applications, *Charles P. MacDermott*
6. Engineering with Rigid PVC: Processability and Applications, *edited by I. Luis Gomez*
7. Computer-Aided Design of Polymers and Composites, *D. H. Kaelble*
8. Engineering Thermoplastics: Properties and Applications, *edited by James M. Margolis*
9. Structural Foam: A Purchasing and Design Guide, *Bruce C. Wendle*
10. Plastics in Architecture: A Guide to Acrylic and Polycarbonate, *Ralph Montella*
11. Metal-Filled Polymers: Properties and Applications, *edited by Swapan K. Bhattacharya*
12. Plastics Technology Handbook, *Manas Chanda and Salil K. Roy*
13. Reaction Injection Molding Machinery and Processes, *F. Melvin Sweeney*
14. Practical Thermoforming: Principles and Applications, *John Florian*
15. Injection and Compression Molding Fundamentals, *edited by Avraam I. Isayev*
16. Polymer Mixing and Extrusion Technology, *Nicholas P. Cheremisinoff*
17. High Modulus Polymers: Approaches to Design and Development, *edited by Anagnostis E. Zachariades and Roger S. Porter*
18. Corrosion-Resistant Plastic Composites in Chemical Plant Design, *John H. Mallinson*

Additional Volumes in Preparation

POLYMER DEVOLATILIZATION

EDITED BY

RAMON J. ALBALAK

Massachusetts Institute of Technology
Cambridge, Massachusetts

Marcel Dekker, Inc.   New York•Basel•Hong Kong

Library of Congress Cataloging-in-Publication Data

Polymer devolatilization / edited by Ramon J. Albalak.
    p. cm. — (Plastics engineering ; 33)
    Includes bibliographical references and index.
    ISBN 0-8247-9627-6 (alk. paper)
    1. Polymers. 2. Volatile organic compounds. I. Albalak, Ramon
J. II. Series: Plastics engineering (Marcel Dekker,
Inc.) ; 33.
    TP1092.P64 1996
    668.9—dc20   95-52700
                                                                CIP

The publisher offers discounts on this book when ordered in bulk quantities. For more information, write to Special Sales/Professional Marketing at the address below.

This book is printed on acid-free paper.

Marcel Dekker, Inc.
270 Madison Avenue, New York, New York 10016

Current printing (last digit):
10 9 8 7 6 5 4 3 2 1

PRINTED IN THE UNITED STATES OF AMERICA

Preface

Devolatilization is an industrial process in which low-molecular-weight components such as unreacted monomer, solvents, water, and various polymerization by-products are separated from a polymeric system. These substances, which are often collectively referred to as "volatiles," may be removed to comply with various regulations, to improve the polymer's properties, or for a variety of other reasons. Devolatilization of a polymer is a complex process involving the transport of volatiles to a polymer–vapor interface, the evaporation of the volatiles at the interface, and their subsequent removal by a vacuum system. In addition to simple diffusion of the volatiles to the polymer–vapor interface, devolatilization progresses in many cases through a foaming mechanism, in which bubbles containing vapor of the volatiles to be removed are formed within a polymer melt. To effectively reduce the concentration of volatile contaminants, a wide variety of devolatilizing equipment is used, which may be broadly classifed into nonrotating and rotating devolatilizers.

To date only two books have been dedicated to the subject of polymer devolatilization,† the more recent of which was published in 1983. The need to devolatilize cost-efficiently, together with the increasing number of restrictions on the acceptable volatile content of polymers, has led to the

---

† *Devolatilization of Plastics* (translated from German), VDI-Verlag, Düsseldorf, 1980, and *Devolatilization of Polymers*, J. A. Biesenberger, ed., Hanser, Munich, and Macmillan, New York, 1983.

growing attention that devolatilization has received over the past decade. This attention has resulted in numerous studies conducted by both the manufacturers and end users of devolatilizing equipment and also by several research groups at academic institutes. This ongoing research has contributed to a deeper understanding of polymer devolatilization—on both the fundamental and applied levels—as presented in great detail in this book.

This volume contains 19 chapters. The first four chapters following the introduction provide the background necessary to understand devolatilization. These chapters focus on the thermodynamics of concentrated polymer solutions, solvent diffusion in polymers, and bubble nucleation and growth. Chapters 6 and 7 report on two extensive studies that probe the actual mechanisms by which devolatilization of polymer melts progresses. The next part of the book addresses devolatilization in various geometries and types of equipment: after a general overview of devolatilizers, specific chapters are devoted to the use and analysis of falling-strand, slit, single-screw, and co-rotating and counter-rotating twin-screw devolatilization. The next section (chapters 14–17) demonstrates industrial applications of devolatilization for a variety of polymers and equipment, and contains several worked examples. This section is followed by a chapter that discusses the future of solvents in the polymer industry in view of increasing regulations. The final chapter in the book addresses the analytical methods by which the concentrations of residual monomers and other volatiles are determined.

The first two of the three appendixes feature data on the vapor pressures of pure solvents and on polymer–solvent interaction parameters, which are needed for various calculations relating to devolatilization. The third appendix presents the abstracts of some 60 papers published on polymer devolatilization over the past decade.

This book targets researchers from both industry and universities who will benefit from a comprehensive, up-to-date report on polymer devolatilization, and designers and end users of devolatilizers who, it is hoped, will find guidance from some of the leading figures in this field.

In addition to thanking the authors who have participated in writing this book, I would like to acknowledge the following people for assisting in various ways in the preparation of this volume: Ted Allen, Paul Andersen, Yachin Cohen, Anca Dagan, Tom Daubert, Philip DeLassus, Garry Leal, Ken Powell, Valeri Privalko, Chris Rauwendaal, Tadamoto Sakai, Judith Schmidt, Bertha Shdemati, and Nam Suh.

My special thanks are extended to Professors Zehev Tadmor and Ishi Talmon, to whom I am indebted for introducing me to the study of polymer devolatilization.

Ramon J. Albalak

Contents

Contributors

Ramon J. Albalak, D.Sc. Department of Chemical Engineering, Massachusetts Institute of Technology, Cambridge, Massachusetts

Colin Anolick, Ph.D. Research Fellow, Department of Central Research and Development, E. I. du Pont de Nemours & Co., Wilmington, Delaware

Gianni Astarita, Ph.D. Professor, Department of Materials and Production Engineering, University of Naples—Federico II, Naples, Italy

Eric J. Beckman, Ph.D. Associate Professor, Department of Chemical and Petroleum Engineering, University of Pittsburgh, Pittsburgh, Pennsylvania

Faivus Brauer Consulting Process Engineer, Department of Process Technology, Werner & Pfleiderer Corporation, Ramsey, New Jersey

Leo F. Carter, Ph.D. Professor, Department of Chemical Engineering, University of Puerto Rico, Mayagüez, Puerto Rico

Timothy J. Cavanaugh, M.S. The Isermann Department of Chemical Engineering, Rensselaer Polytechnic Institute, Troy, New York

Tali Chechik, M.Sc.* Technion—Israel Institute of Technology, Haifa, Israel

---

Current affiliation: Reshet-O-Plast Hahotrim, Kibbutz Hahotrim, Israel.

Thomas R. Crompton, M.Sc.*   Head of Analytical Research Department, Shell Research Ltd., Carrington, Cheshire, England

John Curry, M.E. (Che)   Manager, Process Development, Department of Process Technology, Werner & Pfleiderer Corporation, Ramsey, New Jersey

J. L. Duda, Ph.D.   Professor and Head, Department of Chemical Engineering, The Pennsylvania State University, University Park, Pennsylvania

Moshe Favelukis, D.Sc.   Department of Chemical Engineering, Technion— Israel Institute of Technology, Haifa, Israel

Eric A. Grulke, Ph.D.   Professor and Chair, Department of Chemical and Materials Engineering, University of Kentucky, Lexington, Kentucky

Gary S. Huvard, Ph.D.   Huvard Research and Consulting, Chesterfield, Virginia

Giovanni Ianniruberto, Ph.D.   Department of Chemical Engineering, University of Naples—Federico II, Naples, Italy

Helmut M. Joseph, B.Sc.   Senior Research Chemist, Israel Plastics and Rubber Center, Technion City, Haifa, Israel

Shau-Tarng Lee, Ph.D.   Development Engineer, Sealed Air Corporation, Saddle Brook, New Jersey

J. Thomas Lindt, Ph.D.   William Kepler Whiteford Professor, Department of Materials Science and Engineering, University of Pittsburgh, Pittsburgh, Pennsylvania

Martin H. Mack, M. S.   Vice President, Research and Development, Berstorff Corporation, Charlotte, North Carolina

Pier Luca Maffettone, Ph.D.   Department of Chemical Engineering, University of Naples—Federico II, Naples, Italy

Pradip S. Mehta, Ph.D.   Research Associate, Process Development, Hoechst Celanese Corporation, Bishop, Texas

Edward W. Merrill, D.Sc.   C. P. Dubbs Professor of Chemical Engineering, Department of Chemical Engineering, Massachusetts Institute of Technology, Cambridge, Massachusetts

E. Bruce Nauman, Ph.D.   Professor of Chemical Engineering, Isermann Department of Chemical Engineering, Rensselaer Polytechnic Institute, Troy, New York

---

* Retired.

Russell J. Nichols, P.E.   Vice President, Quality Performance, Farrel Corporation, Ansonia, Connecticut

Armin Pfeiffer, M. S.   Berstorff Corporation, Charlotte, North Carolina

Robert H. M. Simon, D. Eng.***   Fellow, Monsanto Chemical Company, Springfield, Massachusetts

Zehev Tadmor, Ph.D.   Distinguished Professor of Chemical Engineering, and President, Technion—Israel Institute of Technology, Haifa, Israel

Yeshayahu Talmon, Ph.D.   Professor, Department of Chemical Engineering, Technion—Israel Institute of Technology, Haifa, Israel

Alexander Tukachinsky, D.Sc.†   Technion—Israel Institute of Technology, Haifa, Israel

Ali V. Yadzi, Ph.D.   University of Pittsburgh, Pittsburgh, Pennsylvania

John M. Zielinski, Ph.D.   Research Engineer, Air Products & Chemicals, Inc., Allentown, Pennsylvania

---

* Retired.

† *Current affiliation*: Institute of Polymer Engineering, University of Akron, Akron, Ohio.

1

An Introduction to Devolatilization

Ramon J. Albalak

Massachusetts Institute of Technology, Cambridge, Massachusetts

I. INTRODUCTION

Most polymers leaving the reactor contain some low-molecular-weight components such as unreacted monomer, solvents, water, and various reaction by-products. These substances are often collectively referred to as *volatiles* and their presence in the polymer is usually undesired. The concentrations at which these volatiles are present may be as low as several ppm or as high as several tens of percent. Separating them from the bulk polymer may be performed for several reasons, such as

To improve the properties of the polymer
To recover monomer/solvent
To fulfill health and environmental regulations
To eliminate odors
To increase the extent of polymerization

The procedure by which volatiles are separated from the bulk polymer is called *devolatilization* and is usually performed with the polymer above its glass transition temperature, or above the melting temperature for crystalline polymers. This process has been recognized as a unit operation of polymer processing (Tadmor and Gogos, 1979), and it is carried out in industry in a large variety of equipment, covered in great detail in other chapters. Biesenberger and Sebastian (1983) have classified this equipment into two main categories: rotating devolatilizers (such as vented extruders) and still, or nonrotating, devolatilizers (such as falling-strand devolatilizers).

The need to devolatilize on a cost-efficient basis together with the increasing number of restrictions on the acceptable volatile content of polymers has led to the growing attention that devolatilization has received over the past decade. This attention has resulted in numerous studies conducted by both the manufacturers and end users of devolatilizing equipment and also by several research groups at academic institutes. The abstracts of many of these studies are presented in Appendix C.

II. THEORETICAL BACKGROUND

Devolatilization of a polymer is a complex process generally involving the transport of volatiles to a polymer–vapor interface, the evaporation of the volatiles at the interface, and their subsequent removal by a vacuum system. In addition to simple diffusion of the volatiles to the polymer–vapor interface, devolatilization progresses in many cases through a foaming mechanism, in which bubbles containing the volatiles to be removed are formed within the polymer melt. These bubbles may then grow, coalesce, and finally rupture at the polymer–vapor interface, where they release their volatile contents to the vapor phase.

The progress of the devolatilization process depends both on the thermodynamic potential for separation and on the means by which that potential may be realized. Among other factors that determine the extent and rate of devolatilization are the thermodynamics of the polymer–volatile system, the nature of the diffusion of the volatile through the polymer, and the nucleation and growth of vapor bubbles in the polymer melt. The importance of these issues to devolatilization will be presented here briefly, since they are dealt with in depth in the next four chapters.

One of the basic parameters of interest in any separation process is the maximum degree of separation that may be obtained for a given system. The equilibrium weight fraction, W_e, for a given polymer–volatile system may be related to the partial pressure of the volatile in the vapor phase, P_1, by Henry's law:

$$W_e = \frac{P_1}{K_w} \tag{1}$$

The Henry's law constant, K_w, depends on the temperature, the pressure, and the volatile in question.

The behavior of polymer–volatile solutions can be described using the Flory–Huggins theory, discussed in Chapter 2. For high-molecular-weight polymers containing low concentrations of volatile material, the following simplified relation may be written:

$$\ln \frac{P_1}{P_1^0} = \ln \phi_1 + 1 + \chi \tag{2}$$

where P_1^0 is the vapor pressure of the pure volatile, ϕ_1 is the volume fraction of the volatile, and χ is the Flory–Huggins interaction parameter. Correlations and graphs of P_1^0 as a function of temperature for 50 solvents and monomers of interest are given in Appendix A. Values of χ for several polymer–volatile systems are given in Appendix B.

Biesenberger and Sebastian (1983) used an approximate relation between volume and weight fractions of the volatile material to evaluate the Henry's law constant as

$$K_w = P_1^0 \frac{\rho_2}{\rho_1} \exp(1 + \chi) \tag{3}$$

in which ρ_1 and ρ_2 are the densities of the volatile and polymer, respectively.

The foregoing equations enable one to calculate the weight fraction of the volatile at equilibrium. However, in real devolatilization processes lasting a finite period of time, equilibrium is never actually reached, and the final volatile concentration obtained will be greater than W_e.

The time it takes for a certain degree of separation to be achieved in a given polymer–volatile system strongly depends on the rate at which the volatile is able to migrate through the polymer. This is true for both the migration of volatiles directly to the surface, where they are removed by the vacuum system, and for the migration of volatiles to bubble nucleation sites and to vapor bubbles that grow within the polymer melt. The diffusion in concentrated polymer solutions is a very slow process, with typical diffusion coefficients in the range 10^{-8}–10^{-12} m^2/sec (several orders of magnitude smaller than the diffusion coefficients in low-molecular-weight liquids). The diffusion coefficient depends strongly on the temperature of the system and on the concentration of the volatile component. This dependence is especially strong in the vicinity of the glass transition temperature of the polymer, and it may be predicted according to the work of Duda, Vrentas, and their coworkers (Duda et al., 1978, 1982; Vrentas and Duda, 1979) based on a free-volume model (see Chapter 3). The increase in free volume at elevated temperatures results in an increase in diffusivity and a decrease in melt viscosity—both beneficial for devolatilization. An additional reason for devolatilizing at high temperatures is the increase in the vapor pressure of the volatile component, P_1^0.

The strong decrease in diffusivity as the concentration of the volatile component approaches zero may be partially overcome by the addition of an inert substance (usually water) that reduces the weight fraction of the polymer in the system (Ravindranath and Mashelkar, 1988). The efficiency of small amounts of water as a stripping agent in polymer devolatilization has been demonstrated by Werner (1980) and more recently by Mack and Pfeiffer (1993). Other advantages of adding an inert substance were noted by Biesenberger and Sebastian (1983): (1) The addition of an inert substance to the system reduces the partial pressure of the volatile component, P_1, and thus (Eq. 1) reduces its weight fraction at equilibrium; (2) The combined vapor pressures of both the inert substance and the volatile component reduce the temperature and volatile concentration needed to bring about boiling of the polymer solution; (3) Boiling of the inert substance creates bubbles that increase the area available for mass transfer from the polymer to the vapor phase.

Bubbles do not necessarily originate from the presence of an additional inert component. They may be generated by boiling of the volatile alone under superheated conditions at which the partial pressure of the volatile is greater than the surrounding pressure. In general, bubble nucleation may be either homogeneous or heterogeneous. Blander and Katz (1975) have presented expressions for the rate of homogeneous nucleation that occurs within the bulk of a liquid, and for the rate of heterogeneous nucleation that takes place on a surface in contract with the liquid. The general form of these expressions is

$$J = A \exp B \tag{4}$$

in which J is the nucleation rate, and A and B are factors that incorporate system parameters such as temperature, surface tension, and the degree of superheat.

Tadmor (1985) applied Eq. (4) to a polystyrene–styrene system and showed that substantial homogeneous nucleation may occur only at temperatures much higher than those at which devolatilization is performed.

Similar calculations were presented by Lee and Biesenberger (1989) who, based on the work of Jemison et al. (1980), also rejected the theory that heterogeneous nucleation on solid surfaces in contact with the polymer melt may account for the high formation rate of bubbles during devolatilization. Tadmor (1985) also argued that is unlikely that heterogeneous bubble nucleation at the surfaces of devolatilizing equipment plays an important role, since bubbles formed at these surfaces are stagnant and not likely to be swept into the bulk of the viscous melt. Biesenberger and Lee (1986a, b) suggested that significant heterogeneous bubble nucleation may take place during devolatilization based on a model originally proposed for bubble formation in animals (Harvey et al., 1944a, b). According to their theory, stable nucleation sites exist in microcrevices on entrained particles, such as dust, that are dispersed throughout the polymer melt. Chapter 4 presents a detailed account of bubble nucleation in polymers and further discussion of the theory of stable nucleation sites is presented in Chapter 6. Han and Han (1990a, b) have studied bubble nucleation in polymeric liquids both theoretically and experimentally, and they have found that for polystyrene–toluene solutions with 40–60 wt % polymer, the initial size of the bubbles is in the range 0.2–0.4 μm.

The growth of bubbles during devolatilization depends mainly on the rate at which the volatile material can diffuse from the bulk of the polymer to the bubble surface and on the resistance of the viscous polymer melt to displacement by the growing bubble. At the low concentrations of volatiles at which devolatilization is usually performed, it can be shown (Powell and Denson, 1983; Tukachinsky et al., 1994) that heat transfer is not a controlling factor. Many studies have appeared in the literature on bubble growth, and some fundamental approaches to this topic are given in Chapter 5. In general, it is found that the effects of viscosity are predominant at the early stages of bubble growth in which the bubble radius increases exponentially with time. However, the availability of the volatile component at the bubble surface may soon become the limiting factor of bubble growth. In this case, the process becomes governed by the rate of diffusion, and the bubble radius increases proportionally to the square root of time.

III. MECHANISMS AND MODELS FOR DEVOLATILIZATION

The understanding of the mechanisms that govern devolatilization and the mathematical models developed to describe these mechanisms have evolved dramatically over the past three decades. Latinen (1962) was the first to study devolatilization quantitatively and addressed the removal of small amounts of styrene (<1 wt %) from polystyrene in a vented single-screw extruder. He assumed that devolatilization occurred only due to molecular

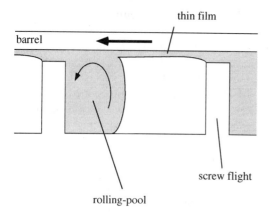

thin film

barrel

screw flight

rolling-pool

Fig. 1   Schematic representation of the rolling pool and thin film formed in a screw extruder.

diffusion of the volatile to the polymer–vapor interface. Latinen took into account diffusion from both the bulk of the polymer film deposited on the extruder barrel and from the bulk of the rolling pool of melt formed at the pushing flight (Fig. 1). Accurate values of the diffusion coefficient of styrene in polystyrene were unavailable at the time, and these were back-calculated from the model based on experimental data. The values thus obtained were at least two orders of magnitude larger than those known today, and even at the time they appeared to Latinen to be too high. He attempted to explain these high values as resulting from the roughness of the polymer surface, which was not taken into account in his model. The possibility of the formation of bubbles within the melt was also suggested, but this explanation was immediately dismissed by Latinen, who claimed that bubble formation was unlikely at the low volatile concentrations he studied.

Coughlin and Canevari (1969) presented two models to describe devolatilization during screw extrusion. The first model assumed that the process was limited by volatile diffusion within the polymer melt, and the second assumed mass transfer resistance at the polymer–vapor interface to be the limiting factor. The diffusion coefficients calculated from the first model based on experiments conducted with polypropylene–xylene and polypropylene–methanol were again orders of magnitude larger than those expected. The possibility of bubble formation was once more suggested, but no evidence of bubbles was detected. Roberts (1970) presented a model for devolatilization in a vented extruder that accounted for surface renewal by both the circular flow of the melt pool and by the continuous formation of a thin film due to flow over the flights of the screw. He found partial support for his model by using the data of Coughlin and Canevari (1969). Roberts discussed the relative importance of volatile removal from these surfaces and concluded that devolatilization from the thin film may account for more than 50% of the total. The diffusion coefficients calculated from this model were also higher than expected.

The reason for the discrepancy between the actual diffusion coefficients and those calculated from models that assume only a mechanism of molecular diffusion became apparent following a study by Biesenberger and Kessidis (1982). In this study, the efficiency of devolatilization in a single-screw extruder at reduced pressure was found to be higher than that of the same process conducted at atmospheric pressure under a nitrogen sweep. From these results, Biesenberger and Kessidis concluded that bubble transport does in fact play an important role in the overall mechanism of devolatilization, and that devolatilization conducted under vacuum is not a simple process of molecular diffusion. Devolatilization under a sweep of nitrogen is, however, a bubble-free process in which mass transfer occurs solely by molecular diffusion. Collins et al. (1985) studied bubble-free devolatilization in a twin-screw extruder under a sweep of nitrogen and derived a model parallel to that developed by Latinen (1962) for a single-screw extruder, which model predicted mass transfer rates in good agreement with experimental data. A diffusional model was presented by Valsamis and Canedo (1989) for devolatilization conducted in a continuous mixer.

Mehta et al. (1984) reported visual observations of bubble formation during devolatilization in a multichannel co-rotating disk processor and postulated that foam devolatilization may be the dominant mass transfer mechanism even at low volatile concentrations. Several models have appeared in the literature to describe bubble transport devolatilization in general and in specific types of devolatilizing equipment. Newman and Simon (1980) modeled the process of falling-strand devolatilization as one of molecular diffusion into a swarm of bubbles, and the expansion of those bubbles against surface tension and viscous forces. Powell and Denson (1983) developed a model for the devolatilization of polymeric solutions containing entrained bubbles in close proximity to each other, and they took into consideration the concentration dependence of the diffusion coefficient. Lindt and Foster (1989; Foster and Lindt 1989, 1990) have presented several studies on foam devolatilization in twin-screw extruders, including a model that describes the transition from a bubbling process to a bubble-free one as the polymer is transported along the extruder and the volatile concentration decreases significantly. Additional models have been presented by others, including Yoo and Han (1984), Chella and Lindt (1986), and Biesenberger (1987).

A thorough study of foam-enhanced devolatilization in rotating equipment was conducted by Biesenberger and Lee (1986a, b, 1987; Lee and

Biesenberger, 1989) in which theories and mathematical models were presented to explain experimental observations (see Chapter 6). Further insight into the actual mechanism of polymer melt devolatilization was presented in a series of studies by Albalak et al. (1987, 1990, 1992) and Tukachinsky et al. (1993, 1994). A systematic study of the morphology created during both falling-strand devolatilization and vented extruder devolatilization led to the discovery that vapor bubbles grow not only due to the diffusion of volatiles from the surrounding melt but also due to the coalescence of each growing bubble with a large number of minute vapor-filled satellite bubbles that surround it (Chapter 7).

IV. EQUIPMENT

There exists a wide variety of devolatilizing equipment used to reduce the concentration of volatile contaminants. As stated earlier, devolatilizers may be broadly classified into nonrotating (Fig. 2) and rotating equipment (Fig. 3). Some of the main units in each category are

Nonrotating (still) devolatilizers	Rotating devolatilizers
Flash evaporators	Thin-film vaporisers
Falling-film devolatilizers	Single-screw extruders
Falling-strand devolatilizers	Multiscrew extruders
	Diskpacks
	Kneaders

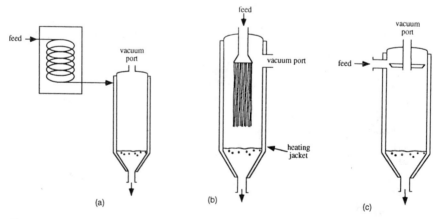

Fig. 2 Schematic representation of several nonrotating (still) devolatilizers: (a) flash evaporator; (b) falling-strand devolatizer; (c) falling-film devolatizer.

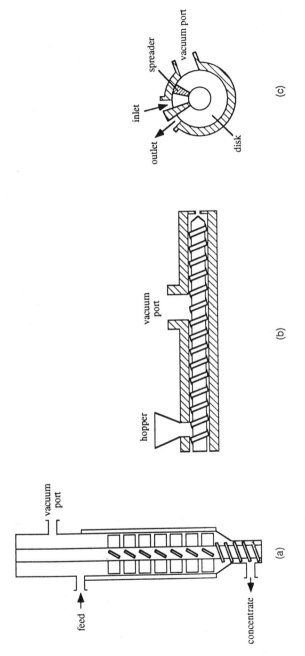

Fig. 3 Schematic representation of several rotating devolatilizers: (a) thin-film vaporizer; (b) vented extruder; (c) diskpack.

Nonrotating equipment relies on gravitational forces to transport the polymer through the devolatilizing zone, whereas in rotating machinery the melt is conveyed by its contact with moving elements. Accordingly, the viscosities that nonrotating devolatilizers may handle are much lower than those processed in rotating equipment. O'Brien (1985) presented operating windows for commercial devolatilizers in which he suggested typical viscosities and volatile levels for each type of equipment. Chapter 8 presents an extensive overview of devolatilizers in general and is followed by five chapters (9–13) that address devolatilization in specific equipment and geometries.

NOMENCLATURE

A parameter in the expression for nucleation rate
B parameter in the expression for nucleation rate
J nucleation rate
K_w Henry's law constant
P_1 partial pressure of the volatile
P_1^0 vapor pressure of the pure volatile
W weight fraction
ρ density
ϕ volume fraction
χ Flory–Huggins interaction parameter

Subscripts

1   volatile
2   polymer
e   equilibrium

REFERENCES

Albalak, R. J., Tadmor, Z., and Talmon, Y. (1987). Scanning electron microscopy studies of polymer melt devolatilization, *AIChE J.*, *33*: 808.
Albalak, R. J., Tadmor, Z., and Talmon, Y. (1990). Polymer melt devolatilization mechanisms, *AIChE J.*, *36*: 1313.
Albalak, R. J., Tadmor, Z., and Talmon, Y. (1992). Blister-promoted bubble growth in viscous polymer melts, *Mat. Res. Soc. Symp. Proc.*, *237*: 181.
Biesenberger, J. A. (1979). "Polymer Devolatilization: Theory of Equipment," 37th SPE ANTEC.
Biesenberger, J. A. (1987). Polymer melt devolatilization: On equipment design equations, *Adv. Polym. Tech.*, *7*: 267.

Biesenberger, J. A., and Lee, S. T. (1986a). A fundamental study of polymer devolatilization. I. Some experiments on foam-enhanced devolatilization, *Polym. Eng. Sci.*, *26*: 982.

Biesenberger, J. A., and Lee S. T. (1986b). "A Fundamental Study of Polymer Devolatilization. II. A Theory for Foam-Enhanced DV," 44th SPE ANTEC.

Biesenberger, J. A., and Lee, S. T. (1987). A fundamental study of polymer devolatilization. III. More experiments on foam-enhanced DV, *Polym. Eng. Sci.*, *27*: 510.

Biesenberger, J. A., and Kessidis, G. (1982). Devolatilization of polymer melts in single-screw extruders, *Polym. Eng. Sci.*, *22*: 832.

Biesenberger, J. A., and Sebastian, D. H. (1983). *Principles of Polymerization Engineering*, Wiley, New York.

Blander, M., and Katz, J. L. (1975). Bubble nucleation in liquids, *AIChE J.*, *21*: 833.

Chella, R., and Lindt, J. T. (1986). "Polymer Devolatilization II. Model for Foaming Devolatilization," 44th SPE ANTEC.

Collins, G. P., Denson, C. D., and Astarita, G. (1985). Determination of mass transfer coefficients for bubble-free devolatilization of polymeric solutions in twin-screw extruders, *AIChE J.*, *31*: 1288.

Coughlin, R. W., and Canevari, G. P. (1969). Drying polymers during screw extrusion, *AIChE J.*, *15*: 560.

Duda, J. L., Ni, Y. C., and Vrentas, J. S. (1978). Diffusion of ethyl–benzene in molten polystyrene, *J. Appl. Polym. Sci.*, *22*: 689.

Duda, J. L., Vrentas, J. S., Ju, S. T., and Liu, H. T. (1982). Prediction of diffusion coefficients for polymer–solvent systems, *AIChE J.*, *28*: 279.

Foster, R. W., and Lindt, J. T. (1989). "Stochastic Simulation of Foaming Devolatilization," 47th SPE ANTEC.

Foster, R. W., and Lindt, J. T. (1990). Twin screw extrusion devolatilization: From foam to bubble free mass transfer, *Polym. Eng. Sci.*, *30*: 621.

Han, J. H., and Han, C. D. (1990a). Bubble nucleation in polymeric liquids. I. Bubble nucleation in concentrated polymer solutions, *J. Polym. Sci. Polym. Phys. Ed.*, *28*: 711.

Han, J. H., and Han, C. D. (1990b). Bubble nucleation in polymeric liquids. II. Theoretical considerations, *J. Polym. Sci. Polym. Phys. Ed.*, *28*: 743.

Harvey, E. N., Barnes, D. K., McElroy, W. D., Whitely, A. H., Pease, D. C., and Copper, K. W. (1944a). Bubble formation in animals. I: Physical factors, *J. Cell. Comp. Physiol.*, *24*: 1.

Harvey, E. N., Whitely, A. H., McElroy, W. D., Pease, D. C., and Barnes, D. K. (1944b). Bubble formation in animals. II: Gas nuclei and their distribution in blood and tissues, *J. Cell. Comp. Physiol.*, *24*: 23.

Jemison, T. R., Rivers, R. J., and Cole, R. (1980). "Incipient Vapor Nucleation of Methanol from an Artificial Site-Uniform Superheat," presented at the 73rd Annual Meeting of AIChE, Chicago, Illinois.

Latinen, G. A. (1962). Devolatilization of viscous polymer systems, *Adv. Chem. Ser.*, *34*: 235.

Lee, S. T., and Biesenberger, J. A. (1989). A fundamental study of polymer devolatilization. IV: Some theories and models for foam-enhanced devolatilization, *Polym. Eng. Sci., 29*: 782.

Lindt, J. T., and Foster, R. W. (1989). "A Comprehensive Model of Devolatilization in Twin Screw Extruders," 47th SPE ANTEC.

Mack, M. H., and Pfeiffer, A. (1993). "Effect of Stripping Agents for the Devolatilization of Highly Viscous Polymer Melts," 51st SPE ANTEC.

Mehta, P. S., Valsamis, L. N., and Tadmor, Z. (1984). Foam devolatilization in a multichannel corotating disk processor, *Polym. Process Eng., 2*: 103.

Newman, R. E., and Simon, R. H. S. (1980). "A Mathematical Model of Devolatilization Promoted by Bubble Formation," presented at the 73rd Annual Meeting of AIChE, Chicago, Illinois.

O'Brien, K. T., (1985). Devolatilization, *Developments in Plastics Technology* (A. Whelan and J. L. Craft, eds.), Elsevier, London.

Powell, K. G., and Denson, C. D. (1983). "A Model for the Devolatilization of Polymeric Solutions Containing Entrained Bubbles," presented at the 75th Annual Meeting of AIChE, Washington, D.C.

Ravindranath, K., and Mashelkar, R. A. (1988). Analysis of the role of stripping agents in polymer devolatilization, *Chem. Eng. Sci., 43*: 429.

Roberts, G. W. (1970). A surface renewal model for the drying of polymers during screw extrusion, *AIChE J., 16*: 878.

Tadmor, Z. (1985). "Polymer Melt Devolatilization Mechanisms," presented at the 1st Annual Meeting of the Polymer Processing Society, Akron, Ohio.

Tadmor, Z., and Gogos, C. (1979). *Principles of Polymer Processing*, Wiley, New York.

Tukachinsky, A., Tadmor, Z., and Talmon, Y. Ultrasound-enhanced devolatilization of polymer melt, *AIChE J., 39*: 359.

Tukachinsky, A., Talmon, Y., and Tadmor, Z. (1994). Foam-enhanced devolatilization of polystyrene melt in a vented extruder, *AIChE J., 40*: 670.

Valsamis, L. N., and Canedo, E. L. (1989). "Devolatilization and Degassing in Continuous Mixers," 47th SPE ANTEC.

Vrentas, J. S., and Duda, J. L. (1979). Molecular diffusion in polymer solutions, *AIChE J., 25*: 1.

Werner, H. (1980). Devolatilization of polymers in multi-screw devolatilizers, *Devolatilization of Plastics*, Verrin Deutscher Ingenieure, Düsseldorf, Germany, pp. 99–131.

Yoo, H. J., and Han, C. D. (1984). Development of a mathematical model of foam devolatilization, *Polym. Proc. Eng., 2*: 129.

2

Thermodynamic Aspects of Devolatilization of Polymers

Edward W. Merrill

Massachusetts Institute of Technology, Cambridge, Massachusetts

I. STATES OF POLYMERS

Polymers under environmental conditions exist in two broad states: semi-crystalline and noncrystalline/amorphous. Semicrystalline polymers include the polyethylenes, polypropylene, poly(vinyl chloride), nylons, polyethylene terephthalate, polyacrylonitrile, and numerous others. Noncrystalline/amorphous polymers include polystyrene, poly(vinyl acetate), poly(methyl methacrylate), and numerous others. The distinction bears on thermo-dynamics: the volatiles to be removed exist only in amorphous/noncrystalline polymer. The crystal lattice of any crystalline polymer is so efficiently packed that no micromolecule larger than hydrogen could be incorporated.

One can further state that a special amorphous state is the molten state, i.e., there is no crystalline order, and the temperature T of the material lies both above the glass transition T_g and above the melting point T_m^0 (if any melting point exists). In this chapter we are particularly concerned with volatile micromolecules in molten polymer. Before addressing that subject, it is worth noting that the rate of removal of volatiles will depend not only on the thermodynamics but also on surface area (m^2/g) and on whether the polymer is above or below its glass transition. If below, the diffusion coefficient of any micromolecular species is reduced by several orders of magnitude from the value that pertains to the same polymer above its glass transition temperature. Poly(vinyl chloride) as produced by different processes can have widely variable but large surface area (owing to microporosity generated in the process of crystallization.

II. PROPAGATION–DEPROPAGATION EQUILIBRIUM: THE IMPLICATIONS OF "THE CEILING TEMPERATURE"

There is at least one class of polymers—the methacrylates—for which the propagation—depropagation equilibrium may be relevant in the course of thermal processing. Polyethers such as poly(methylene oxide) and polyoxolane offer other examples. Underlying this phenomenon is the ultimate balance between propagation (a growing chain adds a monomer unit) and depropagation (an active chain end throws off a monomer unit). This can be written as

$$+\frac{d[P]}{dt} = 0 = k_p[M^*][M]_e - k_u[M^*] \tag{1}$$

where $[P]$ = molar concentration of polymer units, $[M]$ = molar concentration of monomer at equilibrium, $[M^*]$ = molar concentration of active polymer chain ends, k_p = propagation constant, k_u = depropagation constant, and t = time.

In turn, the two constants can be written in Arrhenius form:

$$k_p = A_p e^{-E_p/RT} \tag{2}$$

$$k_u = A_u e^{-E_u/RT} \tag{3}$$

where E_p and E_u are the respective activation energies per mole and A_p and A_u are the frequency factors.

Upon inserting Eqs. (2) and (3) into Eq. (1), we have

$$\ln\left(\frac{A_p}{A_u}[M]_e\right) = \frac{E_p - E_u}{RT} \tag{4}$$

The difference in energies of activation is the enthalpy of polymerization:

$$\Delta H_p = E_p - E_u \tag{5}$$

The left-hand side of Eq. (4) is rearranged to introduce a standard state monomer concentration $[M_S]$, usually taken as unity (mol/L), and the standard state entropy of polymerization ΔS_p^0 becomes defined:

$$R\ln\left(\frac{A_p}{A_u}[M]_e\right) = R\ln\left(\frac{A_p}{A_u}[M]_S\right) + R\ln\frac{[M]_e}{[M]_S} = \Delta S_p^0 + R\ln\frac{[M]_e}{[M]_S} \tag{6}$$

Insertion of (5) and (6) into (4) defines the temperature T_e at which equilibrium occurs:

$$T_e = \frac{\Delta H_p}{\Delta S_p^0 + R\ln([M]_e/[M]_S)} \tag{7}$$

or

$$\ln\frac{[M]_e}{[M]_S} = \frac{\Delta H_p}{RT_e} - \frac{\Delta S_p^0}{R} \tag{8}$$

Insertion of relevant values of ΔH_p and ΔS_p^0 for poly(methacrylate) leads to the conclusion that at $T_e = 132°C$ (405 K) the equilibrium monomer content should be 0.05 mol/L, thus about 5 weight parts per thousand.

Propagation–depropagation has been most intensively studied in living anionic polymerizations, which are not relevant to the general purpose of this book. However, what is implied with respect to devolatilization is the following: *If* the polymer chain undergoes "unzipping" (i.e., throwing off one monomer after another, rather than reacting with the environment, for example, by oxidation), and *if* by some accident, e.g., mechanical stress in thermal processing equipment, polymer chains are broken leaving free radical ends, then these will unravel (unzip), liberating monomer until a concentration is reached as given by Eq. (8).

Thus, if mechanical stress is sufficient to break existing polymer chains during processing, poly(methyl methacrylate) will generate methyl methacrylate monomer, poly(methylene oxide) will generate formaldehyde, and polyoxolane will generate the cyclic ether oxolane (tetrahydrofuran). The concentration of monomer produced will in general be much less than the equilibrium value predicted by Eq. (8) and depends on the rate of scission of chains by the mechanical processing as well as on the intervention of competitive processes such as oxidation.

Nonetheless, one should be aware that devolatilization by extrusion machinery could *generate* volatile monomer under certain conditions.

III. THE ESTIMATION OF PARTIAL PRESSURE OF A VOLATILE COMPOUND PRESENT AT LOW CONCENTRATION IN A POLYMER: RECOMMENDED EQUATION AND TABULATED DATA

The standard vapor pressure P_1^0 of a pure liquid (component 1) is known as a function of temperature for most volatile compounds of interest (see Appendix A). Only if the external pressure is significantly greater than about 100 bar is the value of P_1^0 at a specified temperature T significantly increased.

The problem of paramount interest here is, what is the relation of the *partial* pressure P_1 *to* the vapor pressure P_1^0, given that the temperature T is specified and therefore P_1^0 is known, and given that the content of volatile in the polymer is known?

We shall show that an appropriate expression is, for many practical cases of a low concentration of volatile compound 1 in a polymer,

$$P_1 = P_1^0 \phi_1 e^{1+\chi} \tag{9a}$$

where ϕ_1 = volume fraction of volatile component, e = Euler number, and χ = a dimensionless factor (Flory–Huggins interaction parameter) that is a measure of the compatibility of the volatile in the polymer. Another form of Eq. (9a) is evidently

$$\ln \frac{P_1}{P_1^0 \phi_1} = 1 + \chi \qquad (9b)$$

Generally speaking, high positive values of χ indicate very poor compatibility, and negative values (rarely more negative than -1.0) represent avidity of polymer for the volatile.

In view of the imprecision with which χ is known, one can in most cases substitute the weight fraction W_1 for the volume fraction without introducing great error.

In practically all cases, the temperature T will be such that the polymer is above its glass transition temperature T_g, and for crystallizing polymers the temperature T will be above the crystalline melting T_m^0.

In Eqs. (9a, b), χ itself is a function of temperature, and this functionality is often expressed by the equation

$$\chi = \alpha + \frac{\beta}{T} \qquad (10)$$

In Table 1, representative values of χ for a stated polymer, stated volatile compound, and at stated temperature (or temperatures) are shown. The extensive data in Bandrup and Immergut (1989) (reproduced in Appendix B) from which this has been excerpted should be consulted. Table 2 exemplifies data giving coefficients α and β for χ, as used in Eq. (10), taken from DiPaola-Baranyi and Guillet (1978). These data have been obtained by inverse gas-phase chromatography (IGC), a technique discussed in Section VI and ideally suited to the determination of χ, since the volume fraction of volatile compound is always low and the temperature usually in the range appropriate to devolatilization.

In general, the values of χ as cited in Table 1 and Appendix B should apply with fair accuracy, for the estimation of partial pressure, over a range of at least $\pm 20°C$ of the quoted temperature.

While the preceding suggests that χ is an empirical factor to be found only by experiment, there is a formal theoretical basis for χ and for the form of Eq. (9) as relevant to low volatile fractions (ϕ_1), and this is presented in the following section. There we will see that, theoretically, χ for any named solvent–polymer pair should be independent of the amount of solvent relative to polymer, i.e., independent of volume fraction of solvent (volatile). In fact, this is rarely the case. Values of χ obtained in dilute polymer solutions (2% polymer or less) may be very different from values obtained under conditions of interest in devolatization, i.e., around 2% volatile–98% polymer, for which the data of Tables 1 and 2 were obtained.

Table 1 Values of χ for Low-Volume Fraction of the "Volatile"
(i.e., the "Solvent"), High-Volume Fraction of Polymer $\phi_2{}^a$

Polymer	Volatile	χ	Temp., °C
Polydimethyl siloxane	cyclohexane	0.47	25–70
	toluene	0.80–0.75	25–70
Polyethylene, linear	n-decane	0.31	149
		0.12	185
	cis-decalin	0.07	149
Polystyrene	acetone	1.30–0.56	162–229
	benzene	0.33	120
	water	4.40–3.10	162–229
Polyisoprene	benzene	0.34–0.30	25–55
	n-octane	0.49–0.46	25–55
Poly(butene-1)	n-hexane	0.38	115–135
Poly(methyl acrylate)	benzene	0.51–0.46	90–110
	n-octane	2.38–2.19	90–110

^{a} See also Appendix B.
Source: Bandrup and Immergut (1989).

Table 2 α and β in $\chi = \alpha + \beta/T$ (Kelvin)

Polymer: Polystyrene Volatile (solvent)	α	β
n-decane	−5.60	724.2
benzene	−0.234	227.8
cis-decalin	−0.706	536.1
naphthalene	−0.3313	210.1

Polymer: Poly(methyl acrylate) Volatile (solvent)	α	β
n-octane	−1.661	1469
benzene	−0.3055	295.2
naphthalene	−0.0969	218.2

Source: DiPaola-Baranyi and Guillet (1978).

IV. THE THERMODYNAMIC ACTIVITY OF A VOLATILE COMPONENT DISSOLVED IN A POLYMER: FLORY–HUGGINS THEORY

There is only one case that we can deal with and have reasonable confidence that the model is relevant, namely, that the polymer be completely non-crystalline, thus in an amorphous state. Furthermore, to be of practical interest, the amorphous polymer should be above its glass transition; otherwise, the rate of diffusion of volatiles out of the polymer under a thermodynamic gradient will be very low.

We start with the proposition that the thermodynamic activity a_1 of a volatile species 1, e.g., monomer, water, dissolved in polymer, species 2, is related to the chemical potential:

$$RT \ln a_1 = \mu_1 - \mu_1^0 = \left(\frac{\partial \Delta F_{\text{mix}}}{\partial N_1} \right)_{N_2, P, \dots} \tag{11}$$

where $\mu_1 - \mu_2^0 =$ difference in chemical potential from reference (pure) state, $T =$ absolute temperature, $R =$ gas constant, $N_1 =$ mols of volatile solvent, $N_2 =$ mols of polymer, $\Delta F_{\text{mix}} =$ total Gibbs free energy of mixing species 1 and polymer 2, and $P =$ pressure.

In the ensuing analysis, we are concerned only with single-phase homogeneous solutions with a minor part of 1 in a major part of 2. We are not, for example, treating the case of water as an adsorbed layer on, but not in, emulsion polymerized particles.

Thus, an expression for ΔF_{mix} is required, and the two components thereof, the enthalpy of mixing ΔH_{mix} and the entropy of mixing ΔS_{mix} must be discovered:

$$\Delta F_{\text{mix}} = \Delta H_{\text{mix}} - T \Delta S_{\text{mix}} \tag{12}$$

We take as a measure of activity the ratio of partial pressure P_1 to standard vapor pressure P_1^0, i.e.,

$$a_1 \approx \frac{P_1}{P_1^0} \tag{13}$$

We now give a summary of the Flory–Huggins liquid lattice theory by which the activity of a solvent (volatile), 1, is predicted to vary with its volume fractions in the mixture, and with an interaction parameter χ that accounts (imperfectly) for the relative attraction of polymer for the solvent (Flory, 1953).

A. Definition of the Lattice

The starting point is a lattice, shown as Fig. 1. It is assumed a priori that the total number of lattice sites n_0 is filled either with polymer segments

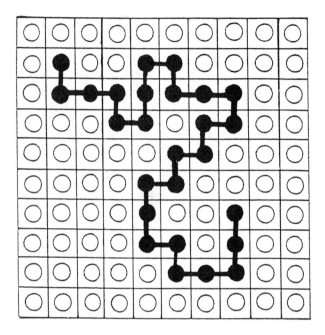

Fig. 1   Segments of a chain polymer molecule located in the liquid lattice. (From P. J. Flory, *Principles of Polymer Chemistry*, 1953. By permission Cornell U. Press.)

or with solvent molecules. Furthermore, the size of the lattice, at least on the first cut, is scaled to accomodate a single solvent molecule. The polymer molecules are modeled as strings of x segments wherein each segment has the size (volume) of a solvent molecule. The total number of lattice sites n_0 remains constant (i.e., volume is constant). Thus, the lattice when filled has the following mass balance:

$$n_0 = n_1 + n_2 x \tag{14}$$

wherein n_1 = number of solvent molecules, n_2 = number of polymer molecules, and

$$x = \frac{M_2}{M_1} \frac{\bar{v}_2}{\bar{v}_1} = \frac{M_2 \bar{v}_2}{V_1} \tag{15}$$

M_2 = molecular weight of polymer, M_1 = molecular weight of solvent, $\bar{v}$ = specific volume of pure amorphous solvent or polymer, and V_1 = molar volume of pure amorphous solvent.

The lattice has a coordination number z; for example, in a cubic lattice (of which Fig. 1 is a two-dimensional representation) $z = 6$.

From the way the lattice is defined, and the size of the polymer chain in relation to the solvent, the volume fractions of solvent ϕ_1 and of polymer ϕ_2 are given by

$$\phi_1 = \frac{n_1}{n_0} \tag{16}$$

$$\phi_2 = \frac{n_2 x}{n_0} \tag{17}$$

B.   Determination of the "Configurational" Entropy of Mixing ΔS^*

The polymer chains are introduced into the initially empty lattice seriatim until they occupy the intended volume fraction ϕ_2. In our context, ϕ_2 is going to be of order 0.95 to slightly less than unity.

The analysis includes a step such as the following, which computes the number of ways v the $(i + 1)$-st polymer chain can be introduced to the lattice after i have already been placed on it:

$$v_{i+1} = (n_0 - ix)z(z - 1)^{x-2}(1 - f_i)^{x-1} \tag{18}$$

In Eq. (18), f_i is the expectancy that a given lattice site is occupied by the segment of a previously inserted polymer chain.

If there are no polar forces that would favor specific associations between polymer segments (i.e., hydrogen bond donor–acceptor pairs), the expectancy that a site is *not* occupied, $1 - f_i$, is taken as

$$1 - f_i = \frac{n_0 - ix}{n_0} = 1 - \frac{ix}{n_0} = 1 - \phi_i \tag{19}$$

where ϕ_i = volume fraction of lattice already occupied by previously inserted polymer chains.

With the assumption of constant total volume (n_0 = constant) and of $1 - f_i$ being correctly given by Eq. (19), the total number of ways, Ω, of introducing n_2 macromolecules is

$$\Omega = \frac{1}{n_2!} \prod_{i=1}^{n_2} v_i \tag{20}$$

The entropy S is given by

$$S = k \ln \Omega \tag{21}$$

where k = Boltzmann constant.

Flory wrote: "If each solvent molecule may occupy one of the remaining lattice sites and in only one way, Ω [as in eq. (20)] represents also the total number of configurations for the solution, from which it follows that the configurational entropy of mixing of the perfectly ordered pure polymer and the pure polymer is given by" Eq. (21), where Ω is given by Eq. (20).

Thus, introduction of (19) into (20) and (20) into (21) leads finally, after slight mathematical simplification, to the result

$$S_c = -k\left[n_1 \ln \frac{n_1}{n_1 + n_2 x} + n_2 \ln \frac{n_2}{n_1 + n_2 x} \right.$$
$$\left. + n_2 \ln x - n_2 \ln x - n_2(x - 1) \ln \frac{z - 1}{e} \right] \qquad (22)$$

To arrive at the net entropy enchange for mixing *disordered* pure polymer with solvent, one deducts the entropy of disorientation S_{dis}, which is given by Eq. (19) in which $n_1 = 0$, $n_2 x = n_0$, leading to

$$S_{dis} = +kn_2 \ln x + (x + 1) \ln \frac{z - 1}{e} \qquad (23)$$

The final result, the entropy of mixing ΔS_m is obtained by subtracting Eq. (23) from Eq. (22), with the result

$$\Delta S_m^* = -k\left(n_1 \ln \frac{n_1}{n_0} + n_2 \ln \frac{n_2 x}{n_0} \right) = -k\left(n_1 \ln \phi_1 + n_2 \ln \phi_2 \right) \qquad (24)$$

Flory placed the asterisk on ΔS_m to indicate that the ΔS_m calculated represents only the configurational entropy computed with the assumption that specific segment–segment interactions do not exist—the assumption that underlies Eq. (17).

C.   The Enthalpy of Mixing: Simplest Version

As a first cut, Flory proposed the following evaluation. The n_2 macromolecule having been introduced onto the lattice so as to occupy $n_2 x$ sites, there remain $n_0 - n_2 x = n_1$ sites to be occupied by solvent molecules.

When a solvent molecule is placed in the lattice in a cell next to a polymer segment, a net exchange of heat ΔW_{12} will take place.

Schematically, we envision the following:

Two solvent molecules come into contact in adjacent cells. *Heat exchanged*

$$1 + 1 \rightarrow 1 \ 1 \qquad\qquad W_{11}$$

Two polymer segments come into contact in adjacent cells.

$$-2- + -2- \rightarrow 2 \ 2 \qquad\qquad W_{22}$$

One solvent molecule comes into contact with one polymer segment.

$$1 + 2 \rightarrow 1 \ 2 \qquad\qquad W_{12}$$

Thus, the net heat exchange per 1, 2 contact pair made from 1/2 of a 1, 1 and 1/2 of a 2, 2 contact pair is

$$\Delta W_{12} = W_{12} - \frac{1}{2}(W_{11} + W_{22}) \tag{25}$$

As formulated, W_{11}, W_{22}, and W_{12} represent "condensations." Thus, each is negative (less than zero).

If W_{12} is more negative than the average of W_{11} and W_{22}, ΔW_{12} will be negative; i.e., heat will be given off, and the mixing is exothermic. Conversely, if W_{12} is less negative than the average of W_{11} and W_{22}, ΔW_{12} will be positive, and hence mixing is endothermic.

The calculation of the enthalpy of mixing proceeds as follows: it is the number of solvent to polymer segment contacts multiplied by ΔW_{12}. The number of solvent to polymer contacts is the following product:

Number of (1) polymer segments × number of (2) cells around each segment × probability (3) that the cell is *not* already occupied by a polymer segment.

Reference to the lattice model shows us (1) $= n_2 x$; (2) $= z - 2$ (neglecting the two ends of each chain, for which there are $z - 1$ adjacent cells); (3) $\approx$ volume fraction *not* occupied by polymer segments; hence (3) $\approx \phi_1$.

Item (3) invokes the same assumption that lies behind Eq. (17), viz., that there is no bias or prejudice for certain preferred polymer segment to polymer segment contacts owing to forces such as hydrogen bond donor–acceptor interactions.

Finally, therefore, collecting and rearranging the foregoing:

$$\Delta H_{\text{mix}} = \Delta W_{12} \cdot n_2 x \cdot (z - 2)\phi_1 \tag{26}$$

and since

$$\frac{n_2 x}{n_0} = \phi_2, \qquad \frac{n_1}{n_0} = \phi_1 \tag{16, 17}$$

we get

$$\Delta H_{\text{mix}} = (z - 2) \, \Delta W_{12} \, n_1 \phi_2 \tag{27}$$

D. Original Definition of the Flory–Huggins Interaction Parameter χ

Flory proceeded to define the interaction parameter χ by

$$\chi \equiv \frac{(z - 2) \, \Delta W_{12}}{kT} \tag{28}$$

which means that if ΔW_{12} is constant, χ should vary inversely as absolute temperature T.

Acceptance of Eq. (28) leads to an expression for ΔH_{mix} (from Eq. 27):

$$\Delta H_{\text{mix}} = kT \chi n_1 \phi_2 \tag{29}$$

and thus to the complete expression for free energy of mixing ΔF_{mix} by Eq. (13) and Eq. (24) for ΔS_{mix}^*:

$$\Delta F_{\text{mix}} = kT(n_1 \ln \phi_1 + n_2 \ln \phi_2 + \chi n_1 \phi_2) \tag{30}$$

E. The Chemical Potential of the Volatile (Solvent) Derived from the Free Energy of Mixing

From Eq. (11) to get the chemical potential change of the solvent *per molecule*, we carry out the operation:

$$\mu_1 - \mu_1^0 = \left(\frac{\partial \Delta F_{\text{mix}}}{\partial n_1} \right)_{T,P,n_2}$$

$$= kT \left[\frac{\partial (n_1 \ln \phi_1)}{\partial n_1} + \frac{\partial (n_1 \ln \phi_2)}{\partial n_1} + \chi \frac{\partial (n_1 \phi_2)}{\partial n_1} \right]_{n_2, T} \tag{31}$$

Note in the last term of the right-hand side of Eq. (31) that χ *is removed from the differentiation* and is taken as a constant, *independent* of n_1 and of ϕ_2.

The result of the manipulation of Eq. (31) is

$$\mu_1 - \mu_1^0 = kT \left[\ln \phi_1 + \phi_2 \left(1 - \frac{1}{x} \right) + \chi \phi_2^2 \right] \tag{32}$$

Since x is presumed large, $x^{-1} \ll 1$. The chemical potential on a *molar* basis (rather than molecular basis) is obtained by replacing the Boltzmann constant k with the gas constant R.

For the purposes of this book, we are dealing with solutions in which ϕ_2 approaches unity and ϕ_1 approaches zero. We can therefore return to Eq. (11), and combine it with Eq. (32), with R replacing k and $\phi_2 \approx 1$, to show

$$\ln a_1 \approx \ln \phi_1 + 1 + \chi \tag{33}$$

or

$$\ln \frac{a_1}{\phi_1} \approx 1 + \chi$$

or

$$a_1 \approx \phi_1 e^{1+\chi}$$

or

$$P_1 \approx P_1^0 \phi_1 e^{1+\chi} \tag{9a}$$

F. Observations About χ: Early Comparisons of Experiment with Theory

From the earliest studies aimed at experimental test of Eq. (32), and all other equations derived from Eq. (30), it has been found that the interaction parameter χ, supposedly independent of composition, usually does in fact vary with ϕ_1 (ϕ_2). Significant variation has been noted in the *dilute polymer* (*concentrated solvent*) range, i.e., $\phi_2 < 0.2$, $\phi_1 > 0.8$. Further, while Eq. (28) suggests that χ should vary inversely with temperature (if ΔW_{12} is only a heat of mixing), it is found that such is not generally the case. In some cases χ decreases with increasing temperature, but not as T^{-1}. In other cases, especially when hydrogen and donor–acceptor pairs exist, χ may increase with increasing temperature.

Some of the discrepancy arises from oversimplification of the original model, for example, the assumptions that

1. There is only one ΔW_{12}, regardless of what "side" of a solvent molecule is in contact with a polymer segment.
2. The probability that a lattice site selected at random is not occupied by a polymer segment is identically the volume fraction $1 - v_2$.
3. On mixing polymer segments and solvent, no volume change occurs, i.e., n_0 is constant.
4. The same lattice is appropriate both to solvent molecules and polymer segments. (Note that the lattice coordination number z never can be determined with any degree of confidence by experiment, and thus the *product* $z\,\Delta W_{12}$ is unknown a priori.)

Recognizing that χT is not usually the constant predicted from the simple model, Flory (1953) postulated that ΔW_{12} must be interpreted as a standard state *free energy* change, and he wrote

$$\Delta W_{12} = \Delta W_\mathrm{h} - T \Delta W_\mathrm{S} \tag{34}$$

where ΔW_h = enthalpy change (i.e., the original concept of ΔW_{12}) and ΔW_S = entropy change. This leads to a revision of the definition of χ:

$$\chi = \frac{z\,\Delta W_\mathrm{h}}{kT} - \frac{z\,\Delta W_\mathrm{S}}{k} \tag{35}$$

From this, it is possible to interpret two more recent methods of expressing χ:

$$\chi = \chi_\mathrm{h} + \chi_\mathrm{s} \tag{36}$$

wherein

$$\chi_\mathrm{h} = \frac{z\,\Delta W_\mathrm{h}}{kT}, \qquad \chi_\mathrm{s} = -\frac{z\,\Delta W_\mathrm{S}}{k}$$

or the form

$$\chi = \alpha + \frac{\beta}{T} \tag{10}$$

wherein

$$\alpha = -\frac{z\,\Delta W_\mathrm{S}}{k}, \qquad \beta = \frac{z\,\Delta W_\mathrm{h}}{k}$$

For many polymer–solvent systems a value of χ_s in Eq. (36), α in Eq. (10), has been found to range between 0.27 and 0.35, with a mean of about 0.30 (Magat, 1949; Ito and Guillet, 1979).

In the system polystyrene–toluene, over the range of polymer volume fraction 0.5 to 0.9 (which might be extrapolated to $\phi_2 = 1$), χ was found to be around 0.3. If χ_s is 0.3, Eq. (36) would suggest that $\chi_\mathrm{h} \approx 0$, i.e., there is no heat given off and taken on when styrene–styrene segments and toluene–toluene pairs exchange to form styrene–toluene pairs. In this case, one would suppose that the same cell on the liquid lattice would serve a styrene segment and a toluene molecule and that their chemical similarities would account for the lack of heat of mixing.

If, in any system, χ_s is positive at about 0.3, ΔW_S in Eqs. (34) and (35) must be negative, and thus there is some small *loss* of randomness of the system upon making 1,2 pairs from 1,1 and 2,2 pairs.

Another example of near constancy of χ over the range $\phi_2 = 0.4$ to 0.99 is the system polyisobutylene–n-pentane, for which χ was found to be 0.63.

If $\chi_s \approx 0.3$, $\chi_h \approx 0.33$, indicating an endothermic process.

We are particularly interested in systems in which the volume fraction of solvent is low: 0.05 or less. Thus, since polymer segments outnumber solvent molecules by a number of the order 20 or greater, it is reasonable to suppose that each solvent molecule will find itself in a "cage" of polymer segments. As long as this situation holds, the free energy of interchange ΔW_{12} (in Eq. 34) should be constant, independent of the number of solvent molecules previously added, and thus independent of ϕ_1.

V. THE PARAMETER χ_1 IN RELATION TO COHESIVE ENERGY DENSITY AND SOLUBILITY PARAMETER

Cohesive energy density CED and solubility parameter δ for any vaporizable pure liquid are defined [Hildebrand and Scott (1950)] as:

$$\text{CED} = \delta^2 = \frac{\Delta E_{\text{vap}}}{V} = \frac{\Delta H_{\text{vap}} - RT}{V} \tag{37}$$

where V is the molar volume of the liquid (at its vaporization temperature), $\Delta E_{\text{vap}} = $ internal energy change upon vaporization, per mol, $\Delta H_{\text{VAP}} = $ enthalpy change upon vaporization, per mol. The usual dimensions of δ are $(\text{cal/cm}^3)^{1/2}$ or $(\text{J/cm}^3)^{1/2}$.

Not long after the appearance of Flory–Huggins theory, the solubility parameter concept was applied to solvent (1)–polymer (2) systems by expressing the enthalpic part of χ, i.e., χ_h in Eq. (36), as

$$\chi_h = \frac{V_1}{RT}(\delta_1 - \delta_2)^2 \tag{38}$$

The form of Eq. (38) suggests that the β of Eq. (10) is to be taken as $(V_1/R)(\delta_1 - \delta_2)^2$, and that $\chi_h T$ should be a constant.

The value of δ_2 cannot evidently be determined *experimentally* via Eq. (37), since polymer cannot in general be vaporized (decomposition in general results upon heating).

Classically, δ_2 was determined by indirect methods such as swelling of cross-linked polymer. Since one of the consequences of Flory–Huggins theory applied to polymer networks in the presence of a solvent is that the degree of swelling should be maximum when the value of χ is a minimum, it was common to measure swelling ratios of a given cross-linked polymer in a series of solvents, i, j, k, etc. Upon plotting swelling ratio as a function of δ_1 ($\delta_i, \delta_j, \delta_k$), a maximum is observed at some δ, and this is taken as δ_2.

A comparable thermodynamic test would be the osmotic pressure of polymer solutions. If one made solutions of a given polymer at a stated concentration (e.g., $\phi_2 = 0.50$) in solvents i, j, k, the maximum osmotic pressure would be developed for the minimum value of χ, thus when the solvent δ_i matched the polymer's δ_2.

By Eq. (38), when $\delta_1 = \delta_2$, $\chi_h = 0$, and this is the minimum value. An obvious shortcoming of Eq. (38) is that it does not allow for values of χ_h less than zero, i.e., when the heat of mixing solvent, and polymer 2 is *exothermic*, as in the case of formation of strong hydrogen donor–acceptor bonds as 1,2 pairs, which were weak or nonexistent as 1,1 or 2,2 pairs.

An example would be acetone (1)–polyvinyl chloride (2). The acetone cannot hydrogen bond to itself. The polyvinyl chloride has the polar dipole

by which it can interact by weak H bonding with like dipoles (2,2 bonding). But the 1,2 bond is stronger, i.e.,

Equation (38) may correctly predict temperature dependence of χ_h for non-hydrogen-bonding systems.

Whether swelling experiments or osmotic experiments are conducted to assess δ_2, the problem of variation of χ with volume fraction polymer (nowhere predicted by Eq. (38) or any preceding equations for χ_h or χ) is *not* resolved. Thus, values of χ_h determined at volume fractions of polymer appropriate to swelling experiments to which χ_s taken as 0.3 is added do not necessarily lead to the correct value of χ.

To the practical end of evaluating χ at $\phi_1 \to 0$, which is needed for devolatilization, the experimental technique of IGC (inverse gas-phase chromatography) has proven ideally suited for two reasons: (1) ϕ_1 is low, close to zero; (2) the temperature is usually high, in a range relevant to devolatilization.

VI. INVERSE GAS CHROMATOGRAPHY (GAS–LIQUID CHROMATOGRAPHY)

Patterson et al. (1971) give an extensive account of the application of inverse gas chromatography (IGC; gas–liquid chromatography) to the determination of parameters characterizing the thermodynamic interaction between a volatilized solvent (i.e., a vapor), component 1, passing over a deposited layer of "liquid," i.e., a polymer, component 2.

They show that the thermodynamic activity a_1 of the solvent vapor, taken as P_1/P_1^0 (cf. Eq. 11), divided by the *weight* fraction W_1 of the solvent dissolved in the polymer is given by

$$\lim_{W_1 \to 0} \frac{a_1}{W_1} = \frac{273.2R}{P_1^0 M_1 V_g^0} \tag{39}$$

where R = gas constant, M_1 = molecular weight of solvent, and V_g^0 = specific retention volume corrected to 0°C (273.2 K). In turn, the specific retention volume V_g^0 corrected to 273.2 K is given by

$$V_g^0 = \frac{Q(t_S - t_a)}{m_2} \frac{273.2}{T} j \tag{40}$$

where Q = volumetric flow rate, T = experimental temperature, m = mass of liquid phase (polymer), t_a = retention time of air (carrier gas helium), and t_S = retention time of solvent (1) (carrier gas helium).

In Eq. (40), j corrects for the pressure drop through the column from inlet P_{in} to outlet P_{out} by the relation

$$j = \frac{3}{2} \frac{(P_{in}/P_{out})^2 - 1}{(P_{in}/P_{out})^3 - 1} \tag{41}$$

An additional term usually appears on the right-hand side of Eq. (41) correcting for nonideality of the gas phase by invoking the second virial coefficient B_{11} of the solvent vapor. As Patterson et al. (1971) note: "This correction (i.e., B_{11}) makes only a relatively small difference in the calculation of χ." Hence, in this account, we omit reference to it. It becomes important if one is seeking accurate values of the partial molal enthalpy of the solvent, but the purpose of this account is to show how values of χ are obtained in order that these can be used to estimate the partial pressure P_1 of a solvent dissolved in a polymer at a low weight concentration W_1. We now refer back to Eqs. (33) and (9a):

$$a_1 = \phi_1 e^{1 + \chi}$$

and

$$P_1 = P_1^0 \phi_1 e^{1+\chi}$$

The volume fraction ϕ_1 is related to the weight fraction W_1 when $W_1 \to 0$, and the total volume is to first approximation the volume of the polymer, by the relation

$$W_1 = \frac{v_1 \bar{v}_2}{\bar{v}_1}$$

where $\bar{v}_1$, $\bar{v}_2$ = partial specific volumes (cm^3/g), respectively, of solvent and of polymer. Thus,

$$\frac{a_1}{W_1} \frac{\bar{v}_1}{\bar{v}_2} = e^{1+\chi} \tag{42}$$

$$\lim_{W_1 \to 0} \chi = \ln\left(\frac{a_1}{W_1} \frac{\bar{v}_1}{\bar{v}_2}\right) - 1 = \ln\left(\frac{273.2R}{P_1^0 M_1 V_g^0} \frac{\bar{v}_1}{\bar{v}_2}\right) - 1 \tag{43}$$

Thus χ is readily determined at the temperature T from determination of V_g^0 as corrected from the specific retention time actually determined at temperature T via Eq. (40).

The literature on χ determined by IGC† contains primarily good to "fair" solvents for the respective polymers. In general, for obvious reasons, very poor solvents have not been examined. One such ubiquitous "solvent" is adventitious water. Except for hydrogen-bonding polymers like the nylons, one would expect χ for water: polymer to be positive and large.

We can estimate the magnitude of χ if we take the limiting case that water at 1 vol % in the polymer has a partial pressure equal to its vapor pressure; then,

$$\frac{P_1}{P_1^0} = 1 = 0.01 e^{1+\chi}$$

which leads to the result: $\chi = 4.605 - 1 = 3.605$.

Further insight into the importance of the value of χ is seen by the calculations in the following table:

---

† The appended list of references *except* Flory (1953), Magat (1949), Hildebrand and Scott (1959), and Scatchard (1949).

Assumption: The volume fraction of the volatile (the solvent) $\phi_1 = 0.005$.

Case I:	$\chi = +1$	(poor solvent)	$P_1/P_1^0 = 0.005e^2 = 0.04$
	$\chi = 0$	("athermal" solution)	$P_1/P_1^0 = 0.005e^1 = 0.0142$
	$\chi = -1$	(exothermic, $\Delta H_{mix} < 0$)	$P_1/P_1^0 = 0.005e^0 = 0.005$

  As one would expect, the more favorable the interaction between solvent and polymer, the lower will be the ratio of partial pressure to vapor pressure, or roughly speaking, the lower will be the volatility, and the more difficult it will be to devolatilize the polymer.

  Finally, the case of several different solvents dissolved in a polymer can lead to extensive calculation, as is true for any multicomponent system (Ruff et al., 1986a). Practically speaking, since we assume that the solvents are present each in low concentration, it is reasonable to treat each one as if it alone were in the polymer (except in the very rare case where there is strong interaction between solvents, for example acetone–chloroform via hydrogen bonding leading to a maximum boiling azeotrope).

  Although it is not an issue of thermodynamics, it is worth mentioning that adventitious (or purposefully added water) upon vaporizing from a polymeric melt during devolatilization will effectively carry out a form of steam distillation, sweeping away organic volatiles i, j, k, according to their P_i, P_j, P_k, set in turn by their volume fractions ϕ_i, ϕ_j, ϕ_k in the polymer, their respective vapor pressures P_i^0, P_j^0, P_k^0 at the devolatilization temperature T, and their respective χ factors χ_i, χ_j, χ_k.

NOMENCLATURE

a	activity
A_p, A_u	frequency factors
E_p, E_u	molar activation energies
ΔE_{vap}	molar internal energy change upon vaporization
f_i	the expectancy that a given lattice site is occupied by the segment of a previously inserted polymer chain
ΔF_{mix}	total Gibbs free energy of mixing species 1 and polymer 2
ΔH_{mix}	enthalpy of mixing
ΔH_p	enthalpy of polymerization
ΔH_{vap}	molar enthalpy change upon vaporization
j	correction factor in Eq. (40)
k	Boltzmann constant
k_p	propagation constant
k_u	depropagation constant

m	mass of liquid phase
M	molecular weight
$[M]$	molar concentration of monomer at equilibrium
$[M^*]$	molar concentration of active polymer chain ends
n	number of molecules
n_0	number of lattice sites
N	number of moles
P_i	partial pressure of component i
$[P]$	molar concentration of polymer units
P_{in}	inlet pressure
P_i^0	standard vapor pressure of component i
P_{out}	outlet pressure
Q	volumetric flow rate
R	universal gas constant
S	entropy
S_{dis}	entropy of disorientation
ΔS_{mix}	entropy of mixing
ΔS_p^0	standard state entropy of polymerization
t	time
t_a	retention time of air
t_s	retention time of solvent
T	temperature
T_g	glass transition temperature
T_m^0	melting temperature
V_1	molar volume of pure amorphous solvent
V_g^0	specific retention volume corrected to $0°C$
W	weight fraction
W	heat exchange
ΔW_h	enthalpy change
ΔW_s	entropy change
z	lattice coordination number
α, β	coefficients in Eq. (10)
δ	solubility parameter, square root of cohesive energy density
μ	chemical potential
ν	number of ways that the $(i + 1)$-st polymer chain can be introduced to the lattice
ϕ	volume fraction
χ	Flory–Huggins interaction parameter
χ_h	enthalpic part of the Flory–Huggins interaction parameter
χ_s	entropic part of the Flory–Huggins interaction parameter
Ω	total number of ways of introducing macromolecules to the lattice

Subscripts

1   solvent
2   polymer

REFERENCES

Bandrup, J., and Immergut, E. H. (eds.) (1989). *Polymer Handbook*, 3rd ed., Wiley-Interscience, New York, Section VII, "Solution properties."

Biros̃, J., Zeman, L., and Patterson, D. (1971). Prediction of the χ parameter by the solubility parameter and corresponding states theories, *Macromolecules*, *4*(1): 30.

DiPaola-Baranyi, G. (1982). Estimation of polymer solubility parameters by inverse gas chromatography, *Macromolecules*, *15*, 622.

DiPaola-Baranyi, G., and Guillet, J. E. (1978). Estimation of polymer solubility parameters by gas chromatography, *Macromolecules*, *11*(1): 228.

DiPaola-Baranyi, G., Guillet, J. E., Klein, J., and Jeberien, H.-E. (1978). Estimation of solubility parameters for poly(vinyl acetate) by inverse gas chromatography, *J, Chromatography*, *166*, 349.

Flory, P. J. (1953). *Principles of Polymer Chemistry*, Cornell U. Press, Ithaca, New York, Chapter 12, "Statistical thermodynamics of polymer solutions."

Glover, C. J., and Lau, W. R. (1983). Determination of multicomponent sorption equilibria using perturbation gas chromatography, *AIChE J.*, *29*(1): 73.

Hildebrand, J. H., and Scott, R. L. (1950). *The Solubility of Non-Electrolytes*, 3rd ed., Reinhold, New York.

Ito, K., and Guillet, J. E. (1979). Estimation of solubility parameters for some olefin polymers and copolymers by inverse gas chromatography, *Macromolecules*, *12*(6): 1163.

Lipson, J. E. G., and Guillet, J. E. (1981). Studies of polar and nonpolar probes in the determination of infinite-dilution solubility parameters, *J. Polymer Sci.*, *19*: 1199.

Lipson, J. E. G., and Guillet, J. E. (1982). Measurement of solubility and solubility parameters for small organic solutes in polymer films by gas chromatography, *J. Coatings Tech.*, *54*: 89.

Magat, M. (1949). Contribution à la thermodynamique des solutions de hauts polymères. II. Solubilité des hauts polymères, *J. Chim. Phys.*, *46*: 344.

Merk, W., Lichtenhaler, R. N., and Prausnitz, J. M. (1980). Solubilities of 15 solvents in copolymers of poly(vinyl acetate) and poly(vinyl chloride) from gas–liquid chromatography. Estimation of polymer solubility parameters, *J. Phys. Chem.*, *84*: 1694.

Patterson, D., Tewari, Y. B., Schreiber, H. P., and Guillet, J. E. (1971). Application of gas–liquid chromatography to the thermodynamics of polymer solutions, *Macromolecules*, *4*(3): 356.

Ruff, W. A., Glover, C. J., and Watson, A. T. (1986a). Vapor–liquid equilibria from perturbation gas chromatography, Part I: Multicomponent parameter estimation, *AIChE J.*, *32*: 1948.

Ruff, W. A., Glover, C. J., Watson, A. T., Lau, W. R. and Holste, J. C. (1986b). Vapor–liquid equilibria from perturbation gas chromatography, Part II: Application to the polybutadiene/benzene/cyclohexane ternary system, *AIChE J.*, *32*: 1954.

Smidsrød, O., and Guillet, J. E. (1969). Study of polymer–solute interactions by gas chromatography, *Macromolecules*, *2*(3): 272.

3

Solvent Diffusion in Polymeric Systems

John M. Zielinski

Air Products & Chemicals, Inc., Allentown, Pennsylvania

J. L. Duda

The Pennsylvania State University, University Park, Pennsylvania

I. INTRODUCTION

Transport behavior of small molecules in polymers affects many industrial applications due to the escalating emphasis on reducing emissions from commercial processes. Although this book centers on the removal of volatiles from polymeric materials, possessing knowledge of the chemical migration rates within polymers is prerequisite in the design of many consumer products and furnishes the information necessary to design environmentally safer operations.

Besides devolatilization, molecular diffusion is a key step in assorted design processes and often governs the utility and manufacture of commercial products. A few appropriate examples include

1. *Time-controlled release products.* Diffusion in polymers is often the controlling mechanism for new medicinal products for the time-controlled release of medication. Similarly, controlled release of other products such as pesticides and fertilizers can be based on microencapsulation of chemicals within polymeric materials.

2. *Paints and coatings.* The physical properties and appearance of coatings can be modified by trace amounts of volatile components (Eaton, 1980; Ellis, 1983; Guerra et al., 1983). Knowledge of organic solvent diffusion rates is therefore vital for the successful market-ability of these products.

3. *Polymerization reactions.* The Trommsdorf (or gel) effect evident in free radical polymerizations arises from the strong dependence of termination reactions on molecular translational mobility, and is undoubtedly an engineering concern (Hanley et al., 1985). Gerbert and coworkers (1990, 1992) have demonstrated that the rate constants for diffusion-limited reactions may be related to the sum of the diffusion coefficients of the molecules undergoing interactions.

4. *Controlled morphology.* Macroscopic material properties are directly affected by microscopic characteristics (Creton et al., 1991; Brown et al., 1989). In one detailed study, the bulk morphology of block copolymer–homopolymer blends was systematically modified through the rate of the solvent diffusion (Löwenhaupt and Hellmann, 1991).

6. *Film production.* The design and control of many processes for producing films involves analysis of the diffusion of solvents in polymer systems.

7. *Packaging or barrier membrane synthesis.* The key property of various membranes is the control of transport through polymeric materials.

8. *Dyeing of fabrics.* The color of apparels is often a primary factor in selecting clothing. Thus, the effectiveness of dyeing fabrics is critical in textile sales (Cegarra and Puente, 1967).
9. *Membrane separation processes.* Most of the data available in the literature, concerning diffusion of low-molecular-weight materials in polymers are related to the development of polymeric membranes as separation processes for vapor and liquid systems.

The literature in these seemingly unrelated fields can provide information concerning the thermodynamics and transport phenomena in polymer–solvent systems that can be useful in the design and analysis of devolatilization processes. Analyses in previous chapters indicate that one of the fundamental physical properties required to design and optimize devolatilization processes is the mutual binary diffusion coefficient, D. Despite the importance of polymer–solvent mutual-diffusion coefficients, little reliable data of this type have been measured. Furthermore, a significant portion of these experimental investigations have focused on the dilute concentration regime, far from the region of interest in devolatilization. Theoretical models that accurately predict the diffusional characteristics of polymer–solvent systems are, therefore, highly sought after.

Although statistical mechanical analysis can be successfully applied to predict the conformational and migrational tendencies of single polymer chains, the overwhelming majority of industrial processing concerns require detailed information of diffusion within *concentrated* polymer solutions or polymer melts. Statistical mechanical analyses have not yet been successful in modeling transport when the coordinated motion of neighboring polymer chains and the existence of intermolecular interactions govern molecular transport.

No general theory currently exists for describing diffusional behavior over the complete concentration range. For the concentrated solutions of interest in polymer devolatilization, however, the theories based on system free volume have been the most useful for correlating and predicting the diffusivity. The most notable models incorporating free-volume concepts were developed by Cohen and Turnbull (1959), Fujita (1961), and Vrentas and Duda (1977a, b).

Currently, the most effective transport theory (that proposed by Vrentas and Duda) describes diffusion both above and below the glass transition temperature, T_g, by employing the notion that all transport processes are governed by the availability of free volume within a system. In this chapter, the fundamental principles associated with the Vrentas–Duda models for self- and binary mutual diffusion shall be developed for solutions containing uncross-linked amorphous polymer chains and low-molecular-weight solvents. Furthermore, since devolatilization equipment typically operates under conditions above the glass transition temperature, T_g, of the polymer–solvent mixture, we will emphasize concentrated rubbery polymer solutions that undergo *classical, Fickian* diffusion.

Besides temperature and composition, diffusion in polymers is influenced by morphological features such as crystallinity and chemical cross-linking, which tend to reduce molecular mobility. Fortunately, these complexities are often absent in devolatilization processes, and the free-volume theory, which is most appropriate for amorphous polymer systems above the glass transition temperature, can be directly applied.

II. FREE-VOLUME CONCEPTS

The motion of molecules within a liquid, relative to one another and with respect to stationary reference points, is greatly affected by temperature, pressure, and composition of the fluid. One approach to elucidate molecular transport within a fluid is rigorous and requires quantification of the frictional drag between molecules as they move past one another. Another strategy that is generally adopted relates the resistance to microscopic motion to macroscopic properties, which are subsequently determined via relatively simple experiments. Viscosity and diffusion coefficients are among such measurable properties providing tremendous insight into the microscopic nature of fluids, in particular, the resistance to molecular translational motion.

Because polymer viscosity is a very sensitive function of temperature, devolatilization processes are often designed for elevated temperatures to take advantage of this characteristic. Similarly, the rate of diffusion in polymers is also greatly enhanced by increases in temperature. The influence of temperature on both of these properties can be related to the amount of free space or free volume betweeen the polymer chains. As the temperature is raised, the kinetic energy of the polymer chains increases, causing the material to expand and enhancing the available free volume. From the free-volume point of view, this expansion of the volume not directly occupied by the polymer chains facilitates reorientational motion and migration of species through the material, and it reduces the friction between chains during flow. A key concept in the free-volume theory presented here is the analysis of viscosity data to predict diffusion rates within polymeric systems.

The specific volume of a condensed fluid ($\hat{V}$) is comprised of two parts: (1) the volume occupied by the molecules themselves ($\hat{V}_0$), and (2) the unoccupied space between the molecules, which is termed the *free volume* ($\hat{V}_{FV}$). In the early versions of free-volume theory (Cohen and Turnbull, 1959), $\hat{V}_{FV}$ was deemed to govern solution diffusion. Molecular transport as perceived by current free-volume theory, however, considers translational mobility to be dictated by only a portion of the free volume, which is continuously redistributed by thermal fluctuations.

As two molecules enter within close proximity to one another, their electron clouds overlap and repulsive forces are no longer negligible. Consequently, not all of the free volume within a system is regarded as equally accessible for molecular transport, and in reality, a distribution of attainable free-volume sites exists throughout the fluid. To account for this distribution, Vrentas and Duda adopted the view originally proposed by Berry and Fox (1968), that the total free volume could be arbitrarily divided into two components: (1) one available for molecular transport and (2) the other unavailable. These portions of the total free volume are termed the *hole free volume* ($\hat{V}_{FH}$) and the *interstitial free volume* ($\hat{V}_{FI}$), respectively.

According to the Vrentas–Duda theory, molecules diffuse by successive discrete jumps. A molecule is expected to take a diffusive step if (1) a vacancy of sufficient size appears adjacent to the molecule and (2) the molecule possesses enough energy to break nearest-neighbor contacts. Assuming that the vacancy and energy availabilities can be represented by Boltzmann probability functions, the expression for solvent self-diffusion (D_1) is obtained (Vrentas and Duda, 1977a)

$$D_1 = D_0 \exp\left(\frac{-E}{RT}\right) \exp\left(\frac{-\gamma(W_1 \hat{V}_1^* + W_2 \xi \hat{V}_2^*)}{\hat{V}_{FH}}\right) \tag{1}$$

where W_1 and W_2 are the weight fractions of the solvent and polymer, respectively. D_0 is a constant preexponential term, E is the molar energy a molecule needs to overcome attractive forces between nearest neighbors, and γ is an overlap factor (between $\frac{1}{2}$ and 1), which is introduced because the same free volume is available to more than one molecule. The specific volumes, $\hat{V}_1^*$ and $\hat{V}_2^*$, are the smallest holes that need to form before a solvent and a polymer, respectively, can take diffusive steps, and they are estimated as the specific volumes of the solvent and polymer at absolute zero temperature. Finally, ξ is the ratio of molar volumes for the solvent and polymer units involved in discrete diffusive jumps.

The total hole free volume available for diffusion, $\hat{V}_{FH}/\gamma$, has contributions from both the solvent and the polymer and is assumed to be a weight fraction average of the pure components

$$\frac{\hat{V}_{FH}}{\gamma} = W_1\left(\frac{K_{11}}{\gamma}\right)(K_{21} - T_{g1} + T) + W_2\left(\frac{K_{12}}{\gamma}\right)(K_{22} - T_{g2} + T) \tag{2}$$

where subscripts 1 and 2 refer to the solvent and polymer, respectively, and the K_{ij} parameters reflect the amount of hole free volume within the pure components (Vrentas and Duda, 1977b).

Solvent self-diffusion is the process by which a molecule undergoes a random walk in a homogeneous medium (i.e., without concentration gradients). Self-diffusion is synonymous with Brownian motion and, as such, is a function of both temperature and concentration. In devolatilizers, polymer–solvent solutions are subjected to conditions that impose concentration gradients. Where a concentration gradient exists, one must account for the spatial variation of penetrant activity. The effect of the imposed gradient on molecular motion is characterized by the mutual-diffusion coefficient (D). For the case of polymer–solvent solutions, D can be written as (Duda et al., 1982)

$$D = \frac{D_1 W_1 W_2}{RT} \left(\frac{\partial \mu_1}{\partial W_1} \right)_{T, P} = D_1 (1 - \phi_1)^2 (1 - 2\phi_1 \chi) \tag{3}$$

where μ_1 is the chemical potential of the solvent, ϕ_1 is the solvent volume fraction, and χ is a polymer–solvent interaction parameter, which reflects the thermodynamic compatibility of the polymer and solvent (Flory, 1942; Huggins, 1942).

Although Eq. (3) is an approximate relationship, it is a reasonable description of mutual diffusion provided that (1) the theoretical expression developed by Bearman (1961) accurately relates the polymer and solvent self-diffusion coefficients to the mutual-diffusion coefficient by means of the change of penetrant activity with concentration, (2) the polymer self-diffusion coefficient is negligible compared with the solvent self-diffusion coefficient, and (3) the change in the penetrant activity with respect to concentration can be modeled by the Flory–Huggins thermodynamic model. These assumptions are quite accurate in most devolatilization processes where the concentration of the volatile species is low.

The two exponential terms in Eq. (1) indicate the energetic and free-volume contributions to the self-diffusion process of a solvent. Since the availability of free volume in polymer systems limits transport much more than the energy required to break free from nearest neighbors (E), the preexponential and energetic terms are often approximated as a single constant, $D_{01}[= D_0 \exp(-E/RT)]$. Procedures for estimating each of the parameters in Eqs. (1) and (2) will be highlighted later in this chapter.

III. APPLICABILITY OF FREE-VOLUME THEORY

The free-volume concepts discussed here may be accurately applied to modeling molecular transport only when a system is free-volume-limited. At high temperatures, all known transport processes (e.g., viscosity, electrical and mechanical relaxation, and diffusion) are described well by the Arrhenius rate expression. However, at lower temperatures, when the availability of free volume is low, deviations from the Arrhenius behavior are observed, and the free-volume formalisms become applicable. The temperature up to which free-volume concepts can accurately be applied has been approximated as $T_g + 100°C$ (Ferry, 1970), although a recent study by Lomellini (1992) demonstrates that the melt viscosity of polycarbonate exhibits classical free-volume-controlled behavior up to $T_g + 185°C$. Clearly, though, at sufficiently high temperatures free-volume models fail as Arrhenius behavior (and the energy associated with nearest-neighbor interactions) dictates the transport mechanism.

Solvents typically contribute significantly more free volume to a polymer solution than do polymer chains themselves, since solvents have lower T_g values. Consequently, free-volume theory is most applicable in the concentrated polymer solutions ($W_1 \rightarrow 0$) realized in most devolatilization processes. However, solvent self-diffusion coefficients are often accurately predicted up to the pure solvent concentration limit ($W_1 \rightarrow 1$) (Zielinski and Duda, 1992a; Vrentas and Vrentas, 1991). Transport of smaller penetrant molecules, e.g., O_2 and N_2, may depend less on the availability of free volume than organic solvents. The motion of small molecules will be less restricted than that of larger molecules, such as monomers or organic solvents; consequently, smaller molecules will find cavities (holes) in which to jump during diffusion more easily.

IV. PARAMETER ESTIMATION FOR DIFFUSION ABOVE T_g

To facilitate use of the Vrentas–Duda free-volume theory (Eqs. 1–3) to predict solvent self- or polymer–solvent binary mutual-diffusion coefficients, a flow chart indicating the step-by-step methodology required is provided in Fig. 1. Based on this figure, only three pathways are conceivable from start to finish:

$$\text{Path 1} = \text{Step 1} \rightarrow \text{Step 2} \rightarrow (\text{Step 2a}) \rightarrow \text{Step 3a} \rightarrow \text{Step 4a} \qquad (4a)$$

$$\text{Path 2} = \text{Step 1} \rightarrow \text{Step 2} \rightarrow (\text{Step 2a}) \rightarrow \text{Step 3b} \rightarrow \text{Step 4b} \qquad (4b)$$

$$\text{Path 3} = \text{Step 1} \rightarrow \text{Step 2} \rightarrow (\text{Step 2a}) \rightarrow \text{Step 3b} \rightarrow \text{Step 4c} \qquad (4c)$$

Step 2a is necessary only if mutual-diffusion coefficients are to be predicted as a function of concentration. Since polymer–solvent thermodynamic interactions do not influence mutual diffusion at the infinite dilution limit ($W_1 \rightarrow 0$) or solvent self-diffusion at any concentration, Step 2a may be bypassed if predictions of these properties are exclusively of interest. In many

The Procedure for the Prediction of Diffusion Coefficients (T > T_{g2})

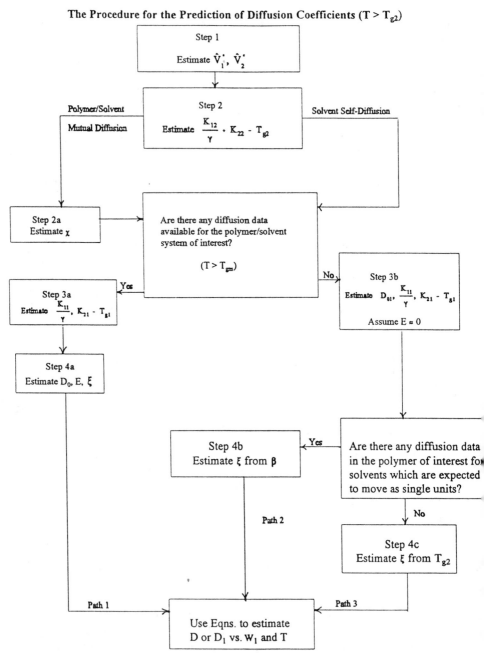

Fig. 1 Procedure for determining parameters needed in free-volume diffusion model. Effectiveness of predictions is rated as follows: Path 1 ≫ Path 2 > Path 3.

devolatilization processes, the concentration of the volatile impurity is very low, and the self-diffusion coefficient is effectively equal to the mutual binary diffusion coefficient.

As one might expect, the availability of a limited number of diffusion coefficient data points, particularly for the mixture of interest, significantly enhances the theory's predictive capabilities. Consequently, the effectiveness of the three pathways is rated as Path 1 $\gg$ Path 2 > Path 3. In the following sections, methods are discussed for completing each step in Fig. 1. In addition, the effectiveness of the predictions from each of the pathways is compared. To facilitate utilization of the free-volume theory, available free-volume parameters for polymers and solvents are presented in Tables 1 and 2.

Step 1: As mentioned earlier, $\hat{V}_1^*$ and $\hat{V}_2^*$, represent the minimum specific hole free volumes required to enable a solvent or a polymer to accomplish a diffusive step. It is reasonable to assume that the *smallest* possible void corresponds to a molecule's dimensions at 0 K. Sugden (1927) and Biltz (1934) have developed group contribution techniques that enable estimations of molar volumes at absolute zero temperature. Therefore, by knowing molecular weight and structure, one can easily estimate $\hat{V}_1^*$ and $\hat{V}_2^*$. Haward (1970) has summarized the Sugden and Biltz group contribution techniques, as do we now, for convenience, in Table 3.

Step 2: The polymer free-volume parameters, K_{12}/γ and $K_{22} - T_{g2}$, can be estimated by correlating the nonlinear temperature dependence of any material attribute that mimics the translational friction coefficient with the Vogel–Fulcher–Tammann (VFT) equation (1921, 1925, 1926)

$$\ln \zeta = \ln A_2 + \frac{\gamma \hat{V}_2^*/K_{12}}{K_{22} - T_{g2} + T} \tag{5}$$

Here ζ is the translational frictional coefficient. Zero-shear-rate viscosities (η_2) (Ferry, 1970), correlation times associated with ^{13}C spin-lattice relaxation (τ_2) (Dekezian et al., 1985; Blum et al., 1986; Spiess, 1990) and dielectric relaxation times (ψ_2) (Bidstrup and Simpson, 1989) are the three most common physical properties correlated with the VFT equation. Any one of these properties may replace ζ in Eq. (5) and be used to estimate K_{12}/γ and $K_{22} - T_{g2}$.

Williams, Landel, and Ferry (WLF) (1955) supplied a theoretical basis to the empirical VFT equation and proposed a slightly different functional form. If the glass transition temperature of the pure polymer, T_{g2}, is known,

K_{12}/γ and $K_{22} - T_{g2}$ can be calculated directly from the WLF parameters using

$$\frac{\gamma \hat{V}_2^*}{K_{12}} = 2.303 C_{12}^{\mathrm{WLF}} C_{22}^{\mathrm{WLF}} \tag{6}$$

$$K_{22} = C_{22}^{\mathrm{WLF}} \tag{7}$$

An extensive list of polymer WLF parameters has been published (Ferry, 1970).

Step 2a: Besides being widely available in the literature (Sheehan and Bisio, 1966; Brandrup and Immergut, 1989; Orwoll, 1977), the Flory–Huggins polymer–solvent interaction parameter, χ, can be determined from equilibrium solubility data in which the volume fraction of solvent, ϕ_1, is known as a function of the solvent activity. Implementing the assumption that the polymer molar volume is much larger than that of the solvent yields the form of the Flory–Huggins expression commonly employed

$$\text{Activity} \approx \frac{P_1}{P_1^0} = \phi_1 \exp(\phi_2 + \chi \phi_2^2) \tag{8}$$

where P_1 and P_1^0 are the experimental and saturation solvent vapor pressures, respectively.

An alternative χ-estimation technique assumes that polymer–solvent compatibility can be appraised through comparisons of solubility parameters. Bristow and Wilson (1958) developed a semiempirical relationship, given as

$$\chi = 0.35 + \frac{\tilde{V}_1}{RT} (\delta_1 - \delta_2)^2 \tag{9}$$

where $\tilde{V}_1$ is the solvent molar volume at temperature T. Although solubility parameters are readily available at 25°C for numerous materials, their temperature dependence is not as widely publicized. Since a solubility parameter *difference* is calculated to estimate χ, the issue of temperature dependence is most often sidestepped since the solubility parameter difference tends to be temperature-independent over relatively broad temperature ranges.

It should be noted that the Vrentas–Duda free-volume formalism for mutual diffusion does not require the Flory–Huggins equation to account for thermodynamic interactions. Equation (3) may be easily modified to accommodate more appropriate polymer solution thermodynamic models if (or when) they become available.

Table 1 Polymer Free-Volume Parameters

Polymer	$M_{2\text{mono}}$	M_{2j}	$\hat{V}_2^*$	$(K_{22}/\gamma) \times 10^4$	$K_{22} - T_{g2}$	C_{12}^{WLF}	C_{22}^{WLF}	β	T_{g2}
Butyl rubber	123.10		1.004	2.39	−96.4	16.76	108.60		205
Cis-1,4-poly(isoprene)	68.05	36.37	0.963	4.64	−146.4	16.79	53.60		200
Neoprene	88.53		0.708	3.91	−163.3	12.13	64.70		228
Poly(α-methylstyrene)	118.09		0.859	5.74	−395.7	13.17	49.30		445
Poly(carbonate)	254.29		0.732	15.20	−385.2	5.50	38.00		423
Poly(dimethyl siloxane)	74.15		0.905	9.32	−81.0	6.11	69.00		150
Poly(ethyl methacrylate)	113.86	140.00	0.915	3.40	−269.5	17.62	65.50	20.90	335
Poly(ethylene-propylene)	70.05		1.005	8.17	−175.3	13.11	40.70		216
Poly(ethylstyrene)	132.20		0.956	4.49	−286.9				355
Poly(isobutylene)	56.04		1.004	2.51	−100.6	16.63	104.40		205
Poly(methyl acrylate)	82.03	128.00	0.748	3.98	−231.0	18.13	45.00	19.50	276
Poly(methyl methacrylate)	100.04	187.81	0.788	3.05	−301.0	14.02	80.00	17.46	381
Poly(propylene)	42.03		1.005	5.02	−205.4	18.24	47.60		253
Poly(p-methylstyrene)	118.17		0.860	5.18	−330.0	15.00	48.00		378
Poly(styrene)	104.08	163.60	0.850	5.82	−327.0	13.78	46.00	10.50	373
Poly(styrene-butadiene)	158.13		0.789	6.60	−184.4	25.60	20.27		205
Poly(vinyl acetate)	86.02	134.20	0.728	4.33	−258.2	15.59	46.80	17.20	305

Source: Zielinski and Duda (1992a).

Table 2   Solvent Free-Volume Parameters

Solvent	M_1	$\hat{V}_1^*$	$\tilde{V}_1^0(0)$	$(K_{11}/\gamma) \times 10^3$	$K_{21} - T_{g1}$	$D_{01} \times 10^4$
Acetic Acid	60.05	0.773	46.45	0.546	−12.5	20.9
Acetone	58.08	0.943	54.77	1.393	−43.31	6.643
Benzene	78.11	0.901	70.38	1.21	−86.4	7.73
n-Butylbenzene	134.2	0.944	126.69	1.61	−112.3	3.3
Carbon Tetrachloride	153.84	0.469	72.15	0.642	−87.93	4.83
Chloroform	119.39	0.51	60.89	0.561	−21.8	7.303
Cyclohexane	84.16	1.008	84.8	2.26	−145.3	2.75
Cyclohexanol	100.2	0.882	88.34	0.733	−170.2	28.57
cis-Decalin	138.25	0.928	128.3	0.848	−86.2	7
trans-Decalin	138.25	0.928	128.3	0.775	−58.3	9
n-Decane	142.28	1.082	154	0.96	−42.8	10.6
di-Butylphthalate	278.3	0.737	205.2	0.777	−153.5	4.3
di-Isobutyl-phthalate	278.3	0.737	205.2	0.737	−189.9	1.7
di-Methyl-phthalate	194.2	0.609	118.2	1.19	−193.2	1.6
n-Dodecane	170.3	1.07	182.17	0.846	−46.1	11.8
n-Eicosane	282.6	1.043	294.85	0.783	−82.2	7.8
Ethylbenzene	106.16	0.928	98.52	1.4	−80	4.6
Ethylene Glycol	62.07	0.779	48.37	0.631	−130.9	16.94
Formic Acid	46.03	0.715	32.91	0.872	107.7	9.4
n-Heptadecane	240.5	1.05	252.6	0.691	−49	12.3
n-Heptane	100.2	1.115	117.75	1.33	−39.9	7.9
n-Hexadecane	226.4	1.053	238.51	0.731	−47.9	11
n-Hexane	86.17	1.133	97.66	1.41	−26.7	7.8
2-Hexanol	102.2	0.99	101.19	1.08	−152.6	1.7
n-Hexylbenzene	162.3	0.954	154.86	2.06	−148.6	1.7
Methanol	32.04	0.961	30.8	0.811	−25.6	25.7
Methyl Acetate	74.08	0.855	63.34	1.016	−33.52	9.52
Methylene Chloride	84.94	0.585	49.69	1.05	−62.2	3.9
Methyl Ethyl Ketone	72.11	0.997	71.9	0.643	51.1	30.2
Naphthalene	128.2	0.813	104.2	0.767	−73	7.4
n-Nonadecane	268.5	1.046	280.77	0.793	−79.9	8
n-Nonane	128.3	1.091	139.95	1.06	−42.9	9.6
n-Octadecane	254.5	1.048	266.68	0.773	−72.2	9.2
n-Octane	114.22	1.121	128.08	1.15	−37.4	9.3
n-Pentadecane	212.4	1.057	224.43	0.738	−47.6	12.1
n-Pentane	72.15	1.158	83.57	1.66	−23.6	6.9
n-Pentylbenzene	148.2	0.95	140.78	2.03	−143.3	2

Table 2 (*Continued*)

Solvent	M_1	$\hat{V}_1^*$	$\tilde{V}_1^0(0)$	$(K_{11}/\gamma) \times 10^3$	$K_{21} - T_{g1}$	$D_{01} \times 10^4$
n-Propylbenzene	120.2	0.937	112.61	1.67	− 109	3.1
1,2-Propylene Glycol	76.1	0.815	62.05	0.58	− 144.5	36.4
Styrene	104.2	0.899	93.67	0.876	− 48.4	11.5
n-Tetradecane	198.4	1.06	210.34	0.76	− 46.7	12.8
Tetrahydrofuran	72.1	0.899	64.82	0.753	10.4	14.4
Tetralin	132.2	0.861	113.84	0.966	− 99.2	5.7
n-Tridecane	184.4	1.064	196.26	0.808	− 47	11.6
Toluene	92.13	0.917	84.48	1.45	− 86.3	4.8
n-Undecane	156.3	1.075	168.09	0.917	− 47.8	10.7
Water	18.02	1.071	19.3	1.739	− 144.5	15.4
o-Xylene	106.16	1.049	111.36	0.899	− 27.7	12.4
p-Xylene	106.16	1.049	111.36	0.673	32	24.2

Source: Hong (1994)

Step 3a: The free-volume parameters, K_{11}/γ and $K_{21} - T_{g1}$, for the volatile species can be estimated by a procedure analogous to Step 2 for the polymer. Several alternative solvent properties such as zero-shear viscosity, NMR relaxation times associated with ^{13}C spin-lattice relaxation (Zielinski et al., 1992), and dielectric relaxation times can be used with the VFT equation (Eq. 5) to determine solvent free-volume characteristics. To obtain good free-volume parameter estimates, data must be available over a broad range of temperature, including low temperatures where the free volume controls the solvent behavior.

Step 3b: In the absence of any diffusivity data, a reasonable simplifying assumption is to neglect the activation energy ($E = 0$) required for the migrating volatile species to disassociate itself from its neighboring molecule. With this approximation, the self-diffusion expression for pure solvents as developed by Dullien (1972) can be coupled to the free-volume theory to obtain

$$\ln \frac{0.124 \times 10^{-16} \tilde{V}_c^{2/3} RT}{\eta_1 M_1 \hat{V}_1} = \ln D_{01} - \frac{\gamma \hat{V}_1^*/K_{11}}{K_{21} - T_{g1} + T} \tag{10}$$

In this expression, $\tilde{V}_c$ and M_1 are the solvent critical molar volume and molecular weight and 0.124×10^{-16} is a constant in Dullien's expression, which has the units of $mol^{2/3}$. The only temperature-dependent parameters in this expression are the viscosity and specific volume of the pure solvent,

η_1 and $\hat{V}_1$, respectively. A three-parameter regression of the pure solvent viscosity and specific volume as a function of temperature can be used to determine D_{01}, K_{11}/γ, and $K_{21} - T_{g1}$. Table 2 presents various solvent parameters that have been determined by this method (Hong, 1994).

Step 4a: The most difficult parameters in the theory to estimate from pure component properties are D_0, E, and ξ. Consequently, any available diffusivity data for the polymer–solvent system of interest should be used to determine these three key parameters. If the data are limited in scope, E may be set equal to zero as a reasonable approximation. If all the steps in the procedure preceding this one have been followed, all the parameters in Eqs. (1)–(3) will have been estimated except D_0, E, and ξ. The preferred approach is to use the available diffusivity data in a nonlinear regression analysis of Eqs. (1)–(3) to determine these three remaining parameters. Previous studies have shown that this is the most judicious use of diffusivity data to predict diffusion coefficients outside the temperature and/or concentration regions where the experimental measurements were performed.

Step 4b: The parameter that has been the most evasive and controversial in the theory is ξ, the ratio of the molar volume of a solvent jumping unit, $\tilde{V}_{1j}$, to the molar volume of a polymer jumping unit, $\tilde{V}_{2j}$. For many common low-molecular-weight volatile impurities in polymers, it is expected that these volatile species or solvents will move as complete units, and the molar volume of the solvent jumping unit is considered to be equal to the total molar volume of the pure solvent at absolute zero temperature: $\tilde{V}_{1j} = \tilde{V}_1(T = 0 \ K) = M_1 \hat{V}_1 \ (T = 0 \ K)$. In contrast, polymer molecules exhibit segmental motion, and the volume of the polymer jumping unit constitutes only a fractional part of the total volume of the polymer chain. $\tilde{V}_{2j}$ can be estimated using any available polymer–solvent diffusivity data for the polymer of interest. In this approach, it is assumed that the molar volume of the jumping unit of a specific polymer is independent of the characteristics of the solvents or low-molecular-weight species that may be present.

Following this line of reason, Ju et al. (1981b) have shown that diffusivity data for several solvents in a given polymer can be correlated by

$$\frac{\gamma \hat{V}_2^* \xi}{K_{12}} = \beta \tilde{V}_1(T = 0 \ K) \tag{11}$$

The success of this correlation indicates that the size of the polymer jumping unit is independent of the solvent and is polymer specific. It should be emphasized that this correlation is valid only for solvent or low-molecular-weight species that can be expected to jump as single units. The behavior of relatively large or chainlike solvents will deviate from this correlation. Values of β as well as other free-volume characteristics of several polymers as compiled by Zielinski and Duda (1992a) are presented in Table 1. Unfortunately, diffusivity data are scarce, and reliable values of β are available for only a limited number of polymers.

Step 4c: Until recently, there were no techniques available to estimate ξ if no diffusivity measurements were available for the polymer of interest. Zielinski and Duda (1992a) have successfully determined, however, that a correlation exists between the molar volume of a polymer's jumping unit and the polymer's glass transition temperature. Such a correlation is qualitatively reasonable since one expects polymers with high glass transition temperatures to have stiff chains and, consequently, relatively large segments of the chain to move together in any motion. It also seems logical that the size of a polymer segment involved in molecular diffusion would be related to the size of polymer units involved in dielectric or mechanical relaxation processes.

Matsuoka and Quan (1991) have presented a theory that can be used to predict the size of the polymer chain segment involved in the α-relaxation process at the glass transition temperature of the polymer. This theory also shows that the volume of a polymer segment leading to α-relaxation is related to the glass transition temperature. In Fig. 2, polymer segment sizes estimated from the Matsuka and Quan model are compared with $\tilde{V}_{2j}$ values obtained by free-volume theory correlations of diffusivity data (Faridi, 1995). The line in this figure represents a correlation of the size of the jumping unit as an exponential function of T_g. This figure shows reasonable agreement between the size of the polymer jumping unit obtained from the correlation of diffusivity data and the size of the polymer chain segment involved in the α-relaxation at the glass transition temperature. The most significant discrepancy occurs for polymers with low glass transition temperatures. These polymers are typically rubbers that contain a large amount of free volume at temperatures where experiments are generally conducted. Consequently, free-volume correlations of diffusion in such systems is precarious since free-volume availability may not be controlling the diffusion process. The scatter in the diffusivity data for these materials is clearly represented by the two data points presented in Fig. 2 for natural rubber. The natural rubber data point above the correlating line was obtained for relatively large solvents diffusing in rubber, and the point falling below the correlating line was obtained from diffusivity measurements of small fixed gases in the rubber.

Although the correlation between the jumping unit of the polymer and the glass transition temperature as represented in Fig. 2 is imprecise, this

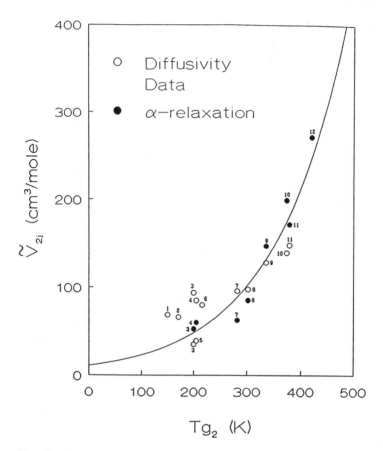

Fig. 2 Correlation of volume of polymer jumping unit with T_g. Open circle symbols refer to values determined from correlation of diffusivity data (Faridi, 1995). Closed symbols refer to volume of polymer chain segments involved in α-relaxation process at T_g as predicted by Matsuoka–Quan (1991) theory. Numbers identify polymers: 1—poly(dimethyl siloxane); 2—poly(butadiene); 3—natural rubber; 4—poly(isobutylene); 5—styrene–butadiene rubber; 6—ethylene–propylene rubber; 7—poly(methyl acrylate); 8—poly(vinyl acetate); 9—poly(ethyl methacrylate); 10—poly(styrene); 11—poly(methyl methacrylate); 12—poly(carbonate).

technique and the one proposed by Zielinski and Duda (1992a) are the only methods available for estimating the jumping unit size of a polymer if no diffusivity data are available for a specific polymer. An estimate of $\tilde{V}_{2j}$ from Fig. 2 can be combined with an estimate of the molar volume of a solvent as determined from the group contribution technique of Haward to estimate ξ for any polymer–solvent system.

Prediction and/or correlation of diffusion coefficients utilizing the free-volume theory presented in this chapter is particularly applicable for devolatilization. Devolatilization usually occurs at elevated temperatures where anomolous effects associated with many polymeric diffusion processes are not observed (Vrentas et al., 1975). Furthermore, the complications associated with semicrystallinity, cross-linking, and polymer chain orientation are usually absent in devolatilization processes. The critical devolatilization stage often occurs at low solvent or penetrant concentrations where the free-volume characteristics of the polymer dominate the process and the mutual binary diffusion coefficient for the polymer–solvent system is equal to the solvent self-diffusion coefficient in a limit of pure polymer. In this critical devolatilization limit of $W_1 \rightarrow 0$, Eqs. (1)–(3) can be approximated as (Ju et al., 1981b)

$$D = D_{01} \exp\left(\frac{-\gamma \hat{V}_2^* \xi}{K_{12}(K_{22} - T_{g2} + T)}\right) \tag{12}$$

Table 3 Group Contribution Methods for Estimating Molar Volumes at 0 K

Component	Sugden (cm³/mol)	Biltz (cm³/mol)
H	6.7	6.45
C (aliphatic)	1.1	0.77
C (aromatic)	1.1	5.1
N	3.6	—
N (in ammonia)	0.9	—
O	5.9	—
O (in alcohol)	3	—
F	10.3	—
Cl	19.3	16.3
Br	22.1	19.2
I	28.3	24.5
P	12.7	—
S	14.3	—
Triple bond	13.9	16
Double bond	8	8.6
3-membered ring	4.5	—
4-membered ring	3.2	—
5-membered ring	1.8	—
6-membered ring	0.6	—
OH (alcoholic)	—	10.5
OOH (carboxyl)	—	23.2

Source: Haward (1970).

This approximation facilitates use of the theory since the number of parameters required is reduced to six. Five of these parameters are readily estimated for most systems. Polymer viscosity data or WLF parameters can be used to determine K_{12}/γ and K_{22} (Step 2). $\hat{V}_2^*$ can be estimated from the group contribution technique of Table 3, and an estimate of ξ can be obtained from T_{g2} (Step 4c). Consequently, an estimate of D_{01} is the most critical step for predicting diffusion in many devolatilization processes. Estimating D_{01} through Step 3b is very approximate, and it is highly desirable that at least one diffusion coefficient measurement be available for an accurate analysis of devolatilization processes. Since it is often difficult to obtain diffusion coefficient measurements at the conditions of devolatilization, the free-volume theory is particularly effective for estimating diffusion coefficients at high temperatures from more easily available low-temperature diffusivity measurements. Many studies, such as Ju et al. (1981a), have shown that the free-volume theory is useful for determining the influence of temperature on diffusivity.

V. QUALITATIVE CHARACTERISTICS FOR POLYMER–SOLVENT DIFFUSION ABOVE T_g

Application of the free-volume theory for diffusion in polymer systems can occur on several levels. The theory unquestionably provides an excellent framework for the correlation of diffusivity data and, with the use of limited data (Paths 1 and 2), enables one to accurately predict diffusion coefficients at temperatures and concentrations where data are not available. At the present state of development, however, the completely predictive theory (Path 3) is inadequate for some applications. Nonetheless, this is the only theory available that can predict mutual binary diffusion coefficients in concentrated polymer solutions. The theory can also provide a basis for a qualitative understanding of the influence of such variables as temperature, concentration, and solvent molecular size on diffusional behavior in devolitilizers. The following confirmed experimental observations are consistent with the free-volume formalism:

1. The apparent activation energy of the mutual binary diffusion process, E_D, is a strong function of temperature and concentration and increases as the solution approaches the glass transition temperature of the system. Stated another way, the concentration dependence of the mutual-diffusion coefficient is greatest near the glass transition temperature. Consequently, forcing a system further away from T_g by altering experimental

conditions (e.g., increasing solvent concentration or system temperature) results in a decrease in the apparent activation energy.

These trends reflect that free volume is gained as solutions are removed from T_g, and that solvents typically contribute more hole free volume to polymer solutions than do the polymers themselves.

2. Molecular diffusion within polymer solutions exhibits greater temperature and concentration dependence as solvent size increases. This is particularly true for large, spherical rigid molecules that are expected to move as a single units, since the probability of a large molecule locating a sufficiently large free-volume pocket enabling it to take a diffusive step is quite low. Small changes in the available free volume, caused by increases in temperature or solvent concentration, can considerably modify the diffusion coefficient.

The diffusivities of small gas molecules within polymeric systems, on the other hand, are either very weak functions, or independent of concentration, because of their small size. For these systems, the apparent activation energy is virtually constant and the effect of temperature on the diffusion process can be accurately described by the Arrhenius expression.

3. Elastomers or rubbers at conditions far above T_g possess a large amount of hole free volume. Solvent diffusion, therefore, occurs rapidly and exhibits a weak dependence on both concentration and temperature.

4. In contrast to behavior in dilute polymer solutions, mutual binary diffusion in concentrated polymer solutions is a very weak function of polymer molecular weight. Although the free volume associated with the terminal chain segments is greater than that associated with interior chain segments, the increase in the number of end units does not significantly affect the overall free volume until relatively low-molecular-weight polymers are involved. Consequently, variations of the molecular weight distribution and the average molecular weights of most commercial polymers do not significantly influence diffusion in concentrated polymer systems.

5. Additives directly influence molecular diffusion through their contribution to system hole free volume. Plasticizers, for example, significantly enhance the availability of free volume and thus, enhance rates of translational motion. Addition of impermeable solids, such as fillers, may not significantly modify the available free volume, but it causes a significant decrease in the diffusivity through modification of the path along which a molecule can travel (i.e., increased tortuosity).

6. In most cases, the free volume associated with the solvent is significantly greater than the free volume of the pure polymer, and the mutual binary diffusion coefficient will increase with the addition of solvent. Thus, the rate of change of the diffusivity with increasing solvent concentration is

greatest when a system is close to T_g. In addition, the larger the solvent, the more dramatic is the influence of concentration.

7. Polymer–solvent thermodynamic interactions serve to decrease the rate of mutual diffusion. This is particularly true for poorly compatible solvent–polymer mixtures. The competing effects of increasing free volume and thermodynamic interactions render a maximum in the mutual binary diffusion coefficient as a function of solvent concentration as a net result. This maximum shifts to lower solvent concentrations as temperature is increased and may ultimately reach the pure polymer limit. For these rare cases, the thermodynamic forces dominate the mutual-diffusion process, and thus, the mutual-diffusion coefficient decreases with increasing solvent concentration over the entire concentration range. The solvent self-diffusion coefficient, which is completely unaffected by thermodynamic interactions, typically increases with solvent concentration.

One concern often expressed regarding free-volume theory is that molecular interactions between the polymer chain and the solvent molecules are not included in the model. This point of view appears to be substantiated by the fact that all the parameters in the theory can be estimated from pure component properties. However, specific polymer–solvent interactions are introduced in two ways: (1) the thermodynamic term in Eq. (3) reflects molecular interactions through the chemical potential of the solvent in the solution; and (2) through the activation energy, E, which reflects the interaction between the solvent jumping unit and its neighbor. E can usually be neglected since its contribution to the activation energy for diffusion is generally small compared with the apparent activation energy associated with free-volume availability. There is evidence (Zielinski and Duda, 1992b), however, that E is concentration-dependent, suggesting that the energy required for a solvent molecule to break the interactions with its neighbors is different depending on whether the neighbors are other solvent molecules or segments of the polymer chain.

From all available experimental studies, it appears that for concentrated polymer solutions in the vicinity of the glass transition ($T_{g2} < T < T_{g2} + 100°C$), the diffusion process is dominated by the scarcity of free volume, while specific molecular interactions are of secondary importance. This conclusion may reflect that the great majority of accessible diffusivity data are for systems that do not exhibit strong interactions such as hydrogen bonding. However, a recent study (Yapel et al., 1994) of the gelatin–water system (which exhibits strong polymer–solvent interactions) over a broad range of concentration indicates that free-volume theory does an excellent job of describing diffusion in this system. Even for this system, of strong interactions, very good correlations are attained by neglecting the energy interactions between the diffusing water molecule and its surrounding polymer matrix (i.e., $E = 0$).

The free-volume concepts implemented to derive the previous equations permit a phenomenonological description of the diffusion process. The success the WLF and VFT models have enjoyed in correlating viscosity data as a function of temperature suggests that the free-volume approach is, at the very least, consistent with empirical evidence.

Several studies have investigated the correlative and predictive capabilities of the free-volume theory for concentrated amorphous polymer systems above T_g, which are of interest in devolatilization. Zielinski and Duda (1992a) present the most recent review and evaluation of the theory.

For illustrative purposes, the influence of concentration and temperature on diffusion in the toluene–polystyrene system is presented in Fig. 3. The parameters presented by Zielinski and Duda (1992a) were utilized for this

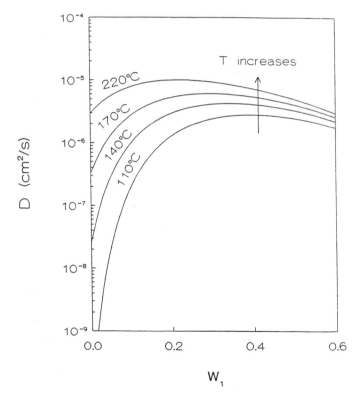

Fig. 3 Free-volume theory prediction of mutual binary diffusion coefficient for the toluene–polystyrene system based on parameters of Zielinski and Duda (1992a).

system, and these predictions are in good agreement with the available data. Figure 3 clearly shows that at relatively low temperatures near the glass transition, diffusivity is a very strong function of temperature and concentration. However, at elevated temperatures, which are characteristic of most devolatilization processes, the influence of temperature and concentration is much more moderate. In fact, the free-volume theory predicts and experimental results by Iwai et al. (1989) confirm, that at temperatures far above the glass transition, the diffusion coefficient can actually decrease with increasing solvent concentration. Significant free volume is available in polymers at these elevated temperatures, and the enhancement of free volume associated with the addition of a low-molecular-weight solvent can be somewhat insignificant. At elevated temperatures, the thermodynamic term in Eq. (3) can start to dominate the free-volume contributions to the solvent self-diffusion coefficient, and the thermodynamics can, in extreme cases, cause the diffusivity to decrease slightly with solvent concentration. This extreme behavior is not realized in most devolatilization processes. However, it is common for the diffusion coefficient to be only a weak function of concentration over the range of conditions utilized in many devolatilization processes.

VI. DIFFUSION OF TRACE AMOUNTS OF SOLVENT BELOW T_g

Most polymers are devolatilized at temperatures above T_g. In fact, the temperature in devolatilizers is often pushed up to the limit of the polymer's degradation temperature to reduce the viscosity of the system and enhance the diffusion process. In a few relatively rare cases, polymers are devolatilized below T_g. The most notable case is the removal of monomer residuals from small, glassy polymers of polyvinyl chloride, as discussed by Berens (1981). As amorphous polymers are cooled, the motion of individual polymer chains becomes so constrained that the cooling rate becomes faster than the rate at which the polymer sample can volumetrically relax. The resulting *nonequilibrium* condition is referred to as the glassy state, and the point at which a material is no longer viewed as rubbery, but rather glassy, is denoted the glass transition temperature, T_g.

Ideally, the T_g transition manifests itself as a step change in the slope of the specific volume–temperature ($\hat{V}$ vs. T) relation, reflecting a finite change in heat capacities. In practice, however, the dynamic phenomenon of the glass transition occurs over a temperature range whose width is influenced by the polymer's thermal and mechanical history, as well as the rate of quiescent cooling and/or frequency of the experiment and the molecular characteristics of the fluid.

Above T_g, polymer chains are capable of achieving equilibrium configurations, whereas polymer segments in the glassy state have inadequate mobility to attain equilibrium conformations within commonly referenced time scales. Consequently, extra hole free volume becomes trapped within the polymer as it is cooled through the glass transition. Although the rate of molecular motion within glassy polymers prevents volume relaxation from reaching equilibrium, molecular motions are not completely eliminated. Density fluctuations persist in the glass and necessitate, consequently, redistributions of the hole free volume.

The formalisms presented in the preceding section would be appropriate for modeling transport below T_g in a polymer that was given ample time to relax to its equilibrium state. Due to the lengthy time required for glassy polymers to attain equilibrium, however, diffusion occurs under non-equilibrium conditions in which a polymer sample has more hole free volume than it would at equilibrium. Predictions of the availability of hole free volume from transport measurements made in rubbery (equilibrium) state are always appreciably low and, consequently, lead to diffusion coefficient predictions in the glassy state that are lower than measured experimentally.

We emphasize that although additional hole free volume is *trapped* in a glassy polymer, this volume continues to be redistributed throughout the sample. Trapping suggests that, within the time frame of interest, the rate of volumetric collapse is not appreciable.

To accurately predict diffusion in the glassy state, the extra hole free volume associated with this state must be quantified. To avoid the complications associated with the reduction in the glass transition temperature with increasing solvent content, the ensuing discussion will focus on devolatilization of trace amounts of solvent molecules.

In the pure polymer limit, the free-volume formalism developed for diffusion in the rubbery state can be modified (Vrentas and Vrentas, 1992) to address this entrapment of excess free volume through a simple adjustment of Eq. (2), i.e.,

$$\frac{\hat{V}_{FH}(W_1 \to 0, T)}{\gamma} = \frac{K_{12}}{\gamma}(K_{22} + \lambda[T - T_{g2}]) \tag{13}$$

where λ is defined as

$$\lambda = 1 - \frac{(\alpha_2 - \alpha_{2g})\hat{V}_2^0(T_{g2})}{\gamma(K_{12}/\gamma)} \tag{14}$$

Here, α_2 is the thermal expansion coefficient of the equilibrium liquid, α_{2g} is the thermal expansion coefficient of the glassy polymer, and $\hat{V}_2^0(T_{g2})$ is the specific volume of the equilibrium liquid polymer at the glass transition temperature of the pure polymer. Therefore, λ delineates the nature of the volumetric contraction as a polymer is cooled through T_{g2} and by definition is found in the interval $0 \leq \lambda \leq 1$. A λ value of unity implies that a polymer is at its equilibrium state.

Laurence and coworkers (Arnould, 1989) applied capillary column inverse gas chromatography (CCIGC) to make extensive infinite dilution diffusivity measurements of solvent transport in polymers over a range of temperatures. Their results, as well as the data obtained by Coutandin et al. (1985), who employed a holographic grating technique, clearly indicate that the temperature behavior of the diffusion coefficient is consistent with the predictions of Eqs. (1) and (2); even when T_g is traversed.

Although λ can be calculated directly from knowledge of a polymer's density as a function of temperature, it may also be regressed from diffusion coefficient data. Equation (14) specifies that λ is a characteristic quantity of the pure polymer and, therefore, must be independent of the penetrant. The available data (Pawlisch et al., 1987) confirms this result since a single value of λ can be used to characterize the diffusion coefficients of several different solvents in a specific polymer. As far as we know, however, no study has ever compared λ values deduced from volumetric data with values estimated from diffusion studies.

VII. QUALITATIVE CHARACTERISTICS FOR POLYMER–SOLVENT DIFFUSION BELOW T_g

Although free-volume transport theory has not been extensively evaluated for finite solvent concentrations at temperatures below T_g, it continues to lend itself to qualitative interpretations consistent with experimental observations. As for the case of diffusion above the glass transition, we summarize the behavior of solvent diffusion in glassy systems suggested by the free-volume formalisms:

1. Contrary to intuition the rate of diffusion (or the diffusion coefficient) does not abruptly decrease as a system cools to the glassy state. Rather, free-volume theory predicts that at a fixed solvent concentration, the diffusion coefficient is a continuous function of temperature as the glass transition temperature is traversed. On the other hand, the apparent activation energy for diffusion, E_D, exhibits a discontinuity at T_g.

2. The step change that E_D undergoes at T_g increases as penetrant size increases. Consequently, it is entirely possible that measurements of the diffusion of small molecules, such as fixed gases, will not register any apparent E_D change at T_g, while larger molecules such as solvent vapors will exhibit significant decreases in E_D as temperature is lowered through T_g. In addition, the change in E_D at T_g will decrease as solvent concentration increases; i.e.,

the influence of the glass transition on the diffusion process is most apparent for trace amounts of large solvent molecules in the polymer matrix.

3. The addition of hole free volume from the solvent generally outweighs the loss of excess hole free volume associated with depressing T_g. Solvents or penetrants that significantly relax the glassy matrix structure without bringing substantial free volume to the mixture are called *antiplasticizers*. Thus, the diffusion coefficient can actually decrease with increasing solvent concentration (Vrentas and Vrentas, 1992; Vrentas et al., 1988).

4. The rate of diffusion in a glassy polymer system is influenced by the polymer's history. Aging or annealing causes a densification of the polymer and a reduction in excess hole free volume. Correspondingly, the free-volume model suggests that as a glassy polymer ages and relaxes toward its equilibrium state, diffusional rates become slower.

VIII. SUMMARY AND CONCLUDING REMARKS

As preceding chapters have demonstrated, information concerning mutual-diffusion coefficients is required to analyze, design, and optimize devolatilization processes. In most cases, a low vapor pressure at the surface of the polymer melt creates a concentration gradient that drives the diffusion of the volatile residual. Consequently, thermodynamics or phase equilibria determine the ultimate theoretical obtainable residual concentration, and the diffusion process dicates the rate of approach to this limiting concentration. The volatile residual mass transfer is usually facilitated by convective flow or mixing of the polymer melt. Convection carries new polymer melt to the vapor interface and minimizes the diffusion path required for the migration of the volatile residuals. Consequently, the complete analysis of devolatilization processes presented in other chapters is a complex coupling of fluid mechanics, thermodynamics, and molecular diffusion. This chapter has concentrated on the behavior and prediction of the diffusion coefficients for volatile residuals–polymer systems. Although it has been shown that it is possible to theoretically predict this coefficient without the utilization of any diffusivity measurements, such predictions are quite precarious; it is not advisable to design a devolatilization process without the guidance of any diffusivity measurements. Quite often, devolatilization processes operate at elevated temperatures where polymer degradation inhibits the direct measurement of the required diffusion coefficients. The free-volume theory presented in this chapter can be quite useful in predicting diffusion coefficients at devolatilization conditions from lower temperature measurements.

Comparisons often show that the apparent effective diffusion coefficient in the devolatilizer is higher than the direct experimental measurement of the coefficient. Molecular diffusion in a polymer melt is a very slow process, and any relatively low flow or convection can completely swamp the diffusion process that would be realized in a stagnant pool of molten polymer. Consequently, inadequacies in the modeling of devolatilization processes that underestimate convective flow will result in effective diffusion coefficients greater than those observed in well-controlled diffusivity measurement experiments.

The preponderance of diffusivity data that were obtained under conditions realized in devolatilizers have been produced by gravimetric sorption techniques. In these techniques, the amount of solvent that diffuses into a molten pool of polymer exposed to solvent vapor is measured as a function of time. Although these experimental techniques appear to be relatively straightforward, it is difficult and time consuming to obtain accurate diffusion coefficients. These experimental difficulties are no doubt the reason that there is a paucity of data in this area. Most recently, experimental procedures based on capillary column techniques have been developed for the measurement of diffusion coefficients in polymer systems and may significantly add to the future data pool (Arnould, 1989; Pawlisch et al., 1987; Gray et al., 1973; Senich, 1979; Senich and Sanchez, 1981; Hadj Romdhane and Danner, 1993). The chromatographic technique utilizes capillary columns coated with the polymer. Problems may arise with the flow of the polymer film down the walls of the capillary at the elevated temperatures characteristic of devolatilization processes. However, these techniques show great promise for the relatively rapid production of diffusivity data at modest temperatures and the infinitely dilute limit ($W_1 \rightarrow 0$). The free-volume formalism presented here can be used to extrapolate the data to the operating conditions within devolatilizers.

The focus of this chapter has been limited to diffusion in binary systems consisting of a solvent or volatile residual and a polymer. Many devolatilization operations involve several volatile species, and the question of diffusion in multicomponent systems naturally arises. In most devolatilization processes, the concentrations of the various volatile residuals is quite low, and as a good first approximation, the free-volume characteristics of the system are controlled by the free-volume characteristics of the polymer. Also, under such dilute conditions, there would be very little interaction between the diffusive fluxes of the various volatile species, so that ternary diffusion effects such as cross-diffusion coefficients can be neglected. Consequently, for most devolatilization processes, the techniques presented in this chapter can be used to predict effective binary diffusion coefficients for the individual volatile species in the polymer. In systems containing two or more solvents in a polymer, the pseudo binary approximation will become invalid when the concentration of the solvents is high enough to significantly influence the amount of free volume available in the system. Under such conditions, increase in the available free volume provided by solvent A will not only enhance the diffusion of solvent A, it will enhance the diffusion of other species in the system. Vrentas et al. (1984) extended the free-volume theory to describe diffusion in polymer–solvent–solvent systems when the free volumes associated with the solvents significantly enhance the total free volume of the system. It has been suggested that the addition of a second solvent can significantly enhance the devolatilization of an undesirable volatile residual. Vrentas et al. (1985) utilized the free-volume model for ternary systems to investigate the enhancement of impurity removal from polymer films by the addition of a second solvent. The enhancement they considered is based on the increase in the diffusion coefficient of the undesirable impurity due to the addition of free volume associated with a second solvent.

It is clear from the free-volume theory that any compound added to the polymer phase that increases the free volume of that phase will have a positive influence on the removal of impurities. Consequently, if possible, additives such as plasticizers and lubricants should be added before the devolatilization process to facilitate impurity removal. As a first approximation, the polymer and additive can be lumped together and treated as a polymer species with specific free-volume characteristics through which the volatile impurity is diffusing. Similarly, the addition of plasticizers can enhance diffusion below the glass transition temperature.

In some cases, a nonsolvent such as water is added to devolatilization processes to enhance the mass transfer. In many of these cases, the enhancement realized is not due to an increase in the effective diffusion coefficient; it is related to a reduction in the effective diffusion path associated with the formation of new surface area by bubbling or foaming of the added volatile material.

The influence of the polymer molecular weight on devolatilization is often of interest. To appropriately account for the influence of the molecular weight and molecular weight distribution, one must obtain the free-volume parameters or WLF coefficients for the specific polymer system to be devolatilized. However, for polymer systems of conventional commercial molecular weights, the dependence on molecular weight is insignificant. If the polymer being devolatilized has a relatively low-molecular-weight additive such as a lubricant or plasticizer, it is advisable to obtain viscosity versus temperature data for the actual mixture to provide the appropriate free-volume parameters.

Devolatilization processes can be significantly enhanced by changes and modifications that influence the convection, mixing, or formation of new surface areas. In contrast, it is difficult to significantly modify the diffusion coefficient, which is a fixed property of the polymer–solvent system at a specific temperature and concentration. In the analysis of devolatilization processes, it is implicitly assumed that the appropriate diffusion coefficient is only a function of the state of the system and is not related to shear rates or levels of strain. There is a possibility that diffusion through a stagnant pool of molten polymer is different from diffusion through polymer chains that have been oriented by the shear associated with flow. To date, however, we have no conclusive experimental evidence that the rates of shear in devolatilization processes significantly influence the mutual binary diffusion coefficients. For most systems, the magnitude of such an effect is probably well within the error associated with experimental measurements or the theoretical predictions.

NOMENCLATURE

A_2 proportionality constant for the VFT equation (Eq. 5)

C_{12}^{WLF} WLF parameter

C_{22}^{WLF} WLF parameter (K)

D polymer/solvent binary mutual-diffusion coefficient (cm^2/sec)

D_0 constant preexponential factor (cm^2/sec)

D_{01} constant preexponential factor when E is presumed to be equal to 0 (cm^2/sec)

D_1 solvent self-diffusion coefficient (cm^2/sec)

E energy required to overcome attractive forces from neighboring molecules (cal/mol)

E_D effective activation energy for diffusion (cal/mol)

K_{ij} free-volume parameters for the Vrentas–Duda formalism

M_1 solvent molecular weight (g/mol)

P_1 solvent vapor pressure (atm)

P_1^0 solvent vapor pressure at saturated conditions (atm)

R gas constant (e.g., 1.9872 cal/mol-K)

T absolute temperature (K)

T_g glass transition temperature (K)

T_{g1} solvent glass transition temperature (K)

T_{g2} polymer glass transition temperature (K)

T_{gm} glass transition temperature of polymer–solvent mixture (K)

$\hat{V}$ specific volume (cm^3/g)

$\hat{V}_{FV}$ specific free volume (cm^3/g)

$\hat{V}_{FH}$ specific hole free volume (cm^3/g)

$\hat{V}_{FI}$ specific interstitial free volume (cm^3/g)

$\hat{V}_0$ specific occupied volume (cm^3/g)

$\hat{V}_1$ solvent specific volume (cm^3/g)

$\hat{V}_2$ polymer specific volume (cm^3/g)

$\hat{V}_1^*$ solvent specific occupied volume at 0 K (cm³/g)

$\hat{V}_2^*$ polymer specific occupied volume at 0 K (cm³/g)

$\tilde{V}_C$ solvent critical molar volume (cm³/mol)

$\tilde{V}_1$ solvent molar volume (cm³/mol)

$\tilde{V}_{1j}$ molar volume of solvent jumping unit (cm³/mol)

$\tilde{V}_{2j}$ molar volume of polymer jumping unit (cm³/mol)

$\hat{V}_2^0$ specific volume of polymer at equilibrium (cm³/g)

W_1 solvent weight fraction

W_2 polymer weight fraction

Greek Letters

α_2 volume expansion parameter for an equilibrium rubber (K^{-1})

α_{2g} volume expansion parameter for a glassy polymer (K^{-1})

β polymer-specific proportionality constant (Eq. 11)

γ overlap factor, which accounts for shared free volume

δ_1 solvent solubility parameter (cal/g)$^{1/2}$

δ_2 polymer solubility parameter (cal/g)$^{1/2}$

ζ translational friction coefficient (dyne-sec/cm)

η_1 solvent viscosity (g/cm-sec)

η_2 polymer viscosity (g/cm-sec)

λ free-volume parameter for glassy polymer (Eq. 14)

μ_1 chemical potential of solvent

ξ ratio of solvent and polymer jumping units

τ_2 correlation times for T_1 relaxation determined by ^{13}C NMR (sec)

ϕ_1 solvent volume fraction

ϕ_2 polymer volume fraction

χ Flory–Huggins polymer–solvent interaction parameter

ψ_2 correlation times associated with dielectric relaxation (sec)

REFERENCES

Arnould, D. D. (1989). *Capillary Column Inverse Gas Chromatography (CCIGC) for the Study of Diffusion in Polymer–Solvent Systems*, Ph.D. thesis, U. of Massachusetts.

Bearman, R. J. (1961). *J. Phys. Chem.*, 65: 1961.

Berens, A. R. (1981). *Pure and Applied Chem.*, 53: 365.

Berry, G. C., and Fox, T. G. (1968). *Adv. Poly. Sci.*, 5: 261.

Bidstrup, S. A., and Simpson, J. O. (1981). Proceedings of 18th North American Thermal Analysis Society, vol. 1, p. 366.

Biltz, W. (1934). *Rauchemie der Festen Stoffe*, Voss, Leipzig.

Blum, F. D., Durairaj, B., and Padmanabhan, A. S. (1986). *J. Poly. Sci.: Part B: Poly. Phys.*, 24: 493.

Brandrup, J., and Immergut, E. H. (1989). *Polymer Handbook*, 3rd. ed., Wiley, New York.

Bristow, G. M., and Watson, W. F. (1958). *Trans. Farad. Soc.*, *54*: 1731.

Brown, H. R., Argon, A. S., Cohen, R. E., Gebizlioglu, O. S., and Kramer, E. S. (1989). *Macromolecules*, *22*: 1002.

Cegarra, J., and Puente, P. (1967). *Text. Res. J.*, *37*: 343.

Cohen, M. H., and Turnbull, D. (1959). *J. Chem. Phys.*, *31*(5): 1164.

Coutandin, J., Ehlich, D., Sillescu, H., and Wang, C. H. (1985). *Macromolecules*, *18*: 587.

Creton, C., Kramer, E. J., and Hadziioannou, G. (1991). *Macromolecules*, *24*(8): 1846.

Dekmezian, A., Axelson, D. E., Dechter, J. J., Borah, B., and Mandelkern, L. (1985). *J. Poly. Sci.: Part B: Poly. Phys.*, *23*: 367.

Duda, J. L., Vrentas, J. S., Ju, S. T., and Liu, H. T. (1982). *AIChE J.*, *28*(2): 279.

Dullien, F. A. L. (1972). *AIChE J.*, *18*(1): 62.

Eaton, R. F. (1980). *J. Coatings Tech.*, *52*(660): 63.

Ellis, W. H. (1983). *J. Coatings Tech.*, *55*(695): 63.

Faridi, N. (1995). To be published.

Ferry, J. D. (1970). *Viscoelastic Properties of Polymers*, 2nd ed., Wiley, New York.

Flory, P. J. (1942). *J. Chem. Phys.*, *10*: 51.

Fulcher, G. S. (1925). *J. Am. Ceram. Soc.*, *77*: 3701.

Fujita, H. (1961). *Fortschr. Hochpolym. Forsch*, *3*: 1.

Gerbert, M. S., and Torkelson, J. M. (1990). *Polymer*, *31*(12): 2402.

Gerbert, M. S., Yu, D. H. S., and Torkelson, J. M. (1992). *Macromolecules*, *25*(16): 4160.

Gray, D. G., and Guillet, J. E. (1973). *Macromolecules*, *6*(2): 223.

Guerra, G., Paolone, C., and Nicolais, L. (1983). *J. Coatings Tech.*, *55*(701): 53.

Hadj Romdhane, I., and Danner, R. P. (1993). *AIChE J.*, *39*: 625.

Hanley, B., Tirrell, M., Balloge, S., and Tulig, T. (1985). *Polymer Preprints*, *26*(1): 299.

Haward, R. N. (1970). *J. Macromol. Sci. Rev. Macromol. Chem.*, *C4*: 191.

Hong S. U. (1994). To be published.

Huggins, M. L. (1942). *J. Am. Chem. Soc.*, *64*: 1712.

Iwai, Y., Maruyama, S., Fujimoto, M., Miyamoto, S., and Arai, Y. (1989). *Poly. Eng. Sci.*, *29*.

Ju, S. T., Duda, J. L., and Vrentas, J. S. (1981a). *Ind. Eng. Chem. Prod. Res. Dev.*, *20*: 330.

Ju, S. T., Liu, H. T., Duda, J. L., and Vrentas, J. S. (1981b). *J. Appl. Poly. Sci.*, *26*: 3735.

Lomellini, P. (1992). *Makromol. Chem.*, *193*: 69.

Löwenhaupt, B., and Hellmann, G. P. (1991). *Polymer*, *32*: 1065.

Matsuoka, S., and Quan, X. (1991). *Macromolecules*, *24*: 2770.

Orwoll, R. A. (1977). *Rubber Chem. and Tech.*, *50*: 451.

Pawlishc, C. A., Macris, A., and Laurence, R. L. (1987). *Macromolecules*, *20*: 1564.

Senich, G. A. (1981). *Chemtech*, *11*(6): 360.

Senich, G. A., and Sanchez, I. C. (1979). *Org. Coat. Plast. Chem.*, *41*: 345.

Sheehan, C. J., and Bisio, A. L. (1966). *Rubber Chem. and Tech.*, *39*: 149.

Spiess, H. W. (1990). *Polymer Preprints*, *31*: 103.

Sugden, S. (1927). *J. Chem. Soc.*: 1786.

Tammann, G., and Hesses, W. Z. (1926). *Z. Anorg. Allg. Chem.*, *156*: 245.

Vogel, H. (1921). *Z. Physik*, *22*: 645.

Vrentas, J. S., and Duda, J. L. (1977a). *J. Poly. Sci. Part B: Poly. Phys.*, *15*: 403.

Vrentas, J. S. and Duda, J. L. (1977b). *J. Poly. Sci. Part B: Poly. Phys.*, *15*: 417.

Vrentas, J. S., and Vrentas, C. M. (1991). *Macromolecules*, *24*: 2404.

Vrentas, J. S., and Vrentas, C. M. (1992). *J. Poly. Sci. Part B: Poly., Phys.*, *30*: 1005.

Vrentas, J. S., Jarzebski, C. M., and Duda, J. L. (1975). *AIChE J.*, *21*(5): 894.

Vrentas, J. S., Duda, J. L., and Ling, H. C. (1984). *J. Poly. Sci.: Part B: Poly. Phys.*, *22*: 459.

Vrentas, J. S., Duda, J. L., and Ling, H. C. (1985). *J. Appl. Poly. Sci.*, *30*: 4499.

Vrentas, J. S., Duda, J. L., and Ling, H. C. (1988). *Macromolecules*, *21*: 1470.

Williams, M. L., Landel. R. F., and Ferry, J. D. (1955). *J. Am. Chem. Soc.*, *77*: 3701.

Yapel, R. A., Duda, J. L., Lin, X., and Von Meerwall, E. (1994). *Polymer*, *35*: 2411.

Zielinski, J. M., and Duda, J. L. (1992a). *AIChE J.*, *38*(3): 405.

Zielinski, J. M., and Duda, J. L. (1992b). *J. Poly. Sci.: Part B: Poly. Phys.*, *30*: 1081.

Zielinski, J. M., Benesi, A. J., and Duda, J. L. (1992). *Ind. Eng. Chem. Res.*, *31*: 2146.

4

Bubble Nucleation in Polymer Mixtures

Ali V. Yazdi and Eric J. Beckman

University of Pittsburgh, Pittsburgh, Pennsylvania

I. INTRODUCTION

In situations where a polymer melt is saturated with a volatile solvent, bubble nucleation is a key step in the overall process of devolatilization. Significantly, as we describe in this chapter, those material parameters on which the process of bubble nucleation depends heavily are themselves strong functions of concentration in polymer systems. As such, the ease of bubble formation in a polymer melt will vary significantly as the process of devolatilization proceeds.

Bubble nucleation can be described as an event where a second (dispersed) phase is generated from within a metastable single phase. Homogeneous nucleation occurs when bubbles of the dispersed phase, here the gas (or solvent), are generated wholly within the continuous phase, in our case the polymer-rich matrix. Heterogeneous nucleation occurs when bubbles are nucleated at the interface between two phases, generally a solid and a liquid, but is not restricted to that case.

II. HOMOGENEOUS NUCLEATION—THEORETICAL APPROACHES

A. Work Through 1980

Theoretical work on the rate of nucleation of gas bubbles from liquids can be traced to the fundamental thermodynamic treatises of Gibbs in the nineteenth century. Numerous researchers continued to hone the theory throughout the twentieth century, and excellent reviews of their work can be found in the literature by Volmer and Webber (1926), Gibbs (1961), Farkas (1927), Kaischew and Stranski (1934), Becker and Doring (1935), Zeldovich (1943), and Frenkel (1946). This review will begin with the series of papers published in the late 1960s (through 1979) by Katz et al. (1966), Katz and Blander (1973), Katz (1970), Katz and Donohue (1979), Blander et al. (1971), which reviewed the earlier work and made substantial contributions to the theory of bubble nucleation. The work by Kagan (1960) must also be mentioned here, for its important advances to nucleation theory through the inclusion of hydrodynamic and thermodynamic constraints on bubble formation.

1. Thermodynamics

To begin, the thermodynamics of homogeneous nucleation was derived originally for formation of gas bubbles in superheated liquids by Volmer and Weber (1926), based on the fundamental work by Gibbs (1961). In these derivations, nucleation involves an activation process that leads to the formation of unstable initial fragments of the dispersed phase, or embryos.

The embryos form due to small density or concentration fluctuations in the metastable continuous phase, and they grow or disappear due to vaporization or condensation of the dispersed phase (gas) molecules. The thermodynamics of bubble nucleation are consequently based on determination of the reversible work (the change in Helmholtz free energy) required to form an embryo within the metastable continuous phase. To form a bubble of volume V_g from the homogeneous liquid phase at constant temperature requires expenditure of work equal to σA to create the new surface, where σ is the interfacial tension at the gas–liquid interface. Creation of the embryo also involves work owing to gas molecules vaporizing and entering the cavity. The total work involved in creating the bubble embryo is therefore (Blander and Katz, 1975)

$$W = \sigma A - (P_B - P_L)V_G + x(\mu_G - \mu_L) \tag{1}$$

where P_B and P_L are the pressures in the bubble and the liquid phases, x is the number of gas molecules, and $\mu_G - \mu_L$ is the chemical potential difference between molecules in the gas (bubble) and liquid phases. By inserting expressions for the surface area and volume of a spherical bubble, one can express Eq. (1) in terms of r, the radius of the bubble. The work required to form a bubble thus increases initially as the volume increases, then decreases above a critical radius. Blander and Katz (1975) expanded Eq. (1) as a Taylor series around both the critical radius and a critical pressure (pressure was incorporated via conversion of the chemical potentials to a fugacity basis) and employed the Laplace equation,

$$P_B \approx P_L + \frac{2\sigma}{r} \tag{2}$$

to approximate the pressure in the bubble. Thus Eq. (1) is transformed to

$$W \approx \frac{4}{3}\pi\sigma r_c^2 - 4\pi\sigma(r - r_c)^2 B + \cdots \tag{3}$$

where, for bubbles close to mechanical equilibrium,

$$B \approx 1 - \frac{1}{3}\left(1 - \frac{P_L}{P_1}\right) \tag{4}$$

At the critical radius, the bubble is in chemical equilibrium with the continuous phase, and thus $P_B = P_1$, and $\mu_G = \mu_L$. Combining Eqs. (2) and (3) at $r = r_c$ leads to

$$W_{cr} = \frac{16\pi\sigma^3}{3(P_1 - P_L)^2} \tag{5}$$

and

$$P_1 = P_L + \frac{2\sigma}{r_c} \tag{6}$$

for critical bubbles. Bubbles smaller than the critical radius tend to collapse, while those equal to or larger tend to continue to grow. Equations (5) and (6) can also be derived via putting Eq. (1) in terms of the radius, assuming chemical equilibrium ($P_B = P_1$ and $\mu_G = \mu_L$), and then minimizing the work with respect to r (Han, 1988).

2. Kinetics of Bubble Nucleation

The kinetics of bubble nucleation are based on the original derivations of Volmer and Weber (1926) and Frenkel (1946). In summary, we assume that bubbles are formed due to density (or concentration) fluctuations in the initially homogeneous fluid. Such fluctuations generate an area within the fluid that, at least momentarily, may be considered to be a bubble containing x molecules of the gas. This bubble can then grow or shrink upon addition or subtraction of gas molecules. It is assumed that the number of bubbles (n) containing x molecules is related to the minimum work required for bubble formation by

$$n = N \exp \frac{-W}{kT} \tag{7}$$

where N is simply the number density of the liquid. At steady state, the net rate at which bubbles go from containing x to $x + 1$ molecules is presumed to be independent of x. Volmer showed that J, the rate of nucleation, can be derived as

$$J = \frac{\beta}{\int [nA]^{-1} dx} \tag{8}$$

where β is the rate per unit area at which a bubble surface gains or loses molecules, approximated by $\beta = P/(2\pi mkT)^{1/2}$, and where n is defined by Eq. (7). The rate of bubble formation has been evaluated (Katz, 1970) via expansion of the logarithmic term in Eq. (7) as a Taylor's series around the critical value, followed by insertion into Eq. (8). The rate of nucleation then becomes

$$J = \beta A(r_c) n(r_c) Z \tag{9}$$

where the rate of nucleation is found to be proportional to the number of critical size bubbles at equilibrium ($n(r_c)$), and the surface area of a critical bubble $(A(r_c))$. The quantity Z, known as the Zeldovitch factor, and given by

$$Z = \left(\frac{\sigma kT}{B}\right)^{1/2} [P_1 A(r_c)]^{-1} \tag{10}$$

is a correction to the nucleation rate accounting for the number of critical size bubbles at steady state being less than that at equilibrium. Consequently, the rate of nucleation becomes

$$J = N\left(\frac{2\sigma}{\pi m B}\right)^{1/2} \exp\frac{-16\pi\sigma^3}{3kT(P_1 - P_L)^2} \tag{11}$$

where m is the mass of a molecule. This equation has been the starting point for many theoretical explorations of bubble nucleation in liquids.

3. Hydrodynamic and Transport Constraints on Bubble Nucleation

In the derivation of Eq. (11), neither hydrodynamic nor transport effects on bubble nucleation have been included; some researchers have modified the basic derivation for J by including these physical phenomena in the derivation. Recent work has also focused on nucleation from liquid mixtures, and the basic theory has been modified accordingly. In their simplest form, these modifications to the basic theory can be seen as means to allow the driving force for nucleation, $P_1 - P_L$, to be altered by thermodynamic and hydrodynamic effects. For example, expansion of a bubble leads to cooling at the interface, which will thus change the effective pressure at the interface. In addition, generation of a bubble within a viscous fluid will induce a retractive force from the fluid itself, which will change the effective pressure within the bubble as well.

For example, Kagan (1960) derived an extension of the basic theory that accounts for both transport and hydrodynamic constraints on nucleation. In Kagan's analysis, the rate of nucleation is governed by the rate of vaporization of molecules from the liquid–gas interface into the bubble. Vaporization carries enthalpy from the liquid into the gas, and thus a corresponding transfer of enthalpy to the interface from the bulk liquid is required (expansion cooling, in effect). The impedance to this enthalpy transfer (i.e., it is not instantaneous heat transfer) slows the rate of vaporization and thus lowers the rate of nucleation. Kagan's derivation begins with an expression for the net rate of vaporization of molecules into the bubble:

$$\frac{dx}{dt} = A(2\pi m kT)^{-1/2}[P_1(T_s) - P_B] \tag{12}$$

where $P_1(T_s)$ is the vapor pressure of the liquid at the interface. As the bubble expands, T_s (and thus $P_1(T_s)$) can drop, thus reducing the driving force for nucleation. Kagan employed an energy balance at the interface to derive a correction factor for Eq. (12):

$$\frac{dx}{dt} = A(2\pi m k T)^{-1/2}[P_1(T_0) - P_B](1 + \delta_\lambda)^{-1} \tag{13}$$

where T_0 is the temperature of the bulk liquid, and the correction factor, δ_λ, is defined as

$$\delta_\lambda = \left(\frac{\Delta H_v}{kT}\right)^2 \frac{\sigma}{\delta_e}\left(\frac{2k}{\pi m T}\right)^{1/2} \frac{P_1}{P_1 - P_L} \tag{14}$$

where ΔH_v is the heat of vaporization and δ_e is the thermal conductivity of the liquid. After completion of the heat balance, Kagan then derived the rate of nucleation as

$$J_\lambda = \frac{J}{1 + \delta_\lambda} \tag{15}$$

where J is given by Eq. (11), and δ_λ is defined in Eq. (14). Thus, the effect of heat transfer on nucleation may be seen to be affecting P_1 in Eq. (11) through a variation of the temperature at the interface, although heat effects appear explicitly through the correction factor in Eq. (15). In effect, because the driving force for nucleation has been altered through a lowering of the actual vapor pressure at the interface, the nucleation rate is depressed over that calculated using Eq. (11) alone.

Likewise, in bubble nucleation in mixtures at steady state, molecules diffusing into the bubble must be replaced by molecules diffusing to the interface from the liquid (again, at a finite rate), which presents yet another mechanism for the effective lowering of P_1, and thus a depression of the driving force for nucleation. Using a similar approach to that taken by Kagan for solving the heat flow problem, Blander et al. (1971) solved the mass balance at the interface for critical size bubbles and consequently derived a correction to the nucleation rate owing to mass transfer effects:

$$J_D = \frac{J}{2(1 + \delta_D)} \tag{16}$$

where δ_D is derived to be

$$\delta_D = 2\sigma \frac{dP}{dC} D(2\pi m k T)^{-1/2}(P_1 - P_L)^{-1} \tag{17}$$

where D is the diffusion coefficient and C is the concentration. Blander and Katz (1975) have subsequently examined the case where a mixture contains a volatile component in very dilute solution with a much less volatile continuous phase (much like the situation existing in polymer devolatilization). In this example, where Henry's law is expected to be obeyed (thus $dP/dC = P/C$), they show that the correction factor δ_D can be very large, and thus the rate of nucleation is directly proportional to the diffusion coefficient (diffusionally limited nucleation). For such dilute mixtures, if the viscosity is only weakly dependent on temperature, the variation of nucleation rate with temperature is essentially reduced to the effect of the enthalpy of mixing on P_1 (and thus the driving force for nucleation). If the enthalpy is negative, then J will increase with temperature, whereas the opposite will occur for a positive enthalpy of mixing.

Along with restrictions due to heat and diffusional effects, growth is impeded by viscous and inertial forces. These forces can be thought of as retractive forces by the fluid resisting expansion of the embryo. In Kagan's derivation for the net rate of vaporization of molecules into the bubble (as shown in Eq. (13)), the rate was found to be proportional to P_B, the pressure inside the bubble, which was taken to be that derived in Eq. (2). In describing the limitations on nucleation due to hydrodynamic forces, Kagan used the following expression for P_B:

$$P_B = P_L + dr\, r + \frac{3}{2}\, d\dot{r}^2 + \frac{2\sigma}{r} + 4\eta\, \frac{\dot{r}}{r} \tag{18}$$

When inertial and viscous forces are small, P_B will be approximately equal to the expression in Eq. (2); when these forces are large, P_B will be approximately equal to P_1. In cases where the viscous terms are small, and the viscosity itself relatively high, Kagan derived the following corrected expression for J:

$$J = N\frac{\sigma}{\eta}\left(\frac{\sigma}{kT}\right)^{1/2}\left(1 - \frac{P_L}{P_1}\right)\exp\frac{-16\pi\sigma^3}{3kT(P_1 - P_L)^2} \tag{19}$$

Thus, just as heat transfer effects can be seen as lowering the driving force for nucleation through a lowering of P_1, so too can viscous forces be seen as depressing the driving force through an increase in P_B. Given the assumptions made by Kagan in generating Eq. (19), it should be considered as a corrected form of the nucleation rate for Newtonian fluids. Other researchers, most notably Ruengphrathuengsuka and Flumerfelt (1992), have included non-Newtonian effects as well.

B. More Recent Approaches

Whereas the early derivations just described were performed primarily for the case of vapor nucleation within a homogeneous liquid (boiling), other researchers have altered the basic theories for use specifically with polymeric systems. Ruengphrathuengsuka and Flumerfelt conducted an investigation of nucleation in the low-density polyethylene (LDPE)/nitrogen system, to support their work on generation of foamed materials. They began their derivation for the nucleation rate for gases in polymer melts at essentially the same place as Kagan, in a description of the net rate of vaporization of molecules into a bubble. A differential equation (in time) for the concentration of bubbles of size n (where n is the number of molecules) was constructed, then integrated at steady state to yield the nucleation rate, J_S. After some manipulation, their base equation for nucleation becomes

$$J_S = kTN\left(\frac{\partial \dot{n}}{\partial R}\right)_{Rc} \left[\frac{\partial^2}{\partial R^2}(\Delta G_n)\right]^{-1} \left[\int \exp\frac{-\Delta G_n}{kT} dR\right]^{-1} \tag{20}$$

where $\dot{n}$ is defined in essentially the same way as Kagan's dx/dt:

$$\dot{n} = \frac{\alpha v_t \pi R^2}{kT}[P_1(T_s, C_s) - P_B] \tag{21a}$$

$$v_t = \left(\frac{8kT}{\pi m}\right)^{1/2} \tag{21b}$$

where α is the evaporation frequency, and ΔG_n is the free energy associated with dormation of an n-molecule cluster. Ruengphrathuengsuka then proceeded to incorporate the effects of heat and mass transfer into the nucleation rate using an approach similar to that of Kagan. Consequently, Ruengphrathuengsuka derived correction factors to J_s accounting for changes to P_1 from heat effects (δ_λ) and mass transfer effects (δ_D). Again, while the former term is a function of the heat of vaporization and thermal conductivity, the latter depends on the diffusion coefficient of the gas in the matrix.

Ruengphrathuengsuka also incorporated hydrodynamic effects into the nucleation rate, again using the technqiue whereby P_B is modified from the classical expression. Unlike Kagan, who cemployed a Newtonian constitutive equation for the fluid, Ruengphrathuengsuka employed the momentum balance for a bubble growing in an infinite fluid:

$$P_B - P_L = \frac{2\sigma}{R} + 2\int \frac{\tau_{rr} - \tau_{\theta\theta}}{r} dr \tag{22}$$

and the Larsen rheological equation (Larsen, 1988) to describe the non-Newtonian behavior (the first normal stress difference) of the fluid. The resulting value for P_B is then employed in the expression for $\dot{n}$ (Eqs. 21a,b), and in an expression for the free energy, ΔG_n. These contributions are then employed with Eq. (20) to derive the nucleation rate:

$$J_S = \alpha N(1 + \delta_\lambda + \delta_D)^{-1} \frac{2\sigma}{\pi m} \left(1 - \frac{\rho^{(G)}\mu_G^{(L)}R_c}{\sigma M_G}\right) \exp \frac{-\Delta G_c}{kT} \tag{23}$$

where ΔG_c is the free energy change for the bubbles of critical size, given by

$$\Delta G_c = \frac{-16\pi}{3\sigma^3} \left(\frac{P_B - P_L + \rho^{(G)}\mu_G^{(L)}}{M_G}\right)^{-1} \tag{24}$$

and $\rho^{(G)}$ is the density of the gas phase, $\mu_G^{(L)}$ is the chemical potential of the gaseous molecules in the liquid phase, and M_G is the molecular weight of the gas. Heat and mass transfer limitations to nucleation appear explicitly through the terms δ_λ and δ_D, while material effects (hydrodynamic contributions) appear through their effect on P_B, which then governs ΔG_c through Eq. (24).

Han and Han (1988, 1990a,b; Han, 1988), employed a different approach to model nucleation of bubbles in concentrated polystyrene–toluene solutions. After finding that the classical theory, as expressed by a form of Eq. (11), did not predict any nucleation under conditions at which nuclei were experimentally observed, they modified the free energy term in the expression. The original free energy term (Eq. 5) was modified by two additional terms: one representing the change in free energy owing to the presence of macromolecules in the solvent, and another owing to the fact that the polymer solution is supersaturated under nucleation conditions. In general, one can summarize this approach by stating that Han and Han assumed that the critical nuclei are not at chemical equilibrium, as is usually presumed, and modified the free energy accordingly. First, using the Flory–Huggins theory for polymer–solvent thermodynamics, they determined the correction to the free energy owing to the nonideality of the solution as

$$\Delta F_t = n_1(\mu_1 - \mu_1^0) = n_1 kT(\ln \phi_1 + \phi_2 + \chi\phi_2^2) \tag{25}$$

As can be seen by examining Eq. (24), Ruengphrathuengsuka also included the effect of nonidealities in the solution through inclusion of an extra term (over that in Eq. (5)) in the critical free energy derivation. This extra term, however, was left in terms of a chemical potential and thus requires a specific model for the mixture thermodynamics to evaluate the free energy completely.

Next, Han and Han generated a correction to the free energy owing to the fact that the critical nuclei are not at chemical equilibrium, and thus $P_B \neq P_1$. This is an interesting approach, in that while other researchers have assumed that, indeed, $P_B \neq P_1$, this was generally assumed to be due to hydrodynamic effects in the fluid. Ruengphrathuengsuka did, however, include a thermodynamic correction to P_B over that found in the traditional Laplace-type expression, which arose due to the form of the free energy he employed. In Han and Han's derivation, the free energy itself is modified a second time by a term that accounts for the supersaturation in the solution, namely

$$\Delta F_s = nkT \ln s = nkT \ln \frac{C_0 - \Delta C(t)}{C_\infty(t)} \tag{26}$$

where C_0 is the initial concentration of the volatile component in a polymer solution at time $t = 0$, $\Delta C(t)$ is the amount of the volatile component consumed per unit volume of solution at time t, and $C_\infty(t)$ is the equilibrium concentration corresponding to the partial pressure of the volatile component in the vapor phase at time t. Thus, the critical free energy becomes

$$\Delta F = \Delta F_{Eq.~(5)} - \Delta F_t - \Delta F_s \tag{27}$$

Han and Han then used Eq. (27) in an expression for the nucleation rate that resembles Eq. (11), i.e.,

$$J = NB' \exp \frac{-\Delta F}{nkT} \tag{28}$$

where B' is a frequency factor that is inversely proportional to r_c^2, directly proportional to the diffusion coefficient, and exponentially dependent on temperature. Goel and Beckman (1994a,b, 1995) used a similar expression to describe bubble nucleation in polymethyl methacrylate (PMMA)–CO_2 mixtures at high pressure, although without Han and Han's modifications to the critical free energy.

Finally, Han and Han employed the following expression to account for changes to P_B due to supersaturation:

$$P_B(t) = P_1 \left\{ 1 + Q \ln \frac{C_0 - \Delta C(t)}{C_\infty(t)} \right\} \tag{29}$$

where Q is an empirical parameter. Equation (29) was derived in its present form to satisfy the boundary conditions that $P_B = P_1$ at $S = 1$ and P_B/P_1 becomes infinite as S approaches infinity. Q was determined numerically by equating the results of Eq. (29) with a hydrodynamic equation for bubble growth (essentially Eq. (18), employed by Kagan). Thus, P_B is modified via both hydrodynamic and supersaturation factors, and the critical free energy is modified by factors accounting for both nonidealities in the solution and supersaturation of the polymer by the diluent.

In a series of papers, Colton and Suh (1987a–c) derived an expression for homogeneous nucleation in polymers whose approach is somewhat similar to that of Han and Han, in that they also modify the free energy term. Here, though, the free energy is modified to account for increases in free volume owing to the addition of the volatile component. Colton derives a potential energy expression based on a simple 6–12 Lennard–Jones type of relationship between energy and distance. Further, the distance between polymer segments is presumed to increase in a predictable way as the amount of gas in the system, the temperature, or the pressure is varied. The change in this potential energy is then subtracted from the free energy expression in Eq. (5), which is then used in an equation of the general form of Eq. (11) to model homogeneous nucleation. Here, therefore, the diluent's net effect is limited to an increase in free volume of the polymer, which reduces the thermodynamic barrier to nucleation.

C. Experimental Observations—Polymer Systems

Gent and Tompkins (1969a,b) and Denecour and Gent (1968) studied the growth of gas bubbles in cross-linked elastomers under relatively low degrees of supersaturation, finding that the gas supersaturation pressure must exceed $5G/2$ (where G is the shear modulus) for bubbles to form. Stewart (1970) expanded on this work through the study of bubble formation in several elastomers that had been pressurized by argon up to 3000 atm. In general, Stewart found that the cell density (the summation of the nucleation rate over the total growth period) rose dramatically above a certain pressure (in the neighborhood of 500 atm, but varying by polymer type), then quickly leveled off. Stewart, employing a slight variant of Eq. (11) to model the rate of nucleation, found the basic theory to predict the pressure dependence of the cell density quite well, although the theory overpredicted the total number of bubbles by at least an order of magnitude. Jennings and Middleman (1985) examined homogeneous nucleation in polymer solutions and found that the presence of the polymer increased the magnitude of the superheat by approximately 10°C, depending on both concentration and the molecular weight of the polymer.

Han and Han (1990a) employed a light-scattering technique to measure the rate of nucleation of bubbles in polystyrene–toluene solutions. They measured the bubble density versus time, finding that the rate increases sharply and then levels out (see Fig. 1). In addition, the rate of nucleation decreases in this system as the concentration of polymer in the system decreases (Fig. 2). This was attributed to the increase in both the diffusion coefficient (which appears in the prefactor in Han and Han's nucleation rate model) and the degree of supersaturation at 150°C as the polymer concentration increases. This experimental trend agrees with the predictions

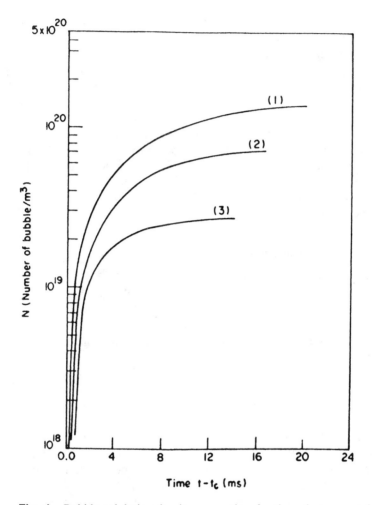

Fig. 1   Bubble polulation density versus time for the polystyrene–toluene solution at 150°C for several polystyrene concentrations (wt %): (1) 60, (2) 50, (3) 40. (From Han and Han, 1990a.)

of their nucleation model (see Eq. 28), although the model overpredicts the number of bubbles by four orders of magnitude. Han and Han also found that as nucleation proceeds, bubble coalescence begins to dominate bubble nucleation.

   In another study, Han and Han (1988b) found that bubble nucleation could be flow induced, as well as induced via pressure quenches and

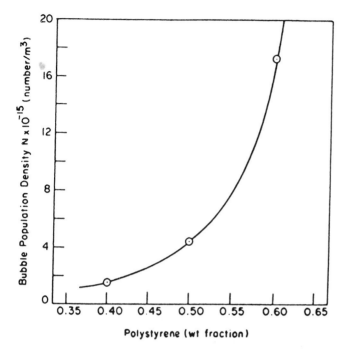

Fig. 2 Average bubble population density versus polystyrene concentration at 150°C and equilibrium pressure 2859 kPa. (From Han and Han, 1990a.)

temperature rises. In a polystyrene–freon 11 mixture, they found that the critical stress for bubble nucleation increases as the amount of diluent in the mixture increases (see Fig. 3). That one can nucleate bubbles via stress-induced mechanisms is not surprising, given that other researchers have shown that polymer–solvent combinations can be induced to both phase separate or mix through the application of stress fields (Rangel-Nafaile et al., 1984; Tirrell, 1986).

Ruengphrathuengsuka and Flumerfelt, in their study of nucleation in LDPE–N_2 mixtures, found trends in bubble number density versus temperature and pressure that are similar to those found by other workers. In particular, Ruengphrathuengsuka observed that the number of bubbles nucleated increased with increasing temperature, owing ostensibly to the drop in surface tension and melt viscosity as temperature increased. The increase occurred despite a drop in gas solubility in the polymer as temperature increased at constant pressure (see Fig. 4). As pressure increases, bubble number also increases, due to an increase in gas solubility with increasing pressure, and the concomitant drops in both surface tension and

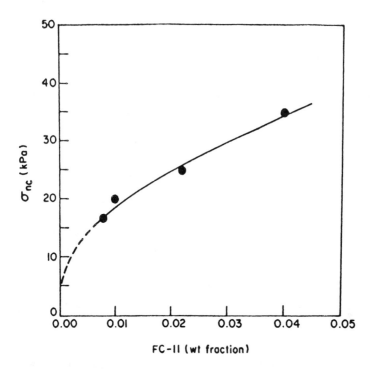

Fig. 3 Critical stress for bubble nucleation (σ_{nc}) versus concentration of FC-11 for STYRON 678/FC-11 Mixtures. (From Han and Han, 1988b.)

melt viscosity (Figs. 5 and 6). Identical trends were reported by Goel and Beckman for the PMMA–CO_2 system, as shown in Figs. 7 and 8.

III. HETEROGENEOUS BUBBLE NUCLEATION

Heterogeneous nucleation occurs when a third phase is generated at the interface between a volatile liquid and another phase in contact with the liquid. The essential difference between homogeneous and heterogeneous nucleation is the existence of two phases initially. This occurs when one of the phases is present at a concentration higher than its solubility limit for the given temperature and pressure, such as a system of a polymer and an additive present in a concentration higher than its solubility limit. In general, heterogeneous nucleation results in the formation of a smaller number of bubbles than homogeneous nucleation, because the number of bubbles in a heterogeneous system is limited by the number of the dispersed particles of the second phase. Because of the tendency of the dispersed phase to

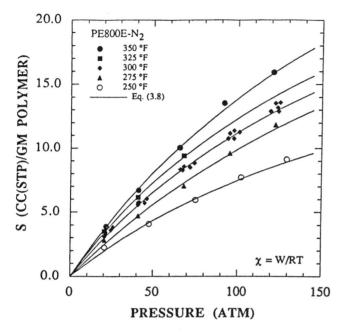

Fig. 4   Solubility of N_2 in LDPE (PE800E). (From Ruengphrathuengsuka, 1992.)

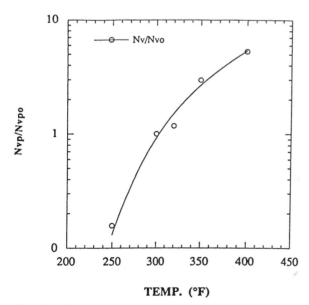

Fig. 5   Effect of temperature on the bubble number density. (From Ruengphrathuengsuka, 1992.)

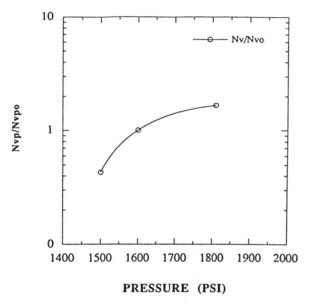

Fig. 6 Effect of pressure on the bubble number density. (From Ruengphrathueng-suka, 1992.)

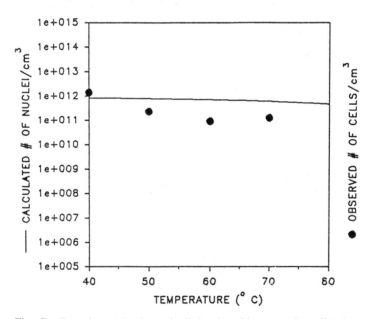

Fig. 7 Experimental values of cell density of foams and predicted number of nuclei as functions of temperature. (From Goel and Beckman, 1994a.)

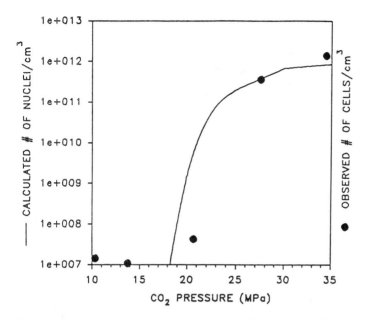

Fig. 8 Experimental values of cell density of foams and predicted number of nuclei as functions of pressure. (From Goel and Beckman, 1994a.)

congregate to form larger particles, this number is usually smaller than the number of sites generated in a homogeneous nucleation process.

The available heterogeneous nucleation theories can be divided into two groups: (1) those which more or less follow the classical nucleation theory principles, which tend to be more theoretical, and (2) those that do not follow the classical nucleation theory, which tend to apply to practical but particular cases.

A. Classical Nucleation Theory

Blander and Katz (1975) proposed an extension to their homogeneous nucleation model applicable to heterogeneous nucleation. The expression for minimum work required for creation of a bubble at the interface of a liquid and a smooth surface is written as

$$W = \sigma_{LG} A_{LG} + (\sigma_{SG} - \sigma_{SL}) A_{SG} - (P_B - P_L) V_G + x(\mu_G - \mu_L) \quad (30)$$

where σ refers to surface tension of corresponding interfaces, A is the surface area, x is the number of gas molecules in the bubble, and μ is the chemical potential. The subscripts LG, SG, and SL refer to the liquid–gas, surface–gas,

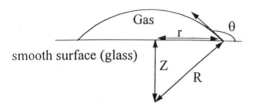

Fig. 9 Geometry of a bubble at a smooth surface (as used by Blander and Katz, 1975).

and surface–liquid interfaces, respectively. Note the resemblance of this equation to Eq. 1.

Using the nomenclature depicted in Fig. 9, they defined $m \equiv Z/R$, and proceeded by substituting volume and area expressions into the foregoing expression. The ideal gas law was used to write the number of molecules in terms of pressure, temperature, and volume, and the chemical potential terms were substituted by the partial pressures of the appropriate phase. Using a similar approach to the development of their homogeneous nucleation expression, the following expression for the number of heterogeneous nucleation sites was derived:

$$J = N^{2/3} S \left(\frac{2\sigma}{\pi m B F} \right)^{1/2} \exp \frac{-16\pi\sigma^3 F}{3kT(P_V - P_L)^2} \tag{31}$$

where $S \equiv (1 - m)/2$, $F \equiv (2 - 3m + m^3)/4$, and the other symbols retain their previous definitions.

The authors took the same approach for heterogeneous nucleation at a liquid–liquid interface. The same expression is obtained for the nucleation rate with the exception of a new definition for F.

$$F = \frac{(2 - 3m_a + m_a^3)\sigma_a^3 + (2 - 3m_b + m_b^3)\sigma_b^3}{4\sigma_a^3} \tag{32}$$

where σ_a, σ_b and are the surface tensions of liquid a and liquid b, and

$$m_a = \frac{\sigma_a^2 - \sigma_b^2 + \sigma_{ab}^2}{2\sigma_a\sigma_b} \tag{33}$$

$$m_b = \frac{\sigma_b^2 - \sigma_a^2 + \sigma_{ab}^2}{2\sigma_b\sigma_{ab}} \tag{34}$$

σ_{ab} is the interfacial free energy per unit area of ab interface.

A more complicated case, heterogeneous nucleation at the interface of a liquid and a nonsmooth surface is discussed by Cole (1974).

Colton and Suh (1987a,b) also developed a model for heterogeneous nucleation at the interface of two phases such as an additive and a polymer. In this model, the number of nucleation sites is expressed by

$$N_1 = C_1 f_1 e^{-\Delta G^*_{het}/kT} \tag{35}$$

where C_1 = concentration of heterogeneous nucleation sites, f_1 = frequency factor of gas molecules joining the nucleus, and ΔG^*_{het} = Gibbs free energy (activation energy barrier) for heterogeneous nucleation.

A balance of surface tensions at the interface of two phases and the nucleated bubble gives

$$\sigma_{ap} = \sigma_{bp} + \sigma_{ab} \cos \theta \tag{36}$$

where σ_{ap}, σ_{bp}, and σ_{ab} are the interfacial surface tensions of the particle–polymer, bubble–polymer, and particle–bubble, respectively, and θ is the wetting angle of the interface, as illustrated in Fig. 10. The excess free energy for the formation of this bubble is given by

$$\Delta G_{het} = -V_b \Delta P + A_{bp}\sigma_{bp} + A_{ab}\sigma_{ab} - A_{ap}\sigma_{ap} \tag{37}$$

Upon substitution of volume and surface area terms,

$$\Delta G_{het} = \left(\frac{-4\pi r^3 \Delta P}{3} + 4\pi r^2 \sigma_{bp} \right) f(\theta) \tag{38}$$

$$f(\theta) = \frac{(2 + \cos \theta)(1 - \cos \theta)^2}{4} \tag{39}$$

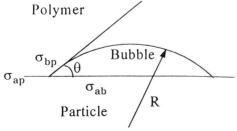

Fig. 10 Geometry of heterogeneous nucleus used by Colton and Suh (1987b).

The critical radius of a bubble in this system, r^*, can be calculated by setting the differential of the free energy to zero:

$$r^* = \frac{2\sigma_{bp}}{\Delta P} \tag{40}$$

and the free energy requirement for a critical nucleus can be calculated by substituting the critical radius into the free energy expression:

$$\Delta G_{het}^* = \frac{16\pi\sigma_{bp}^3}{3\,\Delta p^2}\,S(\theta) \tag{41}$$

A nearly identical expression for homogeneous nucleation is also proposed by Colton and Suh (1987b). The only difference between the two expressions is the absence of $f(\theta)$ in the expression for free energy of homogeneous nucleation, and thus, $\Delta G_{het}^* = \Delta G_{hom}^* f(\theta)$. For a typical wetting angle of $20°$ (Cherry, 1981), $f(\theta)$ is of the order of 10^{-3}, and so ΔG_{het}^* is three orders of magnitude smaller than ΔG_{hom}^*. Hence, a lower interfacial surface energy means a lower activation energy barrier, which in turn translates to a higher nucleation rate. Consequently, a poorly bonded nucleation site provides a lower activation energy and a higher nucleation rate than a well-bonded site. The free energy barrier for heterogeneous nucleation (i.e., in the presence of an interface) is much lower than that for homogeneous nucleation and therefore heterogeneous nucleation proceeds at rates much higher than homogeneous nucleation. As a result, in the presence of heterogeneous nucleation sites, heterogeneous nucleated bubbles form before the homogeneously nucleated bubbles and grow faster.

B. Mixed-Mode Nucleation

Colton and Suh (1987b) allow for the simultaneous occurrence of hetero-geneous and homogeneous bubble nucleation. In many cases, especially when the insoluble phase is present at a concentration just above its solubility limit, both heterogeneous and homogeneous nucleation processes can take place. However, due to the significantly lower activation energy of the heterogeneous nucleation, the rate of heterogeneous nucleation is higher than that of homogeneous nucleation. In this case, by consuming the gas dissolved in the polymer, heterogeneous nucleation reduces the concentration of the gas in the surrounding polymer. Due to the larger size of the heterogeneously nucleated bubbles (because they grow faster), diffusion of gas into a heterogeneously formed bubble is thermodynamically favored, as evidenced by the negative slope of the free energy curves beyond the critical radius point (Fig. 11). The resulting reduction of concentration of gas in the polymer affects the rate of homogeneous bubble nucleation, and therefore a time-

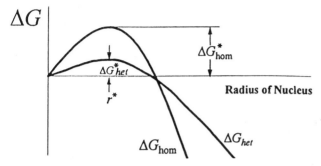

Fig. 11 Excess free energy functions for heterogeneous and homogeneous bubble nucleation. (Colton and Suh, 1987b.)

dependent concentration expression needs to be used in the place of a constant concentration. The time-dependent concentration of gas is approximated by

$$C_t = C_0 - N_{het} t n_b \tag{42}$$

This equation is only a first approximation and, for simplicity, does not take into account the consumption of gas by homogeneously nucleated bubbles (which form at a much smaller rate) or a correction for the volume occupied by the bubbles. Insertion of this equation into Colton and Suh's equation for homogeneous nucleation yields

$$N_{hom} = f_0 C_t e^{-\Delta G^*_{hom}/kT} \tag{43}$$

and the total rate of nucleation can be written as a sum of homogeneous and heterogeneous nucleation rates:

$$N = N_{hom} + N_{het} \tag{44}$$

C. Experimental Results

Colton and Suh compared the results of a series of experiments with the theoretical prediction of their model, allowing homogeneous, heterogeneous, and mixed-mode nucleation. They have reported results on experiments using a system of polystyrene polymer and zinc stearate, stearic acid, and carbon black additives (1987c) and also systems of styrene–butadiene and Teflon (Colton, 1985). Some conclusions of these experiments are as follows:

1. Experimental results show that three distinct nucleation regimes exist, as predicted by the model. Below the solubility limit of the additive, the nucleation is homogeneous. At concentrations much higher than the

solubility limit, heterogeneous nucleation prevails, and at concentrations just above the solubility limit, both homogeneous and heterogeneous nucleation take place. However, the quantitative predictions of the model for the nucleation rate are not very accurate, sometimes in error by several orders of magnitude. The authors suggested that the errors could be due to some inappropriate assumptions and shortcomings of the models used in the calculation of molecular interactions. Inaccuracies in the thermodynamic data used in the calculations may also be partly to blame.

2. An increase in the saturation pressure in the homogeneous nucleation region has a direct effect on the number of nucleation sites, due to the direct effect of ΔP (the driving force) on the activation energy. In the heterogeneous region, however, saturation pressure has little effect on the number of nucleation sites, because the activation energy barrier is already very small. Therefore, increasing the saturation pressure affects only the final size of the bubbles.

3. The number of nucleation sites in the heterogeneous region is controlled by the number of interface sites between the additive and the polymer. The number of sites is controlled by the concentration of the additive and the degree of its dispersity in the polymer. Therefore, to increase the number of bubbles in a heterogeneous nucleation system, higher concentration of the additive and/or better mixing can be employed.

4. Molecular weight and orientation of the polymer do not make a significant contribution to the number or size of cells in any of the three regions. This is probably because nucleation takes place in much smaller scale than the size of polymer molecules.

5. As predicted by the theory, addition of carbon black, a strongly bonded additive, has no effect on the nucleation, and homogeneous nucleation remains the prevailing mechanism after addition of various concentrations of carbon black.

D. Effect of Shear Rate on Heterogeneous Nucleation

Lee (1993) studied the effect of shear rate on the nucleation rate in a low-density polyethylene (LDPE)–dichlorodifluoromethane (CFC-12) system using magnesium silicate (talc) as the nucleating agent. He performed a series of experiments by varying the amount of nucleating agent and shear rate in a counter-rotating twin-screw extruder. As expected, using more nucleating agent produced more cells. More importantly, experiments showed that increasing the shear rate results in a larger number of cells. Lee postulated that shear rate affects the rate of nucleation by (1) breaking up the clumps of nucleating agent and therefore producing more nucleation sites, and (2) by helping "pull" the expanding bubble out of the crevice(s) of the nucleating agent. Based on simple conical geometry of the crevice, Lee then derived a relationship between the capillary number $Ca = R\eta\dot{\gamma}/4\sigma$ and the number of cells generated, which is in fair agreement with his experimental results. In this model, the shear force acts as a "catalyst" that, by reducing the energy barrier, provides a quicker path from stable gas cavity to unstable bubble phase.

E. Nonclassical Theories

The classical theory's inability to provide accurate predictions for heterogeneous nucleation processes in some cases, such as low saturation pressures, has motivated efforts to provide more accurate models. Ramesh et al. (1994a) have proposed a model based on the survival of microvoids. These microvoids are created when a polymer melt consisting of a continuous phase containing particles of another polymer is cooled. The difference between the thermal expansion coefficients of the two components results in thermal and mechanical stresses that induce a microvoid. Due to mechanical limitations, only microvoids that are larger than a critical size survive. Within the nucleating particle, the magnitude of the tensile stresses is given by (Timoshenko and Goodier, 1970; Beck et al., 1968)

$$\Sigma_{rr} = \Sigma_{\theta\theta} = \Sigma_{\phi\phi} = \frac{4(\beta_{\text{particle}} - \beta_{\text{polmer}})(1 + v_{\text{particle}})G_{\text{particle}}G_{\text{polymer}}\,\Delta T}{6(1 - 2v_{\text{particle}})G_{\text{polymer}} + 3(1 + v_{\text{particle}})G_{\text{particle}}}$$

(45)

and within the matrix by

$$\Sigma_{rr} = \left(\frac{R_p}{r}\right)^3 \frac{4(\beta_{\text{particle}} - \beta_{\text{polymer}})(1 + v_{\text{particle}})G_{\text{particle}}G_{\text{polymer}}\,\Delta T}{6(1 - 2v_{\text{particle}})G_{\text{polymer}} + 3(1 + v_{\text{particle}})G_{\text{polymer}}}$$

(46)

where β, v, ΔT, and G are thermal expansion coefficient, Poisson ratio, temperature drop, and modulus, respectively. R_p is the particle size, and r is the radial distance from the center of inclusion. The subscripts polymer and particle denote the continuous-phase polymer and the distributed polymer particle, respectively.

After saturation of this polymer mix with a gas and depressurization, the gas molecules diffuse into the surviving microvoids and produce bubbles. As one might expect, the size and number of these gas bubbles depend largely on the size and number of the microvoids that are present before the foaming procedure starts. The expression derived by Ramesh et al. (1994) for the number of surviving microvoids is

$$N = \frac{N_0}{2}\,\text{erfc}\,\frac{\log(R/\xi)}{\sigma\sqrt{2}}$$

(47)

where N_0 is the total number of potential microvoids (which can be estimated by the number of particles present in the continuous phase), R is the instantaneous bubble radius, ξ is the mean of the distribution of the void size, and σ is the standard deviation of the distribution.

Predictions from this theory were compared with experimental results obtained from nitrogen foaming of polystyrene and high-impact polystyrene matrices nucleated with rubber (polybutadiene) (Ramesh et al., 1994b). The theory provides excellent predictions of the effect of saturation pressure, foaming temperature, rubber particle concentration, and rubber particle size on the cell density in the foamed polymer. The following trends were observed in the experiments:

1. Initially, the number of cells increases sharply with increasing saturation pressure, but eventually it reaches a plateau where all the nucleation sites are filled with the gas. After reaching the plateau, the saturation pressure seems to have little effect on the number of cells. This is as expected from the theory, since the assumption is that there is a finite number of nucleation sites and no new nucleation sites are generated regardless of the saturation pressure.

2. Increasing the foaming temperature increases the cell size and cell density, which are predicted correctly by the theory when appropriate corrections to parameters such as surface tension, viscosity, modulus, and Henry's law constant are made.

3. As predicted by the theory, experiments show that increasing the number of particles increases the number of nucleation sites, and in the presence of sufficient gas pressure, these sites become bubbles. Therefore, the number of particles has a direct effect on the number of bubbles.

4. Increasing the average size of the particles increases the number of cells at first, and once a critical size is reached, the particle size has no effect on the cell density. As theory predicts, this is related to the number of cells that have a large enough radius to create a microvoid. When the particle size approaches atomic sizes, it is conceivable that the stresses due to contractions are not sufficient to cavitate the particle and, therefore, no nucleation site is created. As the average size of the particles increases, more and more particles are cavitated and create nucleation sites for bubbles.

Lee and Beisenberger (1989) have also developed a theory for explaining bubble nucleation based on "pre-existence of heterogeneous germ nuclei located within the cracks and crevices of microscopic particulate matter believed to be present in all liquids." For further explanation of this theory, the reader is referred to Beisenberger and Lee (1986) and Harvey et al. (1944). The theory is discussed by Lee and Beisenberger to explain the inadequacy of both homogeneous and heterogeneous nucleation theories for the modeling of trends in bubble number density, yet no actual predictions are reported. A further discussion of this theory is provided in Chapter 6 of this book.

Finally, Albalak et al. (1992) examined the nucleation behavior accompanying the devolatilization of a polymer melt by quickly quenching a polymer strand extruded into an evacuated tank. They found that heterogeneous nucleation plays a strong role in the rate of devolatilization, although the mechanism for bubble formation is not straightforward. Once a bubble has been nucleated, tensile stresses occurring during bubble growth contribute to an increase of superheating locally, and thus to an effective reduced pressure at the polymer–gas interface. Consequently, the interface acts as a site for enhanced bubble nucleation, which takes the form of blisters on the bubble surface. These blisters eventually coalesce with the growing primary bubble, enhancing the overall devolatilization rate.

IV. DETERMINATION OF PHYSICAL PROPERTIES OF GAS–POLYMER MIXTURES

Regardless of the model chosen to represent bubble nucleation, both viscosity and interfacial tension appear as material parameters in the final equation for the nucleation rate. In addition, as nucleation and growth proceed, the composition of the melt phase will change and so, therefore, will both interfacial tension and viscosity. Thus, in any attempt to model bubble nucleation in a polymer–gas mixture, one will require knowledge of the behavior of both interfacial tension and viscosity as functions of concentration, and hence of temperature and pressure. Finally, if one corrects the nucleation rate to account for finite mass transfer rates, then the solvent diffusivity must also be known as a function of concentration in the polymer–solvent mixture. As shown by Blander and Katz (1975), under certain circumstances nucleation can become diffusionally limited, which could become very important in polymer devolatilization efforts. This short summary of the effect of concentration on material parameters is not meant to be comprehensive; it serves merely to provide the reader with a useful introduction to a subject of great importance in bubble nucleation in polymer systems.

The thermodynamics of polymer systems has been addressed in an earlier chapter—here we will focus on only the concentration dependence of interfacial tension, viscosity, and diffusivity. Little work has been conducted on either viscosity or interfacial tension in concentrated, high-pressure polymer–gas (or polymer–solvent) systems. Clearly, this is a fertile area for further work to support development in polymer devolatilization in general. One could write an entire book simply on a discussion of the composition dependence of the physical parameters of a polymer–diluent mixture; here, we present merely a short review of work on the relevant parameters.

A. Interfacial Tension

Clearly, the interfacial tension of a polymer–solvent mixture will be lower than that of the pure polymer; Ruengphrathuengsuka and Flumerfelt (1992), for example, found that exposure of a low-density polyethylene (LDPE) melt at 300°F to 110 atm of N_2 reduced the interfacial tension from approximately 27 dynes/cm to 10 (Fig. 12), even though they found nitrogen to be relatively poorly soluble in LDPE under these conditions (approx. 11 cm^3 STP/per gram of polymer). Because interfacial tension is a strong function of composition in gas–polymer systems, the value of the interfacial tension will change as the solvent migrates from the polymer matrix into the bubbles during bubble formation and growth. Consequently, because the nucleation rate depends strongly on interfacial tension, the rate will vary as the process of bubble nucleation and growth proceeds. In fact, as shown by the equations in earlier sections, the rate of nucleation will drop sharply as the magnitude of the interfacial tension rises.

Methods by which to predict the interfacial tension of a polymer–gas (or polymer–solvent) mixture can be divided into empirical and thermodynamic expressions. Note that generally the interfacial tension of a mixture is less than a mole fraction average of the individual interfacial tensions of

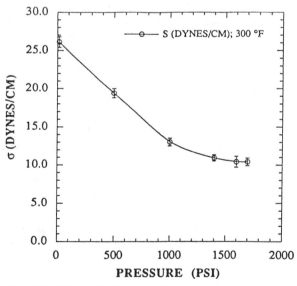

Fig. 12 Effect of dissolved N_2 on surface tension of PE800E. (From Rueng-phrathuengsuka, 1992.)

the components. Macleod (1923) suggested a relationship between the interfacial tension and the liquid and vapor densities:

$$\sigma^{1/4} = P(\rho_L - \rho_V) \tag{48}$$

where ρ_L and ρ_V are the liquid and vapor densities, and P is the parachor, which is thought to be a function of the cohesive energy density of the material (Quayle, 1953). Applying Eq. (48) to mixtures, one obtains

$$\sigma^{1/4} = \sum [P_i](\rho_{Lm}x_i - \rho_{Vm}y_i) \tag{49}$$

where x_i and y_i are the mole fractions of component i in the liquid and vapor phases, and the subscripts i and m refer to pure component and mixture properties. If we assume that the vapor-phase density is negligible compared with that of the liquid, then one may combine these two equations to arrive at

$$\sigma^{1/4} = \rho_{Lm} \sum \frac{x_i \sigma_i^{1/4}}{\rho_{Li}} \tag{50}$$

Goel and Beckman (1994a,b, 1995) used this expression to predict the interfacial tension of a carbon dioxide–polymethyl methacrylate (PMMA) mixture as a function of pressure and temperature. This prediction was subsequently used in the modeling of nucleation and growth of bubbles in PMMA resulting from a pressure quench. Using a form of Eq. (28), they showed that when the interfacial tension of pure PMMA is employed, the model predicts no nuclei at any experimentally accessible CO_2 pressure. On the other hand, generation of the interfacial tension using Eq. (50) predicts a pressure dependence for the cell density closely mirroring that of the experimental data.

Although Eq. (50) is a good first approximation, it must be noted that this expression is empirical. Indeed, if one examines the data of Ruengphrathuengsuka on the nitrogen–LDPE system, one finds that Eq. (50) greatly overpredicts the surface tension of the mixture and, thus, would contribute to an underprediction of the rate of nucleation in this system. Several researchers have applied classical and statistical thermodynamics to the problem of the surface tension of mixtures; many of these results are similar in form, leading to an expression such as (Reid et al., 1986):

$$\sum \frac{x_i^B \gamma_i^B}{\gamma_i^\sigma} \exp \frac{A_i(\sigma_m - \sigma_i)}{RT} = 1 \tag{51}$$

where x_i^B is the mole fraction of i in the bulk liquid, γ_i^B is the activity coefficient of i in the bulk liquid (normalized so that γ_i^B approaches 1.0 as the mole fraction approaches 1.0), γ_i^σ is the activity of i in the surface phase, A_i is the partial molar surface area of component i, and σ_m and σ_i are the interfacial tensions of the mixture and individual components. If one assumes the liquid mixture to be ideal, then Eq. (51) reduces to the more manageable

$$\sigma_m = x_A \sigma_A + x_B \sigma_B - \frac{A}{2RT}(\sigma_A - \sigma_B)^2 x_A x_B \tag{52}$$

for a binary mixture. Here A is the average surface area for the mixture. This simplified form clearly shows that the surface tension of the mixture is less than a mole fraction average of the individual components.

B. Viscosity

Just as in the case of the interfacial tension, the presence of a low-molecular-weight diluent (gas or solvent) will dramatically lower the viscosity of a polymer melt. Gerhardt et al. (1994) and Garg et al. (1994), for example, found that 100 atm pressure of CO_2 lowers the viscosity of a high-molecular-weight poly(dimethyl siloxane) polymer by an order of magnitude. The viscosity of the polymer melt influences nucleation both explicitly and through its influence on the hydrodynamic (viscous) limitations to nucleation, exerted through an influence on the pressure within the growing bubble.

In concentrated polymer solutions, it has long been accepted that the viscosity can be considered as the product of two terms: a structure-sensitive factor, dependent primarily on the number of atoms in the main chain, and a density-dependent fraction factor per chain atom. As shown by Berry and Fox (1968), the structure factor will vary as either $\phi_2^{3.4}$ or ϕ_2 (where ϕ_2 is the volume fraction polymer in the mixture), depending on the molecular weight of the polymer. The friction factor, on the other hand, depends to great extent on the free volume of the system, and as such, can be modeled using either a Williams–Landel–Ferry (WLF) (Williams et al., 1955) or Vogel (1921) type of relationship. The latter is described by

$$\ln \zeta = \ln \zeta_0 + \frac{1}{\alpha(T - T_0)} \tag{53}$$

where ζ is th friction factor, and α and T_0 are the Vogel parameters (T_0 is sometimes assumed to be the glass transition temperature, but it is rather a reference temperature and usually not T_g). The parameter α has been shown to vary linearly with the volume fraction of polymer, and thus in very concentrated solutions can be approximated as $\phi_2 \alpha_2$. On the other hand, T_0 varies with both ϕ_2 and α_2, as shown by Berry and Fox. However, the variation of T_0 with volume fraction polymer likely mirrors the dependence of the glass transition temperature with concentration, which has previously been evaluated for a variety of polymer–diluent and polymer–gas systems. This assumption was employed by Goel and Beckman (1994a,b, 1995) in their modeling of bubble nucleation and growth in the PMMA–CO_2 mixture, given that the effect of CO_2 concentration on T_g was known through earlier work by both Condo and Johnston (1992) and Wissinger and Paulaitis (1991). Manke and colleagues (Gerhardt et al., 1994; Garg et al., 1994) used a variation of this approach to correlate the viscosity of PDMS–CO_2 and PEG–CO_2 mixtures as a function of temperature and CO_2 pressure. They used an equation for the shift factor of the system, i.e.,

$$a = \ln \frac{\eta}{\eta_0} = W_1^n \left(\frac{\rho}{\rho_0}\right)^n \exp\left(\frac{1}{f} - \frac{1}{f_0}\right) \tag{54}$$

where f is the fractional free volume of the system, W is the weight fraction, ρ is the density, and the subscript 1 refers to the pure polymer. Note that in Eq. (54), the weight fraction and density prefactors are, in effect, the structure factor for the system as described by Berry. Further, the exponential is equivalent to the (α, T_0)-containing term in the Vogel equation. Manke used the Panayiotou–Vera (Panayiotou and Vera, 1982) equation of state to calculate the specific volume of each polymer–gas mixture, from which the fractional free volume was then derived. This method provided excellent agreement with the experimental data (see Fig. 13).

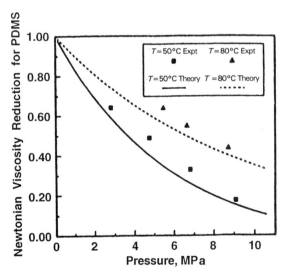

Fig. 13 Newtonian viscosity reduction for CO_2 in PDMS—experimental data and Panayiotou–Vera EOS predictions at $T = 50°C$ and $80°C$. (From Gerhardt et al., 1994.)

C. Diffusion Coefficient

Each of the theoretical approaches to homogeneous nucleation presented here employs the diffusion coefficient of the diluent in the calculation of the rate of nucleation. Many books and reviews have been published concerning diffusion in polymers (Crank and Park, 1951; Dean, 1947; Duda et al., 1982; Frenkel, 1946; Ventras and Duda, 1977, 1979; Ventras et al., 1983); here, we present only those approaches used to model diffusion during nucleation processes. For a detailed discussion of solvent diffusion in polymeric systems, see Chapter 3.

Han and Han employed the free-volume model developed by Vrentas and Duda (Ventras and Duda, 1977, 1979; Ventras et al., 1983), which describes diffusion of a solvent molecule in a polymer–solvent mixture:

$$D(T) = D_0(1 - 2\chi\phi_1)(1 - \phi_1)^2 \exp\frac{-E}{RT}\exp\frac{W_1 V_1^* + W_2 V_2^* \zeta}{V_{FH}\gamma^{-1}} \tag{55}$$

where D_0 is the preexponential factor, W_i is the weight fraction of component i, ϕ_1 is the volume fraction of the solvent, γ is the overlap factor, which indicates how the molecules share the free volume, ι is the ratio of the critical molar volume of the jumping unit of solvent to the critical molar volume of the jumping unit of the polymer, V_i^* is the specific hole volume of component i required for a jump, and V_{FH} is the average hole free volume per gram of mixture. Han and Han employed literature values of the material parameters and Eq. (55) to calculate the diffusion coefficient for toluene in polystyrene to support modeling of their nucleation experiments. Ruengphrathuengsuka and Flumerfelt (1992) measured the diffusivity of nitrogen in an LDPE melt and then used this model to describe diffusion in the system. They found that the measured dependence of diffusivity on concentration was not as strong as the model suggested; yet the scatter in their data prevented a concrete conclusion in this regard.

Beckman and Goel employed a model for diffusivity suggested by Vieth (1991) to describe diffusion behavior in the CO_2–PMMA mixture:

$$D = D_0 \exp\frac{Ac}{B + c}\exp\frac{-E}{RT} \tag{56}$$

where D_0, A, and B are material constants, and c is the concentration of the diluent. The material parameters were determined through fitting of the model to data for diffusion of CO_2 in PMMA measured by Berens and Huvard (1989).

In summary, it can be seen that the material parameters of importance to nucleation in polymer–diluent mixtures—the interfacial tension, the viscosity, and the diffusivity—can all be strong functions of concentration in these systems. Further, as bubbles nucleate and grow within the polymer continuous phase, thus depleting the diluent, it is clear that the viscosity and interfacial tension will rise, and the diffusion coefficient will drop. Consequently, the very process of nucleation will become increasingly difficult in a polymer melt as the process of devolatilization proceeds. Thus, to propel devolatilization to as high an extent as possible, one must consider how to affect these physical properties advantageously, to maintain reasonable nucleation rates throughout the process. One possibility lies in the injection and then removal of a highly volatile plasticizing agent such as carbon dioxide or nitrogen. Because the development of models for the effect of diluent on the physical properties at high pressure and temperature and their supporting data lag behind the development of models for nucleation, it is difficult to recommend any particular nucleation model without improvements in the accuracy of prediction of physical properties such as diffusion constant, activity coefficient, and chemical potential as concentration varies.

NOMENCLATURE

A area

A material constant in Eq. (56)

B frequency factor

B material constant in Eq. (56)

C concentration

Ca capillary number, $R\eta\dot{\gamma}/4\sigma$

$\Delta C(t)$ amount of volatile component consumed per unit volume of solution at time t

c_∞ equilibrium concentration

D diffusion constant

D_0 preexponential factor in Eq. (55)

D_0 material constant in Eq. (56)

ΔF_s Han and Han's correction factor to supersaturation (defined by Eq. (26)

ΔF_t Han and Han's correction factor to free energy (defined by Eq. (25)

F $(2 - 3m + m^3)/4$ (as defined in text)

$f(\theta)$ defined by Eq. (39)

f fractional free volume of the system

f_1 frequency factor of gas molecules joining the nucleus

ΔG free energy

ΔG_n free energy associated with the formation of an n-molecule cluster

G modulus

ΔH heat of vaporization
J rate of nucleation
J number of bubbles per unit area that grow beyond the critical size
k Boltzmann constant
M_G molecular weight of the gas
m mass of a molecule
m Z/R (see Fig. 9)
m_a defined by Eq. (33)
m_b defined by Eq. (34)
N number density of liquid
N_0 total number of potential microvoids
$\dot{n}$ defined by Eq. (21)
P pressure
P parachor
P_1 vapor pressure
P_B pressure in the bubble
Q empirical parameter used in Eq. (29)
R radius
r radius of bubble
S $(1 - m)/2$ (as defined in text)
T absolute temperature (K)
T_0 Vogel parameter
V volume
V_{FH} average hole free volume per gram of mixture
V_i^* specific hole volume of component i required for a jump
W work
W_i weight fraction of component i
x number of gas molecules
x_i mole fraction of component i in the liquid phase
y_i mole fraction of component i in the vapor phase
Z Zeldovich factor (defined by Eq. (10))

Greek Symbols

α evaporation frequency
α Vogel parameter
β rate per unit area at which a bubble surface gains or loses molecules
β thermal expansion coefficient
δ_D correction factor (defined by Eq. (17))
δ_e thermal conductivity of liquid

δ_λ correction factor (defined by Eq. (14))

$\dot{\gamma}$ average shear rate

γ overlap factor; indicates how molecules share the free volume

γ_i^B activity coefficient of i in the bulk liquid

η viscosity

$\imath$ ratio of critical molar volume of the jumping unit of solvent to the critical molar volume of the jumping unit of polymer

μ chemical potential

v_t defined by Eq. (21b)

v Poisson ratio

ρ density

σ standard deviation of the pore size distribution

σ interfacial tension

σ_{ab} interfacial free energy per unit area of ab interface

$\tau_{\theta\theta}$ theta component of the stress tensor

τ_{rr} radial component of the stress tensor

ζ mean distribution of void size

ζ friction factor

$\Sigma_{rr}, \Sigma_{\theta\theta}, \Sigma_{\phi\phi}$ tensile stress in r, θ, ϕ directions, respectively.

Subscripts

1 solvent

2 polymer

ab particle–bubble

ap particle–polymer

bp bubble–polymer

c critical

cr critical

G gas

het heterogeneous

hom homogeneous

i property pertaining to component i

L liquid

LG liquid–gas

SG solid–gas

SL solid–gas

v vapor

Superscripts

* critical property

B bulk liquid

REFERENCES

Albalak, R. J., Tadmor, Z., and Talmon, Y. (1992). Blister-promoted bubble growth in viscous polymer melts, *Mater. Res. Soc. Symp. Proc.*, *237*: 181–186.

Beck, R., Gratch, S., Newman, S., and Rusch K. C. (1968). *J. Polym. Sci. Lett.*, *6*.

Becker, R., and Doring, W. (1935). *Ann. Physik*, *24*: 719.

Beisenberger, J. A., and Lee, S. T. (1986). *SPE ANTEC Tech. Papers*, *32*: 846.

Berens, A. R., and Huvard, G. S. (1989). "Interaction of Polymers with Near-Critical CO_2," *Supercritical Fluid Science and Technology* (K. P. Johnston and J. M. L. Penninger, eds.), ACS Symp. Ser. 406, Washington, D.C.

Berry, G. C., and Fox, T. G. (1968). The viscosity of polymers and their concentrated solutions, *Adv. Polym. Sci.*, *5*: 261–357.

Blander, M., and Katz, J. L. (1975). Bubble nucleation in liquids, *AIChE J.*, *21*(5): 833.

Blander, M., Hengstenberg, D., and Katz, J. L. (1971). Bubble nucleation in *n*-pentane, *n*-hexane, *n*-pentane + hexadecane mixture, and water, *J. Phys. Chem.*, *75*: 3613.

Cherry, B. W. (1981). *Polymer Surfaces*, Cambridge University Press, Cambridge, U.K., Chapter 2.

Cole, R. (1974). Boiling nucleation, *Adv. Heat Transf.*, *10*: 85.

Colton, J. S. (1985). Ph.D. thesis, Massachusetts Institute of Technology, Cambridge, Massachusetts.

Colton, J. S., and Suh, N. P. (1987a). Nucleation of microcellular foam: Theory and practice, *Polymer. Engin. Sci.*, *27*(7): 500–503.

Colton, J. S., and Suh, N. P. (1987b). The nucleation of microcellular thermoplastic foam with additives: Part I: Theoretical considerations, *Polym. Engin. Sci.*, *27*(7): 485–492.

Colton, J. S., and Suh, N. P. (1987c). The nucleation of microcellular thermoplastic foam with additives: Part II: Experimental results and discussion, *Polym. Engin. Sci.*, *27*(7): 493–499.

Condo, P. D., and Johnston, K. P. (1992). Retrograde vitrification of polymers with compressed fluid diluents: Experimental confirmation, *Macromolecules*, *25*: 6730–6732.

Crank, J., and Park, G. S. (1951). Diffusion in high polymers, *Research (London)*, *4*: 515.

Dean, R. B. (1947). The effects produced by diffusion in aqueous systems containing membranes, *Chem. Rev. 41*: 503.

Denecour, R. L., and Gent, A. N. (1968). *J. Polym. Sci.*, *A-2*(6): 1853.

Duda, J. L., Vrentas, J. S., Ju, S. T., and Liu, H. T. (1982). Prediction of diffusion coefficients for polymer–solvent systems, *AIChE J.*, *28*: 279.

Farkas, L. (1927). *A. Phys. Chem.*, *125*: 236.

Frenkel, J. (1946). *Kinetic Theory of Liquids*, Oxford University Press, New York.

Garg, A., Gulari, E., and Manke, C. W. (1994). *Macromolecules*, *27*(20): 5643.

Gent, A. N., and Tompkins, P. A. (1969a), *J. Appl. Phys.*, *40*: 2520.

Gent, A. N., and Tompkins, P. A. (1969b). *J. Polym. Sci.*, *A-2*(7): 1483.

Gerhardt, L. G., Garg, A., Manke, C., and Gulari, E. (1994). "Supercritical Fluids as Polymer Processing Aids," Proceedings of the Third International Symposium on Supercritical Fluids, Strasbourg, France, October.

Gibbs, W. (1961). *The Scientific Papers*, Vol. 1, Dover, New York.

Goel, S. K., and Beckman, E. J. (1994a). Generation of microcellular polymeric foams using carbon dioxide. 1. Effect of pressure and temperature on nucleation, *Polym. Engin. Sci., 14*: 1137.

Goel, S. K., and Beckman, E. J. (1994b). Generation of microcellular polymeric foams using carbon dioxide. 2. Cell growth and skin formation, *Polym. Engin. Sci., 14*: 1148.

Goel, S. K., and Beckman, E. J. (1995). Nucleation and growth in microcellular materials: Supercritical CO_2 as foaming agent, *AIChE J., 41*(2): 357–367.

Jennings, J. H., and Middleman, S. (1985). *Macromolecules, 18*(11): 2274.

Han, J. H. (1988). Bubble Nucleation in Polymeric Liquids, Ph.D. thesis, Polytechnic Institute of New York.

Han, J. H., and Han, C. D. (1988). A study of bubble nucleation in a mixture of molten polymer and volatile liquid in a shear flow field, *Polym. Engin. Sci., 28*: 24.

Han, J. H., and Han, C. D. (1990a). Bubble nucleation in polymeric liquids. I. Bubble nucleation in concentrated polymer solutions, *J. Poly. Sci. Part B, 28*: 711–741.

Han, J. H., and Han, C. D. (1990b). Bubble nucleation in polymeric liquids. II. Theoretical considerations, *J. Poly. Sci. Part B, 28*: 743–761.

Harvey, E. N., Barnes, D. K., McElroy, W. D., Whitely, A. N., Pease, D. C., and Cooper, K. W. (1944). *J. Cell. Comp. Physiol., 24*: 1.

Kagan, Y. (1960). The kinetics of boiling of a pure liquid, *Russ. J. Phys., Chem., 34*: 42.

Kaischew, R., and Stranski, I. N. (1934). *Z. Phys. Chem., 26B*: 317.

Katz, J. L. (1970). *J. Statistical Phys., 2*(2): 137.

Katz, J. L., and Blander, M. (1973). *J. Coll. Interface Sci., 42*(3): 496.

Katz, J. L., and Donohue, M. D. (1979). *Advances in Chemical Physics*, John Wiley and Sons.

Katz, J. L., Saltsburg, H., and Reiss, H. (1966). *J. Coll. Interface Sci., 21*: 560.

Larsen, R. G. (1988). *Constitutive Equations for Polymer Melts and Solutions*, Butterworth Series, Stoneham, England.

Lee, S. T. (1993). Shear effects on thermophastic foam nucleation, *Polym. Engin. Sci., 33*(7): 418–422.

Lee, S. T., and Beisenberger, J. A. (1989). A fundamental study of polymer melt devolatilization. IV: Some theories and models for foam-enhanced devolatilization, *Polym. Engin. Sci., 29*(9): 782–790.

Macleod, D. B. (1923). *Trans. Faraday Soc., 19*: 38.

Panayiotou, C., and Vera, J. H. (1982). *Polymer J., 14*: 681.

Quayle, O. R. (1953). *Chem. Rev., 53*: 439.

Ramesh, N. S., Rasmussen, D. H., and Campbell, G. A. (1994a). The heterogeneous nucleation of microcellular foams assisted by the survival of microvoids in polymer containing low glass transition particles. Part I: Mathematical modeling and numerical solution, *Polym. Engin. Sci., 34*(22): 1685–1697.

Ramesh, N. S., Rasmussen, D. H., and Campbell, G. A. (1994b). The heterogeneous nucleation of microcellular foams assisted by the survival of microvoids in polymer containing low glass transition particles. Part II: Experimental results and discussion, *Polym. Engin. Sci.*, *34*(22): 1698–1706.

Rangel-Nafaile, C., Metzner, A. B., and Wissbrun, K. F. (1984). Analysis of stress-induced phase separations in polymer solutions, *Macromolecules*, *17*: 1187–1195.

Reid, R. C., Prausnitz, J. M., and Poling, B. E. (1986). *The Properties of Gases and Liquids*, 4th ed., McGraw-Hill, New York.

Ruengphrathuengsuka, W. (1992). Bubble Nucleation and Growth Dynamics in Polymer Melts, Ph.D. thesis, Texas A&M University.

Stewart, C. W. (1970). Nucleation and growth of bubbles in elastomers, *J. Polym. Sci. Part A-2*, *8*: 937–955.

Timoshenko, S., and Goodier, J. N. (1970). *Theory of Elasticity*, McGraw-Hill, New York.

Tirrell, M. (1986). Phase behavior of flowing polymer mixtures, *Fluid Phase Equil.*, *30*: 367–380.

Ventras, J. S., and Duda, J. L. (1977). Diffusion in polymer–solvent systems, I. Reexamination of the free-volume theory, *J. Polym. Sci. Polym. Phys.*, *15*: 403.

Ventras, J. S., and Duda, J. L. (1979). Molecular diffusion in polymer solutions, *AIChE J.*, *25*: 1.

Ventras, J. S., Duda, J. L., and Hsieh, S. T. (1983). Thermodynamic properties of some amorphous polymer–solvent systems, *Ind. Eng. Chem. Prod. Res. Dev.*, *22*: 326.

Vieth, W. R. (1991). *Diffusion in and Through Polymers: Principles and Applications*, Hanser Publishers, New York, p. 98.

Vogel, H. (1921). Das Temeraturabhangigkeitsgesetz auf die Viscositat von Flussig-keiten, *Z. Physik*, *22*: 645.

Volmer, M. (1939). Kinetics of Phase Formation, from the Clearinghouse for Federal and Technical Information, ATI No. 81935 (F-TS-7068-RE).

Volmer, M., and Weber, A. (1926). *Z. Phys. Chem.*, *119*: 277.

Williams, M., Landel, R., and Ferry, J. (1955). *J. Am. Chem. Soc.*, *77*: 3071.

Wissinger, R. G., and Paulaitis, M. E. (1991). Glass transition in polymer/CO_2 mixtures at elevated pressures, *J. Polym. Sci: Part B: Poly. Phys.*, *29*: 631–633.

Zeldovich, J. B. (1943). *Acta Physicochem. USSR*, *18*: 1.

5

Fundamentals of Bubble Growth

Moshe Favelukis

Technion—Israel Institute of Technology, Haifa, Israel

Ramon J. Albalak

Massachusetts Institute of Technology, Cambridge, Massachusetts

I. INTRODUCTION

A crucial stage in the devolatilization process is that of the growth of bubbles containing vapor of the volatile material to be separated from the polymer. Rather than to review the vast literature on bubble growth, we have chosen in this chapter to guide the reader through several fundamental approaches to the subject, focusing our attention on the growth of a single stationary spherical bubble in a stationary viscous liquid.

We address the growth of a bubble due to diffusion of volatile material from the bulk of the liquid and its vaporization into the gaseous phase at the bubble surface. In Section II, we discuss bubble growth controlled by mass transfer. We start out by addressing steady and unsteady-state mass transfer to a bubble with a constant radius, and then we incorporate a quasi-steady-state approach to provide a simplified expression for the growth rate. We then present a solution to the general unsteady-state problem. In Section III, we develop an expression for growth in a Newtonian liquid controlled by viscous forces. The combination of both mechanisms is briefly referred to in Section IV. Section V discusses some limitations of the models presented in this chapter as applied to real devolatilization processes.

Bubble growth controlled by heat transfer is not discussed, since it can be shown that at the low volatile concentrations at which devolatilization processes are often conducted, the heat of vaporization to be supplied to the system usually has a negligible effect on the temperature profile (Powell and Denson, 1983).

II. GROWTH CONTROLLED BY MASS TRANSFER

A. Theoretical Background

A differential mass balance for a nonreactive binary system consisting in our case of a polymer melt and solvent, assuming Fickian diffusion, and constant density (ρ) and diffusivity (D) is of the form (Clift et al., 1978)

$$\frac{DC}{Dt} = D \nabla^2 C \tag{1}$$

in which

$$\frac{DC}{Dt} = \frac{\partial C}{\partial t} + v \cdot \nabla C \tag{2}$$

C is the molar concentration of the solvent, v is the velocity vector, and t is time.

The cases presented in this section will be limited to those in which resistance to mass transfer is only in the liquid phase.

B. Boundary and Initial Conditions

The discussion here will also be limited to uniform and constant solvent concentrations at both the bubble surface and far away from the bubble. These conditions may be expressed as

$$C = C_s \ @ \ r = R \tag{3}$$

$$C = C_\infty \ @ \ r \to \infty \tag{4}$$

R is the bubble radius (see Fig. 1).

The initial concentration is given by

$$C = C_\infty \ @ \ t = 0 \tag{5}$$

We assume that the solvent vapor inside the bubble is in chemical equilibrium with the liquid at the bubble surface, so that Henry's law is applicable:

$$P_1 = K_C C_s \tag{6}$$

where P_1 is the partial pressure of the solvent within the bubble and K_C is Henry's constant on a molar base.

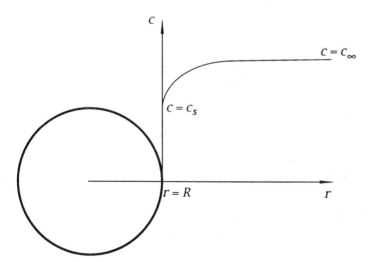

Fig. 1 A schematic representation of a vapor bubble growing in an infinite liquid medium due to mass transfer of the solvent to the bubble surface.

C. Steady-State Diffusional Mass Transfer

At steady state, in the absence of a velocity field in the liquid phase (i.e., $v = 0$) and after assuming spherical symmetry, Eq. (1) is simplified and expresses only radial diffusion, we have

$$\frac{\partial}{\partial r}\left(r^2 \frac{\partial C}{\partial r}\right) = 0 \tag{7}$$

Integrating twice and substituting the boundary conditions given in Eqs. (3) and (4) results in

$$C = C_\infty - (C_\infty - C_s)\frac{R}{r} \tag{8}$$

The local molar flux (N) of the solvent to the bubble at its surface is given by

$$N = D\left(\frac{\partial C}{\partial r}\right)_{r=R} = \frac{D}{R}(C_\infty - C_s) \tag{9}$$

Note that in this case the flux is in the $-r$ direction and that the local and average fluxes are equal.

It is common to define the average flux as

$$\bar{N} = k_m (C_\infty - C_s) \tag{10}$$

in which k_m is the overall average mass transfer coefficient.

From Eqs. (9) and (10), the solution may be written dimensionlessly in terms of the Sherwood number (Sh):

$$\text{Sh} = \frac{k_m R}{D} = 1 \tag{11}$$

The Sherwood number describes the ratio of the total mass transfer to the diffusional mass transfer.

D. Unsteady-State Diffusional Mass Transfer

For zero velocity in an unsteady state, Eq. (1) becomes

$$\frac{\partial C}{\partial t} = \frac{D}{r^2}\frac{\partial}{\partial r}\left(r^2 \frac{\partial C}{\partial r}\right) \tag{12}$$

Using a similar approach to that applied in Section C, the solution obtained is (Clift et al., 1978)

$$Sh = 1 + \frac{1}{\sqrt{\pi \tau}} \tag{13}$$

where τ is a dimensionless time, given as

$$\tau = \frac{Dt}{R^2} \tag{14}$$

Note that for long times ($\tau \to \infty$), Eq. (13) reduces to Eq. (11).

E. Growth Rate

Until now we have discussed in Sections C and D a somewhat imaginary situation in which the bubble size is unaffected by the mass transfer. We shall now demonstrate a simple approach to the problem of determining the change in the bubble radius as a function of time.

We shall make the following assumptions:

1. The bubble contains only solvent vapor, which is assumed to be an ideal gas.
2. The pressure inside the bubble is constant.
3. The process is isothermal.
4. The bubble is large enough so that surface tension does not contribute to the pressure inside the bubble.
5. Viscous normal stresses are negligible.

The ideal gas law for the vapor in the bubble is

$$PV = nR_g T \tag{15}$$

where P, according to our assumptions, is the pressure throughout the system, V is the volume of the bubble, n is the number of moles inside the bubble, R_g is the universal gas constant, and T is the absolute temperature. The time derivative of (15) is

$$P \, 4\pi R^2 \frac{dR}{dt} = \frac{dn}{dt} R_g T \tag{16}$$

Rearranging yields an expression for the average flux:

$$\bar{N} = \frac{1}{4\pi R^2} \frac{dn}{dt} = \frac{P}{R_g T} \frac{dR}{dt} \tag{17}$$

The first equality in Eq. (17) expresses a mass balance for the solvent over the bubble volume.

The average flux was previously defined as

$$\bar{N} = k_m (C_\infty - C_s) \tag{10}$$

Equating (10) and (17) leads to an expression for the growth rate of the bubble:

$$\frac{dR}{dt} = \text{Fm Sh} \frac{D}{R} \tag{18}$$

where Fm is a dimensionless number that indicates the tendency of a given system to foam. It is similar to the Jakob number for mass transfer, and it describes the role of thermodynamic parameters in the growth process:

$$\text{Fm} = \frac{R_g T (C_\infty - C_s)}{P} \tag{19}$$

Bubble growth is enhanced by large values of Fm, i.e., high temperatures, low pressures, and large differences between the concentrations at infinity and at the bubble surface.

Equation (18) describes a general relation for the growth rate of a spherical bubble as a function of the system parameters. Using a quasi-steady-state approach by substituting the steady-state solution of Section C (Eq. 11) into the equation for the unsteady state (Eq. 18), we get

$$\frac{dR}{dt} = \text{Fm} \frac{D}{R} \tag{20}$$

which upon integration from $r = R_0$ at time zero to $r = R$ at some final time t gives

$$R(t) = \sqrt{R_0^2 + 2 \text{ Fm } Dt} \tag{21}$$

Using the dimensionless parameters

$$R^* = \frac{R(t)}{R_0} \tag{22}$$

$$\tau_0 = \frac{Dt}{R_0^2} \tag{23}$$

Eq. (21) may be rewritten as

$$R^* = \sqrt{1 + 2 \text{ Fm } \tau_0} \tag{24}$$

Plots of R^* vs. τ_0 for various values of Fm are given in Fig. 2.

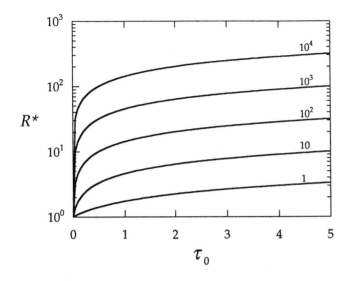

Fig. 2 Plots of the dimensionless bubble radius (R^*) as a function of dimensionless time (τ_0) according to Eq. (24), for typical values of Fm.

Inserting the unsteady-state solution of Section D (Eq. 13) into the expression for the growth rate (Eq. 18) leads to

$$\frac{dR}{dt} = \text{Fm}\,\frac{D}{R}\left(1 + \frac{1}{\sqrt{\pi\tau}}\right) \tag{25}$$

A complete solution for (25) may be derived from the work of Epstein and Plesset (1950). They used a similar approach to this problem and presented their solution in parametric form.

Asymptotic solutions for Eq. (25) may be obtained for two cases. For short times ($\tau \ll 1/\pi$ or $\tau_0 \ll R^{*2}/\pi$), the bubble radius may be expressed as

$$R(t) = R_0 + \frac{2}{\sqrt{\pi}}\,\text{Fm}\,\sqrt{Dt} \tag{26}$$

or in dimensionless form,

$$R^* = 1 + \frac{2}{\sqrt{\pi}}\,\text{Fm}\,\sqrt{\tau_0} \tag{27}$$

Plots of R^* *vs.* τ_0 according to Eq. (27) are given in Fig. 3 for various values of Fm.

For long times ($\tau \gg 1/\pi$ or $\tau_0 \gg R^{*2}/\pi$), Eq. (25) reduces to Eq. (20).

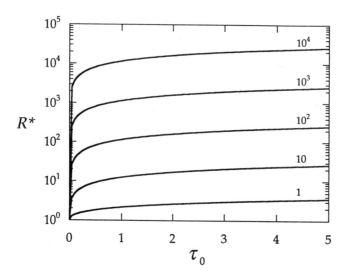

Fig. 3 Plots of the dimensionless bubble radius (R^*) as a function of dimensionless time (τ_0) according to Eq. (27), for typical values of Fm.

F. Unsteady-State Diffusional and Convective Mass Transfer

Although we consider here a stationary liquid, a velocity profile does of course develop in the liquid due to the expansion of the bubble. The cases presented so far neglected the velocity in the liquid phase and consequently disregarded convective mass transfer.

We present the problem and solution following Scriven's classic paper of 1959, in which a detailed solution for the case of bubble growth controlled by heat transfer was developed and in which the final solution for the analogous case of growth controlled by mass transfer was also given. A similar approach and solution were presented by Birkhoff et al. (1958). The present account differs somewhat from Scriven's work since we refer here to molar rather than mass concentrations and assume the density of the liquid phase to be much greater than that of the vapor in the bubble ($\rho_G \ll \rho_L$).

The equation of continuity in the liquid phase in spherical coordinates assuming constant density and spherical symmetry is given by

$$\frac{1}{r^2} \frac{\partial}{\partial r} (r^2 v_r) = 0 \tag{28}$$

which upon integration gives

$$r^2 v_r = f(t) \tag{29}$$

Since r^2v_r is a function of time alone, we may choose to calculate it at any radius. Doing so at $r = R$, at which the radial velocity is approximately dR/dt, leads to

$$v_r = \frac{dR}{dt}\left(\frac{R}{r}\right)^2 \tag{30}$$

Equation (30) may also be derived using an overall integral mass balance. Substituting the velocity profile into the mass balance of Eq. (1) and assuming spherical symmetry leads to

$$\frac{\partial C}{\partial t} + \frac{dR}{dt}\left(\frac{R}{r}\right)^2 \frac{\partial C}{\partial r} = D\left[\frac{1}{r^2}\frac{\partial}{\partial r}\left(r^2 \frac{\partial C}{\partial r}\right)\right] \tag{31}$$

In addition to the boundary and initial conditions given in Eqs. (3)–(5), another boundary condition is required to solve the problem. This requirement is satisfied by a mass balance over the solvent in the bubble, assuming that the bubble contains only solvent vapor that behaves as an ideal gas:

$$\frac{P_B}{R_g T}\frac{dR}{dt} = D\left(\frac{\partial C}{\partial r}\right)_{r=R} \tag{32}$$

which is similar to the expression developed in Eq. (17).

Following Scriven, we now introduce a dimensionless concentration (x), a dimensionless surface concentration (λ), and a dimensionless parameter (ξ):

$$x = \frac{C - C_\infty}{C_\infty} \tag{33}$$

$$\lambda = \frac{C_\infty - C_s}{C_\infty} \tag{34}$$

$$\xi = \frac{P_B}{R_g T C_\infty} \tag{35}$$

Equation (31) may now be written as

$$\frac{\partial x}{\partial t} + \frac{dR}{dt}\left(\frac{R}{r}\right)^2 \frac{\partial x}{\partial r} = D\left[\frac{1}{r^2}\frac{\partial}{\partial r}\left(r^2 \frac{\partial x}{\partial r}\right)\right] \tag{36}$$

The boundary and initial conditions given in Eqs. (3)–(5) and (32) become

$$x = -\lambda \ @ \ r = R \tag{37}$$

$$x = 0 \ @ \ r \to \infty \tag{38}$$

$$x = 0 \ @ \ t = 0 \tag{39}$$

$$\xi \frac{dR}{dt} = D\left(\frac{\partial x}{\partial r}\right)_{r=R} \tag{40}$$

Rather than to solve for $x(r, t)$, Scriven presented a solution for $x(s)$, where s is given by

$$s = \frac{r}{2\sqrt{Dt}} \tag{41}$$

Equation (36) may be presented as

$$\frac{d^2 x}{ds^2} = 2\frac{dx}{ds}\left(-s - \frac{1}{s} + \frac{\beta^3}{s^2}\right) \tag{42}$$

where the dimensionless growth constant β is given by

$$\beta = \frac{R}{2\sqrt{Dt}} \tag{43}$$

Note that although R changes with t, β remains constant for a given system.

Integrating twice and substituting the conditions in Eqs. (38)–(40) leads to

$$x = -2\xi\beta^3 \exp(3\beta^2) \int_s^\infty \frac{1}{\eta^2} \exp\left(-\eta^2 - 2\frac{\beta^3}{\eta}\right) d\eta \tag{44}$$

Applying Eq. (37), the growth constant β may be evaluated:

$$\frac{\lambda}{\xi} = \frac{R_g T(C_\infty - C_s)}{P_B} = \phi(\beta) \tag{45}$$

where

$$\phi(\beta) = 2\beta^3 \exp(3\beta^2) \int_\beta^\infty \frac{1}{\eta^2} \exp\left(-\eta^2 - 2\frac{\beta^3}{\eta}\right) d\eta \tag{46}$$

Note that $\phi(\beta)$ is equal to the dimensionless number Fm defined in Eq. (19).

For a given system, ϕ is first evaluated from Eq. (45), after which β may be evaluated from Eq. (46) or from tables given in Scriven (1959).

Finally, the time dependence of the bubble radius may be found from the rearrangement of Eq. (43):

$$R(t) = 2\beta\sqrt{Dt} \tag{47}$$

Scriven presented approximations for ϕ for both very small and very large values of β:

For $\beta \to 0, \quad \phi \approx 2\beta^2$ (48)

For $\beta \to \infty, \quad \phi \approx \sqrt{\dfrac{\pi}{3}}\,\beta$ (49)

Substituting these asymptotic expressions into (47) we obtain, for small values of β (slow growth),

$$R(t) = \sqrt{\frac{2R_g T(C_\infty - C_s)}{P_B}\,Dt} = \sqrt{2\,\mathrm{Fm}\,Dt} \tag{50}$$

which is equivalent to Eq. (21) developed for long times, since Scriven assumes R_0 to be zero.

For large values of β (fast growth), the asymptotic solution is

$$R(t) = 2\sqrt{\frac{3}{\pi}\frac{R_g T(C_\infty - C_s)}{P_B}}\sqrt{Dt} = 2\sqrt{\frac{3}{\pi}\,\mathrm{Fm}}\sqrt{Dt} \tag{51}$$

which is similar to the solution obtained for short times in Eq. (26).

III. GROWTH CONTROLLED BY VISCOUS FORCES

Bubble growth during devolatilization may be also controlled by the viscosity of the polymer melt, which is typically orders of magnitude higher than that of water. In these cases, the volatile material is always available at the bubble surface and the rate-determining factor is the resistance of the liquid to displacement by the growing gaseous phase.

A. Unsteady-State Viscosity-Controlled Growth

The velocity profile in this case may be derived as in Section II.F using the equation of continuity to again give

$$v_r = \frac{dR}{dt}\left(\frac{R}{r}\right)^2 \tag{30}$$

A differential form of Newton's second law of motion for a Newtonian liquid of constant viscosity (μ) and density (ρ) in laminar flow is given by the well-known Navier–Stokes equation:

$$\rho \frac{Dv}{Dt} = -\nabla P + \mu \nabla^2 v + \rho g \tag{52}$$

The flow of polymer melts of high viscosity is usually in the region of the so-called creeping flow, which is defined as having a low Reynolds number (Re $\ll$ 1). In this region, inertial forces are negligible in comparison with viscous forces. In practice, this implies that the left hand side of Eq. (52) may be set to zero. If we also neglect gravitational forces, Eq. (52) for spherical symmetry becomes

$$0 = -\frac{\partial P}{\partial r} + \mu \left[\frac{1}{r^2} \frac{\partial}{\partial r} \left(r^2 \frac{\partial v_r}{\partial r} \right) - \frac{2}{r^2} v_r \right] \tag{53}$$

Incorporating the velocity profile (30) into Eq. (53) and rearranging leads to

$$\frac{\partial P}{\partial r} = 0 \tag{54}$$

Integrating Eq. (54) over r from the bubble surface to infinity, one obtains

$$P_\infty - P_R = 0 \tag{55}$$

in which P_∞ is the ambient pressure and P_R is the pressure of the liquid at the bubble surface.

We now write a force balance at the bubble surface:

$$P_B - P_R + \tau_{rr} = \frac{2\sigma}{R} \tag{56}$$

in which P_B is the uniform pressure in the bubble, σ is surface tension, and τ_{rr} is the normal radial component of the viscous stress tensor in the liquid, which is given by

$$\tau_{rr} = 2\mu \frac{\partial v_r}{\partial r} \tag{57}$$

Equation (56) is similar in form to the Young–Laplace equation with additional terms for normal viscous stresses.

The radial component of the stress tensor within the bubble was neglected in Eq. (56) since we take the gas viscosity to be zero. Substituting the expressions for P_R and τ_{rr} at the bubble surface into Eq. (56), we obtain

$$P_B - P_\infty - \frac{2\sigma}{R} = \frac{4\mu}{R} \left(\frac{dR}{dt} \right) \tag{58}$$

Equation (58) is a particular case of the well-known extended Rayleigh equation, presented by Scriven (1959) and others, in which the inertial terms have been omitted.

At equilibrium ($dR/dt = 0$), the bubble does not grow and the radius of the bubble remains constant. This critical radius may be found from Eq. (58) as

$$R_{cr} = \frac{2\sigma}{P_B - P_\infty} \tag{59}$$

For the bubble to grow (i.e., $dR/dt > 0$), both sides of Eq. (58) have to be positive, so that

$$P_B > P_\infty + \frac{2\sigma}{R} \tag{60}$$

which is equivalent to stating that the radius of the bubble at $t = 0$ must be greater than the critical radius.

We may evaluate the partial pressure of the solvent within the bubble using Henry's law:

$$P_1 = K_C C_s \tag{6}$$

For a bubble containing only solvent vapor, the pressure in the bubble (P_B) is equal to P_1. In general, P_B is time-dependent; however, the literature indicates that during the initial stages of growth the pressure within the bubble does not vary significantly with time (Barlow and Langlois, 1962).

Under this assumption, for constant ambient and bubble pressures Eq. (58) may be integrated using the initial condition $R = R_0$ at $t = 0$, giving

$$R(t) = R_{cr} + (R_0 - R_{cr}) \exp \frac{(P_B - P_\infty)t}{4\mu} \tag{61}$$

We find that for growth controlled by viscous forces the bubble radius changes as an exponential function of time, whereas for the cases presented for growth controlled by diffusion we saw that the radius changes as the square root of time.

We define the following dimensionless parameters:

$$R^* = \frac{R(t)}{R_0} \tag{22}$$

$$R_{cr}^* = \frac{R_{cr}}{R_0} \tag{62}$$

$$t^* = \frac{t}{4\mu/(P_B - P_\infty)} \tag{63}$$

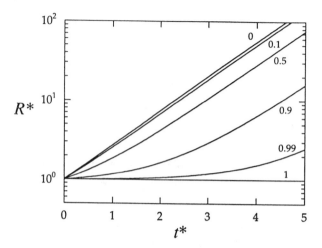

Fig. 4 Plots of the dimensionless bubble radius (R^*) as a function of dimensionless time (t^*) according to Eq. (64), for various values of R_{cr}^*.

Note that $R^* \geq 1$ and $0 < R_{cr}^* < 1$.

Equation (61) may now be written as

$$R^* = R_{cr}^* + (1 - R_{cr}^*) \exp t^* \tag{64}$$

which is plotted in Fig. 4 for various values of R_{cr}^*. We find that the maximum growth and growth rate are obtained for $R_{cr}^* \to 0$, which is the value farthest from equilibrium. From Eq. (59) we see that smaller values of R_{cr}^* indicate a larger driving force for growth ($P_B - P_\infty$).

For very short times ($t^* \ll 1$), Eq. (64) may be approximated as

$$R^* = 1 + (1 - R_{cr}^*)t^* \tag{65}$$

which indicates that the growth rate is constant during the initial stages.

IV. GROWTH CONTROLLED BY MASS TRANSFER AND BY VISCOUS FORCES

So far we have shown two distinctly different situations in which we assumed bubble growth to be controlled exclusively by either mass transfer or viscous forces. However, in a more general case, we can expect bubble growth to be affected by both mechanisms. The problem requires the simultaneous solution of the mass and momentum transfer equations, which along with the appropriate boundary conditions have already been presented in the previous sections of this chapter. This results in mathematical complications that lead to the application of numerical approaches and additional simplifying assumptions and approximations.

A general problem of bubble growth governed by both mass and momentum transfer was presented by Barlow and Langlois (1962) assuming that the concentration changes only in a thin shell surrounding the bubble, and that the thickness of this shell is much smaller than the bubble radius at all times. The resulting equation was solved numerically for a polymeric system.

Barlow and Langlois also derived analytical expressions for the extreme cases of the initial and final stages of bubble growth. Their solution for the early stages of growth, neglecting inertial effects, is identical to Eq. (65), developed here for the initial stages of growth controlled by viscous forces. For growth at long times, the treatment follows that of Birkhoff et al. (1958) and is of the same form as that given by Scriven (1959) (see Section II.F).

The thin-shell approximation was also employed by Patel (1980), who assumed a concentration profile and presented numerical results for systems characteristic of polymer melts that were in good agreement with the results of Barlow and Langlois (1962).

Recently, Tukachinsky (1993) compared bubble growth governed by both viscous forces and mass transfer with that governed by viscous forces alone as described by an equation similar to Eq. (61). His results indicate that the growth rate at early stages may be adequately described using the equations of a viscous-controlled process, whereas at long times the well-known parabolic profile is obtained.

V. LIMITATIONS IN APPLYING THE MODELS TO DEVOLATILIZATION

The models that we have presented here for bubble growth have been developed under assumptions that are not always fully applicable to the devolatilization of polymer melts. In spite of these limitations, these models are sufficient to indicate the trends in bubble dynamics as a function of system parameters and operating conditions and to yield quantitative predictions that may be taken as first approximations. Some of the limitations in applying these models to polymer devolatilization appear to be as follows.

1. For reasons of simplicity we have discussed in this chapter bubble growth in a Newtonian liquid. However, polymer melts that undergo devolatilization are usually non-Newtonian in nature, and they exhibit shear thinning and viscoelastic behavior. The non-Newtonian effects are directly encountered in momentum transfer for which the Navier–Stokes equations

are no longer valid, and a suitable constitutive equation needs to be substituted into the equation of motion. A recent publication by Chhabra (1993) refers to a large number of studies on bubble dynamics in non-Newtonian liquids.

2.   We refer here to a constant diffusion coefficient, although it is well established (e.g., Duda et al., 1982) that the diffusion coefficient of solvents in a polymer depends on the solvent concentration and may change by several orders of magnitude.

3.   Polymer melts processed in rotating devolatilizing equipment are continuously sheared so that any bubbles formed within the melt are deformed, resulting in an increase in their surface area with respect to spherical bubbles of the same volume. The models presented here for spherical bubbles do not account for the increase in mass transfer due to both the increase in surface area and to external convection. The effects of bubble deformation are discussed, for example, by Canedo et al. (1993).

4.   The models presented here are restricted to the growth of a single bubble in an infinite liquid medium. This situation is unrealistic in actual devolatilization processes in which a large number of bubbles form, grow, and interact in a limited amount of polymer melt. A model for the growth of a swarm of bubbles in a polymer melt has been presented by Amon and Denson (1984).

5.   It has been shown (Albalak et al., 1990) that bubble growth during devolatilization is a complex process involving the nucleation of a very large number of minute secondary bubbles around each growing bubble. These secondary bubbles continuously coalesce with the primary bubble, and their volatile content contributes to its growth. This phenomenon is clearly not accounted for in the models presented in this chapter.

NOMENCLATURE

C molar concentration
D diffusivity
Fm   dimensionless number
g gravitational acceleration
k_m overall average mass transfer coefficient
K_C Henry's constant
n number of moles inside the bubble
N local molar flux
$\bar{N}$ average molar flux
P_1 partial pressure of the solvent
P pressure
r radial coordinate

R bubble radius
Re   Reynolds number
R_g universal gas constant
s dimensionless coordinate
Sh   Sherwood number
t time
T absolute temperature
v velocity
V bubble volume
x dimensionless concentration
β dimensionless growth constant
λ dimensionless surface concentration
μ viscosity
ξ dimensionless parameter
ρ density
σ surface tension
τ_{ij} ij component of the viscous stress tensor
τ dimensionless time
ϕ dimensionless parameter

Superscripts

$*$ dimensionless

Subscripts

B   bubble
cr   critical
G   gas
L   liquid
r   radial direction
R   at the bubble surface
s   bubble surface
0   at $t = 0$
∞ far away from the bubble

REFERENCES

Albalak, R. J., Tadmor, Z., and Talmon, Y. (1990). Polymer melt devolatilization mechanisms, *AIChE J.*, *36*: 1313.
Amon, M., and Denson, C. D. (1984). A study of the dynamics of foam growth: Analysis of the growth of closely spaced spherical bubbles, *Polym. Eng. Sci.*, *24*: 1026.

Barlow, E. J., and Langlois, W. E. (1962). Diffusion of gas from a liquid into an expanding bubble, *IBM J. Res. Dev.*, *6*: 329.

Birkhoff, G., Margulies, R. S., and Horning, W. A. (1958). Spherical bubble growth, *Phys. Fluid*, *1*: 201.

Canedo, E. L., Favelukis, M., Tadmor, Z., and Talmon, Y. (1993). An experimental study of bubble deformation in viscous liquids in simple shear flow, *AIChE J.*, *39*: 553.

Chhabra, R. P. (1993). *Bubbles, Drops, and Particles in Non-Newtonian Fluids*, CRC Press, Boca Raton, Florida.

Clift, R., Grace, J. R., and Weber, M. E. (1978). *Bubbles, Drops and Particles*, Academic Press, New York.

Duda, J. L., Vrentas, J. S., Ju, S. T., and Liu, H. T. (1982). Prediction of diffusion coefficients for polymer–solvent systems, *AIChE J.*, *28*: 279.

Epstein, P. S., and Plesset, M. S. (1950). On the stability of gas bubbles in liquid–gas solutions, *J. Chem. Phys.*, *18*: 1505.

Patel, R. D. (1980) Bubble growth in a viscous Newtonian liquid, *Chem. Eng. Sci.*, *35*: 2352.

Powell, K. G., and Denson, C. D. (1983). "A Model for the Devolatilization of Polymeric Solutions Containing Entrained Bubbles," presented at the 75th Annual Meeting of the American Institute of Chemical Engineers, Washington, D.C.

Scriven, L. E. (1959). On the dynamics of phase growth, *Chem. Eng. Sci.*, *10*: 1.

Tukachinsky, A. (1993). D.Sc. thesis, Technion—Israel Institute of Technology.

6

A Fundamental Study of Foam Devolatilization

Shau-Tarng Lee

Sealed Air Corporation, Saddle Brook, New Jersey

I. INTRODUCTION

The removal of volatile components from concentrated polymeric melt constitutes an important separation step in the processing of plastic materials; to name a few examples: unreacted reactants in reactive extrusion, moisture in post-consumer recycling extrusion, unreacted monomer or solvent in polymerization. This process is known as devolatilization (DV). Basically, it is a diffusion-operated separation process, and increase of diffusion area is a logical way to enhance separation. A thin-film evaporator with pitched scraper is a typical example. The rate of diffusion and the rate of effective surface generation become the principal mechanisms for DV. It is common practice to form foams to effect separation. Foam formation in the molten polymer by either reducing pressure or purging with inert gases allows substantial increase of interfacial mass transfer area to maximize separation efficiency. In this case, foam growth and rupture are also necessary steps for DV.

The thermodynamic instability of polymeric solutions known as supersaturation causes bubble nucleation. Temperature increase or pressure reduction are common ways of producing supersaturated systems. Park and Suh (1992) reported nucleation as a result of a sudden temperature rise through finely divided tubes in their microcellular experiments. In fact, a high processing temperature raises degradation concerns, limiting the range of application. Substantial pressure reduction across the die in thermoplastic foam extrusion processes has long been used to make foam products (Lee, 1993). Nonetheless, foam formation through the application of vacuum or inert gas purging are practical methods in DV (Biesenberger, 1983).

In many industrial processes the volatile components are continually removed from molten polymer through foam formation, foam growth, and rupture into a contiguous gas phase. Since foaming can be accomplished in a limited period, DV becomes a unit operation in the current processing equipment. The huge space inherited in the conventional stripping process can thus be avoided. Examples are single-screw extruders, counter-rotating and co-rotating twin-screw extruders. In such equipment the basic process elements are rotating melt pools and stationary melt films (Biesenberger and Lee, 1986a), as indicated in Fig. 1.

A bubble-free mass transfer model for DV in a single-screw extruder was first proposed by Latinen (1962). This model includes longitudinal diffusion, axial convective mass transfer, and evaporation from rejoined film. Down-channel traveling when a film reunites with the melt pool was considered through Taylor's expansion by Roberts (1970). This model is

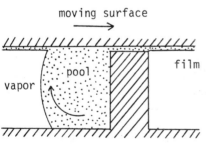

Fig. 1   Partially filled extrusion channel.

based on classical penetration theory, in which the rate-controlling step for evaporation is Fickian diffusion in the melt phase toward the melt–vapor interphase. The model, in essence, consists of the well-known dispersion equation, which in dimensionless form is

$$\mathrm{Pe}^{-1} \frac{d^2\hat{C}}{d\hat{z}^2} - \frac{d\hat{C}}{d\hat{z}} - \mathrm{Ex}\,\hat{C} = 0 \tag{1}$$

where $\hat{C}$ is a dimensionless concentration, $\hat{z}$ is a dimensionless down-channel ordinate, Pe is the longitudinal Peclet number, and Ex is the extraction number. Based on a combined-stage approach, Biesenberger and Kessidis (1982) accounted for the contributions from the film and pool, respectively, to develop the general expression of $\mathrm{Pe_L}$ and Ex including down-channel film–pool reunion.

In certain cases, the reentering distance is minimal when compared with the DV length, L. It is reasonable to ignore backmixing, so that $\mathrm{Pe}^{-1} = 0$. Then, Eq. (1) becomes a simple exponential equation. It has been widely used as a design equation. Moreover, the straightforward solution allows process variables and transport properties to be correlated through system parameters:

$$\frac{d\hat{C}}{d\hat{z}} = -\mathrm{Ex}\,\hat{C} \tag{2}$$

This equation allows prediction of a diffusion coefficient from scattering of separation data. Limited success on a single-screw extruder (Biesenberger and Kessidis, 1982) with nitrogen sweep and on a counter-rotating twin-screw extruder (Collins et al., 1983) has been reported. Note that separation efficiency does not decrease appreciably with increasing machine speed (N_R) despite concomitantly shorter residence time. Evidently, increase of renewal rate of the melt pool offsets the decline inherent with less residence time.

The estimated mass transfer coefficients using Eq. (2) are one or more orders of magnitude less than those measured for polystyrene–styrene in single-screw or twin-screw extrusion under the application of vacuum. This simply suggests a more effective mass transfer mechanism than diffusion for foam DV.

It has long been reported that pressure-reduction-induced DV is accompanied by foaming phenomena, even at very low levels of the volatile component (Lee, 1986). The pressure reduction such as by application of vacuum easily develops a positive superheat (SH) scenario, that is, a system vapor pressure, P_e, greater than surrounding pressure, P_∞:

$$\mathrm{SH} = P_e - P_\infty \tag{3}$$

As a result, a bubble comes into existence. Evidently, the presence of bubbles in the molten polymer substantially enlarges mass transfer area. The phase change via formation, growth, coalescence (if the bubble number is high enough), and rupture of bubbles consists of an effective separation. This process is of "flash" nature when compared with a foam-free diffusion-controlled process. This time benefit can be advantageously transferred to a space benefit. Instead of an independent operation, foam DV can be implemented as part of processing equipment.

Since bubbles become active loci in foam DV, insights into the detailed foaming mechanisms can correlate foaming with separation, and ultimately guidelines on design can be proposed. The existence of a bubble is governed by the force balance equation, which yields the critical bubble radius as

$$R_{Bcr} = \frac{2\sigma}{P_B - P_\infty} \tag{4}$$

in which σ represents the surface tension of the melt phase. Foam will grow when the radius is greater than this; otherwise, it is suppressed into negative radius of curvature. To sustain a bubble, bubble pressure is to equilibrate surrounding pressure and surface tension confinement. This is defined as mechanical equilibrium:

$$P_{me} = P_\infty + \frac{2\sigma}{R_B} \tag{5}$$

In chemical equilibrium, the system pressure corresponds to concentration through an equilibrium constant:

$$P_{ce} = K_W \, W \tag{6}$$

K_W is known as Henry's law constant. Selected polymer–solvent K_W values are listed in Table 1. A detailed expression of K_W can be obtained through the Flory–Huggins partition equation:

$$K_W = P_1^0 \left(\frac{\rho_2}{\rho_1}\right) \exp(1 + \chi) \tag{7}$$

Constant ratio between vapor pressure and volatile concentration in the melt should be assumed with caution for high levels of volatile, as indicated in Fig. 2. However, it is justifiable to use a constant Henry's law constant in the bubble growth of low-volatile vacuum DV.

As a bubble comes into existence, it tends to expand against a viscous medium; and the size increase results in a greater concentration gradient across the bubble–melt interface. Diffusion of volatiles into the bubble is thus encouraged. However, the rheological nature of molten polymer tends to

Table 1 Henry's Law Constant for Selected
Polymer–Solvent Systems

Polymer/Solvent	K_w, atm	Reference
PS/S	30	Werner (1981)
	50	Biesenberger (1983)
PMMA/MMA	85	Werner (1981)
LDPE/Ethylene	2000	Werner (1981)
LDPE/CHClF$_2$	595	Durrill and Griskey (1966)
	170	Gorski et al. (1983)
LDPE/CCl$_2$F$_2$	118	Gorski et al. (1983)
HDPE/Hexane	115	Biesenberger (1983)
	125	Werner (1981)

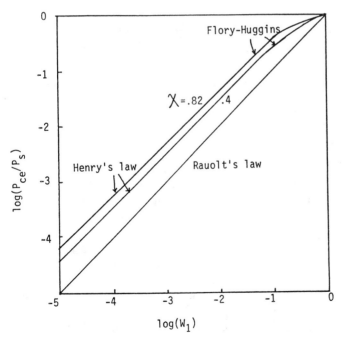

Fig. 2 Comparison between various thermodynamic equations for PDMS/MeCl.

restrict bubble expansion. Diffusion and melt rheology become the controlling factors for foam growth. Duda et al. (1982) developed a solvent–polymer diffusion model including free-volume effects in the polymer. Free-volume enhancement out of the presence of solvent molecules becomes more pronounced when the system is close to the glass transition temperature and has a low solvent level. In the molten state, available sites for solvent molecules to jump into are sufficient, and the amount of solvent has less pronounced effects on the total free volume. It seems reasonable to assume a constant diffusivity for low-volatile transport in molten polymer.

From a separation viewpoint, bubble shrinking in the rolling pool due to pressure variation is not desirable. But bubble rupture at an interface becomes another necessary mechanism to accomplish separation of the contaminant into a contiguous vapor phase. Evidently, foaming appears as a different, but unique, separation process from a conventional diffusion-dominated operation. The mechanisms involved are different from diffusion and interfacial area enhancement. Foam formation, growth, coalescence, and rupture become the loci of foamed DV. Insights into these mechanisms will facilitate upgrading future design and optimizing existing devices.

II. FOAM-ENHANCED SEPARATION

A. Overall Separation

Removal of volatile components from viscous polymer is generally indicated by separation efficiency, defined as

$$E_f = \frac{W_0 - W_f}{W_0 - W_e} \tag{8}$$

W denotes the weight fraction of the contaminant. Subscripts 0, f, and e denote initial, final, and equilibrium. It should be pointed out that in multicomponent system the efficiency could exceed 1 (Werner, 1981). Under high vacuum or spacious vapor space, or both, the ultimate equilibrium composition with respect to the volatile contaminant is very small ($W_e \sim 0$), and separation efficiency is effectively equal to the fractional separation, F_s.

Gas chromatography (GC) is a readily available technology to detect the contaminant level in the polymer. Head space becomes a favorable method for DV, partly because of its convenience and partly because of its operational benefits (Chin, 1982).

A specially designed apparatus was constructed to measure mass transfer rates during foam DV and how they are affected by key parameters.

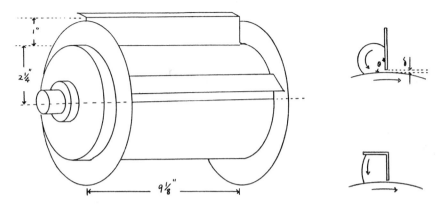

Fig. 3 Schematic drawing of rotating drum apparatus and blade configurations.

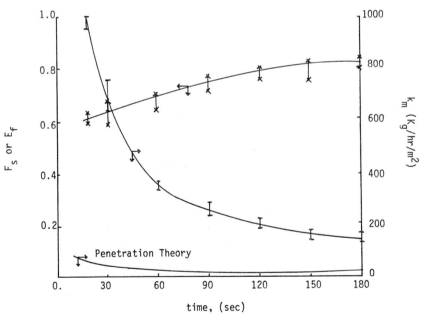

Fig. 4 Separation results ($\times$) and mass transfer rate ($-$) of PDMS/MeCl; η: 1000 poise, W_0: 3600 ppm, N_R: 18 min^{-1}, P: 18 torr. Penetration theory is included for comparison.

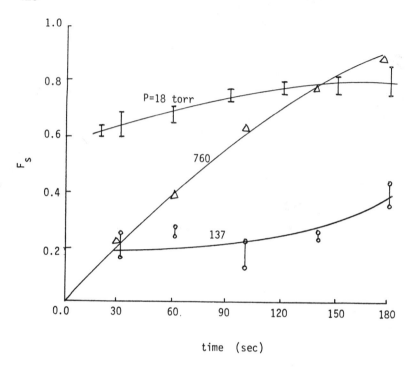

Fig. 5 Separation results of PDMS/MeCl at various times with total pressure as a parameter; N_R: 92 min^{-1}, W_0: 3000 ppm, P: 18 (−), 137 (○), and 760 torr (△).

It consisted of a rolling pool and thin film (Fig. 3) created by rotating a melt-loaded drum against a stationary blade. The apparatus was placed and operated in a spacious vacuum box having a side window for observation. Batch-mode experiments were performed at room temperature to avoid the complexity resulting from thermal degradation.

Methyl chloride (MeCl) and other solvents dissolved in polydimethylsiloxane (PDMS) were selected for the experiments (Lee, 1986). The MeCl/PDMS system has been found similar to molten styrene–polystyrene in viscosity (1000–6000 poise) and Henry's law constant (47 atm). Incorporated with head space GC technology, systematic experiments were performed in such a way that all interesting parameters were independently controlled, including pool time, pool rolling speed, melt viscosity, pressure level, feed composition.

Vigorous foaming appeared when vacuum was applied over the pool.

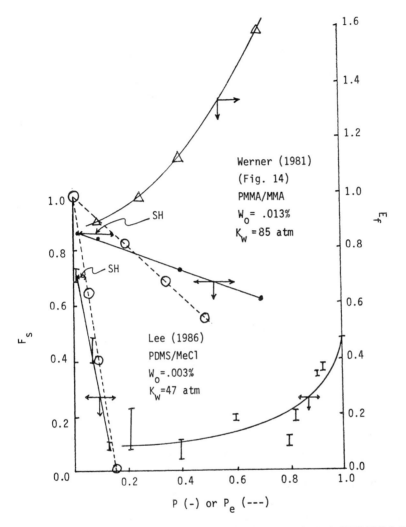

Fig. 6 Various separation results. (△): PMMA/MMA, (−): PDMS/MeCl, (○): system equilibrium pressure, and associated SH plotted against system pressure.

Foaming decreased as DV progressed. Eventually, a bubble-free rolling pool appeared. Figure 4 indicates a fast initial removal followed by a leveling off. Mass transfer coefficients were also computed according to

$$K_M = \rho_2 k_m = \frac{E_f m}{tS} \tag{9}$$

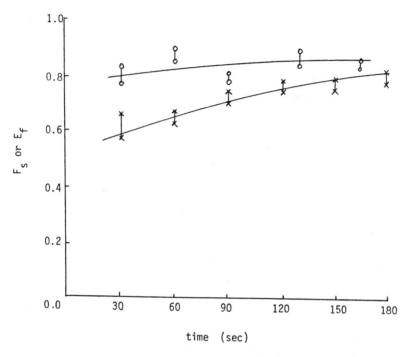

Fig. 7 Separation comparison between degassed ($\times$) and non-degassed ($\bigcirc$) PDMS before loaded with MeCl; P: 18 torr, N_R: 92 min^{-1}.

where m and S are the mass flow rate and surface area of exposure, respectively. K_M was plotted to compare with penetration theory. The latter is by far inferior to the former. It confirms the dominant role of foaming in foam DV. In all low-pressure experiments, SH > 0, yet foaming occurred also in the higher-pressure experiments with SH < 0. The latter is due to air entrainment. Figure 5 shows different separation patterns with various pressure levels. Note the poor separation associated with a moderate pressure drop, in which few bubbles appeared. An exponential curve appears to fit the experiments with $P_\infty = 1$ atm. Despite negative superheat, increase of E_f with rising pressure was reported by Werner (1981) on co-rotating twin-screw DV of MMA/PMMA. The collection was plotted for comparison in Fig. 6.

Figure 7 shows the separation results of degassed and non-degassed PDMS. Normally, PDMS was thoroughly degassed under vacuum before MeCl was introduced to make controlled MeCl/PDMS samples for testing. The extra foaming out of non-degassed PDMS evidently resulted in a better

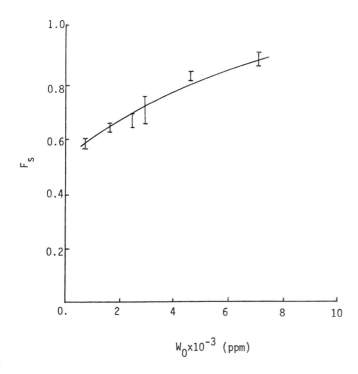

Fig. 8 MeCl separation versus feed concentration; P: 15 torr, separation time: 60 sec.

removal rate. Feed composition and rpm-effects are presented in Fig. 8 and Fig. 9, respectively. Varied F_s suggests that mass transfer is not a constant, as generally applied for DV equipment. Both cases (feed composition and rpm) reinforce the earlier conclusion that Fickian diffusion is not the controlling factor. At the bottom of Fig. 10, separation seems to correlate well with associated SH values. Figure 11 further illustrates the critical role of initial SH. A higher degree of superheat appeared favorable in foam DV.

A modified version of the devolatilizer was established for further foaming investigations. The setup, illustrated in Fig. 12, consists of a scribed line blade affixed to a shaft, mounted with two end disks that fit in a glass tube. Evacuation was effected through a bore in the shaft. This setup allows visualization and videotaping for careful analysis. As illustrated in Fig. 13, in the absence of deformation, vacuum alone is not adequate for bringing bubbles into existence. But slight rotating

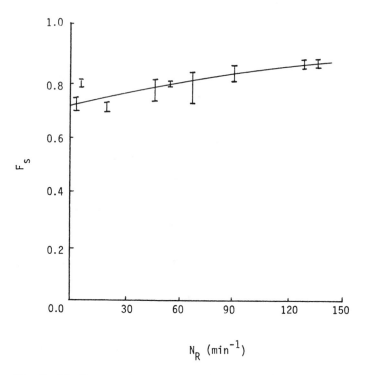

Fig. 9 MeCl separation versus drum rotating speed; P: 10 torr, W_0: 3500 ppm, time: 60 sec.

results in swarms of bubbles, which quantity decreases as rotating continues. The presence of deformation is complementary to vacuum in the low-level solvent foam DV.

Since the limited vapor space is contaminant-enriched during testing, E_f cannot be considered equal to F_s. Figure 14 illustrates rpm (N_R) effects. Increased rpm favors removal rate. During the initial phase, when SH was largest, high rpm caused the rolling pool to shrink in volume. This suggests that a higher internal pressure, as a result of high rpm, reduces SH. At low rpm, relatively high SH encouraged foaming. In spite of this, more bubbles were formed at high rpm, and the time required to remove the contaminants appeared to be shorter. This indicates that a higher rpm enhances both bubble formation and rupture rates, which more than offsets its diminishing effect upon SH, which is a driving force for bubble growth.

Gas bubbles appeared in the rolling pool under atmospheric pressure, and their quantity increased as time passed, up to a saturation

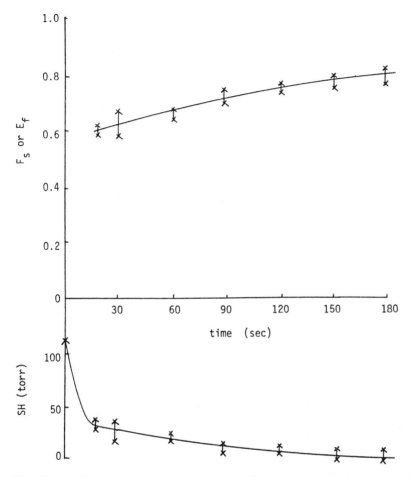

Fig. 10   MeCl separation, same conditions as Fig. 4, and associated superheat (SH).

point. The entrained gas bubbles remained in the pool, a clear contrast to the bubble-free rolling pool after enough time under vacuum.
As indicated in Fig. 5, its separation follows an exponential scattering
pattern.

   The overall separation results have reinforced our belief that foam DV
is a foaming-controlled process. Two types of foaming and the corresponding
removal patterns were distinguishable. The following sections are devoted
to these two foaming mechanisms.

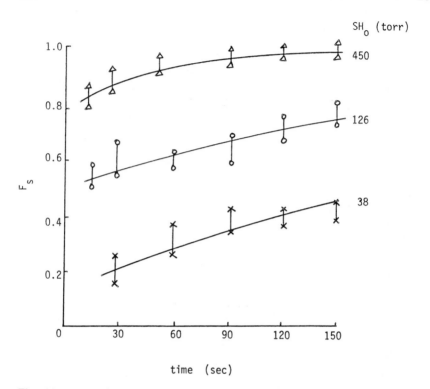

Fig. 11 Separation curves with Henry's law constant (initial superheat) as a parameter; η: 1000 poise, N_R: 75 min^{-1}, P: 7 torr, W_0: 4250 ppm; ($\times$): PDMS/CCl$_2$CF$_4$, K$_w$: 12 atm; ($\bigcirc$) PDMS/CHClF$_2$, K$_w$: 39 atm; ($\triangle$): PDMS/CClF$_3$, K$_w$: 140 atm.

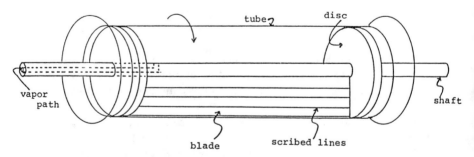

Fig. 12 Schematic drawing of glass tube apparatus.

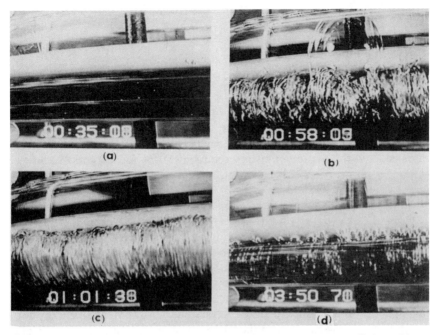

Fig. 13 Photographs taken through glass tube: (a) vacuum applied at 21 sec; (b) low N_R commenced at 50 sec; (c) pool volume increase; (d) bubble disappearance.

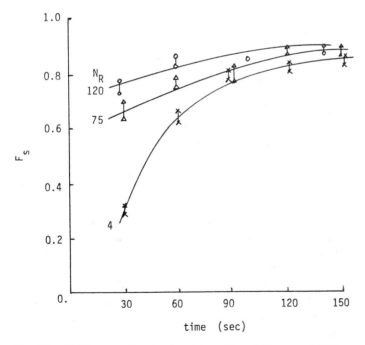

Fig. 14 MeCl separation in glass tube with rotating speed (N_R) as parameter; W_0: 8000 ppm, P: 5 torr, η: 1000 poise, N_R: (○) 120, (△) 75, (×) 4 min^{-1}.

B. Vacuum Foaming

1. Nucleation

Since bubbles are known to be active loci in foam DV, it is a reasonable approach to develop an understanding of bubble formation. Classical nucleation theories assume unstable chemical equilibrium at nucleation so that surface energy and energy to sustain the critical bubble are considered the minimum energy requirement for nucleation. As shown by Blander and Katz (1975), the homogeneous nucleation equation can be expressed by

$$J = A \exp(-B) \tag{10}$$

where A and B are material and system parameters. Blander and Katz (1975) also considered the effects on nucleation rate of thermal, viscous, and diffusion factors. In fact, the energy requirement remained the same. Colton and Suh (1987) presented a heterogeneous model for microcellular foam nucleation for zinc stearate–polystyrene. L–J potential energy for noninteracting molecules was taken into account, but, still, results orders of magnitude under theoretical predictions were observed. Han and Han (1990b) suggested the free-energy change for nucleation in the polymer solution as

$$\Delta F_p = \Delta F - \Delta F_s - \Delta F_t \tag{11}$$

where ΔF denotes the free energy required for single-component phase transformation. ΔF_s and ΔF_t are the free-energy changes due to super-saturation and due to polymer solution, respectively. Incorporating the effects of supersaturation and of free energy of mixing, together with diffusion-controlled mode, the semiempirical nucleation rate ($\#/m^3/\text{sec}$) for polystyrene–toluene can be expressed as

$$J = 4.71 \times 10^{34} M \frac{D}{4\pi r_c} \exp\left[-\left(\frac{42,344}{T} + \frac{\Delta F_p}{nk_B T} \right) \right] \tag{12}$$

where M is the number of molecules per unit volume of the metastable phase, D the diffusion coefficient, k_B the Boltzmann constant, and T the absolute temperature. The discrepancy between Eq. (12) and experimental results is less pronounced. In fact, the solution viscosity Han and Han (1990b) used is at least a magnitude lower than that of molten polymer in the processor. High solvent loading along with low solution viscosity presents favorable conditions for nucleation. Under such circumstances, superheat alone seems sufficient for foam formation.

Lee (1986) reported that neither homogeneous nor heterogeneous theories were adequate in explaining foam formation of low-level volatile in polymer melt. A typical example of polystyrene–styrene is listed in Table 2. Jemison et al. (1980) observed that the favorable conditions to describe heterogeneous nucleation were not realistic. Therefore, these theories must be rejected.

In cavitation theories for simple liquids, Knapp et al. (1970) ascribe bubble formation to vibrations, vortices, or local pressure drops that accompany flow acceleration at high fluid velocities. Polymer melt is known for its viscous and creeping nature, and it is free from significant acceleration. Byon and Youn (1990) observed ultrasonic induced foam formation and enhancement on foam formation in nitrogen/LDPE-PE wax. Similar observations for styrene–polystyrene were reported by Tukachinsky et al. (1993). One could speculate that preexisting cavities in the melt are excited by "disturbances."

Free volume is a well-established concept in polymers. It successfully characterizes polymer melt viscosity by applying "hole theories" of liquids. However, the hole dimension is on the molecular order, and the improbability of finding holes with dimensions comparable with the critical bubble radius is shown in Table 3. In dynamic nucleation, it is not unreasonable to speculate that the distribution of free volume during deformation can shift far enough to give a reasonable probability of the formation of critical radius holes. However, not only is the melt in DV far above the glass transition temperature, but according to our visualization experiments, very low shear was adequate to produce swarms of bubbles, and it is not likely to cause significant variation in free-volume distribution.

Two key aspects of bubble formation were observed in the nucleation experiments. First, as illustrated in Fig. 13, modest superheats alone were necessary, but insufficient, to induce bubbles in the degassed melts (Biesenberger and Lee 1986b). Low rate of deformation was necessary as well. This suggests a reaction path with a metastable state in the presence of positive superheat. Second, Han and Han (1990a) observed bubble formation was in the submicron range. From both observations, it is unlikely that there will be either a sufficiently large force to create a big enough hole or a large enough number of volatile molecules to congregate to form a hole that size. We were thus brought to look into the foam formation theory predicated on the preexisting microscopic particulates, the surfaces of which could accommodate nuclei. Harvey et al. (1944) presented a cavity theory to describe foam formation in organisms under a substantial pressure drop.

Figure 15 illustrates successive geometric configurations for a dispersed gas pocket, idealized as a conical shape cavity. Its radius of curvature determined by the pressure differential across the interface shows its stability.

Table 2 Homogeneous Calculation for Polystyrene–Styrene

Temp. (°C)	Temp. (K)	$K_W{}^a$ (atm)	SH (atm)	ρ^b (g/cm³)	σ^c (erg/cm²)	$A \times 10^{32}$ (#/cm³·sec)	B	J^d (#/cm³·sec)
150	423	5.0	0.012	0.79	18.1	9.66	1.2×10^{10}	0
171	444	7.9	0.026	0.76	15.6	8.62	1.5×10^{9}	0
190	463	10.5	0.039	0.74	13.9	7.91	4.5×10^{8}	0
210	483	15.0	0.062	0.72	12.1	7.18	1.1×10^{8}	0
230	503	21.5	0.095	0.70	10.4	6.47	2.9×10^{7}	0
250	523	34.8	0.161	0.67	8.6	5.64	5.5×10^{6}	0
280	553	52.6	0.25	0.64	6.2	4.58	8.2×10^{5}	0
300	573	109.9	0.54	0.62	4.7	3.86	7.4×10^{4}	0
325	598	181.3	0.89	0.59	2.9	2.88	6.0×10^{3}	0
350	623	403.4	2.00	0.55	1.2	1.70	7.4×10^{1}	2.2

[a] $K_W(T)$ are from Werner (1981).
[b] $\rho(T)$ are found using Watson's expansion method.
[c] $\sigma(T)$ are calculated using Eötvös' equation.
[d] $J = A\exp(-B)$, $A = 3.1 \times 10^{35}(\rho^2\sigma/M)^{1/2}$, $B = 1.2 \times 10^5\sigma^3/(T \cdot SH^2)$ M: molecular weight of styrene.

Table 3 Possibility Calculation on Finding Free-Volume Holes Equivalent to Critical Bubble Radius

$W_s(\%)$	K_W (atm)	P_G (atm)	R_{Bcr} (cm)	$(-\gamma R_{Bcr}^3/\bar{R}_B^3)$	k (prob.)
0.01	47	0.47	8.9×10^{-5}	-7.0×10^{11}	0
	25	0.25	1.7×10^{-4}	-4.9×10^{12}	0
0.005	47	0.24	1.8×10^{-4}	-5.8×10^{12}	0
	25	0.13	3.5×10^{-4}	-4.3×10^{13}	0
0.001	47	0.05	1.0×10^{-3}	-1.0×10^{15}	0
	25	0.025	2.2×10^{-3}	-1.1×10^{16}	0
0.005	47	0.024	2.4×10^{-3}	-1.4×10^{16}	0
	25	0.013	7.0×10^{-3}	-3.4×10^{17}	0

$P = 5$ mm Hg $= 0.007$ atm, $\sigma = 20.9$ erg/cm$^2 = 2.1 \times 10^{-5}$ atm-cm, $\bar{R}_B = 10^{-8}$ cm, $\gamma = 1$, $k = k_0 \exp(-\gamma R_{Bcr}^3/\bar{R}_B^3)$.
Source: Lee (1986).

Harvey et al. (1944) have argued that potentially stable negative curvature can occur in the cavity of a hydrophobic solid surface, since the equilibrium contact angle, θ_e, can be easily established or reestablished. When external pressure decreases to develop superheats, positive radii of curvature can be envisioned. Moreover, the contact angle varies correspondingly. When it is beyond a critical value, θ_r (receding angle) or θ_a (advancing angle), the meniscus tends to slip. This suggests a yield phenomenon. Application of vacuum can easily develop potentially unstable gas cavities with positive radii, as illustrated in Fig. 15b. The continuing decrease of contact angle as a result of external pressure decrease destroys equilibrium; and the meniscus starts to slip and swell toward the cavity mouth (Fig. 15c). In low-volatile DV (i.e., a few percent or lower), superheat is not sufficient to overcome interfacial tension confinement. We postulated that deformation of the surrounding melt provides the necessary impetus via distortion and/or "pulling" to detach the gas cavity from the solid (Fig. 15d). The bubble with radius R_B in (Fig. 15e) will continue to grow in response to positive SH. Han and Han (1990a) reported submicron critical bubble radius as indicated in Fig. 16, in which extrapolation is applied to polystyrene–toluene experimental results.

This cavity model was further tested in thermoplastic foam extrusion. Shear-enhanced foam formation was well correlated with preexisting cavity theory (Lee, 1993). The capillary number, defined as the ratio of shear force

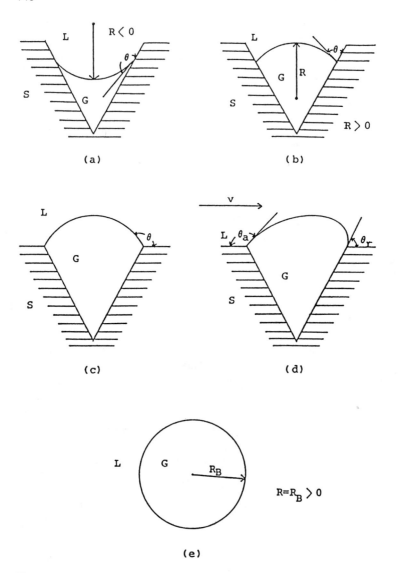

Fig. 15 Successive steps for bubble formation: (a) stable gas cavity; (b) at pressure reduction; (c) metastable cavity; (d) under deformation; (e) unstable gas bubble.

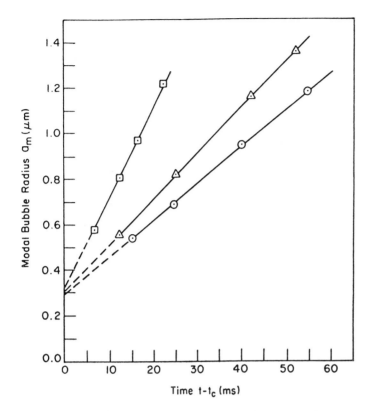

Fig. 16 Bubble radius versus time for polystyrene–toluene at an equilibrium pressure of 2859 kPa and at three temperatures (°C): (○) 150, (△) 170, and (□) 180. (Reprinted with permission from John Wiley & Sons, Inc., Figure 16 of Han and Han, *J. Poly. Sci. B: Poly. Phys.*, *28*: 711, 1990.)

to surface tension force, was established for a metastable gas cavity as illustrated in Fig. 15c:

$$Ca = \frac{R\eta\bar{\dot{\gamma}}}{4\sigma}$$ (13)

where η and $\bar{\dot{\gamma}}$ denote melt viscosity and average shear rate. A straight-line mode in the log-linear plot of Fig. 17 underscores the importance of shear force. More experimental results on thermoplastic foam extrusion support the preexisting cavity nucleation concept (Lee, 1994).

At high loading of volatile solvent, enough pressure reduction can develop sufficient superheats to make bubble nucleation possible. Gas

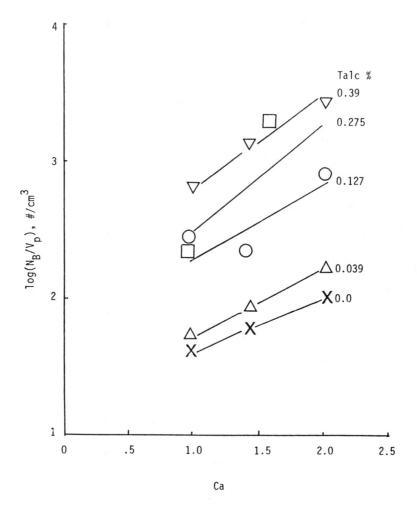

Fig. 17 LDPE/CCl$_2$F$_2$ nucleation in foam extrusion versus capillary number with talc nucleating agent as parameter.

bubble formation is also favored with significantly lower solution viscosity. However, if volatile contents are on the percentage level or lower, only modest superheats can be developed; and that alone is not sufficient to overcome surrounding confinement. Deformation or disturbance becomes essential to detach the gas cavity to create a gas bubble. It will then grow due mainly to its unstable nature. As a result of expansion, diffusion is further encouraged due to the concentration gradient across the bubble melt interface.

2. Bubble Growth

Bubble growth is a rate process, and its limit is determined by thermodynamic parameters. Many research efforts have been devoted to the growth of a spherical gas bubble in a liquid in the absence of flow (e.g., Scriven, 1959). Growth of a swarm of bubbles in a static polymer melt attracted significant research interest in the past decade. Amon and Denson (1984) presented the "cell model," in which a bubble of radius R_B is surrounded by a given amount of polymer to form a spherical shell of radius R, as illustrated in Fig. 18. Assuming an incompressible and Newtonian fluid, isothermal conditions, and a constant diffusion coefficient, Lee (1986) obtained the dimensionless equations as

$$\frac{d\hat{R}_B}{d\hat{t}} = = C_B\left(\hat{P}_b - \hat{P} - \frac{C_A}{\hat{R}_B}\right)\hat{R}_B\left(1 + \frac{\hat{R}_B^3}{\hat{v}_p}\right) \tag{14}$$

$$\hat{R}_B(0) = 1$$

$$\frac{\partial \hat{C}}{\partial \hat{t}} = \frac{\partial}{\partial \hat{r}}\left[\left(\frac{\hat{R}_B^3}{\hat{r}_p} + r\right)^{4/3} \frac{\partial C}{\partial r}\right]$$

$$\hat{C}(0, \hat{r}) = 1, \qquad \hat{C}(\hat{t}, 0) = \frac{P_s}{P_{e0}}, \qquad \frac{\partial \hat{C}(\hat{t}, 1)}{\partial \hat{r}} = 0$$

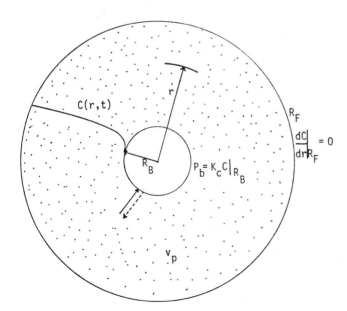

Fig. 18 "Cell" growth model; (—) diffusion direction and (– –) expansion direction.

where

$$C_A = \frac{2\sigma}{R_{B0} P_{b0}} = \frac{P_{me0} - P_\infty}{P_{B0}}$$

$$C_B = \frac{\lambda_D}{\lambda_{bm}}$$

$$\lambda_D = \frac{\hat{v}_p^{2/3} R_{B0}^2}{9D}$$

$$\lambda_{bm} = \frac{P_{b0}}{4\eta}$$

$$\hat{t} = \frac{t}{\lambda_D}$$

$$\hat{P}_{e0} = Kw \frac{W_0}{P_{B0}}$$

The driving forces for growth are either mechanical, $P_B > P_{me}$, or chemical, $P_{ce} > P_B$, or both. As expected, diffusivity and viscosity are the

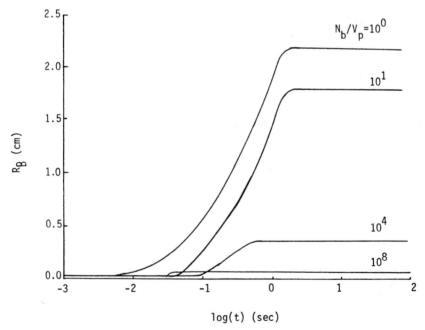

Fig. 19 Computation results based on cell model with cell density as a parameter.

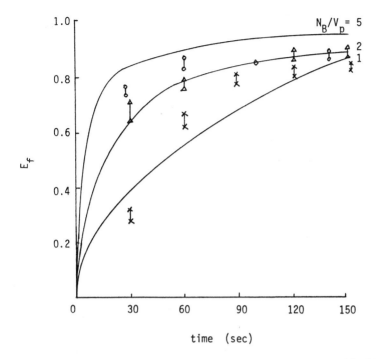

Fig. 20 Modeling results curve-fit separation data as presented in Fig. 14; N_B/V_p: #/cm.

transport and rheological parameters that determine the bubble growth rate. In practice, system parameters are of considerable interest. Figure 19 shows the effect of bubble number density on the growth rate and final size. It agrees with observations that less bubbles reach a larger ultimate size, such as the case of low rpm. In Fig. 20 some of the vacuum foam DV data from the system of MeCl/PDMS have been curve-fitted with the model using bubble number density as a parameter. Despite the aforementioned assumptions such as omission of flow and interactions, the fit is very promising. The values for bubble number density are of the first order, which is somewhat less than estimates from Newman and Simon (1980) (i.e., $500/cm^3$). Coalescence and early rupture are among the possible causes.

3. Rupture

Vigorous foaming appeared after the nucleation threshold was reached. Foaming activity lasted from a few seconds to a few minutes, depending mainly on system parameters. Bubble rupture becomes a necessary step to complete contaminant separation. The rolling pool tends to apply inertial

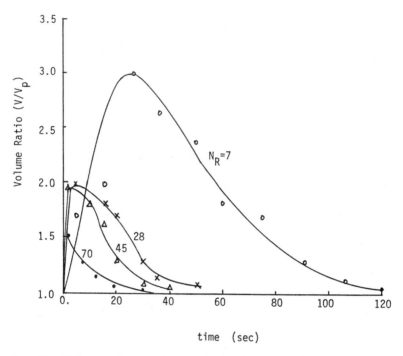

Fig. 21 Volume ratio variation results in glass tube with N as a parameter; P: 5 torr, W_0: 2.3%, N_R: ($\bigcirc$) 7, ($\times$) 28, ($\triangle$) 45, ($\bullet$) 70 min^{-1}.

forces to restrict bubble growth and to terminate bubbles, and an unrealistically low foam–melt density will be otherwise developed. It is of considerable interest to plot pool volume variation during DV. Figures 21 and 22 illustrate the critical roles played by flow and volatile loading. The data distribution suggests competing mechanisms. Pool volume expansion is limited by the flow field. As a result, interfacial bubble rupture occurs.

During visualization experiments, the bubbles stretched near the surface and were squeezed in the nip as illustrated in Fig. 23, and rapid ruptures from the free surface were observed in the early stage of DV. During the later stage, those bubbles that remained attempted to escape at a much slower rate. They extended themselves along the streamlines toward the pool surface. In the final stage, few bubbles reamined rotating in the pool. They were pushed toward the surface due to the pressure gradient in the pool. As a result, bubbles close to the surface are able to escape. Initial vigorous ruptures owing to unproportional growth at the free surface and squeezing at the nip appear to be the main rupture mechanisms. This appears consistent

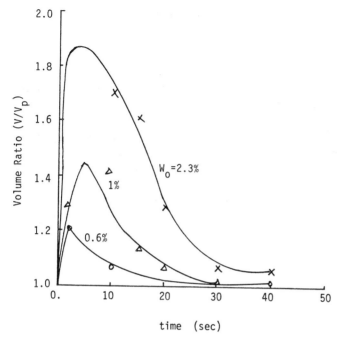

time (sec)

Fig. 22 Volume ratio variation with feed concentration as parameter; N_R: 45 min^{-1}, P: 5 torr, W_0: ($\times$) 2.3%, ($\triangle$) 1%, ($\bigcirc$) 0.6%.

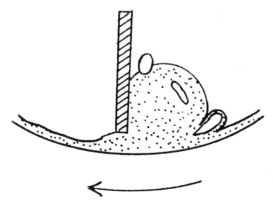

Fig. 23 Sketch of pool with bubbles before bursting.

with the "flash" type separation pattern observed in vacuum DV. The bubble disappearance rate slows down at later stage of DV. In either case, rpm has pronounced effects.

4. Melt Pool Flow Analysis

Owing to the effects of melt flow on the formation, shape, and rupture of bubbles and the separation efficiency observed in our visualization and separation experiments on rolling pools, it is clearly worthwhile to acquire flow insights through fluid modeling. Moreover, the internal pressure profile, a critical parameter for nucleation and growth, will be part of the flow modeling solution. It is a complex flow problem involving two phases and a free surface. As a first approximation, we specified the interface as being circular and applied free surface boundary conditions to it. Other assumptions were two-dimensional flow field, incompressible, Newtonian fluid, and the absence of bubbles. The governing vorticity and streamline function equations, ω and ψ, derived from the Navier–Stokes partial differential equations can be summarized as

$$\frac{\partial \omega}{\partial t} = -\frac{\partial (u_x \omega)}{\partial x} - \frac{\partial (u_y \omega)}{\partial y} + \frac{1}{\text{Re}}\left(\frac{\partial^2 \omega}{\partial x^2} + \frac{\partial^2 \omega}{\partial y^2}\right)$$

$$\frac{\partial^2 \psi}{\partial x^2} + \frac{\partial^2 \psi}{\partial y^2} = \omega \tag{15}$$

where $\text{Re} = \rho a U_0 / \eta$ (Reynolds number).

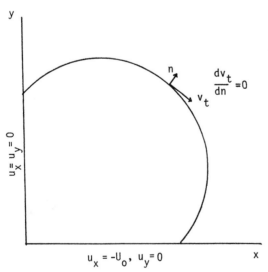

Fig. 24   Rolling pool flow model; v_t: tangent velocity component.

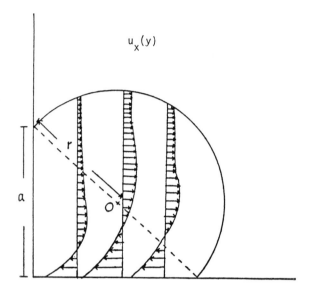

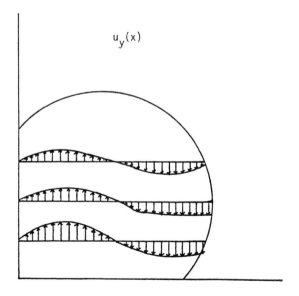

Fig. 25 Velocity profiles: $u_x(y)$ and $u_y(x)$.

The relevant boundary conditions are presented in Fig. 24. Some computed velocity profiles are shown in Fig. 25, in which we find the location of the maximum shear rates to be between the free surface and the moving wall instead of the stationary wall. It is the locus of bubble formation as observed in some of the visualization experiments. The maximum shear rate at the lowest pool rotating speed was of the order of 1 sec^{-1}, which means Newtonian fluid was a reasonable assumption.

Figure 26 illustrates pressure variation as a result of the moving-wall-

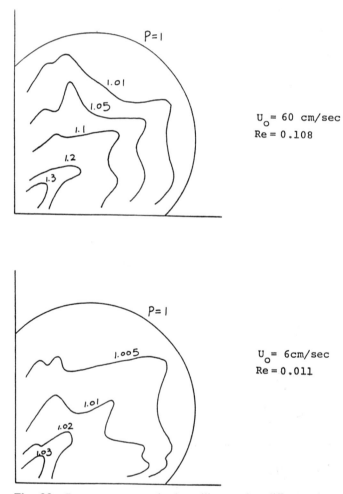

Fig. 26 Pressure contours in the rolling pool at different plate velocities: 6 cm/sec and 60 cm/sec.

induced rolling pool. As expected, a higher moving speed corresponds to a high inernal pressure. As much as a 30% increase was predicted by the model, while the Reynolds number was kept around 0.11. The internal pressure rise is not favorable to growth. It will be sufficient to cause bubbles to collapse and disappear. But at high rpm, the growth restriction on separation is easily offset by nucleation and rupture enhancement. Therefore, rpm has a positive overall effect on separation.

A certain volume was assigned to simulate bubble shape variation in the rolling pool. Lee (1986) reported that computation results had reasonably good agreement with observations. Bubble shape experienced stretching while close to the free surface and contraction at the turning point.

C. Entrained Gas Foaming

Entrained gas bubbles appeared in the rolling pool under atmospheric pressure. These bubbles are different from vacuum bubbles in at least two

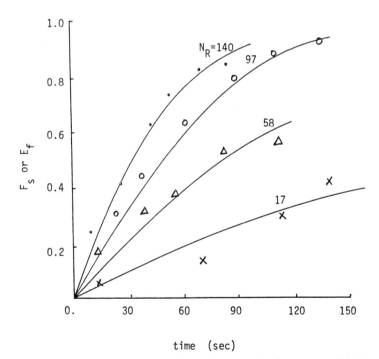

Fig. 27   MeCl separation results under atmospheric pressure with N_R as a parameter; W: 2400 ppm, P: 760 torr, η: 1000 poise, N_R: ($\times$) 17, ($\triangle$) 58, ($\bigcirc$) 97, ($\bullet$) 140 min^{-1}.

respects: low bubble pressure (1 atm), and air content in the bubble. Nevertheless, Fig. 27 illustrates atmospheric separation data from the same setup. The data has been curve-fitted with the exponential model. Clearly the fit is promising, suggesting that the pincipal separation mechanism could be diffusion. However, from the effect of rotating speed illustrated in the same figure, one can speculate that foaming could be the controlling parameter.

Let us investigate diffusion-induced bubble growth. The same growth equations and boundary conditions were employed with different initial conditions. Figure 28 shows the growth results. Noteworthy is the limited expansion and concomitant bubble pressure reduction. It only takes fraction of a second to accomplish the growth, and the corresponding separation is far less than that of experimental results. Diffusion-induced growth is a necessary mechanism, but like vacuum DV, it is not sufficient for separation. One could speculate that a bubble population balance is a good approach to describe foaming in the rolling pool at one atmosphere, with a constant bubble birth and death rate depending on rpm.

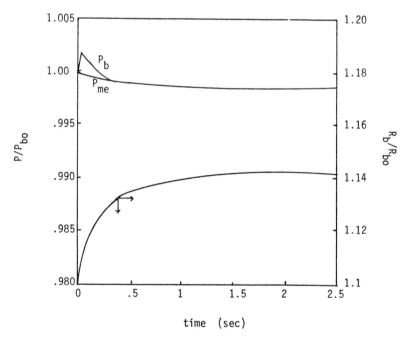

Fig. 28 Computation results on entrained bubble growth; N_B/V_p: 2×10^4 #/cm^3, R_{BO}: 0.01 cm, η: 1000 poise, D: 2.85×10^{-5} cm^2/sec.

Since an entrained bubble reaches chemical equilibrium in a brief period, it is reasonable to assume that a ruptured bubble is chemically saturated. Applying an overall mass balance, we can reach the following equation:

$$-V_p \frac{dC}{dt} = \frac{n_r P_1 V_B}{R_g T} \tag{16}$$

where n_r is the number of interfacial ruptures per unit time, and V_B, V_p, and R_g denote bubble volume, melt pool volume, and the ideal gas law constant, respectively. P_1 represents the solvent partial pressure in the bubble. The familiar exponential expression for the fraction of removal is obtained:

$$F_s = 1 - \exp \frac{-n_r K_C' V_B t}{V_p} \tag{17}$$

K_C' is the Henry's law constant, equal to $P_1/R_g TC$. The rpm effects in Fig. 27 can be understood through its direct relation with n_r. The other parameter,

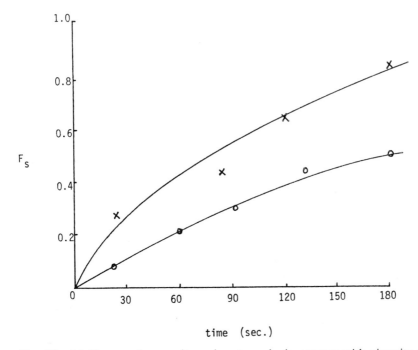

Fig. 29 MeCl separation results under atmospheric pressure with viscosity as a parameter, W_0: 2300 ppm, N_R: 57 min, η: ($\times$) 3000, ($\bigcirc$) 1000 poise.

K_C', also shows good agreement to exponential curve fitting (Lee, 1986). In brief, foaming activity prevails in the entrained gas separation as well. Foam formation appeared while inertial force is greater than surface tension force at the nip. A saturated state on the foam population was established after a certain period, and rupture frequency was supposed to be equal to the nucleation rate from that point on. Again, deformation becomes the key factor behind entrained gas foaming.

Also noted in Fig. 29 is that higher viscosity appears to enhance separation under atmospheric pressure. Higher-viscosity fluid generally possesses a better melt strength, which demonstrates its ability to sustain bubbles (Philips et al., 1992). This effect is not pronounced when bubble lifetime is very brief. However, according to Eq. (17), more bubbles at the steady state are favorable in fast separation.

III. SUMMARY

Devolatilization is primarily a separation process, the limits of which are defined by thermodynamic parameters. Although diffusion is the elementary mechanism, foaming activity has been proved crucial for foamed DV through either application of vacuum or gas entrainment. Foaming in the rolling polymeric melt is an effective mechanism in accelerating separation by generating substantial interfacial areas within limited time. It is useful in separation estimates and in the design of devices to achieve a desired separation.

Foam formation is primarily a superheat-controlled process; however, deformation becomes necessary when low levels of the volatile contaminant cannot provide enough superheat. In fact, rpm effects are more pronounced in the formation and the rate of formation in both vacuum foaming and entrained gas foaming, while the importance of superheat diminishes in the latter. Foam starts to grow right after its birth in the presence of non-equilibrium conditions. In foam DV with positive superheat, the initial bubble number density is primarily determined by superheat, and rpm appears to be one of the major controllable parameters in removing volatiles. Bubble number density decreases as DV time passes, and interfacial bubble rupture and internal shrinking are the main reasons for bubble disappearance. Again, rpm has a strong influence on bubble disappearance. High rpm slows down bubble growth by raising internal pressure, and it can even cause bubble shrinkage in some cases. However, it has an overall positive impact on contaminant removal via vacuum foaming. Moreover, rpm shows principal effects on the separation of entrained gas foaming as well. Even when bubble growth is limited because of a low initial bubble pressure, the bubble population balance indicates qualitative agreement with the exponential separation pattern. The sweep of inert gas is of direct interest in the application of entrained gas. Design for adequate retention time and proper handling of the remaining bubble-saturated melt will make it more practically attractive.

Fundamental study can be used to develop an algorithm that describes DV process components and could be applied to process design, and selection and optimal operation of equipment, which correlates engineering parameters to basic material properties. In addition, we may compute the best possible separation in an existing device or the size and configuration of a device required to achieve a desired separation.

NOMENCLATURE

A material parameter in the nucleation rate equation

a contact length of melt pool on the moving plate

B system parameter in the nucleation rate equation

C solvent concentration in the polymer melt

C_A dimensionless surface tension parameter in cell model

C_B dimensionless constant in cell model

Ca   capillary number, ratio of shear force vs. surface tension force, $R\eta\bar{\dot{\gamma}}/4\sigma$

D diffusion coefficient, cm^2/sec

E_f separation efficiency

Ex   extraction number

F free energy

F_s fraction of removal

J nucleation rate, $\#/(m^3 \cdot sec)$

K_C' dimensionless Henry's law constant

K_W Henry's law constant, atm

k_m local mass transfer coefficient, cm/sec

k_B Boltzmann constant, 1.38×10^{-16} erg/degree

K_M mass transfer coefficient, $kg/(m^2 \cdot hour)$

M number of solvent molecules in a nucleating site

m mass flow rate

N rotating speed, sec^{-1}

N_B number of bubbles

n_r interfacial bubble rupture frequency

P_∞ surrounding pressure, atm

P_B bubble pressure

P_e system equilibrium pressure

P_{ce} chemical equilibrium pressure

P_{me} mechanical equilibrium pressure

P_s solvent vapor pressure, or P_1

Pe longitudinal Peclet number
R curvature of gas cavity
R_B bubble radius
R_g ideal gas constant
r_c critical radius for conventional nucleation
S mass transfer area
SH degree of superheat, atm or torr
T temperature
t time
r_c time at which bubbles appear
U_0 moving plate velocity, cm/sec
u_x x-component velocity, cm/sec
u_y y-component velocity, cm/sec
V_B total bubble volume
V_p melt pool volume
v_p melt volume assigned to each bubble
v_t tangential velocity component
W solvent weight fraction
x x-component
y y component
z down-channel direction

Greek Letters

γ overlapping factor
$\dot{\gamma}$ shear rate, sec^{-1}
Δ difference
η viscosity
λ_{bm} time constant for bubble dynamics
λ_D diffusion time constant in the cell model
ρ density, g/cm^3
σ surface tension, dyne/cm
θ contact angle
ω vorticity function
χ interaction parameter
ψ stream-line function

Superscripts

0 pure state

Subscripts

1 solvent
2 polymer

a   advancing
B   bubble
ce   chemical equilibrium
cr   critical state
e   equilibrium
e0   initial equilibrium
F   melt boundary
f   final
me   mechanical equilibrium
me0 initial mechanical equilibrium
0   initial state
r   receding

Miscellaneous

^   sign for dimensionless symbol
−   average

REFERENCES

Amon, M., and Denson, C. D. (1984). A study of the dynamics of foam growth: Analysis of the growth of closely spaced spherical bubbles, *Poly. Eng. Sci.*, *24*: 1026.

Biesenberger, J. A., and Todd, D. (1983). Section 1: Fundamentals In *Devolatilization of Polymers* (J. A. Biesenberger, ed.), Hanser, New York.

Biesenberger, J. A., and Kessidis, G. (1982). Devolatilization of polymer melts in single screw extruder, *Poly. Eng. Sci.*, *22*: 832.

Biesenberger, J. A., and Lee, S. T. (1986a). A fundamental study of polymer melt devolatilization: I. Some experiments on foam-enhanced DV, *Poly. Eng. Sci.*, *14*: 982.

Biesenberger, J. A., and Lee, S. T. (1986b). *Visualization Experiments on Foam-Enhanced DV of Polymer Melts* [videocassette], Polymer Processing Institute at Stevens Institute of Technology, Hoboken, New Jersey.

Blander, M., and Katz, J. L. (1975). Bubble nucleation in liquids, *AIChE J.*, *21*: 833.

Byon, S. K., and Youn, J. R. (1990). Ultrasonic processing of thermoplastic foam, *Poly. Eng. Sci.*, *30*: 147.

Chin, G. (1982). Simulation of devolatilization in polymer, Master's thesis, Stevens Institute of Technology, Hoboken, New Jersey.

Collins, G. P., Denson, C. D., and Astarita, G. (1983). The length of a transfer unit (LTU) for polymer devolatilization process in screw extruders, *Poly. Eng. Sci.*, *23*: 323.

Colton, J. S., and Suh, N. P. (1987). The nucleation of microcellular thermoplastic foam with additives: Part 1: Theoretical considerations, *Poly. Eng. Sci.*, *27*: 485.

Duda, J. L., Vrentas, J. S., Yu, S. T., and Liu, H. T. (1982). Prediction of diffusion coefficients for polymer–solvent systems, *AIChE J.*, *28*: 279.

Durrill, P. L., and Griskey, R. G. (1966). Diffusion and solution of gases in thermally softened or molten polymer: Part I, *AIChE J.*, *12*: 1147.

Gorski, R. A., Ramsey, R. B., and Dishart, K. T. (1983). "Physical Properties of Blowing Agent Polymer Systems: I. Solubility of Fluorocarbon Blowing Agents in Thermoplastic Resins," Proc. SPI 29th Ann. Tech. Mark. Conf., p. 286.

Han, J. H., and Han, C. D. (1990a). Bubble nucleation in polymeric liquids. I. Bubble nucleation in concentrated polymer solutions, *J. Poly. Sci. B: Poly. Phys.*, *28*: 711.

Han, J. H., and Han, C. D. (1990b). Bubble nucleation in polymeric liquids. II. Theoretical considerations, *J. Poly. Sci. B: Poly. Phys.*, *28*: 743.

Harvey, E. N., Barnes, D. K., McElroy, W. D., Whitely, A. H., Pease, D. C., and Cooper K. W. (1944). Bubble formation in animals: I. Physical factors, *J. Cellular Comp. Physio.*, *24*: 1.

Jemison, T. R., Rivers, R. J., and Cole, R. (1980). "Incipient Vapor Nucleation of Methanol from an Artificial Site: Uniform Superheat," AIChE Annual Meeting, Chicago, Illinois.

Knapp, R. T., Paity, J. W., and Hammit, F. G. (1970). *Cavitation*, McGraw-Hill, New York.

Latinen, G. A. (1962). Devolatilization of viscous polymer systems, *Amer. Chem. Soc. Adv. Chem. Ser.*, *34*: 235.

Lee, S. T. (1986). Study of foam-enhanced devolatilization; Experiments and its theories, Ph.D. thesis, Stevens Institute of Technology, Hoboken, New Jersey.

Lee, S. T. (1993). Shear effects on thermoplastic foam nucleation, *Poly. Eng. Sci.*, *33*: 418.

Lee, S. T. (1994). "More Experiments on Thermoplastic Foam Nucleation," Ann. Tech. Conf. of Soc. Plas. Engr., San Francisco.

Lee, S. T., and Biesenberger, J. A. (1989). A fundamental study of polymer melt devolatilization: IV. Some theories and models for foam-enhanced DV, *Poly. Eng. Sci.*, *29*: 782.

Newman, R. E., and Simon, R. H. S. (1980). "A Mathematical Model of Devolatilization Promoted by Bubble Formation," AIChE Annual Meeting, Chicago.

Park, C. B., and Suh, N. P. (1992). Rapid heating for microcellular nucleation in a polymer melt, *Ann. Tech. Conf. of Soc. Plst. Engr.*, *48*: 1513.

Philips, E. M., McHugh, K. E., Ogale, K., and Bradley, M. B. (1992). Polypropylene with high melt stability, *Kunstsoffe Ger. Plas.*, *82*(8): 23.

Roberts, G. W. (1970). A surface renewal model for the drying of polymers during screw extrusion, *AIChE J.*, *16*: 878.

Scriven, L. E. (1959). On the dynamics of phase growth, *Chem. Eng. Sci.*, *10*: 1.

Tukachinsky, A., Tadmor, Z., and Talmon, Y. (1993). Ultrasound-enhanced devolatilization of polymer melt, *AIChE J.*, *39*: 359.

Werner, H. W. (1981). Devolatilization of polymers in multi-screw devolatilizers, *Kunststoffe*, *71*: 18.

7

The Study of Devolatilization by Scanning Electron Microscopy

Ramon J. Albalak*, Alexander Tukachinsky†, Tali Chechik‡, Yeshayahu Talmon, and Zehev Tadmor

Technion—Israel Institute of Technology, Haifa, Israel

---

* *Current affiliation*: Massachusetts Institute of Technology, Cambridge, Massachusetts.

† *Current affiliation*: Institute of Polymer Engineering, University of Akron, Akron, Ohio.

‡ *Current affiliation*: Reshet-O-Plast Hahotrim, Kibbutz Hahotrim, Israel.

I. INTRODUCTION

Industrially, devolatilization is carried out in equipment belonging to one of two major types, as discussed in length in other chapters. The first type consists of rotating machinery, which is usually a vented modification of classical polymer processing equipment such as single- and twin-screw extruders. The second type of devolatilizing equipment consists of non-rotating equipment such as falling-strand devolatilizers, in which polymer melt containing volatiles to be removed is extruded as thin strands into a vacuum tank. As the strands fall toward the bottom of the tank, their volatile content is reduced (Biesenberger and Sebastian, 1983; Denson, 1985). This chapter discusses the study of devolatilization in both rotating and non-rotating equipment.

It is now recognized that in all types of equipment, devolatilization is a complex process involving boiling, foaming, and foam breaking. Early work (Latinen, 1962; Roberts, 1970) treated devolatilization of polymer melts as a simple surface renewal and diffusion process. However, diffusion coefficients calculated by Latinen (1962) from experimental data obtained from vented single-screw extruders were orders of magnitude larger than what could be expected for the styrene–polystyrene system examined. This indicates that diffusion with surface renewal alone does not govern devolatilization. Further indication that devolatilization under vacuum in single-screw extruders is not a simple diffusive process was given by Biesenberger and Kessidis (1982). Later, Mehta et al. (1984) reported direct visual observation of foaming in co-rotating disk processors. The same holds in strand devolatilization, leading Newman and Simon (1980) to model the process of falling-strand devolatilization as one of molecular diffusion into a swarm of bubbles, and the expansion of those bubbles against surface tension and viscous forces. Bubble nucleation was not considered, as they assumed that the devolatilizer is fed with a melt stream already swollen with vapor bubbles from the previous processing step.

Over the past decade, additional experimental studies have supported the notion that one of the major steps of devolatilization is bubble growth, and we draw the reader's attention to a series of studies conducted by Biesenberger and Lee (1986a,b, 1987; Lee and Biesenberger, 1989). However, previous studies have not probed the actual complexity of devolatilization mechanisms as it is revealed in our work (Albalak et al., 1987, 1990, 1992; Tukachinsky et al., 1993, 1994). Consequently, the many theoretical models suggested over the years for phase growth in both Newtonian and non-Newtonian liquids do not accurately describe the complete mechanism of bubble growth during devolatilization.

Devolatilization has been extensively studied at the Technion over the past 10 years along several parallel supporting lines. Different experimental systems have been used to elucidate the mechanism of devolatilization on a microscopic scale. We present here several aspects of our work.

Scanning electron microscope (SEM) observations of devolatilized polymer strands revealed a previously unknown growth phenomenon in which devolatilization in a falling-strand devolatilizer was seen to proceed through a multigeneration "blistering" mechanism. We discovered that volatile bubbles growing in the melt are fed by the formation of blisters on their inner surfaces. These blisters, which grow in self-similar generations, are formed by the coalescence of a growing bubble and the many satellite microbubbles formed around it, as it expands. We proposed a general mechanism for devolatilization, in which we have shown that heterogeneous bubble nucleation in the core, governed by the degree of superheat, plays a major role in determining the overall rate of devolatilization. Tensile stresses accompanying bubble growth may result in local increase in superheat by reducing the local pressure in the melt. This additional superheat combined with the possible accumulation of impurities on the macrobubble surface may be sufficient to increase the nucleation rate of microbubbles in the melt adjacent to the growing bubble. The result is the large number of blisters formed on the bubble surface.

Subsequent work showed the enhancement of bubble nucleation by ultrasonic treatment of the melt, accompanied by a significant decrease in the level of residual monomer in the polymer strands. Devolatilization was also studied in a vented extruder. Similar morphological findings indicated that a blistering mechanism, similar to that originally proposed for falling-strand devolatilization, also governs devolatilization in rotating equipment.

II. FALLING-STRAND DEVOLATILIZATION

A. Experimental Method

The experimental systems used in this part of our work were built to simulate a falling-strand devolatilizer on a laboratory scale. The initial setup (Albalak et al., 1987) consisted of a melt flow indexer (MFI) that produced the polymer melt, and a heated brass chamber connected to a vacuum system (Fig. 1). The brass chamber was preheated to a set temperature between 170 and 235°C, and evacuated to reduce the pressure to about 100 Pa (0.7 mm Hg). The MFI was fed from above with polystyrene (PS 5E242, British Petroleum) pellets containing 2300 ppm styrene, which were then compressed by a plunger loaded with weights and extruded through a die as a thin strand of melt. The extrusion process was observed through two viewing ports, and stopped after a strand of some 10 cm was formed. The bottom chamber,

Albalak et al.

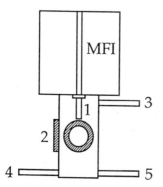

Fig. 1 The initial experimental system. A melt flow indexer was used to extrude polymer strands into a heated evacuated chamber: (1) die; (2) viewing ports; (3) connection to the vacuum system; (4) inlet of cooling water; (5) water drainage.

which was also connected to a source of cooling water, was then flooded to freeze the polymer strand and preserve the microstructure formed during devolatilization.

After freezing, scanning electron microscope (SEM) samples were prepared. The frozen strand, usually with a cross section of 2–4 mm, was detached from the die and fractured in liquid nitrogen to expose its inner structure. The segments of the strands were cemented to specimen holders and sputter-coated with a 25-nm gold layer. The specimens were examined in a JEOL T-300 SEM in the secondary imaging (SEI) mode, using an

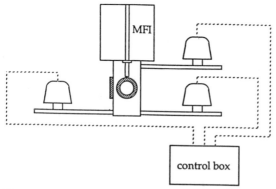

Fig. 2 The modified experimental system. A control system was incorporated to allow for short exposures of the extruded polymer strands to the superheated conditions necessary for bubble nucleation and growth.

electron acceleration voltage of 25 kV. The cross sections and lateral surfaces of the specimens were examined at magnifications up to 35,000 ×.

For the next part of our study (Albalak et al., 1990), in which we focused on the initial stages of bubble formation and growth, the experimental system was modified to allow short exposures of the extruded polymer strands to the superheated conditions necessary for bubble nucleation and growth. Controlling the exposure time enabled us to capture morphologies created at the initial stages of the process and also allowed us to determine the sequence in which the different morphologies appeared. To achieve this goal, the bottom chamber was connected to the vacuum and cooling systems through timer-controlled solenoid valves (Fig. 2). The molten polymer from the MFI was extruded into the bottom chamber kept at 235°C and initially at atmospheric pressure. At these conditions, the volatiles in the polymer were *not* superheated. After a strand of some 10 cm had formed, the control system was activated. As the control system was operated, the solenoid valve connecting the chamber and the vacuum system was opened for a preset time, of 0.6 to 28.0 sec. While the valve was open, the polymer strand within the chamber was subjected to pressure in the range of 130–400 Pa (1–3 mm Hg) and the polymer/volatile solution was immediately superheated, as indicated in Fig. 3, which shows

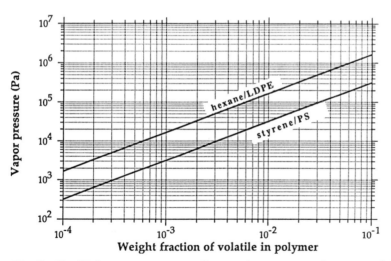

Fig. 3 Equilibrium vapor pressure of styrene in a styrene–polystyrene solution and hexane in a hexane–LDPE solution at 235°C. According to Flory–Huggins theory with an interaction parameter $\chi = 0.42$. (Reproduced from Albalak et al. (1990) by permission of the American Institute of Chemical Engineers. © 1990 AIChE. All rights reserved.)

the equilibrium partial pressure of the volatiles in the polymeric systems used. As the vacuum valve was shut, the valve connecting the chamber to the incoming cooling water was opened, and a split second later so was the drain valve.

The materials used in this part of the study were a general-purpose polystyrene (PS 5E242, British Petroleum) containing either 300 or 10,000 ppm styrene, and a low-density polyethylene (LDPE 111, Israeli Petrochemical Enterprises) enriched with hexane to various concentrations up to 30,000 ppm. Polymers were enriched with volatiles by storing pellets in a closed vessel saturated with vapors of the volatile component.

For the 10,000 ppm styrene–polystyrene system examined, the residual monomer was determined by gas chromatography as a function of exposure time to superheated conditions. A Packard model 430 was used with a Chromosorb 80/100 mesh column, 1 m in length. The temperature of the FID detector used was 200°C, that of the column was 130°C, and the injection port was maintained at 160°C.

B. Results and Discussion

A reference sample for which no boiling-foaming was expected to occur was extruded at 180°C into atmospheric pressure (no superheat), using the unmodified system of Fig. 1. A low-magnification micrograph of this strand is shown in Fig. 4a. As expected, there is no evidence of bubbles, neither on the lateral surface, nor on the cross section.

The first characteristic morphological feature to appear in samples extruded at superheat is a swarm of randomly scattered relatively large voids. These voids, which we have termed *macrobubbles*, are of the order of 100 µm and above, and can be clearly seen on the cross section of a sample extruded at 170°C into 100 Pa pressure (Fig. 4b) and on the lateral surface of a sample extruded at 200°C into 200 Pa pressure (Fig. 4c).

Figure 4d is a high-magnification micrograph of the inner surface of a macrobubble found in the cross section of a sample extruded at 235°C into 300 Pa pressure. As shown in the figure, the inner surface of a macrobubble is sometimes inhabited by *blisters*, which are thin, dome-shaped vapor-filled pockets. Figure 4e is a high-magnification micrograph of the inner surface of a macrobubble on the lateral surface of a sample extruded at 200°C into 200 Pa pressure, exhibiting similar findings.

We have found that there are two distinctly different types of blisters: *microblisters* ranging in diameter from 1 to 3 µm, and *miniblisters* ranging in diameter from 10 to 15 µm. In the bubble growth mechanism we propose, the two types of blisters evolve from each other in a cyclic manner: a first generation of microblisters emerges through the soft molten surface of a

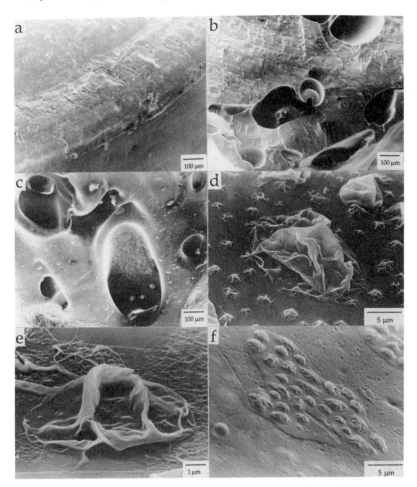

Fig. 4 Scanning electron micrographs of polystyrene strands extruded using the initial experimental system shown in Fig. 1. (a) Reference sample extruded at 180°C into atmospheric pressure (no superheat); (b) macrobubbles on the cross section of a strand extruded at 170°C into 100 Pa pressure; (c) macrobubbles on the lateral surface of a strand extruded at 200°C into 200 Pa pressure; (d) mini- and microblisters on the surface of a macrobubble within the core of a strand extruded at 235°C into 300 Pa pressure; (e) a single collapsed blister on the lateral surface of a strand extruded at 200°C into 200 Pa pressure; (f) microblisters growing on the remains of a collapsed miniblister on the lateral surface of a strand extruded at 170°C into 100 Pa pressure. (Reproduced from Albalak et al. (1987) by permission of the American Institute of Chemical Engineers. © 1987 AIChE. All rights reserved.)

macrobubble after having been formed as tiny *microbubbles* under the surface. These vapor-filled microblisters grow to a maximum diameter of about 3 μm, at which stage the skin containing the vapor is too thin and, therefore, too weak to withstand the pressure difference. The blister bursts, releasing the contained vapor into the growing macrobubble. If microblisters emerge close enough to each other, they may merge to form a larger structure, a miniblister. At some stage the miniblister also bursts and the nearly empty skin collapses, entrapping small vapor-filled pockets. These are the nucleation sites for a second generation of microblisters, growing on the remains of the previous-generation miniblister. Figure 4f shows a cluster of a new generation of microblisters, growing on the visible remains of a miniblister on the lateral surface of a sample extruded at 170°C into 100 Pa pressure.

Not all macrobubbles were found to contain blisters or their remains. It is reasonable to assume that when the polymer is depleted of volatiles, no more blisters form, and surface tension heals the inner surface of the macrobubble to a smooth appearance. The presence of blisters on the surfaces of macrobubbles is therefore related to the length of time that the polymer strand is exposed to superheat, a parameter that was essentially uncontrolled in our initial setup. To capture the morphology before it healed, and to study the initial stages of the process, we modified our experimental system to allow for controlled devolatilization time, as described in the previous section.

The first series of time-controlled experiments was conducted at 235°C using polystyrene with an initial concentration of 300 ppm styrene. Exposing the polymer strand to superheat for 0.6 sec gave no evidence of foaming or blistering on the lateral surface or cross section (Fig. 5a), on both macroscopic and microscopic scales. Microblistering of the lateral surface was the first evidence of devolatilization, and it was encountered on strands exposed to vacuum for as short as 0.8 sec. Figure 5b shows the lateral surface of a strand exposed to vacuum for 2.2 sec, in which overlapping tracks indicate that several generations of blisters have evolved by this time. After 6.0 sec of superheat, the strand is still macroscopically smooth on the lateral surface, and there is no evidence of bubbles in the core of the strands. Although single macrobubbles were observed on the lateral surface of strands as early as 1.4 sec of superheat, only after 7.3 sec of vacuum exposure did we detect macroscopic foaming on the lateral surface (Fig. 5c). At that stage the cross section exhibits only single macrobubbles (Fig. 5d), covered with blisters (Fig. 5e). Complete foaming of the core of the polymer strands occurs only after being exposed to superheat for a least 9.6 sec, as shown in Fig. 5f.

A second series of experiments was conducted using polystyrene containing an initial concentration of 10,000 ppm styrene. Figure 6a shows that already after 0.8 sec of superheating, the lateral surface is swollen with

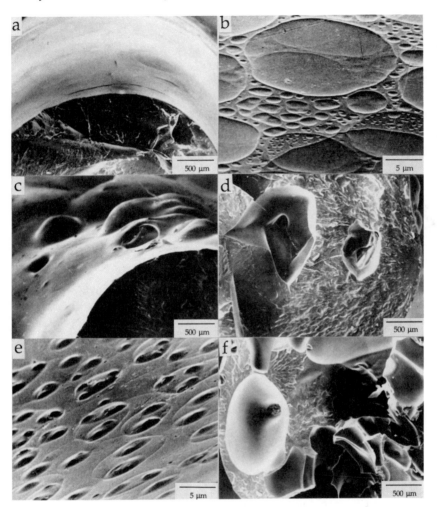

Fig. 5 Scanning electron micrographs of polystyrene strands with an initial concentration of 300 ppm styrene extruded using the modified experimental system shown in Fig. 2: (a) cross section and lateral surface after 0.6 sec; (b) blister remains on the lateral surface after 2.2 sec; (c) foamed lateral surface after 7.3 sec; (d) single macrobubbles on the cross section after 7.3 sec; (e) blister tracks on a macrobubble surface on the cross section after 7.3 sec; (f) foamed cross section after 9.6 sec. (Reproduced from Albalak et al. (1990) by permission of the American Institute of Chemical Engineers. © 1990 AIChE. All rights reserved.)

Albalak et al.

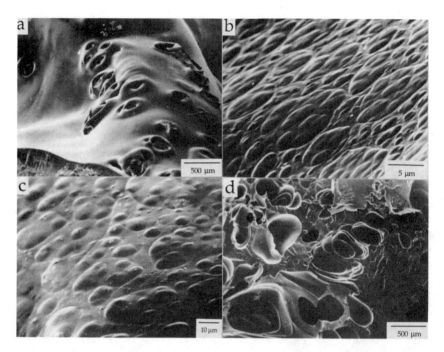

Fig. 6 Scanning electron micrographs of polystyrene strands with an initial concentration of 10,000 ppm styrene extruded using the modified experimental system shown in Fig. 2: (a) foamed lateral surface after only 0.8 sec; (b) blistering on the lateral surface after 0.8 sec; (c) foamed cross section after 2.2 sec. (Reproduced from Albalak et al. (1990) by permission of the American Institute of Chemical Engineers. © 1990 AIChE. All rights reserved.)

macrobubbles. The lateral surface was also found to be covered with blisters at this stage (Fig. 6b). Bubble growth in the core is also found to occur after shorter vacuum exposure than with 300 ppm styrene; single macrobubbles covered with blisters were found in the core after only 0.8 sec superheating, whereas similar features were found on samples with an initial 300 ppm styrene after 7.3 sec of superheating. Intense foaming of the cross section was observed after exposing the strands to vacuum for 2.2 sec (Fig. 6c).

In addition to SEM examination, the strands extruded in this series were also used to determine the level of residual styrene concentrations left in the polymer as a function of exposure time to superheated conditions. The residual monomer concentration, as determined by gas chromatography, for exposure times up to 16.9 sec is plotted in Fig. 7, which exhibits two or three measurements for each exposure time. It appears that, for the system

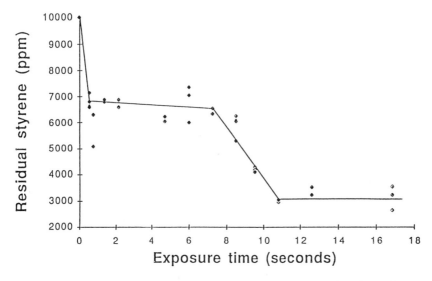

Fig. 7 Residual styrene content as a function of exposure time for polystyrene strands with an initial concentration of 10,000 ppm styrene, as determined by gas chromatography.

examined, the changes in residual styrene concentration could be roughly described by four main regions shown by the straight lines in the figure. In the first region (0–0.6 sec) there is a significant decrease in the content of the styrene from 10,000 to about 6800 ppm in a short period of time. This finding corresponds to the morphology shown in Fig. 6a: after 0.6 sec the lateral surface has already foamed, and apparently released a relatively large amount of volatiles to the surrounding. During the second region of Fig. 7, lasting to about 7.3 sec, there is a much more gradual decrease in monomer content to a level of some 6500 ppm. The micrographs corresponding to these times show the formation of later-generation blisters on the lateral surface, and the simultaneous growth of macrobubbles in the core of the strands. However, since the lateral surface was depleted of most of the volatiles during the initial foaming, and because at this stage the macro-bubbles in the core are still trapped within the strands, there is little reduction in the level of residual styrene. During the third region in Fig. 7, lasting to about 10.8 sec, there is again a significant decrease in the styrene content. We believe that this finding corresponds to the rupture of core macrobubbles that reach the lateral surface and release their contents to the surroundings. The fourth region in Fig. 7 is characterized by a nearly constant level of monomer on the order of 3200 ppm.

Gas chromatography measurements (Chechik, 1993) of polystyrene strands initially containing 20,000 ppm styrene indicated two distinct regions, rather than the four regions shown in Fig. 7. The first region, lasting to about 2 sec of exposure, showed a sharp decrease in the styrene concentration from 20,000 to 7500 ppm. Most of the styrene was apparently removed by intense foaming of the lateral surface, as indicated by micrographs of strands exposed to superheated conditions for 2 sec. Some 40% of the volume of these strands was composed of macrobubbles. In the second region, lasting from 2 to about 17 sec of exposure time, a very gradual decrease in styrene content from 7500 to 4500 ppm was measured.

The final plateau level of volatiles reflects to a large extent the vapors trapped in the bubbles and blisters in the core of the strands that cannot burst and empty their contents. This amount is naturally greater when the original styrene concentration is higher. Extensive mixing of the melt, e.g., in a vented extruder, may release these vapors.

An important conclusion from these findings is that the residence time of the strands in the vacuum chamber significantly affects the final volatile concentration. It would appear that for each polymer/volatile system there is an optimal exposure time for a given set of operating parameters. For the polystyrene–styrene system containing 10,000 ppm styrene, the optimal time is on the order of 11 sec. Devolatilizing for shorter periods of time would not remove all the volatile possible with the apparatus, whereas operating for much longer than 11 sec would appear from these preliminary results not to be beneficial to the efficiency of the process. This study stresses the importance of the parameters that determine the exposure time of the polymer to superheat in a real falling-strand devolatilizer, parameters such as the rate of extrusion, the number and diameter of strands, and the geometry of the vacuum tank.

Comparing Fig. 7 with theoretical predictions given in the literature, such as the plots shown by Newman and Simon (1980), that describe a smooth decrease in monomer content as a function of time, gives further indication that the current mathematical models for polymer melt devolatilization do not provide an accurate description of all devolatilization processes. Possible reasons for this are that current models do not fully address (1) bubble coalescence and rupture; (2) the blistering mechanism; and (3) the differences between the process on the lateral surface and in the core of the strands.

An additional series of experiments was conducted with low-density polyethylene (LDPE) enriched to various concentrations of hexane. Figure 8 consists of micrographs of low-density polyethylene strands with an initial concentration of 4000 ppm hexane that were exposed to vacuum for various times up to 28.0 sec. These micrographs show that the polyethylene strands

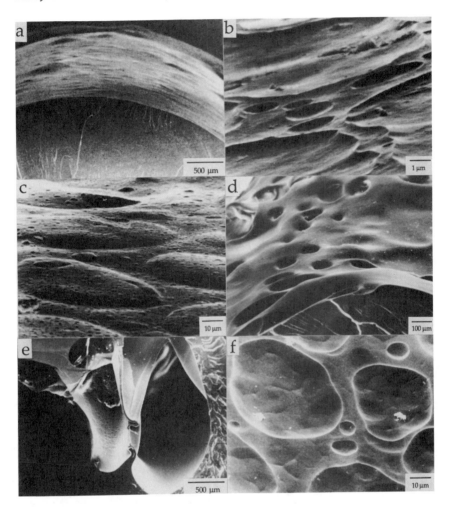

Fig. 8 Scanning electron micrographs of low-density polyethylene strands with an initial concentration of 4000 ppm hexane extruded using the modified experimental system shown in Fig. 2: (a) cross section and lateral surface after 0.6 sec; (b) multigeneration blister remains on the lateral surface after 0.6 sec; (c) blister remains on the lateral surface after 2.2 sec; (d) foamed lateral surface after 9.6 sec; (e) foamed cross section after 28.0 sec; (f) blister remains on a macrobubble on the cross section after 28.0 sec. (Reproduced from Albalak et al. (1990) by permission of the American Institute of Chemical Engineers. © 1990 AIChE. All rights reserved.)

undergo a process similar to that encountered in polystyrene strands: initial blistering and foaming of the lateral surface, followed by the formation of blister-covered macrobubbles in the core, finally leading to total foaming of the cross section.

Figure 8a shows the cross section and lateral surface of a strand after being superheated for 0.6 sec. The macroscopic appearance of the strand is similar to that of the polystyrene strand of Fig. 5a, and it exhibits no evidence of foaming. However, taking a close look at the lateral surface of the strand (Fig. 8b) shows that on the microscopic scale blistering has already begun. Microblistering of the lateral surface becomes more intense with longer exposure to superheated conditions, with additional generations of blisters appearing (Fig. 8c). However, after 6.0 sec of vacuum exposure, both the cross section and lateral surface remain macroscopically featureless, and only after 9.6 sec some foaming of the lateral surface is observed (Fig. 8d). After 9.6 sec some of the extruded strands exhibited single macrobubbles in the core, and after 16.9 sec of superheat all the strands examined had macrobubbles in their cross sections, showing evidence of blistering. After 28.0 sec exposure to vacuum, the entire cross section of the strand is foamed (Fig. 8e), with signs of several generations of blisters (Fig. 8f).

Additional work (Chechik, 1993), in which polymer strands were subjected to superheat for preset periods of time up to 17.0 sec, was conducted using polystyrene containing an initial concentration of 20,000 ppm styrene. The results of this part of the study exhibit similar morphological findings as those previously found, and they give further evidence of the multigeneration blistering mechanism. Moreover, the different generations of blisters formed on the polymer strands indicated the existence of a self-similar system, in which the morphology formed has a similar appearance at different magnifications. This phenomenon is illustrated in Figs. 9a–c, showing micrographs of a strand superheated for 1.4 sec.

Figure 9a, lowest magnification, shows first-generation blisters on the inner surface of a macrobubble. Magnifying the area 10 times (Fig. 9b) reveals additional features in the form of smaller blisters of a later generation. An additional 10-fold magnification reveals yet another generation of still smaller blisters (Fig. 9c).

This fractal-like geometry may be described by a scaling law, which in our case relates the number of blisters per unit area (N) to the scale length of the blisters (r):

$$N \sim r^{-D_f} \tag{1}$$

D_f is the fractal dimension, experimentally calculated from Figs. 9a–c to be 2.3.

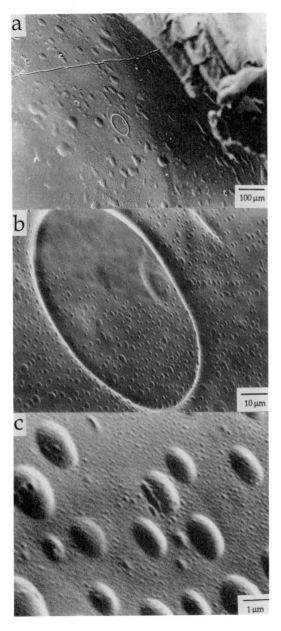

Fig. 9 Scanning electron micrographs of polystyrene strands with an initial concentration of 20,000 ppm styrene extruded using the modified experimental system shown in Fig. 2. Three different magnifications of the same area reveal a self-similar fractal-like geometry.

III. PROPOSED MECHANISM

It is suggested that initial blistering and foaming occur on the lateral surface for a number of reasons. The first is the immediate contact the surface has with the surrounding vacuum, as opposed to an undetermined pressure within the core of the strand. Any bubble growth within the strand must overcome the energy required to displace viscous material to form a void, whereas blistering on the surface does not encounter such resistance. Moreover, blister growth, which is observed to be a rapid process, may occur within the strand only after a microbubble has nucleated, and initially grown by diffusion. A possible additional reason for intensive activity on the lateral surface of the strands is the continuous bombardment of the surface by minute dust particles found in the atmosphere. These particles are present even at the modest vacuum at which industrial devolatilization is performed, and they are certainly present in our experiments in which strands are extruded into the vacuum chamber initially at atmospheric pressure. On impact with the strands, these particles may form heterogeneous nucleation sites for bubble–blister growth.

In the case of the core of the strand, we would expect nucleation of microbubbles to start as soon as the polymer melt is superheated. The generally accepted model for heterogeneous nucleation in the core of the liquid is that of Harvey et al. (1944a,b), who studied bubble formation in superheated biological systems, and who postulated the existence of entrained microparticles (e.g., dust particles), themselves containing acute-angle microcrevices, that may act as stable nuclei. This model is discussed in more detail elsewhere in this book.

The pressure in a gas pocket in such a microcrevice, shown schematically in Fig. 10, is less than that of the liquid as can be seen from the Laplace equation:

$$P_B - P_L = \frac{-2\sigma}{R} \tag{2}$$

where P_B is the pressure within the bubble, P_L is the pressure in the liquid phase (the melt), σ is surface tension, and R is the bubble radius.

The deeper the liquid penetrates into the cavity, the smaller the gas pressure becomes, until a value must be reached at which the gas pressure in the crevice is at equilibrium with the dissolved gas. Hence conditions for stable nuclei exist.

The rate of heterogeneous bubble nucleation in superheated systems was presented by Blander and Katz (1975) and is of the general form

$$J = A \exp\left(\frac{B}{P_1 - P_L}\right) \tag{3}$$

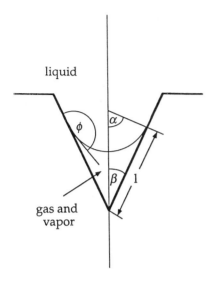

Fig. 10 A stable Harvey-type nucleus in a conical cavity, which may be found on dust or other particles in the system.

where J is the nucleation rate, P_1 is the equilibrium vapor pressure of the volatile, and A and B are factors that incorporate temperature, surface tension, molecular weight, and the geometry of the nucleation sites shown in Fig. 10. Specific factors for conical cavities, such as those discussed by Harvey et al., have been proposed by Cole (1974).

Equation (3) expresses the rate at which *primary* microbubbles nucleate in the melt. Primary microbubbles are those formed in the melt before any blistering occurs, which finally grow to form the macrobubbles found in the core of the strands. Bubbles grow by two mechanisms: (1) diffusion of volatiles from the surrounding melt; (2) formation of blisters on the bubble surface, subsequent blister rupture, and release of volatiles to the interior of the bubble. The blistering process is, in essence, the coalescence of a primary bubble with the many *secondary* microbubbles formed around it. These secondary microbubbles nucleate at a much higher rate than the primary microbubbles. Calculations, based on micrographs of time-controlled devolatilized strands, show that the blister-forming secondary microbubbles may nucleate at rates up to seven orders of magnitude higher than the nucleation rate of primary microbubbles. The increase in the nucleation rate is hypothesized to be due to the additional superheat caused by the buildup of negative pressure during bubble growth.

Raising the volatile content has the effect of increasing the equilibrium vapor pressure, P_1 causing an increase in the degree of superheat expressed as $P_1 - P_L$ in the denominator of Eq. (3). Alternatively, the degree of superheat may be raised by lowering the value of P_L. Subjecting the system to an external vacuum may reduce P_L to a minimum value close to zero. Further decrease in the local value of P_L at the bubble surface may occur by cavitation due to tensile stress generated by the moving boundary of the growing gas macrobubble in the liquid medium.

Street (1968) showed that the angular tensile stresses at the bubble surface in a viscoelastic liquid are given by

$$\tau_{\theta\theta} = \tau_{\phi\phi} = -\left\{ 2\mu_0\alpha + \left(\frac{4\mu_0\alpha^2\lambda}{1 - 2\alpha\lambda}\right)\left[1 - \exp\frac{(1 - 2\alpha\lambda)t}{\lambda}\right]\right\} \quad (4)$$

in which μ_0 is the zero-shear viscosity, λ is the first relaxation time (which is zero for time-independent fluids), and

$$\alpha = \frac{\dot{R}}{R} \quad (5)$$

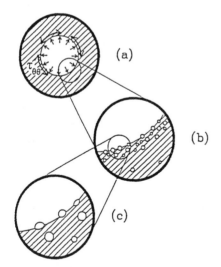

Fig. 11 The proposed mechanism for the core of the strands. (a) A growing macrobubble generates angular stresses in the surrounding melt; (b) secondary microbubbles nucleate around the macrobubble due to a local reduction in the pressure and coalesce with it, forming blisters on the inner surface of the macrobubble; (c) detail of (b). (Reproduced from Albalak et al. (1990) by permission of the American Institute of Chemical Engineers. © 1990 AIChE. All rights reserved.)

Adding the tensile stress contribution to the superheat term in Eq. (3) gives an expression for the rate at which secondary bubbles nucleate.

The buildup of tensile stresses in the liquid adjacent to a growing bubble may activate preferentially oriented nuclei, which remain inactive at the surrounding pressure, thus forming a large number of satellite microbubbles. Because the original bubble continues to grow toward these microbubbles, they eventually coalesce with it in the form of a first generation of microblisters on the surface of the larger bubble. This process is schematically shown in Fig. 11. The next stage is the cyclic repetition of the previously discussed blistering mechanism, forming several generations of microblisters and miniblisters. Another reason for the high rate of secondary microbubble nucleation is the possible accumulation of solid particles by the surface of the growing macrobubble; impurities, which are initially randomly distributed in the polymer melt, tend to be picked up by the advancing bubble front, resulting in more potential nucleation sites at the bubble surface. Once primary bubbles begin to grow via blistering in addition to diffusion, bubble growth is accelerated, and adjacent macrobubbles start to coalesce, thus rapidly foaming the entire cross section.

IV. ULTRASOUND-ENHANCED DEVOLATILIZATION

When the volatile content of a polymer is low, bubble nucleation seems to be the bottleneck of the devolatilization process. Formation of tiny bubbles is suppressed by surface tension of the polymer melt. The driving force of the process is the difference between the equilibrium pressure of saturation and the ambient pressure. At a given concentration of volatiles, saturation pressure can be increased by raising the temperature, the upper limit being determined by thermal stability of the polymer. High vacuum applied to a polymer melt can reduce the ambient pressure almost to zero. Even then, however, the pressure difference often is not sufficient for full-scale bubble nucleation throughout the melt.

Triaxial stretching of a liquid may reduce the ambient pressure to negative values. That can be practically achieved by means of a powerful ultrasound source. The acoustic field causes high-frequency stretch-compression stresses within the liquid; the stresses may upset the continuity of the liquid, and as a result, areas of tiny bubbles arise in the liquid (acoustic cavitation). The phenomenon, thoroughly investigated for low-viscosity liquids (Knapp et al., 1970), can also be observed in polymer melts (Peshkovskii et al., 1983). The idea of intensifying foaming by use of an ultrasonic acoustic field during polymer melt devolatilization was the subject of an additional aspect of our study (Tukachinsky et al., 1993).

A. Experimental Method

The experimental setup, shown schematically in Fig. 12, was designed to extrude an acoustically treated (or untreated) polymer melt into a vacuum chamber. The 20-mm-diameter, single-screw extruder was equipped with a crosshead die incorporating a waveguide attached to an ultrasound generator and acoustic transformer. The waveguide was a longitudinally oscillating 10-mm-diameter titanium rod immersed in the polymer melt. The removable vacuum chamber, attached to the crosshead, was equipped with a special sample holder.

Electric heaters controlled the temperatures of the crosshead and the vacuum chamber. In experiments without ultrasonic treatment, the temperature of the head was set higher, to compensate for the temperature increase of the polymer melt as a result of ultrasonic treatment. Thus, the temperature

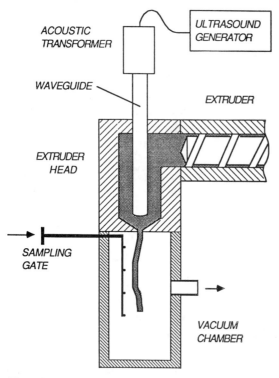

Fig. 12 The experimental setup for studying ultrasound-enhanced falling-strand devolatilization. (Reproduced from Tukachinsky et al. (1993) by permission of the American Institute of Chemical Engineers. © 1993 AIChE. All rights reserved.)

of the extruded strand, measured by a needle thermocouple, was the same for ultrasonically treated and untreated melts. In the experiments reported here the temperature was 255°C and the ultrasound frequency 20 kHz. The material used was commercial polystyrene with a residual styrene content of 750 ppm, enriched with styrene up to 3000 ppm. As in the falling-strand experiments, enriched polymer was prepared by storing the pellets in a closed vessel saturated with styrene vapors. Styrene content before and after devolatilization was determined using a Hewlett Packard 5890 gas chromatograph with a flame ionization detector.

B. Results and Discussion

A series of experiments was carried out to study the effect of the acoustic treatment on the devolatilization process. Melt strands were initially extruded into the open with and without the acoustic treatment. The samples obtained clearly showed that the acoustically treated samples visibly foamed up even at atmospheric pressure, whereas their untreated counterparts showed no signs of foaming. These experiments, therefore, indicated a qualitative difference in devolatilization of acoustically treated and untreated polymer melts.

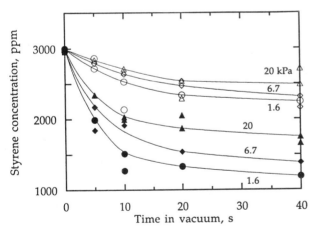

Fig. 13 Residual styrene concentrations in polystyrene extruded at 255°C. The open symbols refer to experiments without ultrasound, the filled ones to experiments with ultrasound treatment. The parameter is the absolute pressure in the vacuum chamber (Pa). (Reproduced from Tukachinsky et al. (1993) by permission of the American Institute of Chemical Engineers. © 1993 AIChE. All rights reserved.)

To study the quantitative effect of the acoustic treatment, the strands were extruded into the vacuum chamber and exposed to pressures of 20,000, 6650, and 1600 Pa for periods of time from 5 to 40 sec. After reaching a steady state, the extruder was switched off, and the melt strand sample was collected by a specially designed ladderlike sample holder. The latter was attached to a rod, that when inserted, shut off the die exit and cut off the melt strand sample. The samples, dissolved in dichloromethane, were then analyzed for residual styrene by gas chromatography.

The results shown in Fig. 13 indicate that acoustic treatment signifi-cantly reduces the level of residual styrene, and that the effect is amplified with increasing vacuum levels. Considering the time scale at which the phenomenon occurs, it can be concluded that acoustic treatment may be an important, practical means for improving polymer melt strand devolatiliza-tion (Tukachinsky et al., 1993).

V. VENTED EXTRUDER STUDIES

Vented extruders are the most frequently used rotating-type devolatilizers and are discussed in detail in other chapters. Shear deformation, mixing, and continuous renewal of the polymer melt surface distinguish devolatilization in a vented extruder from that in falling-strand equipment.

A. Experimental Method

The experimental setup designed for the study of devolatilization in a vented extruder enables us to use the same technique of quenching the polymer melt with water for SEM investigations as previously described (Fig. 14). The installation consists of two single-screw extruders connected in series, so that polymer melt is fed from the feeding extruder into the vented extruder, providing "starve feed" of the vented extruder (Tukachinsky et al., 1994). The outer diameter of the vented extruder screw is 50 mm, the lead is 50 mm, and the channel depth is 2 mm.

A sampling port in the barrel of the vented extruder is located three leads downstream from the melt inlet. The port (sampling window) in the barrel of the vented extruder is used for venting, observation (and video-taping), quenching, and polymer sampling. A three-way nozzle is installed on the port. Its central arm, the widest one, is covered with glass and serves for observing the melt during devolatilization. Vacuum venting is carried out via another arm of the nozzle. Cold water for quenching the melt is fed through the third arm. Solenoid valves are installed on the second and third arms of the nozzle. They are regulated by an automatic control system, which opens or closes the valves at predetermined time and stops the screw of the

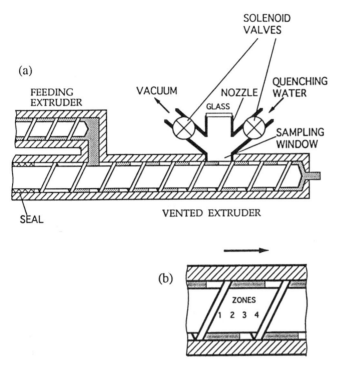

Fig. 14   (a) Schematic drawing of the vented extruder installation. (b) The four zones of melt sampling along the width of screw channel. The arrow shows the direction of melt movement in the extruder. (Reproduced from Tukachinsky et al. (1994) by permission of the American Institute of Chemical Engineers. © 1994 AIChE. All rights reserved.)

vented extruder simultaneously. A magnet transducer of screw angular position is used to stop the screw in a position convenient for sampling. The vacuum system includes a vacuum pump, a 100-liter vacuum buffer tank, a vacuum trap, and a gauge.

For the purpose of comparison with our falling-strand studies, the experimental parameters were similar: the material was polystyrene with a styrene content of 6000 ppm, the temperature was set at 235°C, the absolute pressure during vacuum exposure was 400 Pa, and the vacuum exposures were set between 0.5 and 20 sec. Other parameters were flow rate 0.54 kg/hour, screw speeds of 72 and 20 rpm, with corresponding shear rates in the screw channel of 94 and 26 sec^{-1}.

To quench polymer melt at various stages of devolatilization in a vented extruder, the experiment was conducted as follows. Both extruders were heated and initially run at atmospheric pressure. After steady-state extrusion was reached, the automatic control system first opened the valve connecting the vacuum tank with the inner space of the vented extruder to start devolatilization. After a certain predetermined time lapse (0.5 to 20 sec), the automatic system opened the second valve to feed quenching water, and simultaneously, the vented extruder screw was stopped. The screw angular position transducer assured that every time the screw stopped, its interflight space (from flight to flight) was facing the sampling port.

The extruder was cooled, its inner space was dried, and polymer samples were collected after flooding the sampling port with liquid nitrogen. Polymer chips were chopped off the four zones of the screw channel width, as shown in Fig. 14b. The first zone is located directly in front of the pushing flight, the second and the third zones in the middle part of the interflight space, and the fourth zone behind the rear edge of the following flight. The collected polymer chips were immersed again in liquid nitrogen and fractured into specimens for SEM analysis.

Polymer melt foaming in the vented extruder was also recorded by a video camera. For this, the sampling window was covered with Pyrex glass and sealed. Volatiles were pumped off through another port in the vented extruder barrel. A Sony video-camcorder CCD-F450E was mounted above the window, using shutter speeds up to 4000 sec^{-1}. The camera field of view covered the entire window (40-mm diameter), a little less than one lead of the screw (50 mm). Videotaping started about half a minute before vacuum was applied within the devolatilization zone, and continued for 2 to 3 min, when steady-state vented extrusion conditions were attained.

B. Results and Discussion

Melt foaming started, according to the videotape and micrographs, the instant the vacuum was applied. Foaming developed considerably faster than in falling-strand devolatilization under similar conditions. The stresses and disturbances induced by screw rotation appeared to promote bubble nucleation in a supersaturated polymer. The higher the screw speed, the more intensive the foaming was.

At 72 rpm, the strongest foaming occurred during the first 5 sec: the melt actually turned into foam, filling the whole width of the screw channel. Then foaming subsided, and by the fifth second, a strip of nonfoamed melt appeared in the zone adjacent to the pushing surface of the flight (Fig. 15). The strip, about 7 mm wide, divided by quite a distinct border from the rest of the foamed mass, was observed during the subsequent 15–20 sec. In about half a minute after vacuum had been applied, foaming terminated throughout

Fig. 15 The sampling port with the polymer quenched after 10 sec of exposure to vacuum. (Reproduced from Tukachinsky et al. (1994) by permission of the American Institute of Chemical Engineers. © 1994 AIChE. All rights reserved.)

the observed space. Thus, under steady-state conditions, polymer melt foaming took place only in the starting section of the screw, in no more than two leads of its length.

Similar phenomena were observed in experiments at lower screw rotation speed (20 rpm), but on a different time scale. Intensive foaming developed during 1–2 sec after vacuum was applied, and continued for about 15 sec. At the end of that period, a strip of nonfoamed melt was formed near the screw flight; foaming receded and then ended in about 1 min from the beginning of the process.

Scanning electron micrographs of quenched specimens of a foamed polymer melt showed microstructural features similar to those observed in our studies of falling-strand devolatilization. These are indicative of the multistage nature of the foaming process in rolling-pool-and-film apparatus. We found the free surface of the material and the surface of macrobubbles covered in some sites by traces of micro- and miniblisters, about 1 to 100 μm in size (Fig. 16a). The size and frequency of appearance of such areas depend on the duration of vacuum exposure and on the particular zone of the screw channel (as mentioned, the channel width was subdivided into

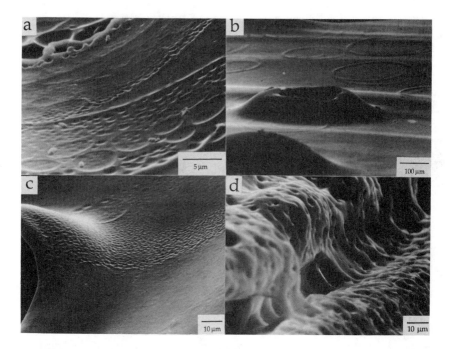

Fig. 16 Scanning electron micrographs of quenched polystyrene samples from the vented extruder system. Screw speed: 72 rpm, t: vacuum exposure time, n: number of the zone. (a) Blister traces on the free surface ($t = 0.8$ sec, $n = 4$); (b) craterlike features and big blister traces ($t = 0.5$ sec, $n = 4$); (c) microblister traces near a macrocrater ($t = 0.5$ sec, $n = 1$); (d) micropores on the macrobubble surface ($t = 0.8$ sec, $n = 2$). (Reproduced from Tukachinsky et al. (1994) by permission of the American Institute of Chemical Engineers. © 1994 AIChE. All rights reserved.)

four zones). Foaming developed considerably faster than in falling-strand devolatilization under similar conditions, where 10 sec were needed for complete foaming of the strand core.

For a screw speed of 72 rpm, the following lengths of vacuum exposure (prior to quenching) were chosen: 0.5, 0.8, 1.0, 2.0, 5.0, 10, and 20 sec. Already at 0.5 sec, large craterlike features were observed (Fig. 16b); also, microblister traces such as shown in Fig. 16c were found in the first and fourth zones of the channel. At $t = 0.8$ sec (one rotation of the screw), the material was fully foamed, both at the macro and micro level. Micro- and miniblisters covered the entire outer surface of specimens and inner surface of macrobubbles (Figs. 16a,d). Similar images of vigorous foaming were also observed at $t = 1$ and 2 sec. At $t = 5$ sec, an essentially different picture was observed: Bubbles were

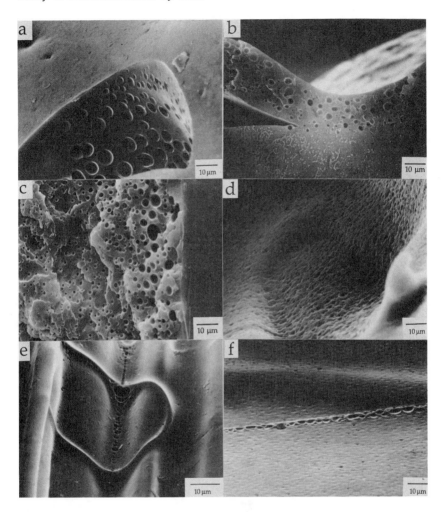

Fig. 17 Scanning electron micrographs of quenched polystyrene samples from the vented extruder system. Screw speed: 72 rpm, t: vacuum exposure time, n: number of the zone. (a) Microblister traces on a macropore surface ($t = 5$ sec, $n = 4$); (b, c) microbubbles on the cross section near a free surface ($t = 1$ sec, $n = 2$); (d) surface covered with micropores ($t = 2$ sec, $n = 2$); (e) blister chain in a gully ($t = 1$ sec, $n = 4$); (f) blister chain ($t = 20$ sec, $n = 4$). (Reproduced from Tukachinsky et al. (1994) by permission of the American Institute of Chemical Engineers. © 1994 AIChE. All rights reserved.)

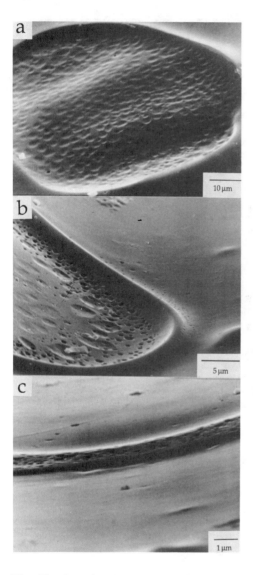

Fig. 18 Scanning electron micrographs of quenched polystyrene samples from the vented extruder system. Screw speed: 72 rpm, t: vacuum exposure time, n: number of the zone. (a) Microblisters cover the surface inside a big blister trace ($t = 2$ sec, $n = 3$); (b) microblister traces near a big blister trace ($t = 1$ sec, $n = 3$); (c) microblister trace on a big blisher trace ($t = 1$ sec, $n = 3$). (Reproduced from Tukachinsky et al. (1994) by permission of the American Insitute of Chemical Engineers. © 1994 AIChE. All rights reserved.)

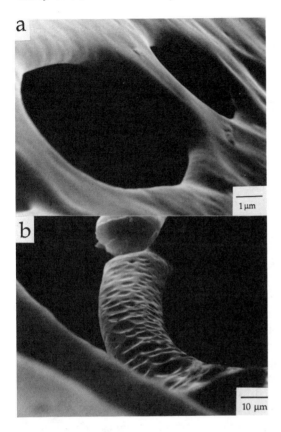

Fig. 19 Scanning electron micrographs of quenched polystyrene samples from the vented extruder system. Screw speed: 72 rpm, t: vacuum exposure time, n: number of the zone. (a) Coalescing microbubbles ($t = 0.8$ sec, $n = 2$); (b) blister traces on the stretched neck between coalescing bubbles ($t = 0.8$ sec, $n = 3$). (Reproduced from Tukachinsky et al. (1994) by permission of the American Institute of Chemical Engineers. © 1994 AIChE. All rights reserved.)

absent from the first zone (adjacent to the pushing flight) and in the three remaining zones, traces of blisters were seen as before, but they were not dispersed so densely; they were mainly located on the inner surface of the pores remaining from macrobubbles (Fig. 17a). After 10 sec had elapsed, there were still many macrobubbles in zones 2, 3, and 4, but notably less fresh traces of microblisters were seen. At $t = 20$ sec, foaming on a macro-level was not observed in any of the four zones.

Shear is the main difference between foaming in a vented extruder and foaming during falling-strand devolatilization. Bubbles of the order of several microns in size were almost spherical in shape; macrobubbles became substantially elongated in shear field, their length to width ratio reaching 10 and more. The increase of the bubble surface together with the convection caused by shear flow appears to enhance mass transfer of volatiles into the bubbles. Additional benefits of the shearing action in rotating devolatilizers are surface renewal, and the accelerated rupture of bubbles. Bubbles within the melt in a vented extruder may burst due to the extensive mixing, thus accelerating the release of their contents to the surroundings. In falling-strand devolatilizers the contents of the bubbles remain trapped within the polymer strand until they can reach the lateral surface.

Microbubbles, several microns in size, were not frequently observed in the bulk of the polymer. When they were found, they were located near the surface (Fig. 17b); their size increased the closer they were to the surface (located at the right-hand side of Fig. 17c). Blisters were mostly located on the surface of the material in groups of various sizes and shapes: a very large group covering a significant part of a specimen surface (Fig. 17d), or chains of blisters either following (Fig. 17e) or not following (Fig. 17f) lines of surface deformation. Quite often the remaining rim of a partially reabsorbed big blister served as a border of a blister-covered patch (Fig. 18). "Bridges," formed as a result of coalescence of neighboring bubbles (Fig. 19a), were also found to be the sites of microblister concentration (Fig. 19b). Both the "bridges" and the blister surfaces are regions of polymer melt stretching.

Calculation of heat released as a result of viscous dissipation near a growing bubble and of heat absorbed as a result of volatile evaporation shows that devolatilization of a polystyrene–styrene system can be regarded as isothermal, if the initial volatile concentration does not exceed approximately 1%. Mechanical stresses appear to be the main factor enhancing nucleation near growing bubbles and at the sites of elongational microflows.

ACKNOWLEDGMENTS

Different parts of this work were supported by grants from the National Council for Research and Development, Israel, the KFA Jülich, Germany and the German–Israeli Foundation for Scientific Research and Development (GIF). The authors wish to gratefully acknowledge the contributions of Judith Schmidt, Bertha Shdemati, and Baruch Paz to this study.

NOMENCLATURE

A parameter in Eq. (3)
B parameter in Eq. (3)

D_f fractal dimension
J heterogeneous nucleation rate
l cavity length, Fig. 10
N number of blisters per unit area
P_L pressure in the liquid medium
P_1 equilibrium pressure of the volatile in the polymeric solution
P_B pressure of vapor in a bubble
r scale length of the blisters
R bubble radius
t time
α angle, Fig. 10
α defined in Eq. (5)
β angle, Fig. 10
θ angular coordinate
λ first relaxation time
μ_0 zero-shear viscosity
σ surface tension
$\tau_{\theta\theta}$ angular normal stress
$\tau_{\phi\phi}$ angular normal stress
ϕ angular coordinate
χ Flory–Huggins interaction parameter

REFERENCES

Albalak, R. J., Tadmor, Z., and Talmon, Y. (1987). Scanning electron microscopy studies of polymer melt devolatilization, *AIChE J.*, *33*: 808.

Albalak, R. J., Tadmor, Z., and Talmon, Y. (1990). Polymer melt devolatilization mechanisms, *AIChE J.*, *36*: 1313.

Albalak, R. J., Tadmor, Z., and Talmon, Y. (1992). Blister-promoted bubble growth in viscous polymer melts, *Mat. Res. Soc. Symp. Proc.*, *237*: 181.

Biesenberger, J. A., and Kessidis, G. (1982). Devolatilization of polymer melts in single-screw extruders, *Polym. Eng. Sci.*, *22*: 832.

Biesenberger, J. A., and Lee, S. T. (1986a). A fundamental study of polymer melt devolatilization. Part I: Some experiments on foam-enhanced devolatilization, *Polym. Eng. Sci.*, *26*: 982.

Biesenberger, J. A., and Lee S. T. (1986b). "A Fundamental Study of Polymer Melt Devolatilization. II. A Theory of Foam-Enhanced DV," 44th SPE ANTEC.

Biesenberger, J. A., and Lee S. T. (1987). A fundamental study of polymer melt devolatilization. III: More experiments on foam-enhanced DV, *Polym. Eng. Sci.*, *27*: 510.

Biesenberger, J. A., and Sebastian, D. H. (1983). *Principles of Polymerization Engineering*, Wiley, New York.

Blander, M., and Katz, J. L. (1975). Bubble nucleation in liquids, *AIChE J.*, *21*: 833.

Chechik, T. (1993). M.Sc. dissertation, Technion—Israel Institute of Technology.

Cole, R. (1974). Boiling nucleation, *Adv. Heat. Transfer*, *10*: 85.

Denson, C. D. (1985). In *Advances in Chemical Engineering*, vol. 12 (J. Wei, ed.), Academic Press, New York.

Harvey, E. N., Barnes, D. K., McElroy, W. D., Whitely, A. H., Pease, D. C., and Cooper, K. W. (1944a). Bubble formation in animals. I: Physical factors, *J. Cellular Comp. Physiol.*, *24*: 1.

Harveey, E. N., Whitely, A. H., McElroy, W. D., Pease, D. C., and Barnes, D. K. (1944b). Bubble formation in animals. II: Gas nuclei and their distribution in blood and tissues, *J. Cellular Comp. Physiol.*, *24*: 23.

Knapp, R. T., Daily, I. W., and Hammit, F. G. (1970). *Cavitation*, McGraw-Hill, New York.

Latinen, G. A. (1962). Devolatilization of viscous polymer systems, *Adv. Chem. Ser.*, *34*: 235.

Lee, S. T., and Biesenberger, J. A. (1989). A fundamental study of polymer melt devolatilization. IV: Some theories and models for foam-enhanced devolatilization, *Polym. Eng. Sci.*, *29*: 782.

Mehta, P. S., Valsamis, L. N., and Tadmor, Z. (1984). Foam devolatilization in a multichannel corotating disk processor, *Polym. Process Eng.*, *2*: 103.

Newman, R. E., and Simon, R. H. S. (1980). "A Mathematical Model of Devolatilization Promoted by Bubble Formation," presented at the 1980 AIChE 73rd Annual Meeting, Chicago, Illinois.

Peshkovskii, S. L., Friedman, M. L., Tukachinsky, A. I., Vinogradov, G. V., and Enikolopian, N. S. (1983). Acoustic cavitation and its effect on flow in polymers and filled systems, *Polym. Composites*, *4*: 126.

Roberts, G. W. (1970). A surface renewal model for the drying of polymers during screw extrusion, *AIChE J.*, *16*: 878.

Street, J. R. (1968). The rheology of phase growth in elastic liquids, *Trans. Soc. Rheol.*, *12*: 103.

Tukachinsky, A., Tadmor, Z., and Talmon, Y. (1993). Ultrasound-enhanced devolatilization of polymer melt, *AIChE J.*, *39*: 359.

Tukachinsky, A., Talmon, Y., and Tadmor, Z. (1994). Foam-enhanced devolatilization of polystyrene melt in a vented extruder, *AIChE J.*, *40*: 670.

8

An Overview of Devolatilizers

Pradip S. Mehta

Hoechst Celanese Corporation, Bishop, Texas

I. INTRODUCTION

Polymer devolatilization is a *separation process* in which one or more undesirable volatile components are removed from a polymer matrix. The majority of the processes used in the manufacture of polymers today incorporate this unit operation in one or more pieces of equipment, as a means to *recover* unreacted monomer or solvent used in polymerization, *puri fy* the product, and/or *enhance* the product via reactive processing. Stringent regulations imposed by governmental agencies often dictate the amounts of tolerable residues of some components in the final commercial product.

A variety of methods and equipment are used industrially for separating volatile components for the base polymer matrix. The separation can be carried out while the polymer is in the solid state, but more frequently, due to the ease of separation, volatiles are removed from molten polymer streams. Examples of the former are (1) solid state polymerization of polyesters and (2) deodorization of a polymer through drying. The removal of a solvent

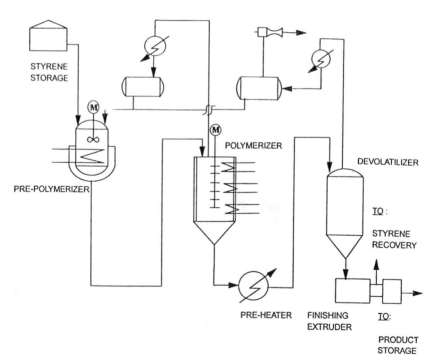

Fig. 1 A simplified flow sheet for bulk polymerization process for polystyrene. (From Fong, 1992; Courtesy of SRI International.)

from the polymer in a vented extruder exemplifies the latter. Though practiced industrially, devolatilization in the solid state is more akin to the drying process and is not discussed in this chapter.

Polymer devolatilization in the molten phase is a more widely used separation process. Melt devolatilization is often the intermediate step between the polymerization and the finishing steps, especially in the larger integrated polymer units. Examples of two integrated processes—the bulk polymerization process for polystyrene and the process for making polyester—are shown in Figs. 1 and 2, respectively. Polymer devolatilization is the post-polymerization step for styrene recovery in the process for manufacturing polystyrene (e.g., see Fong, 1992), whereas devolatilization is carried out as the driving force to enhance the polymerization reaction in the polyester process (e.g., see Ellwood, 1967). Such integrated polymer units are compact, contain minimal in-process inventory, are less maintenance prone, and are less capital and energy intensive. Further, the end product has better properties because the material does not experience a second heat history.

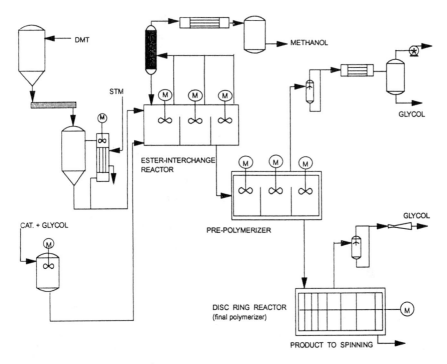

Fig. 2 A simplified flow sheet for manufacturing polyester. (From Ellwood, 1967, reprinted with permission from *Chemical Engineering.*)

Melt devolatilization is also carried out in secondary processes such as (re)compounding of plastics (Mauch, 1981), or processes involving polymer modification, for example, alloying (Hu et al, 1993), and sometimes even in injection molding machines (Nunn, 1980). These operations are frequently carried out in extrusion-type machinery fitted with venting capability. Here, the objective is to enhance the end-product properties.

In Section II we examine the technical requirements of the process— limitations and enhancement mechanisms for separation—prior to reviewing the available equipment. In Section III, we differentiate among broad classes of the commercially available equipment in meeting these needs. For each class, the equipment characteristics and their relative strengths are discussed qualitatively in relation to the process fundamentals. An emphasis is placed on the practical aspects of their operation. More detailed analyses of some of the classes—for example, falling strand devolatilizers and the extruders— are the subjects of other chapters in this book. In Section IV, a few broad criteria for the selection of the equipment are discussed.

This chapter, being an overview, covers a wide variety of subjects and equipment. As one may expect, there is some overlap with descriptions provided in other chapters of this book specifically devoted to a particular equipment or principle.

II. PROCESS FUNDAMENTALS

Devolatilization of molten polymers is a thermodynamically driven, mass-transfer-limited separation process. This unit operation presents one with many technical challenges. The difficulties stem from the very viscous nature of the materials being handled and the severe heat and mass transfer limitations that accompany the process. Frequently, the process requirements are further complicated by simultaneous chemical reactions affecting the very nature of the material being devolatilized. With these limitations, separation equipment needs to be specially designed to induce flow patterns conducive to efficient separation. The following discusses some of the limitations often experienced with the separation equipment, methods at our disposal to enhance separations, and their implications for equipment design.

A. Mass Transfer Mechanisms

1. Diffusing Films

The key assumption by earlier investigators of melt devolatilization was that the bulk of the evaporation occurred at exposed surfaces, created specifically for that purpose, in the existing equipment. They further assumed that during the course of devolatilization, some mechanical means was provided intermittently to renew the exposed interfacial layers as they became depleted. In a partially vented screw extruder, for example, Latinen (1962), Coughlin and Canevari (1969), and Roberts (1970) all assumed that as the screw rotates, a thin film of the polymer solution is deposited on the barrel wall. Figure 3 shows a schematic of this process. Volatile components from the film diffuse through the layers within the film to the vapor phase over it—governed by penetration theory (Crank, 1956). The driving force for diffusion is the gradient between the concentration of the volatile component in the polymer solution and the vapor phase above it. As the screw rotates, the returning flight, presumably, scrapes off the deposited film, mixes it with the bulk of the polymer being conveyed in the screw channel and redeposits a new film, freshly enriched in the lighter components. Due to the increased concentration gradient, process efficiency is enhanced several-fold. The model by Roberts (1970) also accounted for the surface evaporation occurring from the exposed surface of the bulk pool of the polymer as it was being transported (also shown in Fig. 3).

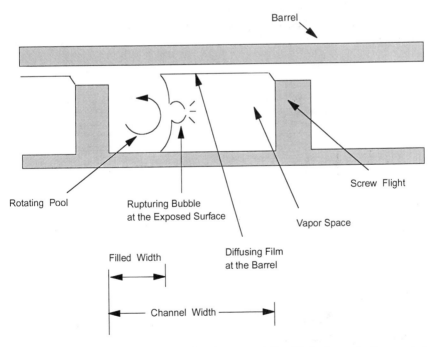

Fig. 3 Surface evaporation occurring in a partially filled channel of a vented extruder.

The two elemental steps in this process,

Evaporation from an exposed surface
Enrichment of depleted layers through surface renewal

occur concomitantly in an extruder. In equipment such as thin-film evaporators or a co-rotating disk processor (discussed in the following section), these steps are carried out sequentially. In analogy with classical chemical unit operations, each sequence of these two steps can be considered as a single separation stage. Figure 4 shows a simplified schematic of the sequential process steps occurring within each stage of a staged model. Commercial equipment typically comprises many such stages in series. Biesenberger (1980) developed a generalized theory that encompasses both types of flow patterns and showed equivalence between them. According to this model, the separation efficiency for most equipment can be summarized in a compact form:

$$E_F = f(\text{Ex,Pe,}N) \tag{1}$$

where Ex is the *extraction efficiency per exposure*, which depends on the volatile diffusivity, the polymer exposure times, and the geometry parameters for the equipment. The Peclet number, Pe, is a measure of the axial (back)

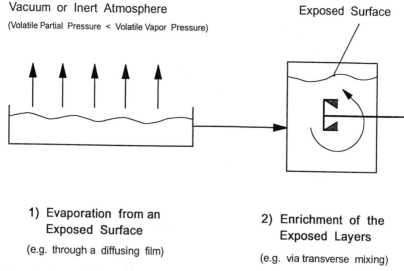

Vacuum or Inert Atmosphere
(Volatile Partial Pressure < Volatile Vapor Pressure)

Exposed Surface

1) Evaporation from an Exposed Surface
(e.g. through a diffusing film)

2) Enrichment of the Exposed Layers
(e.g. via transverse mixing)

Fig. 4   Sequential process steps in a single separation stage of a "staged model."

mixing, and N is the number of separation stages. Two simplified expressions for f take the forms

$$E_F = 1 - (1 - Ex)^N \tag{2}$$

for the staged model where no backmixing occurs, and

$$E_F = 1 - \exp(-Ex) \tag{3}$$

for the flows encountered in extruder-type equipment where backmixing is negligible and mass transfer from the pool exposure dominates. For convenience, the mechanism described so far will be called the *diffusing film* model.

    a. Process Implications for the Diffusing Film Model. If one were to imagine a piece of equipment ideally suited to carry out devolatilization based on the foregoing theory, then it would comprise several *evaporation-surface renewal* stages in series. Some of the key characteristics of this equipment would be

1. All or part of the polymer solution is conveyed through the devolatilization section as thin films spread over large surfaces within the equipment.
2. The freshly created thin films are exposed to an atmosphere where the partial pressure of the component to be devolatilized is less than its vapor pressure. This can be ahieved by lowering the total pressure over the films (e.g., by drawing vacuum) or by exposing the films to an inert atmosphere.
3. The flow is disrupted intermittently to provide transverse mixing for enriching the depleted layers (surface renewal). At the heart of the model is the assumption that the depleted layers are completely mixed with the rest of the fluid in the mixing stage. Since polymer flows are strictly laminar, this assumption can only be met with some means of rigorous reorientation or a stream-splitting/recombinant type of flow.
4. Operation at high speeds to maintain respectable throughputs and still have most of the flow exposed as thin films.

2. Bubble Transport

    Even though the diffusing film model explained the often observed (Todd, 1974; Padberg, 1980) speed dependence in extruder-type devices, it failed to explain other significant observations. For example:

    Dilute polymer solutions are often first flashed as a pre-devolatilization step in stationary equipment such as flash tanks. High separation efficiencies are attained in such equipment even in absence of thin films or surface renewals.

Attempts to back-calculate volatile diffusivities in the base polymer matrix from observed separations in extruding machinery yielded abnormally high numbers (Biesenberger and Kessidis, 1982; Mehta et al., 1984).

Under the assumptions of diffusing film model, separation efficiencies would be similar if the solution was exposed to inert atmospheres (instead of vacuums), contrary to observations (Biesenberger and Kessidis, 1982).

Mehta et al. (1984) exposed polymer solutions coated as thin films of varying thicknesses and surface areas. The diffusing film model predicted higher efficiencies with thinner films. Surprisingly, poorer efficiencies were observed with thinner films coated over larger areas.

Newman and Simon (1980) proposed an alternative mass transfer mechanism of devolatilization promoted by bubble formation that may explain these discrepancies. In this model, they assume that a multitude of bubbles are simultaneously nucleated when the polymer solution is exposed to a pressure lower than the vapor pressure of the volatile components. Many of the nucleated bubbles grow under the imposed pressure differential, while simultaneously being fed by the volatiles diffusing from the surrounding polymer matrix. The grown bubbles travel to the polymer–vapor interface, where they rupture, releasing volatiles to the vapor phase. A schematic of this process is shown in Fig. 5.

Transport via bubbles has been the premise of mathematical models of several investigators (Newman and Simon, 1980; Yoo and Han, 1983;

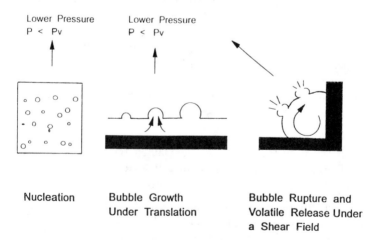

Nucleation   Bubble Growth   Bubble Rupture and
                    Under Translation   Volatile Release Under
                                               a Shear Field

Fig. 5   Sequential process steps occurring during bubble-assisted devolatilization.

Powell and Denson, 1983; Lee, 1982). Lee and Biesenberger (1986, 1987, 1989) built a device specifically designed to visually observe this process and confirmed bubble-dominated transport. Their observations along with those from the author's industrial experience can be summarized as follows:

1. When a polymer solution is exposed to a pressure lower than the vapor pressure of the volatile components, a multitude of bubbles are nucleated (Lee, 1982). Exposure to an inert atmosphere at a higher pressure would fail to nucleate bubbles, resulting in lower separation efficiency. Chapter 4 is devoted to the subject of bubble nucleation. Significant bubble formation even at volatile concentrations as low as 50 ppm has been visually confirmed.

2. Even though transport depends on the rate of volatile diffusion into the bubbles from the surrounding polymer solution, the overall mass transfer rate is significantly enhanced due to the surface generated by a very large number of bubbles. This explains the abnormally large diffusivity numbers observed by Biesenberger and Kessidis (1982) and by Mehta et al. (1984).

3. Initial bubble nucleation is a rapid process. However, the growth rate of the bubbles is determined by both volatile diffusivity in the solution and the mechanical resistance offered by the highly viscoelastic surrounding medium (see Chapter 5). In fact, many mathematical models (Newman and Simon, 1980; Yoo and Han, 1983; Powell and Denson, 1983) have assumed the "bubble growth phase" to be the rate-controlling step.

4. The imposed pressure differential and flow field help convey the growing bubbles to the liquid–vapor interface, where they rupture and release volatiles. Important factors for bubble rupture appear to include the pressure differential, temperature, bubble size, surface tension, polymer viscoelasticity and imposed shear field.

5. Under "still" devolatilization conditions, the bubble transport appears to decay after an initial rapid flash. After the initial swarm of bubbles has grown and ruptured, the separation may then proceed only via diffusion of the volatile to the exposed vapor–liquid interface (i,e., diffusing film mechanism).

6. If the polymer mass is, however, continuously agitated, then bubbles continue to be formed even after the initial flash has died. Under an appropriate shear field, yet unruptured, larger bubbles were observed to fragment, giving rise to many new smaller bubbles, which then grew further as volatiles continued to diffuse into them. The right type of intermittent (or concomitant) agitation is therefore the key to *autonucleation*. Under these conditions, the overall separation efficiency improves significantly.

7. If, however, the method of agitation is such that the polymer mass is simultaneously being pressurized, then the already grown bubbles tend to shrink, governed by the force balance around an individual bubble. The already separated volatiles are then *redissolved* back into the polymer solution. This explains the poorer separation that Mehta et al. (1984) observed when thinner films were deposited.

8. Since the viscosity of the polymer solution increases rapidly as the lower-molecular-weight components are removed, viscoelastic resistance to bubble growth becomes the rate-controlling factor. This effect is more evident in the final phases of the finishing operation, where the last traces of the volatile components are being removed. An increased dependence of devolatilization on the speed of agitation becomes evident here.

a. Process implications with bubble transport as the dominant mechanism. An "ideal" equipment configuration that would enhance bubble transport is very different from the one that relies on diffusion from the surfaces of freshly created thin films. In this equipment:

1. A saturated polymer solution is exposed to an atmosphere where the total pressure is less than the volatile component's vapor pressure over the solution. In the initial phase, a large volume is provided for the sudden expansion of the polymer solution that results from formation of a large number of bubbles.

2. After the initial *flash*, the polymer solution is translated for a short period. Ideally, the translation would be *pluglike* (no shearing or pressurizing flow fields) within which the bubbles could grow freely. The minimum condition for the bubble growth is that the right-hand side of the modified Rayleigh equation (Lee, 1982), given by

$$\frac{dR_B}{dt} = \frac{v_P + (4/3)\pi R_B^3}{4\eta R_B} R_B \left(P_B - P_\infty - \frac{2\sigma}{R_B} \right) \tag{4}$$

is positive. Here, R_B is the bubble radius, P_B the pressure inside it, v_p is the volume of a cell around the bubble over which the mass balance has been made, η the viscosity of the surrounding medium, and σ is the surface tension. Equation (4) gives the rate at which a single bubble will grow under the conditions of the translating melt. To estimate the pressure inside the bubble, one must also solve the diffusion equation (component balance around the bubble; see Biesenberger and Sebastian, 1983), given by

$$\frac{\partial C}{\partial t} = D \left(\frac{\partial^2 C}{\partial r^2} + \frac{2}{r} \frac{\partial C}{\partial r} \right) - v \frac{\partial C}{\partial r} \tag{5}$$

where C is the volatile concentration within the bubble, D its diffusivity, r the instantaneous bubble radius, and v the interface speed. Equations (4) and (5) will form a system of partial differential equations that must be solved simultaneously, along with appropriate boundary conditions that include a vapor–liquid equilibrium relationship. The solution, which may require use of a good numerical technique, would yield the rate of bubble growth and the rate of depletion of the volatile from the surrounding medium to the bubble. This in turn determines the *period of translation*. Note that other approaches to estimating the bubble growth dynamics are also available in the literature (Yoo and Han, 1983; Powell and Denson, 1983).

     3. The solution is then agitated in a melt pool within which the melt experiences an imposed shear field that is devoid of internal pressurization, but one that still generates the stresses that encourage bubble fragmentation (as a means to *renucleate* bubbles). The flow field in this region should be such that bubbles are continually exposed to the vapor–liquid interface, where due to flow stresses bubbles can *rupture* or *disengages* from the fluid and volatiles are released. Within the melt phase, the bubbles continue to fragment, as a means of *renucleation*. Enough shear energy should be dissipated here to provide the latent heat for subsequent evaporation.

     4. Steps 1 through 3 are repeated in the same sequence (or concomitantly) several times within the devolatilization section of the equipment. Analogous to the "diffusing film–surface renewal" model, each such sequence can be considered as a separation stage. The devolatilizing equipment can be viewed as comprising of N stages in series.

     Later, as we review commercially available equipment, their effectiveness in carrying out the foregoing steps is examined.

B. Heat Transfer Limitations

Effective means of transferring heat to and from the polymer solution are essential while carrying out devolatilization. When evaporating large quantities of volatiles, one must supply the latent heat of vaporization. On the other hand, in the final phase of the finishing operation, where polymer viscosity is high, removal of the dissipated heat is of prime concern as polymer would degrade under excessive temperatures. The solution properties, such as the thermodynamic equilibrium, the solution viscosity, and the volatile diffusivity, are also functions of temperature. Good temperature control is therefore necessary in devolatilization equipment.

     Vapor–liquid equilibrium determines the limiting residual concentration of the volatile component at a given temperature (Chapter 2). The equation proposed by Flory (1953) is often used for finding the vapor pressure of a volatile component over a polymer solution:

$$\ln \frac{P_1}{P_1^0} = \ln(1 - \phi_2) + \phi_2 + \chi\phi_2^2 \tag{6}$$

where P_1^0 is the vapor pressure of pure solvent, ϕ_2 the volume fraction occupied by the polymer, and χ is the Flory–Huggins interaction parameter. Figure 6 shows the vapor pressure curves for the polystyrene–styrene system, calculated at different temperatures (Padberg, 1980).

Operating the devolatilizer at the maximum possible temperatures (at a given absolute pressure) has several advantages. As seen in Fig. 6, vapor pressure over a given solution increases with temperature. The concentration gradient between the bulk of the solution and its interface, $c_{bulk} - c_{interface}$, is the driving force for mass transfer, where the latter is essentially equivalent to the equilibrium concentration of the volatile in the melt, as estimated by Eq. (6). At constant pressure, this value drops with increased temperature, thereby driving up the gradient. Hence, higher mass transfer rates are experienced at higher temperatures. In addition, mass transfer is also enhanced due to increased volatile diffusivity and decreased solution viscosity. Mechanical resistance to bubble growth is also lower due to decreased

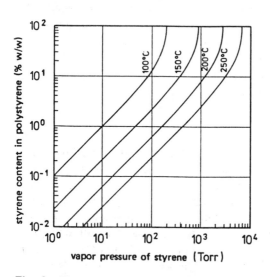

Fig. 6 Vapor pressure curves for polystyrene–styrene system at various temperatures. (From Padberg. 1980; reprinted with permission from VDI—Gesellschaft Kunstofftechnik, VDI—Verlag GmbH.)

polymer viscosity. Devolatilizers are therefore operated at as high a temperature as feasible.

The temperatures at which devolatilization occurs within various sections of the equipment (or subsequent devolatilization equipment) is first mapped out. Once the devolatilization temperature, T_{pol}, is determined, an overall energy balance written around the devolatilization unit (see also Fig. 7) gives an estimate of the amount of external heat needed to be removed or supplied:

$$\dot{m}_p c_p (T_{pol} - T_{in}) + \dot{m}_v c_v (T_{pol} - T_{in}) + \dot{m}_v \, \Delta H_v = \dot{Q}_{ext} + P_w \qquad (7)$$

where $\dot{m}_p$ and $\dot{m}_v$ are the polymer and the vapor mass flow rates, respectively, c_p and c_v are the respective heat capacities, ΔH_v is the heat of vaporization, P_w is the mechanical energy input due to viscous dissipation, and $\dot{Q}_{ext}$ is the required heat input or removal. In Eq. (7), we have assumed that the vapors and the polymer exiting the equipment are at the bulk temperature of the solution. The external heat input, $\dot{Q}_{ext}$, can be estimated from

$$\dot{Q}_{ext} = U_0 A (T_{pol} - T_{ext}) \qquad (8)$$

where U_0 is the overall heat transfer coefficient, A is the available heat transfer area, and T_{ext} is the temperature of the external cooling or heating medium.

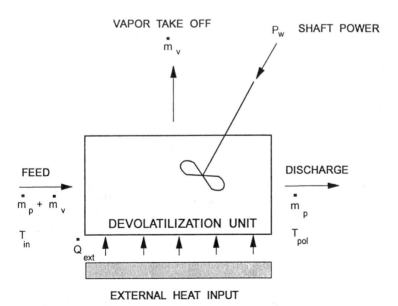

Fig. 7 An overall energy balance around the devolatilization unit.

When Eq. (8) is used to calculate the overall heat transferred, an average value of U_0 (normally supplied by the equipment manufacturers) should be used. U_0 values in equipment processing viscous substances are usually low (5–50 Btu/hour-ft^2-°F or 24–240 Kcal/m^2-hour-°K). Agitated equipment, especially those that scrape and re-create fresh films, exhibit higher U_0 values.

The power dissipated during processing, P_w, depends largely on the solution viscosity and the amount of shear imposed on the polymer. Most of the power is dissipated within close clearances that an agitator makes with the stationary walls of the equipment. Assuming a pure drag flow within the clearance, the power dissipation, P_w, within the clearance may be estimated from

$$P_w = m V_b w L \dot{\gamma}^n = \frac{m V_b^{n+1}}{H^n} wL \tag{9}$$

where m and n are the fluid power law parameters, V_b is the relative velocity of the agitator, H is the clearance between the agitator and the stationary surface, and w and L are the width and the length of the clearance area, respectively. Equation (9) alternatively expresses the dissipated power in terms of the applied shear rate within the clearance, $\dot{\gamma}$, a terminology more familiar to rheologists. Power dissipation within other areas may be calculated using classical chemical engineering methods.

The solution viscosity can change by orders of magnitude during devolatilization. To make a good estimate of the power consumed, it may be necessary to apply Eq. (9) separately to various sections of the equipment. Care must be taken during the design phase that the equipment drive can provide the required power for handling these estimated loads and that its components can withstand the delivered torque.

Most polymer devolatilizers are designed to renew surfaces via close tolerance devices or inserts that scrape previously deposited films. It is prudent to analyze the flow characteristics in these close clearance areas, where shear rates are usually an order of magnitude higher than those in the bulk of the flow. An application of Eq. (9) would reveal that possibly half of the dissipated power could come from shear experienced in these areas. This is especially the case in the final stages of devolatilization, where the polymer viscosity is at its maximum. If the heat dissipated within these areas is not efficiently removed, severe degradation or depolymerization may occur.

C. Chemical Reactions

Devolatilization is sometimes used to enhance chemical reactions such as polycondensation or grafting. On the other hand, most polymer processing steps, including devolatilization, are vulnerable to such chemical reactions as depolymerization or degradation. In both situations, the nature of the material being processed is continually being altered, which further complicates the separation process.

The rate-limiting step in the final stages of most polycondensation reactions is the removal of volatile by-products. If the reactors are engineered to effectively separate the by-products, the reaction rates can be markedly increased. Manufacture of polyester (see also Fig. 2) provides a good example. The polycondensation of polyethylene terephthalate (PET) proceeds by the combination of two 2-hydroxyethyl end groups to form an internal ester group and an ethylene glycol molecule (Pell and Davis, 1973):

$$2(-COOC_2H_4OH) \underset{k_2}{\overset{k_1}{\rightleftharpoons}} -COOC_2H_4OOC- + HOC_2H_4OH \quad (10)$$

As seen from Eq. (10), the balance between the forward and the backward reaction rates can be significantly altered if the volatile ethylene glycol molecule, HOC_2H_4OH, is being continuously removed. In fact, the production of a fiber-grade polyester requires a glycol vapor pressure less than 6 torr (Farney, 1966). Even at vacuum levels of <0.5 torr, the reaction rates are quite low due to diffusional constraints, and polymer exposure as *thin films* is necessary. Figure 8 (taken from Farney, 1966) shows that a combination of high vacuum (0.1 torr) and thin film exposure (0.1 mil or 2.5×10^{-6} m) improves the reaction time from 100 to <10 sec for equivalent molecular weight buildup. This dramatic improvement in the reaction rate should be viewed in light of a modern commercial PET finishing reactor train, where residence times of the order of 3 hours are not uncommon.

Examples of polycondensation reactions include manufacture of various nylons, polybutylene terepthalate, polycarbonate, polyphenylene sulfide, polyarylate, liquid crystal polymers, polyurethane, and so on. Not only is the removal of the by-products important while designing these reactors, but care must also be taken to account for changes in physical properties as these molecules build up. For example, the power consumed within the final polymerizer in the manufacture of PET in Fig. 2 would be much higher than that consumed in the initial stirred reactor, as the average polymer viscosity would have increased several thousand–fold.

Polymer degradation is another chemical reaction that accompanies most melt processing operations. Degradation can be thermal (chain scission, depolymerization), thermooxidative, or other chemical side reactions such as hydrolysis. Thermal and thermotoxidative degradation could lead to polymer chains with more low-molecular-weight segments, formation of undesired by-products, and in many cases, formation of color bodies.

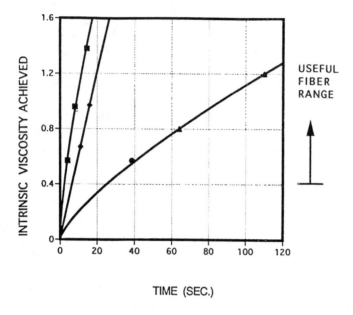

Fig. 8 Polymerization rate of polyethylene terephthalate in thin films. (Based on data from Farney, 1966; reproduced with permission of the American Institute of Chemical Engineers © 1966 AIChE; all rights reserved.)

Depolymerization results in a polymer chain breakdown, leading to increased monomer content and hence increased separation load. Uncontolled cross-linking reaction is another degradation mechanism, common in melt processing, whereby two or more polymer chains are combined giving rise to variations in physical properties of the product (e.g., appearance of gels). This reaction is of particular concern when processing rubber-modified plastics such as ABS (acrylonitrile butyl styrene) or HIPS (high impact polystyrene).

Hydrolysis of polyesters due to presence of moisture,

$$-R-O-CO-R'- + H_2O \quad \rightarrow \quad ROH + R'COOH \qquad (11)$$

is yet another example of a degradation reaction one often faces while reprocessing polyesters (e.g., while recycling polyesters). Thus, it is essential that PET be dried prior to being melt processed. When certain polymers are compounded with acidic pigments, degradation also occurs unless the polymer is protected by acid scavengers. In all these cases, the product properties could be further enhanced by providing a quick devolatilization step in conjunction with the respective main processing steps.

The degradation reactions are temperature sensitive and increase rapidly beyond a certain critical temperature. However, as we saw earlier (Fig. 6), higher temperature provides a greater concentration gradient for increased mass transfer. Thus, we reach an optimum temperature at which a devolatilizer should operate. It may prove useful to use a surrogate molecule for monitoring the rate of depolymerization while evaluating devolatilizing equipment. For example, ethylbenzene was used as an inert surrogate molecule by Blanks et al. (1980) during their investigation of polystyrene depolymerization.

Thermal and thermooxidative degradation usually begins at close clearance areas within the equipment, as heat builds up due to high shear and limited heat transfer. It is not unusual to see a temperature elevation of 20–25°C for material flowing through close clearance areas, for example, the flow over the flight of an extruder. Well-designed equipment either avoids high shear areas or provides means to improve heat transfer in these areas.

D. Multistaged Operation and Devolatilization Aids

Large separation loads and/or stringent requirements for residues in the end product are two extreme situations requiring special considerations while designing the devolatilization system. Two practical approaches to handle such difficult separations are at the disposal of a processor:

1. Operation of several devolatilizers at different pressure (vacuum) levels
2. Use of carrier substances to enhance devolatilization in the final stages

1. Staged Operation

While designing a separation system, one often has the knowledge of feed composition and the required residual volatile level in the exiting stream. The residual solvent concentration may then be used in Eq. (6) to determine an upper bound on the operating pressure for effective separation in a given devolatilizing apparatus.

When the total separation requirements exceed one order of magnitude (e.g., a solvent reduction from 5% to 0.05%), it may be more prudent to use several stages of separation, operating at different absolute pressures, to achieve this separation, especially if the operating pressures required are subatmospheric. There are many advantages to this strategy:

1. By vacuum staging, the bulk of the volatiles are removed at higher absolute pressures. Computation using the ideal gas law reveals that the demand on the vacuum system is considerably alleviated at higher pressures. The final traces of the volatile component can be then removed at high vacuums but with minimal load. Removal of majority of volatiles in the first few stages also has the advantage of reducing the overall refrigeration costs related to condenser coolant, as warmer coolant may be satisfactory in these stages. The optimum ratio (if the ideal gas law is applicable) between the successive stages equals the nth root of the overall pressure ratio (Todd, 1974), where n is the total number of stages,

2. Vacuum staging also reduces polymer entrainment arising from high vapor velocities in the initial stages, where vapor load is high. Entrainment is caused by excessive foaming, which tends to carry the polymer into the vent and plug the vent opening. This is especially the case in volumetrically limited equipment.

3. The polymer can be reheated intermittently, between the stages, to compensate for the evaporative cooling that the polymer experiences during the initial flash.

4. While carrying out polycondensation reactions, it is usually critical to maintain reactant ratios as the polymerization progresses. It is likely that the monomer with lower vapor pressure could inadvertently boil along with the by-product that is being evaporated. This would upset the ratio between the reactants. Vacuum staging controls excessive monomer boil up during the initial period, when the polymer viscosity is lower. The rate of reactant boil up reduces drastically as the polymer viscosity increases, and higher vacuum levels do not pose a problem during successive vacuum stages.

2. Devolatilization Aids

Intentional addition of a devolatilizing aid to the polymer is another practical approach available to the processor for improving separation efficiency, especially during the final stages of vacuum devolatilization. As with the use of steam stripping in distillation operations, a devolatilization aid reduces the partial pressure of the solute in the vapor phase and therefore increases the concentration gradient for devolatilization. A carrier substance can be miscible or immiscible with the polymer matrix. Its effectiveness, however, is increased substantially if it is mixed well or dispersed finely within the polymer matrix.

Other advantages of adding a carrier substance are:

1. Carrier substances have lower vapor pressure than the substances to be devolatilized. Upon exposure to lower pressures, they quickly superheat inside the polymer and nucleate bubbles into which the solute will start to diffuse. Thus carrier substances act as effective nucleation aids.
2. A right choice of the carrier substance could even improve the bubble growth and rupture steps as the interfacial tension and mixture viscosity are both favorably affected.
3. A carrier substance with a large heat of vaporization (e.g., water) can also be used to control the polymer temperature.
4. Carrier substances can sometimes form low boiling azeotropes with the solute (for example, Werner (1981) and Hess (1979) used water to boil styrene from polystyrene), lowering the vacuum requirements even further.

The carrier substance should be of low cost, inert, easily recoverable, with lower vapor pressure (than the component to be devolatilized), and tolerable in trace quantities in the final product. It can be a gaseous substance (e.g., nitrogen), but more frequently a lower boiling liquid such as water or pentane is used. The carrier substance is preferably added in a pressurized zone, mixed intimately with the polymer solution, and then the volatilized mixture is exposed to vacuum.

III. COMMERCIAL EQUIPMENT

A wide range of commercial equipment is available to the processor. The equipment can be classified in a variety of ways—e.g., magnitude of the solution viscosity handled, heat transfer characteristics, equipment cost, or equipment simplicity. In the following, the equipment is broadly classified as still equipment, where mechanical agitation is absent or not critical to the separation, or rotating equipment, where devolatilization is significantly enhanced because of mechanical agitation. A few equipment geometries are described under each category to illustrate the interaction between the machine geometry on the mass and heat transfer effectiveness. This is not an all-inclusive list; many types of equipment are commercially available. This section also points out specific advantages or applications of certain geometries.

A. Still Equipment

In an analogy with chemical flash chambers, polymer solutions containing large amounts of volatiles are frequently devolatilized in still equipment, in part because the equipment is simpler, reliable, and less expensive; but also because the solution viscosities are low enough to allow material movement with ease. The flashing device has a large volume, is devoid of moving parts, and is maintained at a pressure lower than the vapor pressure of the solute, The heat of solute evaporation is obtained either from an external source or at the expense of the sensible heat of the solution. Further, because of the low solution viscosities, heat and mass transfer limitations are small and extensive surface renewal is not necessary.

1. Flash Evaporators

The liquid to be devolatilized is first heated in a long (and sometimes spiral) tube using an external heat transfer medium. The saturated/super-saturated solution, under pressure (as the solutes start evaporating), is then carried into a wide opening—usually a flash tank—where the pressure is relieved. Figure 9 is a schematic of this process. Vapors are removed overhead. The devolatilized solution being discharged from a flash eva-porator is usually not very viscous and can be discharged using a gear pump. One might also use a screw conveyor at the bottom of the tank for discharging the polymer.

Various modifications of standard tanks can be made to aid the separation. Figure 10 shows one specific design by Boucher (1964), which

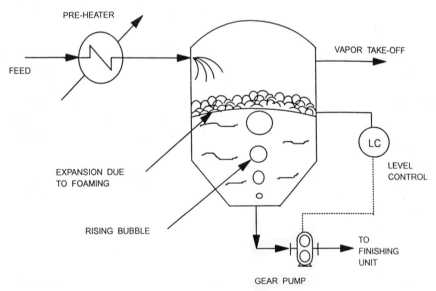

Fig. 9 Rise of a bubble in a flash evaporating unit.

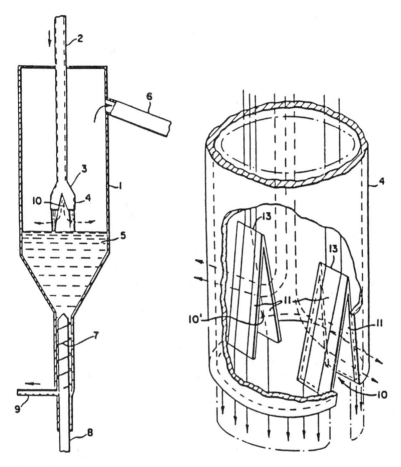

Fig. 10   Use of a deflector baffle to separate steam from polymer solution. (From Boucher, 1964; U.S. Patent 3,134,655.)

he proposed for separating steam from the polymer while preparing Nylon 6,6. It uses deflector baffles to physically separate steam from the solution. Some tanks are designed with a tangential inlet (as in cyclone separators) to minimize liquid entrainment in the vapor discharge.

The mass and heat transfer process begins in the preheating tube itself To enhance the transport, the tube diameter is chosen so that the flow within it is not laminar. Further, the tube is kept under pressure to prevent premature devolatilization. Upon entrance to the flash chamber, the pressure is suddenly released, and a multitude of bubbles are nucleated. Vapors quickly disengage, the solution viscosity increases, and the down-flow becomes laminar. Usually the thermodynamic separation limit is approached, especially when the solution viscosity is low. The partially devolatilized polymer solution then collects in a pool formed at the bottom of the chamber.

Mass transfer within the pool is poor as the entrained bubbles can only rise slowly (see also Fig. 9) against gravity due to the difference between the pressure within the bubble and the pressure in the vapor phase above the pool. One may use Stoke's equation,

$$v = \frac{\Delta \rho \, g_c d^2}{18\eta} \tag{12}$$

which gives the rising velocity, v, for the bubble of diameter, d, in conjunction with Eq. (4), to calculate the bubble growth period (total rise time). The effective pressure over the bubble within the liquid is further increased by the height of the liquid over it (see also Fig. 9). If the chamber is operated at subatmospheric pressures, the liquid height above the bubble could be enough to repressurize the volatiles into the liquid phase. This effect must be accounted for while solving Eq. (4). The bubbles entrapped in the polymer tend to "redissolve" quickly as the polymer travels toward the tank discharge, especially if the liquid level is maintained high.

Care must be taken, especially when the solution in the tank is viscous, to ensure that there are no stagnant areas where the polymer could degrade. Often the entrained polymer/oligomers would deposit overheads and degrade and char. This could well become the source of contamination (black specks) that appear in the final product.

2. Falling-Strand Devolatilizers

One specific configuration of a flash chamber is the falling-strand devolatilizer, where the material enters the flash chamber as a multitude of strands, which fall free due to gravity into the accumulated melt pool below. Figure 11 is a schematic of this design. The free fall of strands allows easy disengagement of the already grown bubbles. During the free fall of the liquid, the configuration allows new bubbles to nucleate and grow free of external shear or pressure gradients. A detailed discussion of the falling-strand devolatilizer appears in the next chapter.

Other configurations such as a falling film (nonagitated) or rising film evaporators also come under the heading of still equipment and can be analyzed similarly.

3. Evaporating Kettles

Continuously stirred tanks are often modified for a vacuum take-off overhead to allow operation at subatmospheric pressures. This is often the

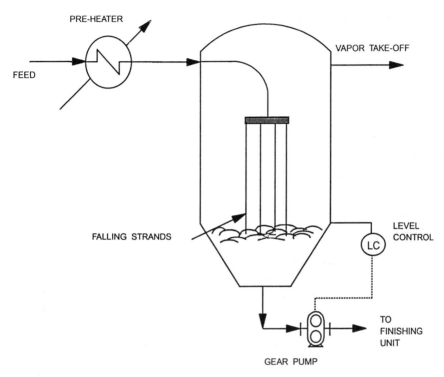

Fig. 11   A schematic of a falling-strand devolatilizer.

case while carrying out polycondensation reactions during the initial reaction stages. For example, evaporating kettles are used in the process for manufacturing polyester, shown in Fig. 2, until the polymer viscosity becomes too high for the kettle to handle. One needs to supply latent heat of vaporization, as the reaction by-products evaporate. The evaporating kettles are therefore usually jacketed to maintain the reaction temperature.

Here again, despite low solution viscosity, the mass transfer rate is poor as the bubble growth is inhibited in the lower layers of the tanks. An agitator is therefore employed to continuously stir the liquid within so that the successive layers are interchanged. In case of a simultaneous reaction, the agitator also serves to mix the reactants intimately. Whenever a reaction is occurring within the kettle, bubble nucleation is automatic as a result of the reaction—only the bubble growth and bubble rupture are the rate-limiting factors for mass transfer.

B. Rotating Equipment

The equipment used to handle polymer solutions with high viscosities usually contains rotating parts that not only aid forward movement of the material within, but also impart surface renewal for heat and mass transfer. The equipment geometries are usually complex, and power requirements to drive the rotating parts are higher. A few of the geometries are chosen for discussion in this section, either because of their widespread use or because the equipment contains some unique features that increase its separation potential.

1. Evaporating Kettles with Special Impeller Designs

The advanced ribbon reactor (ARR) and the vertical cone reactor (VCR) shown in Figs. 12 and 13 are two attractive evaporating kettle designs by Mitsubishi Heavy Industries (Shimada et al., 1985, 1987). Both are fitted with impellers that create desirable circulation patterns (for improved devolatilization) within the kettles. In ARR geometry, the impeller scrapes the walls as it pushes the liquid near the wall down, while the liquid at the kettle bottom moves upward. In the VCR geometry, the impeller scrapes the fluid upward and forces it to rise along the inner wall to form thin films. The forced circulation flow results in higher devolatilization rates as is evidenced by the strong dependence on rotational speed as, seen in Fig. 14 for the VCR. The surface renewal could be aiding disengagement of bubbles that would have been otherwise trapped at the bottom of the vessel. The scraping at the walls enhances the heat transfer at the walls (if the kettle is jacketed). The unique impeller design (a jointly moving parallel plate flow within it) can handle fluids with much higher viscosities. Shimada et al. (1987) claim that VCR type of designs can been used over a wide viscosity range (100–8000 poise) and is especially suitable for batch operations. These and similar kettles where the agitation aids devolatilization are better classified as rotating equipment.

2. Screw Extruders

In the post-reactor polymer processing field, single- and multiscrew extruders possess a unique status because of their capability to carry out several processing steps—melting of plastics, mixing with additives, devolatilization, and finally, pressurization for ultimate shaping—all on the same shaft. They can efficiently handle fluids with viscosities varying over several orders of magnitude on the same shaft(s). Vented, single- or multiscrew extruders are used for devolatilization, stand-alone or in combination with other polymer processing functions, in a wide variety of applications. Single- or multiscrew extruders have been known to concentrate polymer solutions containing highly dilute solutions, and they have also been used

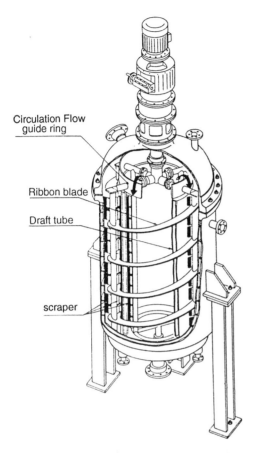

Circulation Flow
  guide ring

Ribbon blade

Draft tube

scraper

Fig. 12 Schematic of MHI's advanced ribbon reactor (ARR) and the fluid flow patterns within. (From Shimada et al., 1987; courtesy of H. Takeda and H. Mori, Mitsubishi Heavy Industries.)

to purify the product with remnant volatile levels in a few parts per million range.

The vent zone of a screw extruder has deep channels to provide a *pressure-free* zone within which the polymer is exposed to lower pressures (or vacuum). Figure 3 shows a typical flow pattern that exists in the typical vented zone of an extruder. The screw channel is partially filled in this section. Polymer melt rotates in the cross-channel direction, as the bulk of the material is conveyed downstream. Due to the clearance between the barrel and the screw flight, a thin layer of polymer is deposited on the barrel as

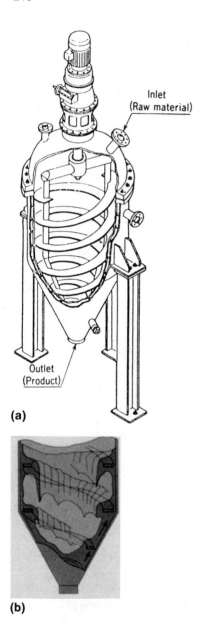

Inlet
(Raw material)

Outlet
(Product)

(a)

(b)

Fig. 13 MHI's vertical cone reactor, VCR, incorporates helical ribbons with an inverted cone base. (b) A high-viscosity fluid creates a large surface area. (From Shimada et al., 1987; courtesy of H. Takeda and H. Mori, Mitsubishi Heavy Industries.)

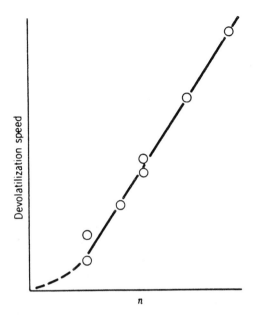

Devolatilization speed

n

Fig. 14   Devolatilization rate dependence on rotational speed in the VCR. (Taken from Shimada etal., 1987; courtesy of H. Takeda and H. Mori, Mitsubishi Heavy Industries.)

the screw flight rotates. Thus evaporation may occur either at the surface of the rotating bulk or at the surface of the deposited films. Earlier investigators postulated that the cross-channel rotation of the pool provided renewal to both these exposed surfaces (as explained earlier in Section II.A.1). However, one may also explain the high separation efficiencies observed in extruders using *bubble transport* theory. Upon exposure, many bubbles are nucleated within the pool. As the pool rotates, some of the bubbles are brought to the exposed pool surface. They grow due to lower pressures near the polymer–vapor interface and are either ruptured at the surface or are drawn back into the pool. The flow pattern in the vented channel of twin-screw extruders is similar to that seen in Fig. 3, except in the apex region between the adjacent screws. The flow in this region is unique and depends on the type of the twin-screw extruder.

Three chapters in this book detail various aspects of screw devolatilizers; one describing single -screw extruders, and the other two addressing twin-screw extruders. The following comments are some additional observations that may prove useful:

1. Many extruder manufacturers base the design of the vented sections on equations similar to those proposed by Latinen (1962), Coughlin and Canevari (1969), and Roberts (1970), derived from the mathematical model describing *diffusing films*. Indeed, in many cases (e.g., Todd, 1974) a strong dependence of devolatilization efficiency on screw speed has been observed, as Eqs. (1)–(3) predict. The strong dependence on screw speed can also be explained by the role that the agitation and surface renewal plays in bubble rupture, disengagement, and bubble *renucleation*. One must, however, be careful while scaling up the data taken on smaller equipment, as the inferences drawn could be vastly different, depending on the mechanism chosen. A safer bet is a scale up based on the elemental flow patterns (described later in Section IV.C).

2. Operation with several vacuum stages is feasible on the same extruder shaft(s). By designing a section of the screw with a small channel depth (metering section), between adjacent vent zones, one would ensure pressurized polymer between successive zones, effectively separating the two vent pressures. Further, by imparting precalculated shear to the polymer between the successive vents, one can efficiently impart dissipative heat required for evaporation, in a progressive fashion.

3. Screw geometry is ideal for addition of carrier substances between adjacent vacuum zones. The intermittent zone can be designed to be a mixing section where the carrier substance is added under pressure and dispersed within the polymer melt as fine droplets. The carrier substance can significantly enhance performance of the next devolatilization zone.

4. Foster and Lindt (1989) identified two separate mass transfer regimes in devolatilizing extruders:

a. The *first regime* is volume constrained and characterized by the existence of a foam that is not allowed to grow freely. The foam density has not been lowered yet to the point that a mechanical equilibrium between the growing bubble and the viscous resistance (from the surrounding medium) is approached. The separation is limited in this regime due to relatively low residence time. This regime typically occurs when either the flow rate is high relative to the screw conveying capacity or while working with solutions containing very large amounts of volatiles. In principle, one could also observe this behavior if the bubbles form a *stable foam* (e.g., with polymers having very high viscoelasticity). The screw design required to remove the mass transfer limitation in this regime is that yielding maximum conveying volume.

b. The *second regime* is one in which the foam is free to grow and the density allowed to approach a mechanical equilibrium. Here, the bubble growth period and right type of agitation are the keys to improving mass transfer.

A judicious use of "staged exposure" with appropriate temperatures and pressures in each stage can remedy uncontrolled foaming.

5. The degree of channel fill in the vent zone is a critical parameter while designing the vented screw sections, especially if the screw is operating in the *first regime*, where conveying is the limiting factor. The degree of channel fill, g, can be estimated from the implicit relationship

$$Q = n_{fl} \frac{\pi N_R D_b w \cos \theta}{2} f_L F_d \left(g, \frac{H}{w}\right) F_{dc}\left(\theta, \frac{R_i}{R_o}\right) C_f(n, \theta) \tag{13}$$

where Q is the volumetric throughput; N_R is screw speed; n_{fl}, w, H, D_b, and θ are the number of flights, the screw channel width, the channel gap, the screw diameter, and the helix angle, respectively. F_{dc} is the shape factor accounting for the curvature (Tadmor and Klein, 1978), which is dependent on the helix angle θ and R_i/R_o, the ratio of the shaft to outer radii; f_L is the factor accounting for the leakage over the flights (Tadmor and Klein, 1978); C_f is the factor accounting for the effect of viscosity reduction due to the cross-channel shear (Mehta, 1989); and F_d is the flight edge shape factor, originally introduced by Squires (1958), but modified to account only for one side channel wall (instead of two—polymer melt in a partially filled channel is not retarded by the other wall). In the form presented in Eq. (13), F_d includes the degree of channel fill factor, g, in its definition. Note that Eq. (13) could have also been written alternatively, with w replaced by gw and with F_d now only a function of H/w.

To obtain the maximum volumetric efficiency or the lowest degree of fill, maximum screw channel depth, H, is usually chosen in the vent zone. However, mechanical considerations limit the channel depth. One prediction from Eq. (13) is that the conveying capacity can be significantly increased when the helix angle is increased well beyond the conventional 17.65° for the square pitched screws. Equation (13) can be used to construct the plot shown in Fig. 15, where the extruder speed required to maintain a given throughput of polymer in an 8-in. (0.203 m) extruder at constant degree of fill (g) is determined at different screw helix angles. Operational parameters, screw geometry, and material property parameters are listed in Table 1. Indeed, the optimum helix angle for conveying, in this example, is seen to lie somewhere between 30° and 40°. High helix angles would therefore be especially effective in the volumetrically limited *first regime*. In addition, higher helix angles also reduce the power dissipation within the extruder

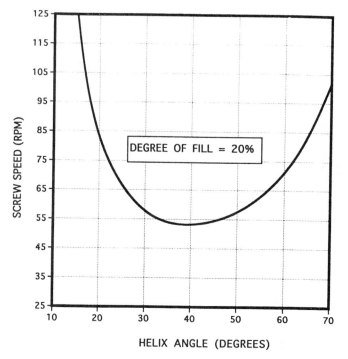

Fig. 15 Effect of screw helix angle on its conveying capacity at constant degree of fill (parameters as listed in Table 1).

dramatically. Figures 16 and 17 show the power dissipation within the screw channel and over the flight as a function of the helix angle. The power dissipation in Figs. 16 and 17 were computed using the same parameters as listed in Table 1 and computational methods described in Tadmor and Klein (1978).

Figure 18 shows the pressure gradient generated due to the cross-channel flow as a function of the helix angle (Mehta, 1989). A typical bubble nucleated upon exposure to the vent atmosphere would tend to redissolve quickly because of the cross-channel pressure gradient, if it follows the cross-channel circulation depicted in Fig. 18. Thus, one needs to balance the benefits of a lower degree of fill against the increased cross-channel pressure gradient as the helix angle is increased.

A multiflighted screw section with a larger helix angle not only can improve material flow and reduce power but may also be more effective in preventing material accumulation in the vent, especially if excessive foaming occurs.

Table 1   Parameters Used for Computation of Figs. 15–1

Screw Geometry

No. of flights   = 1
Diameter   = 8   inch
Channel gap   = 2   inch
Flight width   = 0.8 inch
Flight clearance = 0.01 inch

Operational Parameters

Flow rate	= 3000 lb/hour
Degree of fill, g	= 20%
Average temperature within the channel	= 450°F
Average temperature within flight clearance	= 500°F

Material Properties

Viscosity: Power law $= \dot{\gamma}^{n-1}$
             at 450°F $m = 0.5455$ lbf-sec^n/in.^{2} $n = 0.555$
             at 500°F $m = 0.464$ lbf-sec^n/in.^{2} $n = 0.522$
Density: $\rho = 54.594 - 0.01947213\ T$ lbm/ft^3 (T in °F)
Specific heat: Cp = 0.60 Btu/lbm-°F
Conductivity: $k_m = 0.105$ Btu/hour-°F-ft

a.   *Choosing between single- and twin-screw extruders.*   Processors are often faced with options involving a single-screw and a twin-screw. The processor should evaluate whether the performance of a twin-screw extruder justifies the added costs. Biesenberger et al. (1990) did not see significant improvements in separation efficiency when a co-rotating twin-screw extruder was used in place of a single-screw devolatilizer. These results could have been system-specific.

However, one advantage of the co-rotating twin-screw geometry (over the single-screw geometry) is its ability to wipe the adjacent screw. In a single-screw geometry, the leakage over the flight clearance tends to accumulate in the unoccupied area trailing the flight, which in turn remains stagnant and degrades. Further, in a single-screw geometry, the area under the vent is especially vulnerable to stagnation. Twin-screw geometry is effective in reducing these occurrences. The elemental assembly of the co-rotating twin-screw geometry also offers flexibility for quick material changeovers.

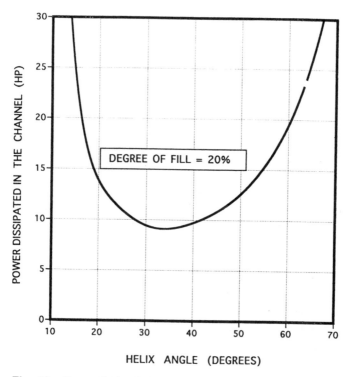

Fig. 16 Power dissipation within the screw channel as a function of the helix angle at constant degree of fill (parameters as listed in Table 1).

The processor should also consider evaluating application of the counter-rotating, nonintermeshing, twin-screw geometry. Figure 19 shows the comparison of the flow near the apex area between the counter-rotating and the co-rotating geometries. Sakai (1991/92) observed that the compression and decompression occurring within the apex region of the counter-rotating geometry resulted in a superior devolatilization due to improved surface renewal effect in this region. Within the counter-rotating geometry, the material experiences a stretching or extensional flow in the area under the vent. Such flows tend to promote bubble nucleation and growth. In the co-rotating geometry, on the other hand, the material experiences a compression in the apex region, under which already grown bubbles tend to *redissolve*. Depending on the application, counter-rotating, nonintermeshing twin screw geometry can be very effective.

It behooves the processor to choose between the various alternative

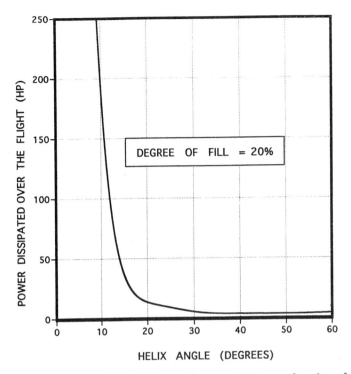

Fig. 17 Power dissipation over the screw flight as a function of the helix angle at constant degree of fill (parameters as listed in Table 1).

geometries only after sufficient experimentation on smaller scales. Further, the processor needs to evaluate the economics of the three options.

We note that extruders with more than two adjacent screws have been successfully used to devolatilize polymer solutions. A recent example is the "Multi-screw extruder (MSE)" (Mack and Pfeiffer, 1991, see also Fig. 20) offered on smaller scales by Berstorff Corporation.

b. Shear in the flight. The material in the area over the flight is vulnerable to high shear rates in all extruder geometries. Figure 21 (Dusey and Mehta, 1995) shows the development of a temperature profile over a screw flight as it wipes the barrel wall. Indeed, this area is the prime source for material degradation and side reactions. In processing shear- or heat-sensitive materials, one may find that the depolymerization occurring over the flights increases the volatile load beyond the separation ability of the extruder. Good cooling within the barrel and proper sizing of the flight land and flight

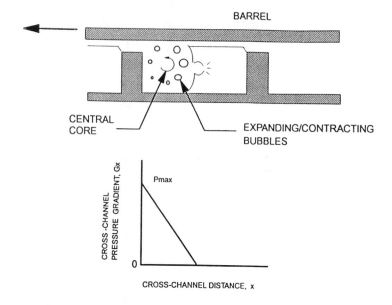

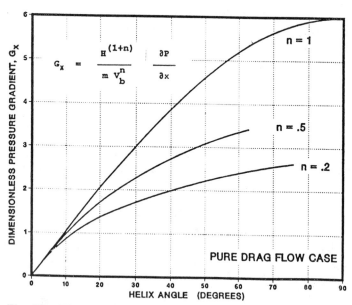

Fig. 18 Dimensionless cross-channel pressure gradient as a function of the helix angle at various power law indices, *n*.

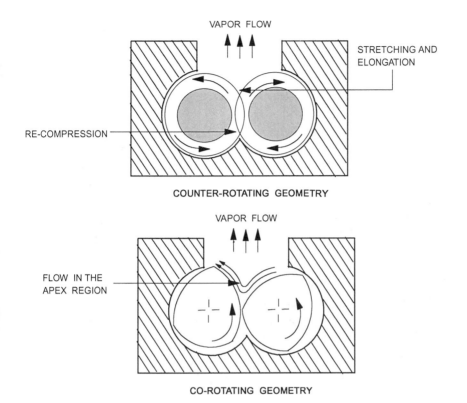

VAPOR FLOW

STRETCHING AND
ELONGATION

RE-COMPRESSION

COUNTER-ROTATING GEOMETRY

VAPOR FLOW

FLOW IN THE
APEX REGION

CO-ROTATING GEOMETRY

Fig. 19 Flow near the apex area between the co-rotating and counter-rotating twin-screw geometries.

clearance are keys for minimizing material degradation. As seen in Fig. 17, higher helix angles can reduce the shear over the flight significantly, primarily as a result of reduction in the flight length.

3. Thin-Film Evaporators

The mechanically agitated thin-film technology has evolved over years as a means to rapidly concentrate a variety of solutions. The technology has been applied to *concentrate polymer solutions* for more than three decades (Mutzenberg, 1965; Widmer, 1971). In addition, this technology has been extensively used to

Remove residual monomers (e.g., removal of residual styrene from polystyrene)

Recover solvents (e.g., manufacture of engineering resins)

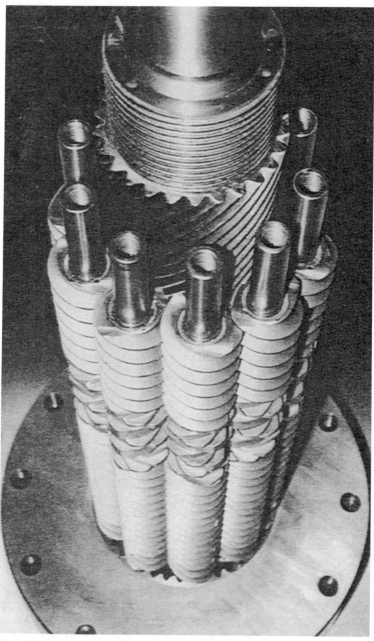

Fig. 20   A cross section of Berstorff's MSE with 10 self-wiping screws. (From Barth, 1993; courtesy of M. Mack, Berstorff Corp.)

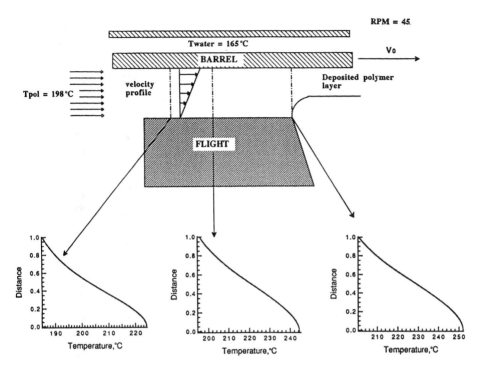

Fig. 21 Development of temperature profile within the clearance over the flight of the screw—an expanded view of the area over the flight; screw diameter = 12 in., screw speed = 45 rpm, bulk polymer temperature = 198°C; barrel cooled at 165°C.

Carry out polycondensation reactions (e.g., final reactors in manufacture of polyesters)

Figures 22a and 22b show cross-sections of a vertical thin-film vaporizer. Its cross section is shown in Fig. 22b. The apparatus comprises of a vertical heated drum within which a mechanical agitator rotates. The incoming liquid is first distributed evenly along the inside of the heated body via a distributor ring. The material flows down due to gravity. The revolving rotor comprises several blades (see Fig. 22b) that spread the down-flowing solution into thin films. Upstream of the newly deposited films, in front of the blades, a *bow wave* is formed, where surface renewal takes place. Evaporated vapors leave the solution through the core of the vaporizer, either in the co-current fashion (high volatile content) or counter-current fashion (for removal of residual volatiles), as shown in Fig. 23.

For processing very high viscosity fluids (e.g., viscosity greater than

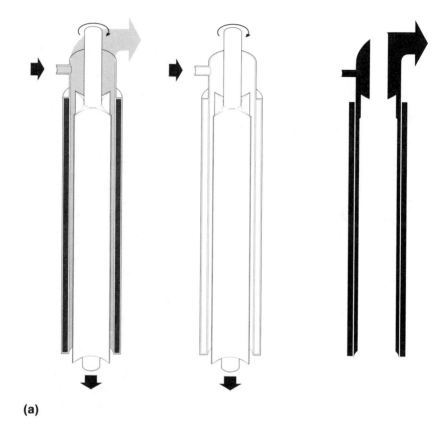

(a)

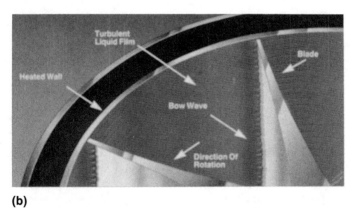

(b)

Fig. 22 (a) Cross section of a thin-film vaporizer. (b) Cross section of a thin-film vaporizer showing formation of bow waves. (Courtesy of W. L. Hyde, LCI Corp.)

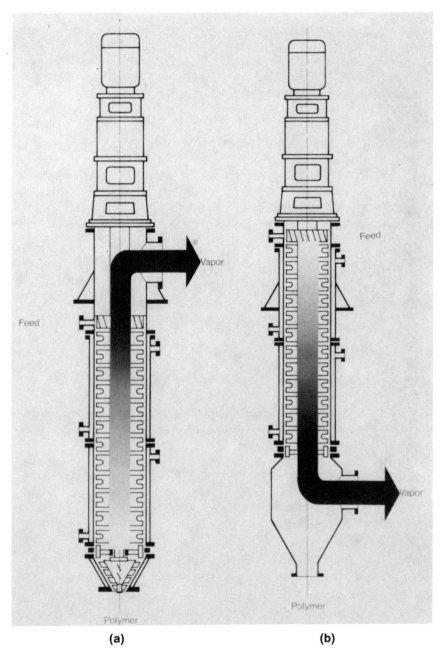

(a)   **(b)**

Fig. 23   (a) Counter-current and (b) co-current vapor flows in filmtruder evaporator. (From Heimgartner, 1980; courtesy of W. L. Hyde, LCI Corp.; reprinted with permission from VDI-Gesellschaft Kunstofftechnik, VDI-Verlag GmbH.)

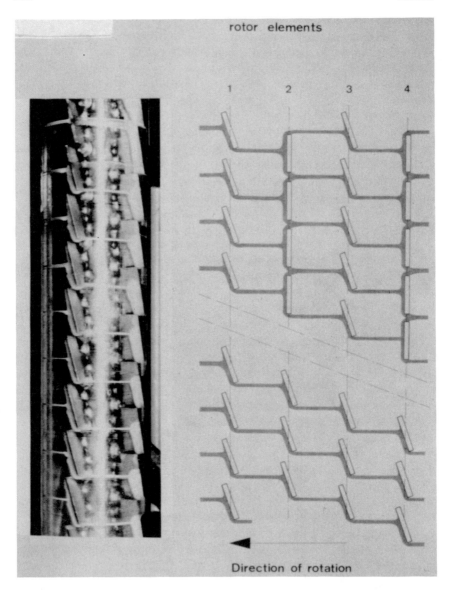

Fig. 24 Flow model of a Filmtruder evaporator with flow-transporting blades. The blades are also arranged to allow a stream-splitting/recombinant flow pattern that induces mixing. (From Heimgartner, 1980; courtesy of W. L. Hyde, LCI Corp.; reprinted with permission from VDI-Gesellschaft Kunstofftechnik, VDI-Verlag GmbH.)

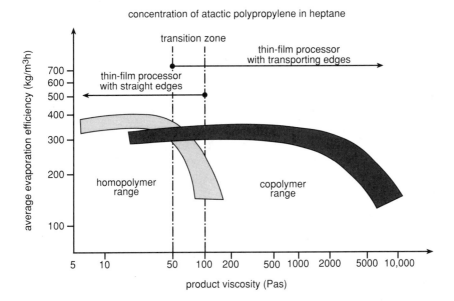

Fig. 25 Comparison of average evaporation efficiency in the thin-film vaporizer with fixed blade rotor and in the Filmtruder evaporator with transporting blades; polypropylene–heptane solutions with varying viscosities. (From Heimgartner, 1980; courtesy of W. L. Hyde, LCI Corp.; reprinted with permission from VDI-Gesellschaft Kunstofftechnik, VDI-Verlag GmbH.)

50,000 centipoise), the downward flow due to gravity is limited. This flow can be aided by replacing the straight blade rotor with one having *pitched* blades. The pitched blades provide positive transport to the fluid in the downward direction along the heated wall and can process fluids with viscosity ranging from 50 to 15 million centipoise. Figure 24 shows the flow within one such device: a Vistran† evaporator from LCI Corporation. Figure 25 (Heimgartner, 1980) gives a comparison between the viscosity ranges over which a thin-film vaporizer and a Filmtruder† evaporator may be used, respectively. The discharge system may consist of a surge tank, level control system, and gear-type discharge pump (also available from LCI Corp.) at the bottom of the unit. Alternatively, an extractor screw may be used to discharge the fluid. The lower bearing in this equipment is designed

---

† LCI Corporation now provides high-viscosity processors under the tradename Vistran.

to ride on a polymer film, and therefore this device does not require bottom mechanical seals.

a. Observations on thin-film devolatilizers.

1. The units are capable of removing large amounts of volatiles. Figure 26 shows the concentration gradient (along with other parameters) across a Luwa (now LCI Corporation in the United States) high-viscosity unit (Lane, 1981, see also Widmer, 1971), as polymer moves across the unit. As the polymer solution is concentrated, the solution viscosity increases by several orders of magnitude. As the material progresses downward, its temperature also increases, due to viscous heat dissipation. A special LCI high-viscosity gear pump or an extraction screw is used at the bottom to

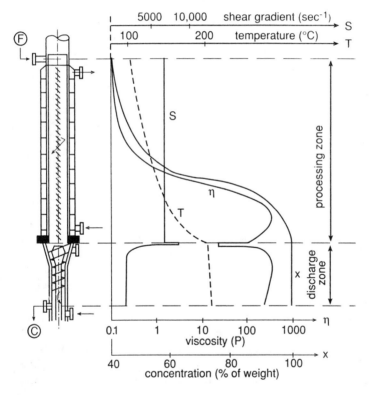

Fig. 26 Gradients of various parameters along the length of a Luwa (now LCI Corp.) high-viscosity unit for PVA–toluene solutions; S: shrear rate gradient; T: solution temperature; x: degree of concentration; η: solution viscosity. (From Lane, 1981; courtesy of W. L. Hyde, LCI Corp.)

pump the devolatilized product out of the unit. The equipment can be designed to process fluids with starting viscosities ranging from a few poise to final viscosities in tens of thousands of poise.

2. The equipment is designed to provide repeated thin film exposures with intermittent surface renewal steps. The flow patterns exemplify a "staged model" as shown in Fig. 4. The polymer is exposed as thin films between each successive *bow wave*. Surface renewal takes place within the *bow wave*. If, indeed, diffusion through films is the governing mechanism, then Eq. (2) predicts very high separation efficiency for this type of equipment, as there are multiple surface renewal steps (Lane, 1980). The assumption of ideal transverse mixing, for good surface renewal in the staged model, is well exemplified in this equipment. Good mixing is achieved as the flow is split into many small discrete rolls in front of the individual blades (see Fig. 22b and Fig. 24) and then recombined. The stream-splitting/recombinant flow pattern reorients the depleted layers, and the layers still rich in volatiles are exposed.

3. If bubble transport is the dominant mechanism, then this geometry allows many successive pluglike flow elements (deposited films) for undisturbed bubble growth. The bubble growth period depends on the speed of rotation, throughput, and the rotor geometry. Bubble release and renucleation would occur at the bow waves. One should note that if the rotor-to-wall clearance is too small (very thin films), a large pressure gradient can develop within the wedge of the blade prior to the film deposition (already-grown bubbles can redissolve).

Measurements by Woschitz (1982), however, indicated that local mass transfer rates peak upstream of the rotor blades in the bow wave region. Although Woschitz built a mass transfer model based on modified penetration theory, his results suggest dominance of bubble transport at the bow waves.

4. A heat transfer analysis, similar to the *diffusing film* model, with assumption of good surface renewal at the bow wave, would predict excellent heat transfer at the walls. High heat transfer coefficients, $\sim 100\text{--}200$ kcal/m²-hour-°C, have been observed (Lane, 1981). However, as in extruders, the polymer could be overheated due to viscous dissipation within the clearance of the rotor blades, which operate at relatively high speeds.

5. In contrast to extruders, devolatilization in this geometry is not volumetrically limited. Deposited films foam inward in the radial direction, where there are no growth restrictions. It is, however, prudent to avoid excessive foaming or high vapor velocities as entrained polymer could deposit on the rotating shaft, stagnate, and degrade—contaminating the product. Lower-molecular-weight oligomers can also vaporize and redeposit on the shaft, causing contamination problems.

6. While designing a polymer line, the processor may need more than one unit, depending on the number of vacuum stages needed. Unlike extruders, intraunit vacuum staging is not feasible. Also, devolatilization aids may be added only in separate equipment between two successive units. It should be noted, however, that use of high vacuums and high mass transfer rates in this equipment may alleviate the need for a devolatilization aid in most circumstances.

7. If operated under high vacuums, the sealing at the unit top is achieved through mechanical seals. The discharge end is seated by the polymer itself. Specially designed condensing and vapor handling systems are used for high vacuum operations (<1 mm Hg abs.).

4. Disk Ring Reactor

This apparatus was introduced as a continuous polymerizer for making polyester resins (Ellwood, 1967). The reactor, shown in Fig. 27, is a large horizontal sectioned vessel. The outer barrel is a heated stationary drum through which high vacuum is drawn. The pre-polymer fed to the reactor

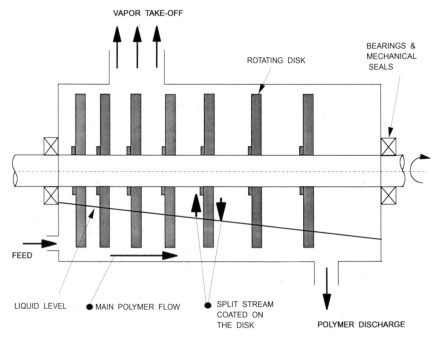

Fig. 27 A schematic representation of a disk ring reactor; disk spacing increases as the polymer viscosity builds.

travels through a pool formed at the reactor bottom. The pool level depends. on the product viscosity and is maintained by controlling the inlet flow rate to the reactor. Many rotating disk elements, mounted on a shaft, pass through the polymer pool and drag the viscous polymer along as they emerge from the pool, resulting in a thick coat of polymer film. As the disks rotate, the excess polymer runs off the disks by gravity, leaving behind a thinner film. The thin pre-polymer film polymerizes further according to Eq. (10), as it is exposed to very high vacuum and as the reactor by-product is being continuously removed. As the polymer coated on the disks is dipped into the polymer pool again, sufficient material exchange occurs to allow a fresh coat of pre-polymer, which gets polymerized during the next disk rotation.

  a. Observations on the operation of the disk ring reactor.

  1. The disk ring reactor is typically the final reactor in a train of polyester reactors. The residence times within the reactor are of the order of a few hours.

  2. The disks rotate very slowly (few revolutions per minute). However, due to large disk diameter, the disk motion should cause sufficient agitation within the pool to invoke *bubble transport*. Higher rotational speeds are avoided to give as much exposure as possible to the film on the disks (this restriction would be relieved if material exchange at the pool is improved).

  3. The spacing between the disks increases in the downstream direction. The spacing is primarily determined by the (anticipated) local viscosity of the polymer. If the gap between consecutive disk is too small, then it could fill completely with the polymer, especially if the polymer is too viscous. Reactors today are therefore designed with shafts that allow several placement positions for the disk, elements.

  4. The disk elements could be customized to meet processor's needs. Disks with holes, spokes, or even screens are not uncommon. These allow the excess polymer to run off more easily. Further, these designs help pool agitation and material interchange.

  5. The level measurement for the pool at the reactor bottom is a critical operational need. A very high level would be detrimental to devolatilization, whereas a very low level would not be sufficient for material flow (which is entirely gravity driven). A nuclear gauge would give a good indication of the reactor level. One must be careful, however, as the pool at the reactor bottom tends to "slosh around" due to the disk motion.

  6. Possible stagnation areas include the disk elements and the shaft. Low-molecular-weight oligomers can also condense on the inner barrel surface near the "vapor take-off" and char.

  7. As mentioned earlier, production of polyester with high intrinsic viscosity, for it to be in a *useful fiber range* (Farney, 1966), requires vacuum

levels below 5 torr. It is not uncommon to see processes operating well below
1 torr. The shafts of the reactors may be fitted with mechanical seals of special
design that do not allow external air leakage into the reactor. These
mechanical seals work well due to low shaft speeds.

5. Co-rotating Disk Devolatilizer

The concept of drag induction through *jointly moving plates* was
introduced by Tadmor et al. (1979). The co-rotating disk processor was
introduced to the market by Farrel Corporation as the DISKPACK polymer
processor (Tadmor et al., 1983, 1985). The co-rotating disk geometry differs
from the disk ring geometry in its ability to convey the material forward
through induced drag.

The processor comprises a number of parallel disks mounted on a
rotating shaft, closely fitted in a barrel, as shown schematically in Fig. 28.
The space formed between two adjacent disks and the barrel constitutes a

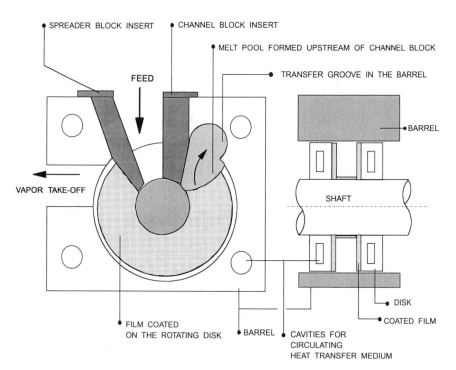

Fig. 28 Cross-sectional views of a co-rotating disk processor showing coated film
on the disks and melt pool formed at the channel block insert; both the disks and
the barrel are temperature controlled. (Courtesy of L. Valsamis, Farrel Corp.)

devolatilizing chamber. A stationary *channel block insert* placed between the two disks separates the inlet and the outlet of each chamber. A heat transfer fluid may be circulated in cavities formed within the stationary barrel and through the cavities in the rotor via a rotary joint for effective temperature control. The processor can be designed to convey material in sequential flow (one disk to the next) arrangement, or the material can be split over several disks in parallel (see Fig. 29). A series configuration is used for increasing separation efficiency (Mehta et al. 1984). A common vapor take-off manifold can be placed (Mehta and Valsamis, 1985b) over several disks to form one devolatilization stage. Specially designed *polymer seals* (Mehta and Valsamis, 1985a) over the disk top may be incorporated into design to allow *staging.* Further, it is possible to incorporate additive chambers between adjacent devolatilization stages, where one may inject *devolatilization aids.* Finally, the material discharge is attained by incorporating a pressurizing chamber as the final pumping chamber, which can pressurize the material for a downstream shaping operation.

Figure 30 shows the developed (unfolded) view of the processor at a

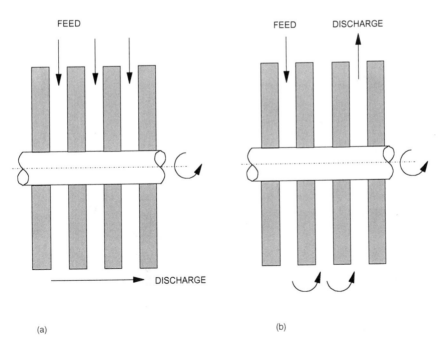

Fig. 29 Multiple-disk assembly: (a) parallel configuration; (b) series configuration. (Courtesy of L. Valsamis, Farrel Corp.)

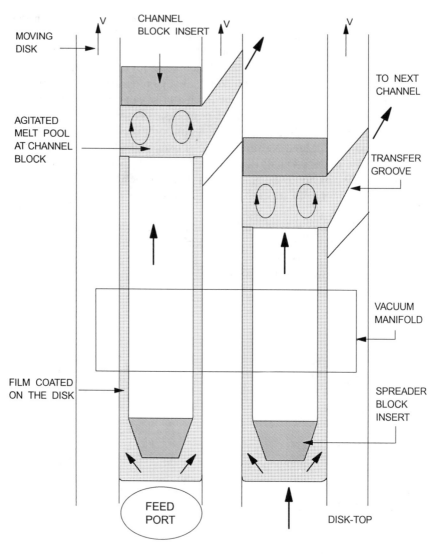

Fig. 30 A developed view of the co-rotating disk processor at radius $r = R_d$, showing a transfer groove over the disk top separating two adjacent chambers; series flow configuration. (Courtesy of L. Valsamis, Farrel Corp.)

constant radius. Polymer melt entering the inlet of the first devolatilizing chamber is dragged forward by the rotating disks, and it is coated on the disks by a stationary *spreading block insert* (Mehta et al., 1984), shaped to form a converging wedge with each of the disk surfaces. The coated films translate with the disks in a pluglike fashion, while vacuum is drawn over the void spaces between the films through a port connected to a vacuum source. The translating films meet a circulating pool formed at the channel block, just large enough to generate pressure for melt transfer into the succeeding chamber through a transfer groove formed in the stationary housing. The polymer melt repeatedly undergoes a similar process in succeeding chambers, until it reaches the last chamber from which it is discharged.

    a.   Observations on the co-rotating disk processor.

    1. With simultaneous drag induction from two surfaces, the processor is *extremely* efficient at conveying the material forward (much more so than conventional extruding equipment; e.g., see Tadmor et al., 1985). Therefore, the flow is not significantly influenced by gravity. Conveying capacities are large, and residence times much smaller. Consider comparison between two processors, a disk ring reactor and a co-rotating disk processor, each with a diameter of 500 mm and a length of approximately 1500 mm, and both with several disk elements. Due to high conveying capacity, the co-rotating disk processor is capable of a throughput of approximately 600–1000 Kg/hour at operating speeds of 30–50 rpm. With the nearly plug-like motion, the residence time distribution is narrow, with an average of ~ 1–2 min. The flow rate within the disk ring reactor, which rotates slowly (~ 5 rpm) and relies entirely on gravity flow, would be a mere 30–50 Kg/hour. Due to a large inventory, the average residence time within the reactor would be several hours.

    2. As with the thin-film evaporator, the flow pattern within the co-rotating disk processor is that of a "staged" model, shown in Fig. 4. The polymer solution is exposed as two thin films in each successive chamber for a finite period of time. The surface renewal occurs in the pools formed upstream of the channel block inserts. In order for the material to exit the chamber, the melt must change its orientation. Thus a high degree of folding, reorientation, and mixing occurs at the block. By properly designing the channel width, one may ensure that pressurization at the pool is minimal. These are necessary ingredients for good surface renewal.

    3. If, indeed, *diffusing films* is the prevalent mass transfer mode, then Eq. (2) applies, with N equal to the number of devolatilization chambers in series. High efficiencies result even with low diffusivities when sufficiently thin films are formed (1–2 mm thick). Good transverse mixing at the pool ensures that depleted layers of the film are renewed efficiently.

4. If, on the other hand, bubble transport is the dominant mass transfer mechanism, then the processor has been shown to be very efficient. An example of its effectiveness was given by Mehta etal. (1984), where greater than 90% separation efficiency was attained in a processor with just three chambers. This corresponded to effective volatile diffusivities in the range of 10^{-3}–10^{-4} cm^2/sec. The processor geometry approaches the "ideal equipment" configuration for bubble transport described earlier. The entering polymer solution is exposed to lower pressures, as thin films; the bubbles within the films can grow in an absence of externally imposed pressure field within the films for a fixed period; and the bubbles are released within the films and at the surface of the pool formed upstream of the channel block. The pressure field at the pool is minimal (no bubble redissolution), and a vigorous agitation occurs there to provide ample opportunities for bubble renucleation.

5. Attempts to coat the disks with very thin films (1 mm or less) resulted in a loss of separation efficiency by as much as 50% (Mehta et al., 1984). The results could only be explained using bubble transport as the dominant mechanism. The foamed polymer was dragged through the narrow spreader wedge clearances, where high hydrodynamic pressures developed and already-grown bubbles redissolved.

6. In an analogy with the "staged" model for mass transfer, Mehta and Donoian (1990) showed that the processor was capable of excellent heat transfer rates. The geometry provides a very large surface-to-volume ratio for efficient heat transfer. Also, both the rotor and the barrel are temperature controlled. In addition to the large surface for heat transfer, the unit also exhibits very high heat transfer coefficients, reflecting good surface renewal characteristics. Experimentally measured polymer–side heat transfer coefficients in a pilot unit were measured at 250 Kcal/m^2-hour-°C, unusually high for polymer systems.

7. The processor can be designed to operate at several pressure levels on the same shaft (staging) through use of *polymer seals* (Mehta and Valsamis, 1985a). The processor can therefore carry out several processing steps, such as addition and mixing of additives, etc., on the same shaft. This technology, however, relies on forming a polymer seat on the disk top separating the adjacent devolatilization chambers at different pressure levels. Even though the polymer seal is formed with only a fraction of the bulk flow, the material within the seals experiences high shear and high temperatures and may degrade. Careful experimentation on smaller-scale units is recommended prior to designing "staged" units.

6. Twin-Paddle Devolatilizers

Apart from twin- or multiscrew extruders, an entirely different category of specially designed devolatilizers/reactors exists in the industry that operate with twin or multiple agitators for good surface renewal. Some pieces of equipment have conveying capabilities, whereas others rely on gravitational flow. Most were developed to carry out polycondensation reactions and therefore are capable of operating at high temperatures and high vacuums. In contrast to single-shaft devices, they are mechanically more complex but provide superior agitation, and sometimes self-cleaning characteristics. Twin-screw extruders, co-rotating or counter-rotating, are not discussed in this section. Rather, examples of a few unique geometries are reviewed.

    a. Co-rotating shafts. One interesting twin-paddle device is the Self Cleaning Reactor (SCR), built by Mitsubishi Heavy Industries (Shimada et al., 1985, 1987). The SCR, schematically illustrated in Fig. 31, comprises many thick eccentric disks mounted on two parallel co-rotating shafts. The eccentricity of the disks is such that the adjacent disks on the two shafts move together, forming a pair. The eccentricities of subsequent disk pairs are staggered, forming a "long helix" angle. The helix angle at 60° allows forward movement, and at 180° (partially filled channels) provides the maximum exposure with little or no forward conveying.

    Figure 32b shows typical sectional flow patterns available for partially filled channels (Shimada et al., 1985). The vapor space moves in pockets,

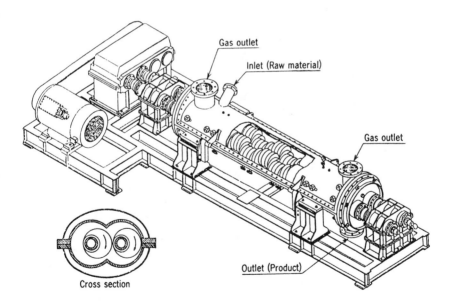

Fig. 31 MHI's SCR (Self Cleaning Reactor) comprises two shafts fitted with many eccentric rotors. (From Shimada et al., 1987; courtesy of H. Tekeda and H. Mori, Mitsubishi Heavy Industries.)

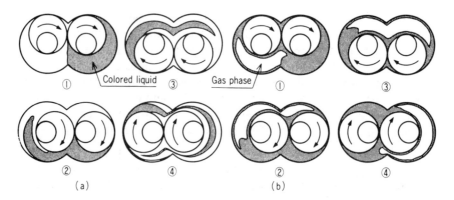

Fig. 32   Flow pattern of high-viscosity fluid in (a) full and (b) half-full volume operation of SCR. (From Shimada et al., 1987; courtesy of H. Takeda and H. Mori, Mitsubishi Heavy Industries.)

varying in configuration with the motion of the rotors, thus providing constant surface renewal. A very large exposure area is created as the polymer is coated as thin films on the sides of each disk by the disks upstream and downstream. The clearance between the successive disk pairs determines the thickness of the film coated on them. Further, a polymer film is also coated on the inside of the "figure eight" barrel.

In the forward conveying mode, the fluid introduced at the upstream end moves forward by drag-induced pressurization in the wedge formed between the eccentric disks and the "figure eight" barrel wall. Each pair of disks produces enough pressure to move the material to the subsequent pair of disks longitudinally in both directions. Staggered eccentricity ensures that the fluid moves preferably in the forward direction.

This device, with a length-to-diameter ratio of about 6, is available in different sizes up to 8 m³, according to the manufacturer. The geometry is designed to take full advantage of a very large exposure area for good devolatilization and has been used extensively for carrying out polycondensation reactions. Because of the large exposure area and the claimed high renewal rate, it has the potential of yielding very high separation efficiencies under the diffusing films mechanism. However, if bubble transport was dominant, it would still provide good devolatilization conditions with a proper choice of the "stagger" helix angle. For viscosities less than 10,000 poise, the manufacturer recommends (Takeda and Mori, 1992) use of N-SCR, which features lower power consumption and higher effective volume rate. Both versions are used extensively in carrying out polycondensation reactions.

b. Counter-rotating shafts. An example of a twin-paddle reactor with counter-rotating shafts is shown in Fig. 33. The horizontal "High Viscosity Reactor," manufactured by Mitsubishi Heavy Industries, Ltd., comprises two shafts rotating at the same speed in opposite directions. Mounted on the two shafts are thin concentric circular disks on which the polymer adheres and that rotate with the shafts (as in the disk ring reactor). The disks are notched and have scraper blades to enhance surface renewal. The flow within the reactor is by gravity. The reactor can be viewed as a series of stirred tanks. According to the manufacturer, each pair of disks can be considered as one tank in the model (Shimada et al., 1987).

A similar concept is used by Hitachi Ltd. of Japan in their Spectacle Blade reactor, shown in Fig. 34, where spectacle-shaped rotating blades are used instead of disks. This type of reactor can also be viewed as a series of stirred tanks, with each pair of blades regarded as a single mixing tank. Another agitator configuration provided by the same manufacturer comprises lattice-type blades (Fig. 35).

The AP-series reactors by List AG provide another version of twin-paddle systems. The reactors are comprised of two parallel intermeshing agitators rotating at different speeds. Both agitators are equipped with

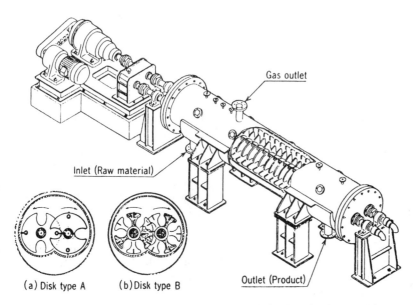

Fig. 33 Mitsubishi Heavy Industries' HVR (High Viscosity Reactor) incorporates two shafts fitted with many disk elements. (From Shimada et al., 1987; courtesy of H. Takeda and H. Mori, Mitsubishi Heavy Industries.)

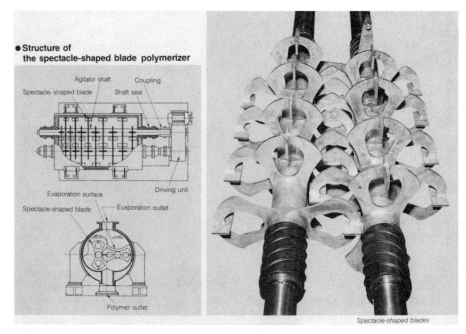

Fig. 34   Reactor from Hitachi Ltd., with spectacle-shaped blades. (Courtesy of T. Kinoshita and O. Kito, Hitachi Ltd.)

radially mounted, heatable or coolable plates with U-shaped mixing/kneading bars welded to their peripheries, arranged in a manner that each rotor completely cleans the other shaft's core, plates, and mixing/kneading bars. The kneading elements provide good lateral mixing and surface renewal, whereas the oblique positioning of the kneading elements transport the polymer forward, yet maintaining a very narrow retention time distribution. Large cross-sections provide for good heat exchange and facilitate removal of gases and vapors. Figures 36(a) and (b) show the LIST-ORP reactor in which the rotors counter-rotate at a speed ratio of 1:4. Figures 37(a) and (b) show LIST-CRP reactor in which shafts co-rotate at a speed ratio of 4:5.

Observations show that counter-rotating shafts produce a "stretching" type of flow pattern at the apex. It is plausible that in addition to the transverse mixing at the apex, the normal stresses so introduced would enhance formation and growth of bubbles.

The foregoing examples of twin-shaft devolatilization devices are not exhaustive. Many other devices, not covered here, are being utilized as devolatilizers. For example, the flow pattern in a properly designed internal

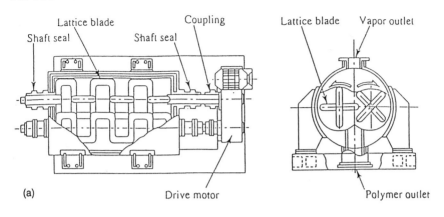

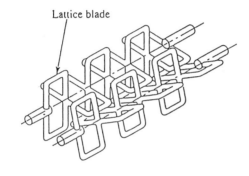

Lattice-Blade Type Polymerizer

Fig. 35 Reactor from Hitachi Ltd., with lattice-type blades. (a) Overall construction; (b) strirring blade (lattice blade). (Courtesy of T. Kinoshita and O. Kito, Hitachi Ltd.)

mixer, such as a Farrel Continuous Mixer (FCM mixer), could provide several stages of diffusing layers followed by rolling pools (Valsamis and Canedo, 1989) where good surface renewal occurs.

C. Vacuum and Condensing Systems

1. Vacuum Systems

Government regulations, product stewardship, and processing requirements at the customer's shops may require that the monomer/solvent

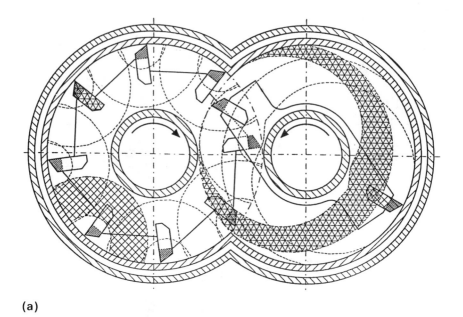

(a)

Fig. 36   LIST-ORP reactor for devolatilizing high viscosity liquids. Counter-rotating rotor shafts are cleaned by the adjacent rotor elements. (a) Cross-sectional view; (b) perspective view. (Courtesy of List AG.)

content in the polymers be reduced to a few parts per million, sometimes even to a few parts per billion. Application of Eq. (6) reveals that the most of commercial devolatilization processes would have to operate at sub-atmospheric pressures to meet these requirements. In addition, recovery of high boiling solvents is possible only when the polymer solution is subjected to a high vacuum. Finally, many polycondensation reactions require vacuums down to 1 torr to achieve the high molecular weights needed for desired product properties. A well-designed vacuum system is therefore critical to process reliability.

Many industrial devolatilizers operate between 1 and 760 torr. The following considerations may prove useful in the selection and design of vacuum systems:

1.   Key components of the load on the vacuum pump are the volatiles removed and the noncondensables (typically air leakage).

A precondenser is the most effective way of reducing the condensable load on the vacuum pumps. A precondenser can be

(b)

a surface condenser or a direct contact condenser. If pre-condensers are used, the coolant for the precondenser must be kept below the dew point of the vapor stream that it is condensing. In some cases, this may necessitate extra refrigeration of the coolant. One must weigh the benefits of lower vacuum loads against increased refrigeration needs.

Noncondensables typically comprise air leaked into the process through vessel joints such as flanges and threads, valves, mechanical seals in case of equipment with agitators, and mechanical and thermal fatigue of gaskets and seals. Methods to detect various leakage sources and to estimate the amounts are discussed by Ryans and Roper (1986).

2. Vacuum pumps available in the market are either mechanical pumps (such as rotary vane) or steam-jet type. The pumps are rated

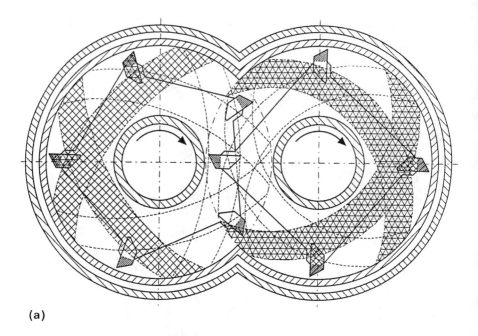

(a)

Fig. 37   LIST-CRP reactor for devolatilizing high viscosity liquids. Co-rotating rotor shafts are cleaned by the adjacent rotor elements. (a) Cross-sectional view; (b) perspective view. (Courtesy of List AG.)

by capacities and the maximum vacuum or the "blind" suction pressure (i.e., at zero load). The required suction pressures are calculated from the process requirements (e.g., see Eq. (6)), and the pressure drops over the condenser and the vacuum line. The required capacity is estimated from the devolatilizer load minus the estimate of recovery in the precondenser. Table 2 (Ryans and Roper, 1986) gives approximate ranges for commercially available vacuum pumping systems. Current industrial practices favor integrated pumping systems.

3.   Selection of vacuum pumps may be based on several criteria:

Capacity and the required suction pressures.

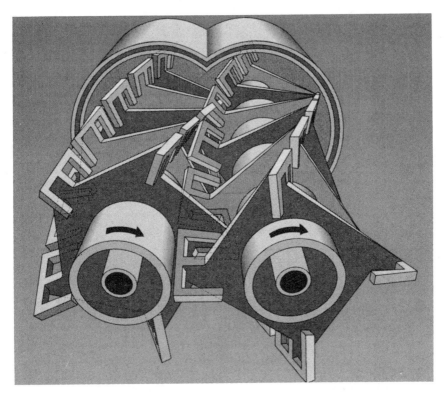

(b)

Process economics including purchase and installation costs versus operating costs such as energy.

Operational reliability including such factors as the tolerance of the system for entrained solids and/or liquids, tolerance for surges in load, response to precondenser failure, response to an increase in the discharge pressure, ability to handle corrosive chemicals, and availability of field maintenance.

Environmental considerations, such as the amount of effluent discharge and the vapor escape with use of the motive fluids.

4. The maximum pressure drop occurs prior to the precondenser. It is therefore advisable to design devolatilizers with nozzles as large as feasible, and to size the piping to the precondenser accordingly.
5. Many industrial devolatilizers are fitted with knockout pots in the vicinity of the vacuum nozzles to minimize entrainment down-

Table 2 Capacity and Operating Range for Vacuum Pumps and Vacuum Pumping Systems Commonly Used in Process Applications

Type	Blind or base pressure	Lower limit for process applications	Single unit capacity range, ft³/min
Steam ejectors:			
One-stage	50 torr	75 torr	10–1,000,000
Two-stage	4 torr	10 torr	
Three-stage	800 micron*	1.7 torr	
Four-stage	100 micron	250 micron	
Five-stage	10 micron	50 micron	
Six-stage	1 micron	3 micron	
Liquid ring pumps:			
60°F water-sealed:			
one-stage	50 torr	75 torr	3–18,000
two-stage	20 torr	40	
Oil-sealed	4 torr	10 torr	
Air ejector first stage	2 torr	10 torr	
Rotary piston pumps:			
One-stage	5 micron	100 micron	3–800
Two-stage	0.001 micron	10 micron	
Rotary vane pumps:			
Operated as a dry compressor	20 torr	50 torr	20–6,000
Oil-sealed, rough-vacuum pump	0.5 torr	2.0 torr	50–800
Oil-sealed, high vacuum pump			
One-stage	5 micron	100 micron	3–150
Two-stage	0.001 micron	10 micron	
Rotary lobe blowers:			
One-stage	100 torr†	300 torr	30–30,000
Two-stage	10 torr†	60 torr	
Integrated pumping systems:			
Ejector–liquid ring pump	1 micron	3 micron	100–100,000
Rotary-blower–liquid ring pump	1.0 torr	5 torr	100–10,000
Rotary-blower–rotary piston pump	0.1 micron	0.10 micron†	100–30,000
Rotary-blower–rotary-vane pump	20 micron	200 micron§	100–30,000

* 1.0 micron = 0.001 torr.
† Based on intercooled design that uses gas admitted to a trapped discharge pocket to cool the blower.
‡ Based on using a two-stage rotary-piston pump as the backing pump.
§ Based on a two-stage rough-vacuum rotary-vane design that exhibits a base pressure of approximately 0.5 torr.
Reprinted with permission from: J. L. Ryans and D. L. Roper, "Process Vacuum System Design and Operation," Copyright © 1986 by McGraw-Hill, Inc.

stream. If the devolatilizer is operated above 10 torr, then the knockout pots should be flanged for occasional cleaning. A parallel spare knockout pot could be installed.

6. For vacuum requirements below 10 torr or for applications where even a slight amount of oxygen is not permissible, "weld rings" at the flanges and in the piping should be considered in lieu of conventional flanges.

2. Condensing Systems

Vapors removed from the devolatilizer are condensed quickly, both to reduce the load on the vacuum system and to recover most of the volatiles back into a process stream. As mentioned before, precondensers are effective vapor pumps and should always be considered whenever significant amounts of condensables are being drawn off. Two types of condensers are commonly used: direct contact condensers and surface condensers.

In the direct contact condensers, heat is removed from the process off-gases (volatiles) by direct physical mixing (contact) with a cooling medium. The coolants are kept at temperatures near those of the condensing gases, and pressure drops across the condensers are low. Whenever feasible, a direct contact precondenser is preferable, as it can better handle entrained solids (e.g., polymer and low molecular weight components). A barometric leg is, however, often needed for the direct contact condensers operating under vacuums. Spray condensers and cascade condensers are examples of direct contact condensers. Sometimes, to reduce recovery costs, the condensed volatiles are subcooled and used as the coolant. In this case, the maximum vacuum level attainable is limited by the vapor pressure of the condensed volatile. Surface condensers do not allow contact between the cooling medium and the condensing volatiles. They are used when direct contact condensers cannot be used or when recovery of the volatiles from the coolant would be expensive. A more complete description of the two types of condensers can be found in Kern (1950) and Ryans and Roper (1986).

Condenser coolant refrigeration needs depend primarily on the total volatile load, operating vacuum levels, and the amount of inert leakage in the devolatilizer and the vacuum system. In cases where a large amount of the vapors needs to be removed, it is beneficial to remove most of the volatiles at higher operating pressures, not only to reduce the vacuum pump load but also to minimize the cooling requirements. The coolant temperature requirements for such systems are moderate because the condensation temperature is moderate. Whenever feasible, vacuum staging within the devolatilizer provides an effective solution.

Air leakage in the devolatilizing equipment and other parts of the vacuum system invariably leads to higher coolant refrigeration needs as the temperature difference between the coolant and the volatiles, ΔT, is often increased to compensate for the resulting ineffective heat transfer coefficients.

IV. CONCLUDING REMARKS

Equipment selection is usually carried out early during the phase of process development of the entire polymer line. In some instances, however, regulations or product enhancement considerations lead to modification of an existing line. For example, upgrade of an existing finishing extruder to provide a vented zone will fall into the latter category, The decision to select certain equipment will be based on technical considerations, scale-up methodology, operational reliability, and process economics.

A. Process Economics

Economic considerations are the prime motive for installing equipment at a manufacturing site. Even regulations, product stewardship factors and environmental considerations can be evaluated in economic terms such as increased (or loss of) sales due to enhanced (or inferior) product, losses caused by shutting down a product line, and so on. Two measures frequently used in industry are the *return on investment* (ROI) and the *payback periods*. The former reflects the percentage of investment recovered per year, whereas the latter is the time it takes from startup until the cumulative cash inflow equals the investment costs. Benchmarks for minimum ROI or maximum payback periods required by various manufacturers differ, but usually fall in the range of 15%–25% or 3–5 years, respectively.

The economic analysis could be cursory ($\pm 50\%$ accuracy) in the early stage of an investigation, or detailed ($\pm 5\%$–10%) prior to equipment purchase. A detailed economic analysis would account for such factors as comparison of process improvement alternatives, cost of technical feasibility trials and evaluation, cost of capital, installation costs (includes modification to existing lines) and project expenses, startup and working capital, interest rate, depreciation, royalties, income due to increased sales, variable and fixed costs (e.g., additional labor required for operating the devolatilizer), and a sensitivity analysis of product volume/price changes.

B. Technical Considerations

Most technical considerations hinge upon a good match between the proces requirements and the capability of the equipment. Given the many choices of equipment available in the market, the processor needs to make a feasibility study to narrow the choices. After examining the properties of the polymer solution over the range of the anticipated process conditions, the processor must evaluate such questions as the following:

1. What is the required degree of separations? How many stages of take-off pressures are required?
2. How many pieces of equipment are necessary for the required degree of separation? What kind of agitation is required for good mass transfer? Is the available surface adequate for mass transfer? Can the equipment under consideration accommodate pressure (vacuum) staging?
3. What is the required heat load for separation (latent heat)? Is the heat transfer adequate? What is the acceptable exit temperature?
4. Are there local hot spots due to high shear? If so, what is the expected local temperature increase? Would the material degrade at these temperatures?
5. Is it necessary to add a devolatilization aid? What kind? Is the equipment suitable for effective dispersion of the aid? What about handling and recovery of the aid?
6. How much residence time is required by the process? For processes with simultaneous chemical reaction, the residence time may be dictated by the rate of reaction even if devolatilization is efficient. Is the residence time distribution narrow or broad?
7. Is the equipment available in large enough sizes to handle the throughput? How about the separation load? If not, how many parallel lines need to be installed?
8. Has the equipment been used in a similar service? What is the maximum size of equipment in commercial usage? Are data available for evaluation?

One of the effective ways of evaluating the technical feasibility is to carry out trials, either in-house or at the vendors' laboratories on a smaller scale. The initial expense of the trials is usually well justified in terms of avoidance of costly mistakes later on. These trials also afford the critical data required for process and mechanical scale up.

C. Scale Up

Critical areas for equipment scale up are (1) fluid mechanics, (2) mass transfer, (3) heat transfer, and (4) mechanical integrity. Traditionally, data taken on smaller (pilot) scale are either fit into statistical models or semitheoretical models. These models, although fairly accurate in predicting responses of a given piece of equipment, sometimes fail to predict accurate responses for the scaled-up sizes. This is especially the case while scaling up mass transfer (because theoretical models for bubble transport are still in their infancy). In the example of an extruder devolatilizer, Biesenberger et al. (1990) found that the exponential model was not sufficiently accurate for scale up.

Manufacturers of commercial equipment develop in-house methods and rules of thumb based on accumulated experience in scaling up equipment. These methods are generally reliable for scale up to the sizes existing in the field and for materials similar to those that the equipment has already been processing. For unfamiliar applications, the following guidelines may prove useful:

1. Design experiments in the pilot trial to identify the dominant mass transfer mechanism and determine the operational regime. For example, the pilot trials may indicate whether bubble transport is dominant, whether the separation is close to thermodynamic equilibrium, and if is there excessive foaming.

2. Determine key parameters that are critical to mass transfer with the help of existing models available in literature. In an extruder with bubble transport dominant, some of the key parameters would be (for each vented zone)

Parameter	As a Measure of
degree of screw fill	effective control of foaming
melt exposure time at the pool surface	bubble growth period in one exposure
total residence time under vent	number of surface renewal stages
cross-channel pressure gradient	bubble redissolution
surface velocity of the rotating pool	bubble rupture
frequency of pool cross-channel rotation	bubble renucleation

3. Ensure similarity of the key parameters while geometrically scaling up the fluid dynamics and operating conditions (e.g., rotational speed, vacuum level, temperature). Some compromise is usually necessary. In some cases it may be more convenient to keep parameter ratios (dimensionless form) constant (for example, the ratio of melt surface exposure time to the total residence time).

4. Heat transfer in most equipment can be calculated through

application of penetration theory separately to different fluid segments.

5. Ensure that the operating speeds do not impose excessive shear in the close clearance areas. This can be best evaluated by estimating the fluid temperature rise within the clearance using an unsteady-state energy balance.

6. Wall shear stresses in critical areas can be calculated using fluid rheology. Equipment components must be able to mechanically withstand these stresses. A complete finite element analysis would be valuable in identifying weak areas.

7. One may need to scale up the equipment in separate segments if fluid properties change drastically over the equipment (see Fig. 26).

8. Include upset conditions or variable throughput rates as possible scenarios in all calculations.

D. Operational Reliability

Operational reliability is critical in a manufacturing environment. Loss of production, even for a small time, can be very expensive. The following considerations may be helpful in improving total operational reliability:

1. Ensure that spare parts for the equipment as well as necessary installed spare equipment, the temperature control system, and the vacuum system are readily available. Spares may include pumps, seals and gaskets, knockout pots, precondenser, inserts such as the agitator within the devolatilizer, and so on.

2. Ensure that the equipment design meets applicable design code requirements for the maximum surge (or upset) that can occur in the line. The surge may be in the form of polymer feed, volatile concentration (the vaccum system should be designed to handle this), solution temperature, or discharge system failure.

3. Ensure that the equipment is able to restart loaded after a power failure.

4. Consider those designs that are effective in minimizing solids entrainment into the vapor take-off systems. This is the most common problem experienced in the industry. A good vacuum line design is critical to mechanical reliability. Rotor designs that wipe vent areas are also effective at containing solids flow into the vapor take-off system.

5. A good predictive and preventative maintenance program is essential for reliable operation. This typically includes proper

vibration analysis, lubrication analysis, and standard preventative
maintenance, especially for lubrication of the rotating parts.

6. Mechanical parts should be designed to withstand deflections
    caused by unbalanced loads created by multiple melt pools in
    mechanically agitated equipment.
7. In many instances, solids deposit on the unwiped surfaces within
    the devolitilizers. These solids char at high operating temperatures
    and occasionally fall back into the product. Consider the use of
    melt filters at the discharge end to eliminate contamination of the
    product.

NOMENCLATURE

A effective heat transfer surface area

C volatile concentration in the polymer solution

C_f correction factor accounting for the cross-channel shear in a partially
filled extruder channel

C_p heat cap, cal/gm °C

d diameter of the rising bubble in a stationary polymer solution

D volatile diffusivity in the polymer solution

D_b inside diameter of the extruder barrel (inches)

E_F overall separation efficiency

Ex   extraction efficiency per exposure

f degree of channel fill in the extruder

F_d wall shape factor accounting for reduced drag due to flight effect in
an extruder channel

g_c gravitational acceleration

H channel depth in an extruder channel (inches)

L length of the clearance (inches)

m fluid power law parameter, lbf-sec^n/in.^{2}

$\dot{m}_p$ polymer mass flow rate

$\dot{m}_v$ volatile mass flow rate

n fluid power law parameter

N number of surface renewal steps

N_R agitation speed of the extruder

P_1 vapor pressure of the volatile component over the polymer solution

P_2^0 vapor pressure of the pure volatile component at the solution
temperature

Pe   Peclet number

P_B pressure within a bubble

P_W power dissipated within the devolatilizer

P_∞ pressure far from the bubble

Q volumetric flow rate in the extruding channel (in.^{3}/sec)

$\dot{Q}_{ext}$ externally supplied heat

r radial position

R_B bubble radius

R_i screw shaft radius

R_o screw order radius

t bubble growth period

T_{ext} temperature of the external heating/cooling medium

T_m temperature of the incoming polymer solution

T_{pol} temperature of the polymer solution

U_0 overall heat transfer coefficient (Btu/hour-ft^2-°F)

v_p volume of the cell over which the mass balance has been made

ΔH_v latent heat of vaporization

$\Delta\rho$ density difference between the vapor and the surrounding polymer solution

ϕ_2 volume fraction of the polymer

χ Flory–Huggins interaction parameter

η polymer solution viscosity

σ surface tension

θ screw helix angle

REFERENCES

Barth, U. (1993). Multi-Screw Extruder MSE for Degassing and Reaction Processes, Technical Literature, Berstorffe Corp., Hanover, Germany.

Biesenberger, J. A. (1980). Polymer devolatilization—theory of equipment, *Polym. Engin. Sci.*, *20*(15): 1015.

Biesenberger, J. A., and Kessidis, G. (1982). Devolatilization of polymer melts in single screw extruders, *Polym. Engin. Sci.*, *22*: 832.

Biesenberger, J. A., and Sebastian, S. (1983). *Principles of Polymerization Engineering*, Wiley, New York.

Biesenberger, J. A., Dey, S. K., and Brizzolara, J. (1990). Devolatilization of Polymer Melts: Machine Geometry Effects, *Proc. SPE's 48th ANTEC*, Dallas, Texas.

Blanks, R. F., Meyer, J. A., and Grulke, E. A. (1980). Mass transfer and depolymerization of styrene in polystyrene, *Polym. Engin. Sci.*, *21*(16): 1055.

Boucher, D. F. (1964). Apparatus and Method for Steam-Polymer Separation, U. S. Patent 3,134,655.

Coughlin, R. W., and Canevari, G. P. (1969). Drying polymers during screw extrusion, *AIChE J.*, *15*(4): 560.

Crank, J. (1956). *The Mathematics of Diffusion*, Oxford University Press, London.

Dusey, M., and Mehta, P. S. (1995). Development of temperature profile over a screw flight, in preparation.

Ellwood, P. (1967). Continuous polyester condensations obtained with two reactors, *Chem. Eng.*, *74*: 98.

Farney, R. C. (1966). *Chem. Eng. Prog.*, *62*(3): 88.

Flory, P. J. (1953). *Principles of Polymer Chemistry*, Cornell University Press, Ithaca, New York.

Fong, W. S. (1992). Polystyrene: Process Economics Program, S.R.I. International, Report 3913, Menlo Park, CA.

Foster, R. W., and Lindt, J. T. (1989). Bubble growth controlled devolatilization in twin screw extruders, *Polym. Engin. Sci.*, *29*(3): 178.

Hess, K.-M. (1979). *Kunstoffe*, *69*(4): 199.

Heimgartner, E. (1980). *Devolatilization in a Thin Film Vaporiser*, VDI-Verlag GmbH, Dusseldorf, pp. 69–97.

Hu, G. H., Chen, L. Q., and Lindt, J. T. (1993). Acceleration of Chemical Reaction in Boiling Polymer Solutions, *Proc. SPE's 50th ANTEC*, New Orleans, pp. 37–40.

Kern, D. Q. (1950). *Process Heat Transfer*, McGraw-Hill, New York.

Lane, J. B. (1980). Mass Transfer in the Luwa Filmtruder®, *Seminar on Polymer Processing*, Luwa Corp., Charlotte, NC.

Lane, J. B. (1981). Design and Performance Factors for Luwa Polymer Machines, *Seminar on Thin Film Technology*, Luwa Corp., Charlotte, NC.

Latinen, G. A. (1962). Devolatilization of viscous polymer systems, *Adv. Chem. Ser.*, *34*: 235, American Chemical Society, Washington, D.C.

Lee, S. T. (1982). Computational Analysis of Bubble Behavior in the Devolatilization of Polymer Melts, Master's thesis, Stevens Institute of Technology, Hoboken, New Jersey.

Lee, S. T., and Biesenberger, J. A. (1986). A fundamental study of polymer melt devolatilization, I, *Polym. Engin. Sci.*, *26*: 982.

Lee, S. T., and Biesenberger, J. A. (1987). A fundamental study of polymer melt devolatilization, II, *Polym. Engin. Sci.*, *27*: 517.

Lee, S. T., and Bisenberger, J. A. (1989). A fundamental study of polymer melt devolatilization, III, *Polym. Engin. Sci.*, *29*: 782.

Mack, M., and Pfeiffer, A. (1991). "The Continuous Thin Film Reactor Using 10 Intermeshing Screws," *Proc. SPE's 49th ANTEC*, Montreal, pp. 109–113.

Mauch, K. (1981). *Kunstoffe*, *71*(5): 266.

Mehta, P. S. (1989). The Influence of Cross-Channel Shear on Isothermal Conveying of Power Law Fluids in Extruding Channels—An Analytical Solution for Pure Drag Flow Case, *Proc. SPE's 47th ANTEC*, New York.

Mehta, P. S., and Donoian, G. (1990). Heat transfer characteristics of a rotary disk processor, *I & EC Res.*, *29*(5): 829.

Mehta, P. S., and Valsamis, L. N. (1985a). Rotary Processor and Seals, U.S. Patent 4,527,900.

Mehta, P. S., and Valsamis, L. N. (1985b). Rotary Processor and Vacuum Systems, U.S. Patent 4,529,478.

Mehta, P. S., Valsamis, L. N., and Tadmor, Z. (1984). Foam devolatilization in a multichannel corotating disk processor, *Polym. Proc. Engin.*, *2*(2/3): 103.

Mehta, P. S., Valsamis, L. N., and Tadmor, Z. (1985). Rotary Processors and Methods for Devolatilizing Materials, U.S. Patent 4,529,320.

Mutzenberg, A. B. (1965). Agitated thin film evaporators; Part 1: Thin film technology," *Chem. Engin.*, *72*: 175.

Newman, R. E., and Simon, R. H. M. (1980). A Mathematical Model of Devolatilization Promoted by Bubble Formation, 73rd Annual AIChE Mtg., Chicago.

Nunn, R. E. (1980). Vented-barrel injection modling has come of age, *Plastics Eng.*, February: 35–39.

Padberg, G. (1980) *Fundamental Principles of Polymer Devolatilization*, VDI-Verlag GmbH, Dusseldorf, pp. 1–23.

Pell, T. M., and Davis, T. G. (1973). Diffusion and reaction in polyester melts," *J. Polym. Sci.*, *11*: 1671.

Powell, K. G., and Denson, C. D. (1983). A Model for the Devolatilization of Polymeric Solutions Containing Entrained Bubbles, 75th Annual AIChE Mtg., November, Washington, D.C.

Roberts, G. W. (1970). A surface renewal model for the drying of polymers during screw extrusion, *AIChE J.*, *16*(5): 878.

Ryans, J. L., and Roper, D. L. (1986). *Process Vacuum System Design and Operation*, McGraw-Hill, New York.

Sakai, T. (1991/92). Report on the state of the art: Reactive processing using twin screw Extruders, *Adv. Polym. Tech.*, *11*(2): 99.

Shimada, T., Omoto, S., and Mori, H. (1985). Evaluation of Mixing Performance of High Viscosity Reactors, *Proc. Indian Chem. Engrg. Congress*, Calcutta.

Shimada, T., Mori, H., Kajimoto, H., and Omoto, S. (1987). Development of high viscosity reactors, *Mitsubishi Heavy Industries Tech. Rev.*, *24*(3): 1.

Squires, P. H. (1958). *SPE J.*, *14*: 24.

Tadmor, Z., and Klein, I. (1978). *Engineering Principles of Plasticating Extrusion*, R. E. Krieger, Huntington, New York.

Tadmor, Z., Hold, P., and Valsamis, L. (1979). A Novel Polymer Processing Machine—Theory and Experimental Results, *Proc. SPE's 37th ANTEC*, New Orleans, 193.

Tadmor, Z., Valsamis, L. N., Yang, J. C., Mehta, P. S., Duran, O., and Hinchcliff, J. C. (1983). The corotating disk plastics processor, *Polym. Engin. Rev.*, *3*(1): 29.

Tadmor, Z., Valsamis, L. N., Mehta, P. S., and Yang, J. C. (1985). Corotating disk pumps for viscous liquids, *I & EC Proc. Des. & Dev.*, *24*(2): 311.

Takeda, H. and Mori, H. (1992). Private communication, July.

Todd, D. B. (1974). Polymer Devolatilization, Proc. SPE's 32nd ANTEC, San Francisco, p. 472.

Valsamis, L. N., and Canedo, E. L. (1989). Mixing, devolatilization, and reactive processing in the Farrel continuous mixer, *Intern. Polym. Proc. IV, 4*: 247.

Werner, H. (1981). *Kunstoffe*, *71*(1): 18.

Widmer, F. (1971). The treatment of viscous substances in thin-film apparatus, *Dechema Monographica*, *66*: 143.

Woschitz, D. (1982). Lokaler Stoffaustausch im Mechanisch Beeinflussten Flussigkeit, Doctoral dissertation, Eidgenossichem Technischen Hochschule, Zurich.

Yoo, H. J., and Han, C. D. (1983). Development of a Mathematical Model of Foam Devolatilization, 75th Annual AIChE Mtg., Washington D.C.

9

Falling-Strand Devolatilization

Robert H. M. Simon†

Monsanto Chemical Company, Springfield, Massachusetts

---

† Retired.

I. INTRODUCTION

A. General Description

The falling-strand devolatilizer (FSD) operates via vapor flashing from a polymer melt at a temperature such that the pressure maintained in the devolatilizer is below the saturation pressure of the volatiles in the melt. In principle, except for feed and discharge pumps, an FSD has no moving parts contacting the melt. Key functions such as melt forwarding, mixing, and surface generation are brought about primarily via combinations of gravity flow with vapor bubble formation, growth, and disengagement. An FSD is often called a flash devolatilizer.

It should be borne in mind that a genuine "falling-strand" condition only exists over a limited range of devolatilization parameters, as will be discussed shortly. More typically, the melt is actively foaming as it descends through the vacuum chamber, and in some instances the falling melt may have been initially formed as sheets. The term *FSD* is therefore largely a misnomer. It originated with some of the early applications (e.g., Stober and Amos, 1950) and was then incorporated into the jargon.

B. Devolatilization Regimes

When considering vapor flashing from a melt, Simon (1983) pointed out that there are basically three regimes, each with separate rate-limiting mechanisms, and each largely dependent on SH, the degree of "superheat," a term often used in industrial practice, and defined here as

$$SH = P_1 - P_\infty \tag{1}$$

where P_1 is the saturation pressure exerted by the volatile in the melt, and P_∞ is the total pressure within the vacuum chamber. With a given system, P_1 is largely a function of temperature and volatile concentration, as discussed in Section I.D. Because of the broad variation possible with key parameters such as melt properties and volatiles' molecular weight, it is difficult to predict which devolatilization regime will result from a given level of superheat. Even so, the qualitative guidance provided can be very useful.

The first of these regimes, termed "free boiling," occurs where $P_1 - P_\infty$ is large and liquid viscosity often relatively low. This usually obtains with volatile-rich conditions. Vapor bubbles rapidly initiate and expand via mass transfer from the liquid phase and by coalescence. The resulting convective mixing reduces the depletion of volatiles at the melt–vapor interfaces and further promotes mass transfer to the vapor phase. Devolatilization rates in the free-boiling regime are ultimately limited by operational problems caused by excessive melt–vapor foam expansion and entrainment into the vapor line. These are tied to bubble coalescence and vapor disengagement mechanisms. With free boiling, melt temperature can rapidly drop in the absence of external energy inputs since the latent heat of vaporization is otherwise provided primarily by sensible heat. For example, under adiabatic conditions a styrene–polystyrene melt at about 250°C will drop approximately 12°C for every 10 wt % of the melt that vaporizes. As the melt temperature falls, P_1 will drop, and superheat will therefore diminish. Melt viscosity will rise due to both temperature drop and volatile removal. These effects cause bubble growth and movement to slow down, and the free-boiling regime gradually fades out.

The second regime is termed "bubble growth" and characterizes devolatilizing conditions where superheat has dropped and/or viscosity has increased to the point where bubble initiation and growth now become the rate-controlling mechanisms. Foam expansion and entrainment are no longer process limitations. Since volatile removal is slower with this regime, cool-down is reduced (even in the absence of external energy inputs). Melt viscosity will still increase due to further volatile depletion even under isothermal conditions.

With the third regime, "diffusion control," superheat is now at a low value due to depletion of volatiles and/or temperature drop. As a result, there is little or no new bubble formation, existing bubbles grow very slowly, and their coalescence and disengagement virtually cease. The rate of volatile loss is now relatively slow, being primarily controlled by molecular diffusion at the melt–vapor interface in the devolatilization chamber. Conditions will therefore be nearly isothermal in the absence of thermal or mechanical energy inputs. Surface regeneration can accelerate devolatilization in this regime, but an FSD inherently has only limited capability here. Thus, with a devolatilization dominated by this regime, such as where very low levels of residual volatiles are sought, equipment designs providing mechanical means for surface generation are likely to be preferred. These machines are discussed in detail elsewhere in this book.

C. Maximizing "Superheat"

It can be seen that the FSD operates most effectively where superheat is maintained as high as possible. There are several ways to accomplish this.

1.   Raise melt temperature so as to increase P_1. The practical extent of this will, of course, be limited by the overall thermal stability of the feed. The effects of excessive cross-linking, depolymerization, polymer cleavage, formation of low-molecular-weight polymers, and other undesirable reactions must be considered.

2.   Minimize evaporative cool-down. This helps limit the drop-off of P_1 to that brought about by depletion of volatiles. The key here lies with preheater design and, to a lesser extent, with providing additional heat exchange surfaces within the FSD chamber. We discuss these in Sections II.B and V.

3.   Add a lower boiling liquid and thereby increase P_1. Even relatively low concentrations of such an additive can be very effective with some systems. In addition, if FSD chamber pressure is not changed when such additives are used, the partial pressure of the higher boiling volatile vapor will be lowered by the presence of the inert vapor. This will lower the equilibrium concentration of the volatile in the melt and thereby increase the devolatilization driving force. Vrentas et al. (1985) point out that if the additive diffuses faster than the volatiles in the melt, the free volume of the system is increased, thus facilitating removal of those volatiles. Volatile additives are discussed further in Section II.D.

4.   Reduce FSD chamber pressure, P_∞. Economic factors will usually set the lower practical limits of chamber pressure. These include (1) higher vacuum equipment capital, operating and maintenance costs, (2) higher costs related to handling larger volumes of vapor, and (3) the possible need to install or increase refrigeration capacity to ensure adequate condensation of

the devolatilized vapors. Of course, where P_1 is already very low, a further reduction of P_∞ will effect only a small increase in superheat. However, the reduction of P_∞ will favor bubble growth as well as diffusion, and these will enhance devolatililization.

D. Operating Guidelines

When considering the devolatilization of a given polymer melt in an FSD, among the first questions to be decided are those relating to FSD temperature and pressure, and whether any reasonable combination of these parameters might achieve the targeted residual volatile levels in the effluent. A useful relationship for this purpose is the Flory– Huggins equation (Flory, 1953), presented in Chapter 2 and shown here for the case where there is a single-component dissolved volatile:

$$\frac{P_1}{P_1^0} = \phi_1 \exp[\phi_2(1 + \chi\phi_2)] \tag{2}$$

where P_1^0 is the vapor pressure of the pure volatile at its temperature in the FSD chamber, ϕ_1 and ϕ_2 are respectively the volume fraction of the volatile and polymer in the melt, and χ is the Flory–Huggins interaction parameter for the volatile–polymer system.

Where the volatile concentration is low, ϕ_2 approaches unity and Eq. (2) simplifies to

$$\frac{P_1}{P_1^0} = \phi_1 \exp(1 + \chi) \tag{3}$$

Rearranging Eq. (3) and shifting to a weight fraction basis, we obtain

$$W_1 = \frac{P_1}{P_1^0} \frac{\rho_1}{\rho_2} \exp[-(1 + \chi)] \tag{4}$$

where W_1 is the weight fraction of the volatile in the melt, while ρ_1 and ρ_2 are respectively the densities of the pure volatile and polymer at the devolatilization temperature.

Using Eq. (4), it is possible to estimate the combinations of FSD temperature and pressure to obtain a desired weight fraction of residual volatile under equilibrium conditions. Besides χ, it is necessary to know vapor pressure versus temperature for the pure volatile, and density versus temperature for the volatile as well as for the polymer. In general P_1 is taken to be equivalent to FSD chamber pressure. Errors with Eq. (4) can arise (1) if there is significant departure from equilibrium conditions, (2) where there is uncertainty in melt temperature measurement, and (3) where the chamber contains quantities of noncondensable gases originating, for example, from air leakage. Errors from the latter normally become significant only with relatively low FSD chamber pressures. As indicated earlier, Eq. (4) assumes a single-component volatile. The use of the Flory–Huggins equation with multicomponent volatiles is discussed by Meister and Platt (1989). Figure 1 is a plot of dissolved volatile concentration in the melt versus saturation pressure of the volatile for the styrene–polystyrene system at selected temperatures ($\chi = 0.33$).

Once a pressure–temperature operating "window" can be defined for the desired effluent residual volatile level, say, via the Flory–Huggins equation, the next recommended step is to specify a maximum melt temperature while still avoiding excessive unreacted monomer polymerization, thermal degradation, undesirable side reactions, and cross-linking. With this melt temperature specified, the required FSD chamber pressure can be estimated. Here, too, there are numerous constraints, particularly as chamber pressure is lowered, usually related to capital and operating costs of the vacuum system and any refrigeration that might be needed. On top of this dropping FSD chamber pressure too far can lead to excessive melt foam expansion and entrainment into the vapor lines because of high volumetric vapor flow rates.

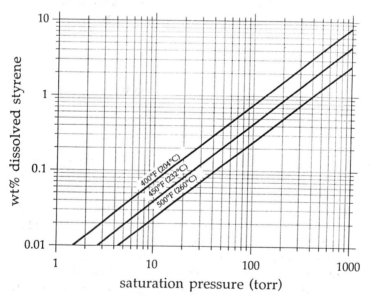

Fig. 1 Styrene concentration in polystyrene and resultant saturation pressure, P_1 (Flory–Huggins equation, $\chi = 0.33$).

In the course of optimizing FSD operating conditions, it may seem desirable to lower temperature. Care should be taken with this, however. Besides requiring lower pressure to achieve equivalent residual volatiles, lowering temperature has the undesirable effect of increasing melt viscosity and promoting the transition to less efficient devolatilization regimes.

It is possible that there may be no practical combination of FSD temperature and pressure to achieve the desired lowering of residual volatiles in a single step. In such cases, a multistage devolatilization system is called for. This will be discussed in Section III.

E. Materials of Construction

For many applications, including polystyrene and rubber-modified polystyrene, carbon steel has frequently been a satisfactory FSD material. With some polymers, stainless steel and sometimes aluminum have been preferred to prevent black specks, product discoloration, or other corrosion-related problems. It is usually a safe bet to use stainless steel or stainless steel–clad material unless it is known that particularly corrosive agents (e.g., HCl) might be produced at devolatilization conditions. When considering a commercial installation where minimum purchase cost is targeted and a novel polymer is being made, a suitable testing program with different candidate materials is strongly recommended.

II. BASIC SINGLE-STAGE DESIGN AND OPERATION

A. Preheater

Many of the principal elements of an FSD that could be used with the free-boiling or bubble growth regimes are shown schematically in Fig. 2. The feed is generally the pumped effluent from the final continuous reactor stage and consists of polymer along with a mixture of unreacted monomers, oligomers, solvents, and/or additives. Undissolved liquids and gases may also be present. This feed passes through a preheater to raise its temperature to a desired level, as discussed in Section I.D. The preheater is typically a single-pass shell-and-tube heat exchanger with the heat transfer fluid, very often hot circulating oil or water, on the shell side. Steam, sometimes at high pressure, is also occasionally employed. This heat exchanger design is simple, and its operation is generally self-correcting. Should flow diminish in one of the tubes due to a perturbation, the melt temperature will rise in the tube due to increased average residence time, causing viscosity to drop relative to the melt in the other parallel preheater tubes. This viscosity drop will tend to reestablish proper melt flow in the tube. The preheater is vital to successful FSD performance and is discussed further in Section VI.

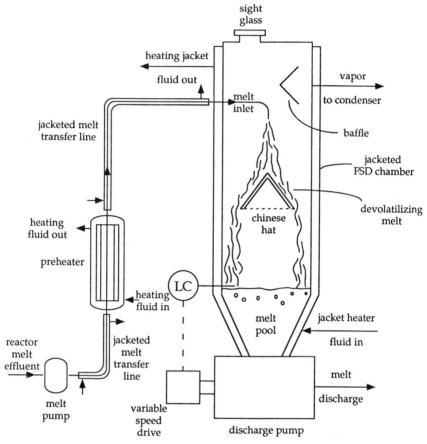

Fig. 2 Principal elements of an FSD.

B. FSD Chamber

In Fig. 2 the preheated melt passes into the FSD chamber through a line
that is usually jacketed and open at the downstream end. In a few instances
a throttle valve (not shown) may be placed in the line downstream of the
preheater if there is a need to prevent vapor flashing in that unit. The
devolatilized vapor passes to a condensing system, described in Section IV,
which connects to a vacuum source. The pressure in the FSD chamber is
controlled via the latter and set according to principles discussed in Section
I.D.

Although the melt to the FSD chamber generally originates from a polymerization line and passes through a preheater as just described, in some applications the FSD melt feed may be the output of a single-screw extruder (Craig, 1990). In these instances there is generally no need for a preheater.

The FSD chamber is normally jacketed for circulating heat transfer fluid or steam, or else electrically heated. This is primarily to achieve the required surface temperatures for startup, and later to make up for heat losses to the ambient.

Under free-boiling conditions, the preheated effluent from the melt inlet line is usually an expanding foam made up of rapidly coalescing and bursting vapor bubbles. Some FSD designs may incorporate a baffle, as shown, to help prevent foam gobs from being entrained into the vapor outlet line.

If feed and FSD conditions promote the bubble growth or diffusion control regimes, melt foaming on exiting the melt inlet line will be limited or absent. Under these circumstances it is advantageous to have the melt feed enter the FSD chamber through a spinneret, a showerhead distributor, or narrow slots rather than the plain open pipe shown in Fig. 2. The melt thus emerges as individual falling strands or thin sheets. Their relatively high surface-to-volume ratios promote flashing of volatile feed components, and there will be some desirable surface generation as they elongate or break up while falling.

Some FSD designs may feature a melt spreading surface, as shown, below the inlet line. It tends to prolong the exposure time to the chamber vacuum. The devolatilizing melt then drops from there into the melt pool maintained at the bottom of the FSD chamber. Although these surfaces, as discussed in Section V.B, are sometimes complex in shape and often internally heated with steam or circulating oil, very frequently they are just "Chinese hat" conical devices, as indicated in Fig. 2. They can be effective especially where melt viscosity is relatively low. With increasing viscosity, say, above 10^4 poise (10^3 Pa sec) the melt layer on the surface becomes thicker, diminishing heat transfer from the circulating oil, if employed, and otherwise interfering with volatile removal. Melt buildup, as would occur on a Chinese hat, is discussed by Bird et al. (1960).

The melt pool should be sized for minimum residence time consistent with stable operation. This means it should at least provide sufficient fluid head to avoid cavitation at the melt pump. Wherever possible, it is good practice to have enough extra room in the FSD chamber to enlarge the melt pool and thereby provide some surge capacity. In this way, short-term upsets upstream or downstream in the production line need not force a line shutdown. The melt pool level is usually controlled by an automatic system, illustrated in Fig. 2 based on a capacitance probe, nuclear gauge, or other level detecting device, operating through the variable-speed drive of the discharge pump. Often, FSDs can be equipped with a sight glass, as shown, to permit observation of vacuum chamber conditions or to assist in manual control of the melt pool level if needed.

Unless there is very active boiling in the melt pool accompanied by low two-phase fluid density and upwelling, which allow rapid transport of bubbles to the surface, any devolatilization taking place there is likely to be relatively minor. If the melt pool is operating in the bubble growth or diffusion-controlled regimes, such rapid fluid movement is not likely, and the actual band where there is active transport of volatiles to the melt pool surface can be quite shallow, often well under 1 mm in thickness. Since fresh material is continuously pouring over this surface band from above, the average lifetime of a fluid element within it is short, typically less than a second. Any bubble in the vicinity of the active band will tend to be "buried" in the melt pool more rapidly than it could normally rise, unless melt viscosity is comparatively low and bubble diameter is large. Once a bubble starts sinking this way in the melt pool, it will contract under the increasing head of melt and may disappear even before entering the discharge pump. This effect becomes more pronounced as chamber pressure is reduced.

Where conditions favor free boiling in the melt pool, heat transfer from the jacket can help reduce temperature drops due to sensible heat losses from vaporization and thereby increase devolatilization there. With some units, heating coils are installed in the melt pool to provide additional heat transfer. However, heat transfer coefficients are likely to be low unless there is very active melt movement from boiling. Thus, wall and coil temperatures may have to be significantly hotter than the melt pool to achieve a satisfactory level of heat transfer. This method of heat transfer to the melt pool is therefore limited to systems essentially free of cross-linking or thermal degradation effects at those elevated wall temperatures.

C. Melt Discharge Pump

Gear pumps with large inlet ports and designed for vacuum operation are generally used for melt discharge because of their high pumping efficiency, reliability, and controllability. Where viscosity is very high and an especially large inlet port is needed to ensure proper feeding, a screw pump may be preferred. Because of the lower efficiency of a screw pump, it will have a greater tendency to heat the melt. This may be useful if there is much evaporative cool-down in the FSD chamber or if further melt heating is needed, say for a second devolatilization stage, as will be discussed in Section III.

D. Volatile Additives

As mentioned in Section I.C, volatile additives are sometimes employed to improve devolatilization performance. Water is frequently the choice for this

(Nauman et al., 1972), although methanol (Fujimoto, 1976) and other organic liquids have been used with some systems. Gases such as nitrogen have also been employed (Hinds, 1972). The benefits of using a volatile additive must be balanced against potential problems with recovery of the additive and possible chemical effects. Moreover, for this technique to be most effective, the additive must be dissolved in the melt or, if not soluble, finely dispersed therein. This is very frequently difficult to accomplish because the volatile additive levels needed are generally quite low, often around 1% or less, and melt viscosity is likely to be several orders of magnitude greater than that of the additive. With the latter condition in particular, extensional flow and flow reorientation rather than shearing are likely to be the key to breakup of the dispersed component. The success of this technique will hinge, therefore, on the design choice and operation of the in-line mixer. Craig (1990) used a 12-element Sulzer SMX in-line motionless mixer to blend a polystyrene melt containing about 4000 ppm volatiles with 1% water, with the mixture then fed at 245°C to an FSD maintained at 50 torr (6.7 kPa). Product volatiles were reported to be about 210 ppm versus about 2200 ppm where the in-line mixer was not used. Pressure drop data indicated that some foaming developed in the mixer.

A mechanically agitated in-line mixer may be preferable to a motionless type, particularly where the volatile additive is not fully soluble in the melt.

E. FSD Performance

The patent literature occasionally provides FSD performance data that may be useful with specific units. Elsewhere in the literature, data are sparse. Meister and Platt (1989) published detailed performance data from a single-stage commercial unit operating with polystyrene melts. Unfortunately, they provided no information on the design and dimensions of the FSD.

III. MULTISTAGE DEVOLATILIZERS

Here, as shown schematically in Fig. 3, the effluent melt from an initial FSD is pumped through a second preheater and then into a second FSD chamber, almost always maintained at a lower pressure than the first. Most multistage FSD systems employ two stages, but occasionally three may be used (McCurdy and Jarvis, 1983). An early example of a two-stage FSD system is found with a patent by Amos and Carroll (1960) in connection with a continuous styrene–acrylonitrile copolymerization process.

As pointed out in Section I.D, single-stage units are often simply unable to achieve the required volatile reduction. Particularly under initially free-boiling conditions, for example, evaporative cool-down will often slow

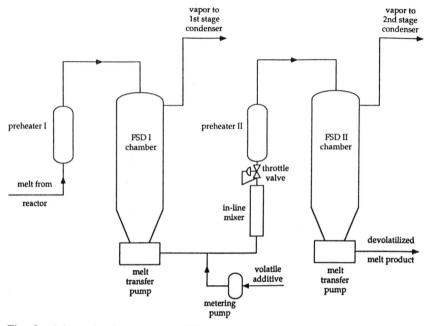

Fig. 3   Schematic of a two-stage FSD.

the devolatilization rate excessively. Reheating in the melt pool, as described in Section II.B, is often impractical. It is frequently preferable, therefore, to employ a second FSD stage.

Often where it is possible to achieve a targeted total volatile reduction with a single FSD stage, very low chamber pressure may be required. If the devolatilization load is high, the FSD chamber and the vapor lines must be sized for high volumetric flow rates so as to avoid excessive foam expansion as well as excessive pressure drop and possible foam entrainment in the vapor line. Also, as discussed in Section IV, refrigeration of the coolant in the vapor condensation system may be necessary, depending on condensate volatility. It may be possible, however, to raise operating pressure considerably with only a modest sacrifice in devolatilizing performance. The higher FSD pressure will reduce the volumetric vapor flow rate, making possible smaller-diameter vapor lines and fewer problems from foaming. In addition, the need for first-stage refrigeration may be avoided. Although the second devolatilization stage may still have to be operated at very low pressure to achieve the desired residual volatile level, any refrigeration need there would be greatly reduced due to the smaller vapor load. The savings accrued from reduced refrigeration requirements and smaller vapor handling lines will often more than compensate for the cost of the additional stage. In addition, since the total devolatilization requirement is handled by two stages rather than by one, each of these stages can be more optimally designed and operated. A two-stage unit can therefore be expected to have greater capacity and flexibility.

It is possible to incorporate features into a two-stage design to further promote flexibility and improve capacity. For example, Fig. 3 also shows a volatile additive stream entering the interstage melt transfer line upstream of an in-line mixer. Where there may be both low and high boiling volatiles to be removed (e.g., acrylonitrile and styrene), the first-stage feed may require little or no preheating for removal of most of the low boiler. The first-stage effluent can then be heated as required for removal of the high boiler in the second stage. This two-stage strategy can significantly reduce high temperature exposure of the devolatilizing melt and promote improved product properties.

IV. VAPOR CONDENSATION SYSTEMS

The design of vapor condensation systems with devolatilizers must generally take the following into account: vacuum operation, superheated vapor feed, and a low pressure drop requirement. The last becomes increasingly critical as FSD chamber pressure is lowered. Figure 4 shows schematically a frequently employed design for such an application.

It is first necessary to desuperheat the incoming vapor in order to maximize the capacity of the condenser. This is accomplished here by passing the vapor feed into a desuperheating chamber, where it is contacted by a spray of recirculating condensate. The now saturated vapor passes next to a condenser, normally of the single-pass shell-and-tube type and designed for rapid condensation and subcooling. In some designs there is no separate desuperheating chamber. Instead, the recirculating condensate is sprayed into the inlet vapor header of the condenser. The condensate drops into a level-controlled pool at the bottom of the condensate receiver. A line from the receiver connects to the vacuum system. Commercial units generally employ multistage steam jets, although mechanical vacuum pumps may also be used. A pump beneath the receiver circulates a constant stream of condensate back to the desuperheater. This recirculating condensate also provides a flushing and solvent action, which helps prevent buildup of plugs and surface deposits should polymerizable monomers be present. Excess condensate accumulating in the receiver is continuously bled off through a control valve for one or more of the following purposes: direct recycling, further treatment such as fractionation or chemical purification, or to be purged from the system.

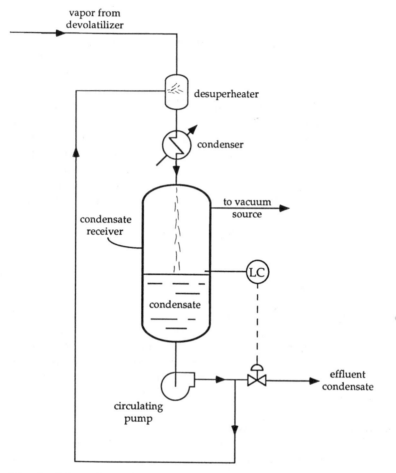

Fig. 4 Schematic of a vapor condensation system.

Where FSD chamber pressure must be very low, a direct contact condenser, also functioning as a desuperheater, may be used instead to minimize pressure drop. Figure 5 shows such an arrangement.

To minimize costs, the preferred coolant for the condenser is process water. However, if condensate vapor pressure is high at temperatures attained this way, vapor losses to the vacuum system may be excessive in terms of raw material costs and environmental problems. Should this be the case, refrigerated coolant is generally required.

McCurdy and Jarvis (1983) describe a method to avoid refrigeration

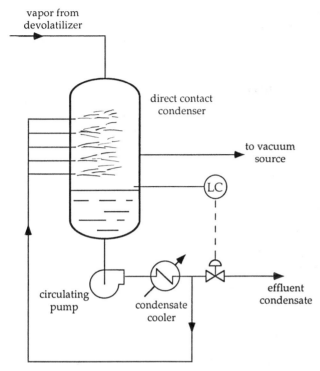

Fig. 5   Schematic of a vapor condensation system for low pressure drop.

where it might be needed beyond the first stage of a multistage FSD system. Assuming the first stage can be operated with a nonrefrigerated condenser coolant, the second and subsequent stages are operated without condensers. With a two-stage system, for example, the second-stage effluent vapor, at lower pressure than that from the first, is drawn directly into a vacuum booster pump, from which it is discharged at a pressure equal to the first stage. This recompressed vapor joins the first-stage effluent vapor, and they pass on together to the condenser.

V. OPERATIONAL DESIGNS

As with most other types of processing equipment, FSDs have evolved considerably since they were first developed in the 1940s. However, the published literature has largely been confined to patents, which are, of course, inherently descriptive in nature and only rarely provide quantitative scale-up criteria or useful design correlations. Where that type of information has been developed, it has largely remained proprietary. We will discuss here some design features that have been utilized in practice and some underlying reasons for them.

A. Single-Stage with Multiorifice Melt Manifold

Stober and Amos (1950) took out one of the earliest patents clearly describing an FSD. It was part of a line for the continuous mass polymerization of styrene, and the chamber is shown in cross section in Fig. 6. The preheated polystyrene melt feed at 255°C contains 4.5–5 wt%

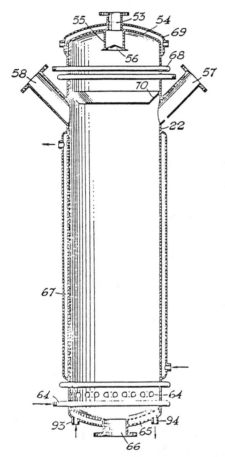

Fig. 6 Cross section of FSD chamber. (From Stober and Amos, 1950).

volatiles (principally styrene monomer). The feed flows at about 9.1 kg/hour (20 lb/hour) through inlet (53), passes through a header (55) and then through a multiple-orifice plate (56), from which it drops as individual strands though the FSD chamber maintained at a pressure of 4 torr (0.53 Kpa). The chamber is 2.44 m (8 ft) tall with an inside diameter of 0.61 m (2 ft), and has three jacketed sections (65, 67, 69) for hot oil circulation. Figure 6 also shows heating coils (64) for the melt pool. Two large-diameter lines are shown attached at the upper third of the FSD. The first (57) is an inert gas inlet. Its practical value is uncertain, but it may have been intended for purging the FSD or perhaps to dilute solvent or monomer vapors, as discussed in Section II.D. The second (58) is the vapor outlet line connecting to the vacuum system. A truncated conical baffle (70) is positioned as shown to help prevent the falling melt strands from being deflected into the vapor outlet line. A pump, not shown, withdraws melt at 255°C from the pool at the bottom of the FSD chamber through outlet (66) and passes it through a strand die, following which the strands are cooled and pelletized. The unidentified residual volatiles content of the product is 0.2 wt %.

This design is still often employed with relatively high-solids melts containing fairly low concentrations of polymerizable monomers, and where the initial operating regime is either bubble growth or diffusion control.

B. Heated Melt Spreading Surfaces

A patent to Kimoto and Yamagisawa (1972) describes an FSD with a melt spreading surface heated with circulating oil. The jacketed FSD chamber is shown in cross section in Fig. 7. Provided that melt viscosity is not too high, the design provides a reportedly effective melt spreading surface as discussed in Section II.B. The hemitoroidal melt inlet manifold has a number of orifices on its flat bottom side, directing the falling melt strands onto the melt spreading surface, which has a similar annular cross section.

In an example where a polystyrene melt feed at 250–260°C containing 10 wt % ethylbenzene and 8 wt % styrene is fed to the FSD tank held at 30–60 torr (4–8 kPa), the effluent melt residual volatiles were reduced to 0.1–0.3 wt%. The residual volatiles increased to 2–4 wt% when the heated melt spreading surface was removed.

It is possible in principle to incorporate the preheating and melt spreading surface functions into a single unit. Hon (1989) describes such an embodiment shown in Fig. 8. Here the vertical preheater tube (4) is divided longitudinally by a pair of crossed vanes (32) welded to one another and to the inside wall of the tube. The tube wall and vanes are heated by conduction from band heater (44) acting through metal jacket members (40, 42). The melt enters through the top, and it coats the heated surface of the vanes and

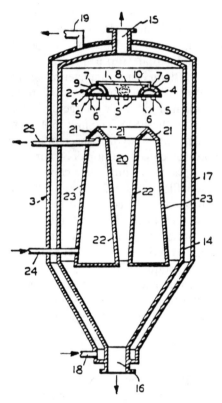

Fig. 7   FSD with heated, melt spreading surface. (From Kimoto and Yamagisawa, 1972).

inside tube wall as it flows downward. At the bottom of the tube, the devolatilized melt drops into the receiver (24) under vacuum, from which it is discharged via pump (50). The capacity of this design is limited, and it appears intended for operation in the bubble growth or diffusion-controlled regimes. It could be effectively used, however, with a bench-scale or pilot plant system.

C.   FSD with Top-Mounted Preheater and Melt Recirculation

The preheater configurations shown in Figs. 2 and 3, where the melt line between the preheater and the FSD chamber might be long, can lead to serious problems. Since the preheater generally raises melt temperature substantially to promote devolatilization, it is possible to have significant

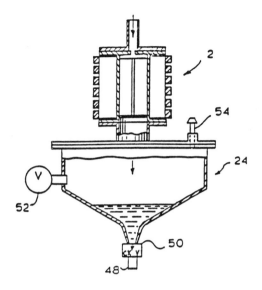

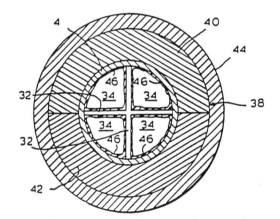

Fig. 8 Preheater with integral melt spreading surfaces. (From Hon, 1989.)

additional polymerization take place in that melt line, especially if conversion is not very high or if conditions favor the gel effect. These conditions can also promote formation of oligomers and low-molecular-weight polymers, often very deleterious to product properties. To avoid the problem, designs have been developed where the preheater is mounted on top of the FSD

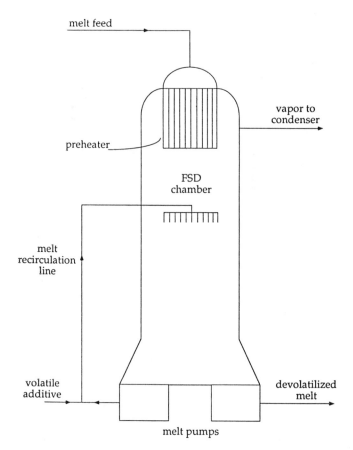

melt feed

vapor to
condenser

preheater

FSD
chamber

melt
recirculation
line

volatile
additive

devolatilized
melt

melt pumps

Fig. 9 FSD with top-mounted preheater and melt recirculation. (From Hinds, 1972.)

chamber and its tubes open directly into the chamber. An example is shown in Fig. 9 (Hinds, 1972).

The tubes of this type of preheater may terminate in small orifices, which reduce flashing upstream and form the exiting melt into falling strands. In general, however, the tube diameter is not changed at the outlet, and the resulting upstream flashing can provide important benefits. The two-phase flow in the preheater tubes reduces average preheater residence time, and it can bring about a significant increase in heat transfer so that the melt cool-down that would normally result from the vapor flashing can be greatly reduced or even eliminated. This more nearly isothermal feature is capable of markedly increasing devolatilization.

The melt recirculation feature illustrated in Fig. 9 allows a portion of the total flow to be exposed to a second fall through the FSD chamber via a showerhead type of melt manifold. Also included are means to introduce a low boiling volatile additive into the melt recirculation stream to promote further devolatilization. However, as discussed earlier, this is not likely to be effective in the absence of a suitable mixer incorporated into the recirculation line. In other embodiments, the recirculated melt also passes through a preheater.

D. FSD with Stacked Stages

With two-stage FSDs it may be desirable to mount the first stage directly on top of the second. This configuration, shown schematically in Fig. 10

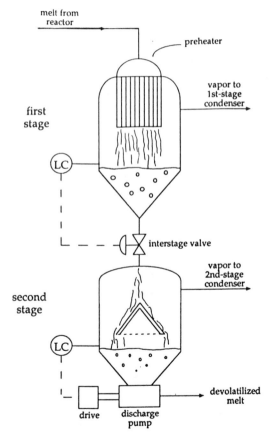

Fig. 10 Schematic of two-stage "stacked" FSD. (From Hagberg, 1976.)

(Hagberg, 1976), in addition to requiring less floor area, virtually eliminates the interstage melt line and thereby shortens total average residence time. Similar to the unit shown in Fig. 9, the first-stage preheater is mounted directly atop its FSD chamber. Melt from the first stage flows by gravity into the second-stage chamber through an interstage valve controlled by the melt pool liquid level detector. The first-stage preheater has vapor flashing within the tubes with reportedly very little melt temperature drop. In this configuration no interstage preheater is considered necessary. The single-pass preheater tubes described as 0.91 m (3 ft) long, with 23 mm (0.9 in.) inside diameter. When devolatilizing a polystyrene melt, the throughput per tube is in the range of 2.7 to 5.5 kg/hour (6 to 12 lb/hour).

If melt viscosity is very high, it may be difficult to achieve sufficient flow through the interstage valve consistently. Substituting a transfer pump for the valve can be an effective solution to the problem.

VI. PREHEATER DEVELOPMENTS

The importance of preheater design and operation to improved FSD performance cannot be overstated. Indeed, patents in recent years related to the FSD have to a considerable extent focused on the preheater. Discussed here are some of the developments in this area since the early 1970s.

A. Tubes Fitted with Motionless Mixer Elements

It has been found advantageous in a number of applications to have the preheater tubes fitted with Kenics or similar spiral motionless mixer elements (Hagberg, 1976; Chemineer, 1979). The radial mixing provided by these elements reduces temperature gradients and improves the inside film coefficient by as much as fivefold compared with conventional open tubes. In addition, plug flow conditions are more closely approximated, leading to a more uniform residence time distribution in the preheater. Equivalent heat transfer performance can therefore be achieved with lower tube wall temperatures, important for reducing polymer degradation. Although these mixer elements increase pressure drop per unit length in the tubes, experience has shown that flashing upstream of the tube exit, as described in Sections V.C and V.D, can still occur and promote near-isothermal vaporization.

B. Preheaters with Shell-Side Temperature Gradients

Metzinger et al. (1975) describe a vertical preheater with a horizontally baffled shell operating with a condensable vapor heating medium such as a biphenyl–biphenyl oxide mixture. Inert gas is metered into the vapor at certain points to bring about progressive reduction of the vapor concentration counter to the flow of the melt, thus achieving a desired temperature gradient in the descending melt.

This preheater type is reported to be of particular use where the melt contains residual monomers differing considerably in volatility (e.g., acrylonitrile and styrene), and where the low boiler is temperature-sensitive.

C. Preheaters with Narrow Longitudinal Melt Channels

The preheater described in a patent by Fink et al. (1979) is intended to minimize the temperature difference between the devolatilizing melt and the heat transfer surface, and to prevent melt cool-down as it flashes. Unlike the case with preheaters based on the conventional shell-and-tube heat exchanger, the melt here flows downward through a large number of shallow channels of rectangular cross section machined into metal blocks, which can be assembled on the unit construction principle. The blocks provide conductive heat transfer to the melt from any desired primary heat source such as hot circulating oil, steam, electrical heating rods, or the like.

Figure 11 provides an exploded view of this preheater design. In

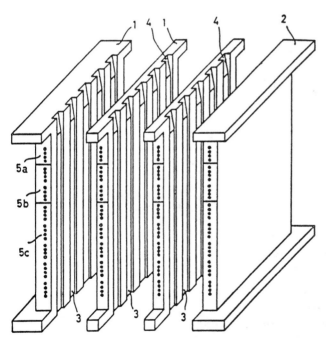

Fig. 11 Exploded view of elements of modular preheater with narrow vertical melt channels. (From Fink et al., 1979.)

practice, the blocks are firmly held together with tension bolts and screws so that the open channels (3) of any block (1) form, with the flat back of the adjacent block, sealed leakproof slit-shaped melt channels. Holes (5) drilled into the blocks at right angles to the melt channels carry the primary heat source. They can be vertically subdivided into groups (e.g., 5a, 5b, 5c), each operable at different temperatures. Although in principle the unit can be readily dismantled for maintenance, there are likely to be practical difficulties in cleaning and reassembling this particular design, and in keeping it leak-free over the long term.

The upstream end of each channel has a short recess (4) to ensure uniform distribution of melt from a feed header (not shown). The recommended channel depth is 0.5–4 mm to minimize temperature gradients across that dimension and thereby permit lower block temperatures. The pressure drop down each channel is intended to have flashing start at some intermediate point, with heat transfer from the block reducing any evaporative

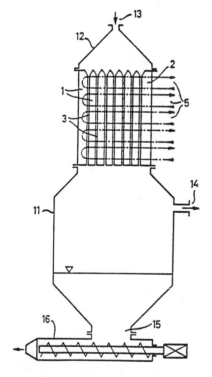

Fig. 12 FSD unit with preheater of Fig. 11. (From Fink et al., 1979.)

cool-down of the melt. It is possible to modify the pressure gradient down the channels by varying their width in some manner rather than maintaining it constant as shown in Fig. 11. The foaming melt exiting the channels enters directly into the FSD chamber.

Figure 12 is a schematic of this preheater with its melt header (12) mounted atop the FSD chamber (11) with its vapor outlet (14), melt pool (15), and discharge pump (16).

In an example, the preheater has a total of 108 channels, each 2 mm deep, 20 mm wide, and 1.1 m long. Heating zones are progressively set at 145, 185, 225, and 260°C. The feed, a styrene–acrylonitrile melt with 55% solids at 193 kg/hour, enters the preheater at 145°C at a pressure of 25 atm (2.53 MPa). It exits as foam at 246°C, entering the FSD chamber maintained at 18 torr (2.4 kPa). The volatile levels are shown in the following:

Wt%	Acrylonitrile	Styrene	Ethylbenzene
in feed	8.8	26.4	19.8
in product	trace	0.04–0.06	0.04–0.06

D. Preheaters with Short, Stacked, Horizontal Melt Channels

Figure 13 shows a devolatilizer with a preheater of this type (Aneja and Skilbeck, 1989). The FSD chamber (111) houses the cylindrical preheater with an inlet (122) for the melt feed, an outlet (123) for the devolatilized melt, a vapor outlet (124), and feed and discharge lines (170, 171) for the heat transfer fluid. The preheater is made up of a series of horizontally stacked plates (130), detailed in Fig. 14, which shows an exploded perspective view. The melt feed is directed downward through the annular zone (150), and then outward through a large number of relatively short radial channels (140) of shallow, rectangular cross section. The melt exits the channels as a foam and flows down through a second annular space between the preheater and the wall of the FSD chamber to the outlet (123). Heating oil circulates through the stacked plates via the inlet and outlet (170, 171), the conduits in parallel (161), and the headers (165, 166, 167).

An example cites a preheater with 7404 channels, 152 mm (6 in.) long, 25 mm (1 in.) wide and 2.5 mm (0.10 in.) deep. The plates are 2.5 mm (0.10 in.) thick, and the stack has an inside and outside diameter respectively of 381 and 686 mm (15 and 27 in.). The cylindrical spacer (200) within the central melt receiving zone has an outside diameter of 305 mm (12 in.). No feed rates are provided, unfortunately. Other embodiments feature stacks with a square rather than circular horizontal cross section.

This design appears to provide very rapid heat-up and short average

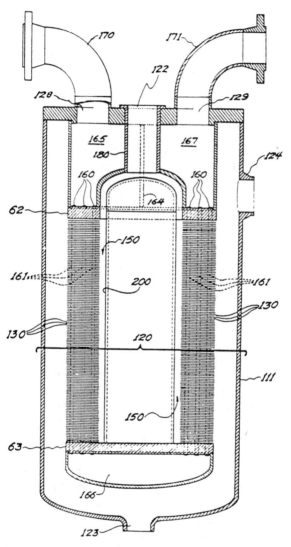

Fig. 13 FSD with preheater comprising short, flat, horizontal melt conduits. (From Aneja and Skilbeck, 1989.)

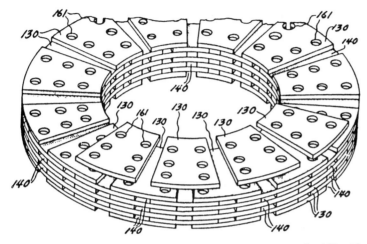

Fig. 14 Exploded perspective view of preheater plate stack of Fig. 13. (From Aneja and Skilbeck, 1989.)

FSD residence times, useful with heat-sensitive materials. However, it would not be expected to have high devolatilization capability. Compared with the more conventional shell-and-tube preheaters, the construction here is more complex, and maintainence is likely to be more difficult. A preheater patented by Mattiussi et al. (1992) has similar design features.

VII. COMPARISONS WITH MECHANICALLY POWERED DESIGNS

It may be useful at this point to list some of the pros and cons of the FSD when compared with mechanically powered designs such as are discussed elsewhere in this book.

A. Advantages

1. Cheaper to build since there are essentially no moving parts and key fabrication steps more generally involve welding rather than machining.
2. Thermal energy is introduced via heat exchange equipment rather than as the result of the degradation of mechanical energy. This is usually cheaper.
3. Designs can easily provide surge capacity, often of practical importance in a continuous process.
4. Simpler operation, less prone to leaks because there are few if any moving seals. Operating problems generally easier to diagnose and

correct. Direct observation of the vacuum chamber operation often possible via sight glasses.

5. Particularly useful when dealing with shear-sensitive polymers.
6. Can be especially well adapted for devolatilizations dominated by the free-boiling regime.

B. Disadvantages

1. Average residence times tend to be longer, and a long residence time "tail" is also more likely. May therefore be less suitable for polymers with reduced thermal stability, and peak operating temperatures may have to be lower. Careful design can minimize these effects, however.
2. More limited at very high viscosities since flow is largely via gravity.
3. Ultimate residual volatiles are likely to be higher.
4. Less adaptable for blending in additives since there are no moving parts.
5. In the absence of moving parts, disengagement of vapor bubbles may be less efficient.
6. Since it has little "self-cleaning" capability, the FSD may be less suitable with polymers that easily cross-link or otherwise form insoluble wall deposits.

VIII. FUTURE TRENDS

As seen from the patents discussed in Section VI, emphasis can be expected to continue in preheater design aimed at minimizing melt exposure to high temperatures, reducing average residence time, and promoting in situ vaporization so as to minimize evaporative cool-down. This will involve further development in melt channel shape and configuration, and innovation in the zonation and circulation of the heating medium. Further design developments to simplify the assembly and maintenance of this type of equipment will also probably occur. We should also expect further development of in-line mixing equipment for volatile additives.

There will be further recognition that, in many instances, a properly, designed and operated FSD can provide performance comparable with or better than many mechanically powered units, and at lower installation and operating costs. The particular suitability of an FSD for operating in the free-boiling regime may prompt further development of hybrid multistage devolatillizers where very low residual levels must be achieved from volatile-rich feed. Here a mechanically powered devolatilizer such as a wiped-film unit or twin-screw vented extruder can function very advantageously as a second stage. This combination should often constitute a very effective, low-cost option. In some instances, the hybrid feature may be merely a mechanical mixing device in the melt pool of the second-stage FSD chamber enhancing devolatilization by generating fresh surface, thereby improving volatile diffusion as well as bubble disengagement. It can also provide a way to blend additives effectively into the melt. An apparatus of this type is described in a patent by Moore and Wessel (1990).

NOMENCLATURE

P_1 saturation pressure of volatile component(s) in the melt
P_1^0 vapor pressure of pure volatile component in the FSD chamber
P_∞ total pressure in the FSD chamber
SH   "superheat" $(P_1 - P_\infty)$
ϕ_1 volume fraction of the volatile in the melt
ϕ_2 volume fraction of the polymer in the melt
χ Flory–Huggins interaction parameter

REFERENCES

Amos, J. L., and Carroll, T. M. (1960). U.S. Patent 2,941,985.
Aneja, V. P., and Skilbeck, J. P. (1989). U.S. Patent 4,808,262.
Bird, R. B., Stewart, W. E., and Lightfoot, E. N. (1960). *Transport Phenomena*, Wiley, New York, p. 121.
Chemineer, Inc. (1979). *Agitation Insights 12*, Chemineer Agitators, Dayton, Ohio.
Craig, T. O. (1990). Application of an enhanced flash-tank devolatilization system to a degassing extruder, *Adv. Polym. Tech.*, *10*: 323–325.
Fink, P., Wild, H., Zizlsperger, J., Reffert, R. W., and Thielen, G. (1979). U.S. Patent 4,153,501.
Flory, P. J. (1953). *Principles of Polymer Chemistry*, Cornell University Press, Ithaca, New York.
Fujimoto, S. (1976). U.S. Patent 3,987,235.
Hagberg, C. G. (1976). U.S. Patent 3,966,538.
Hinds, J. R. (1972). British Patent 1,269,936.
Hon, C. C. (1989). U.S. Patent 4,865,689.
Kimoto, K., and Yamagisawa, K. (1972). U.S. Patent 3,694,535.
Mattiussi, A., Buonerba, C., Balestri, F., Dall'Acqua, D., Matarrese, S., and Borghi, I. (1992). U.S. Patent 5,084,134.
McCurdy, J. L., and Jarvis, M. A. (1983). U.S. Patent 4,383,972.
Meister, B. J., and Platt, A. E. (1989). Evaluation of the performance of a commercial polystyrene devolatilizer, *Ind. Eng. Chem. Res.*, *28*: 1659–1664.
Metzinger, L., Gottschalk, K., and Schwab, J. (1975). U.S. Patent 3,865,672.
Moore, E. R., and Wessel, T. E. (1990). U.S. Patent 4,954,303.

Nauman, E. B., Szabo, T. T., Klosek, F. P., and Kaufman, S. (1972). U.S. Patent 3,668,161.

Simon, R. H. M. (1983). Flash evaporators and falling strand devolatilizers, *Devolatilization of Polymers* (J. A. Biesenberger, ed.), Hanser Publishers, New York, p. 35.

Stober, K. E., and Amos, J. L. (1950). U.S. Patent 2,530,409.

Vrentas, J. S., Duda, J. L., and Ling, H.-C. (1985). Enhancement of impurity removal from polymer films, *J. Appl. Polym. Sci.*, *30*: 4499–4516.

10

Slit Devolatilization

Giovanni Ianniruberto, Pier Luca Maffettone, and Gianni Astarita

University of Naples–Federico II, Naples, Italy

I. INTRODUCTION

Devolatilization of a polymer is typically a two-phase process that takes place when the vapor pressure of the component to be stripped is higher than its partial pressure in the gas phase. The difference between these two quantities represents the thermodynamical driving force for devolatilization, and the design of any stripping apparatus tries to enhance this driving force. Rather obviously, the simplest way to tackle this problem is to increase the volatile vapor pressure, by operating at a high temperature (compatibly with the polymer thermal degradation) and, simultaneously, to reduce the volatile partial pressure, by imposing a low pressure (often by pulling a vacuum) on the gas phase. An apparatus that essentially works by exploiting these design criteria is a flash tank. This operation maximizes the thermodynamical driving force, but it is very inefficient in another aspect of the devolatilization process: the mass transport in the polymeric phase. It is well known that the diffusivity of low-molecular-weight components in a polymeric solution can be orders of magnitude lower than in ordinary solutions. Furthermore, the polymeric phase is likely to be very viscous, so that mass transport is controlled by diffusion. Thus, flash operations would require very long residence times to overcome their scarce mass transport efficiency, but long residence times increase the thermal degradation of the polymer.

The mass transport efficiency can be improved in two ways: by increasing the interfacial area between the gas and the polymeric phase, and by inducing strong surface renewal (Denson, 1983). This latter expedient determines a rapid mixing of the polymeric solution, and thus, mass transfer by diffusion occurs at the favorable conditions of short exposure times.

An increase of the exposed surface is successfully achieved with the falling-strand devolatilizers. Experimental results have in fact shown that this process can obtain high purity, although it is effective only as a refinement stage, that is, when the feed concentration has already been reduced to few percent. As a matter of fact, bulk devolatilization is commonly performed with extruders where both the exposed area and the surface renewal are maximized. The fluid dynamics of this class of devices allows, with a proper screw design, the generation of large wiped films together with a continuous remixing of the polymeric solution and, hence, determines effective separations in a variety of operating conditions.

When dealing with bulk devolatilization, the latent heat of vaporization has to be continuously supplied to the liquid phase to limit the quenching due to adiabatic cooling. A simple ordering argument can be used to clarify this point. Let W_1 be the mass fraction of volatile, and ΔW_1 the mass fraction reduction needed in the devolatilization unit; in an efficient apparatus, ΔW_1 should be of the order of the feed mass fraction, $W_{1,0}$. The amount of heat (per unit time) corresponding to the endothermic devolatilization is $\tilde{Q} \lambda \Delta W_1$, where $\tilde{Q}$ is the mass flow rate fed to the unit, and λ represents the latent heat of vaporization per unit mass. Let ΔT_{int} be the temperature change that the polymeric phase undergoes in the devolatilizer. If no heat is supplied to the system, the solution cools down; the overall heat balance is given by

$c\,\Delta T_{int} = \lambda\,\Delta W_1$, with c being the specific heat per unit mass. Since the typical value of the ratio λ/c is about $200°C$, the adiabatic cooling is negligible for refinement stages where ΔW_1 is much smaller than unity, while it can be quite significant for the case of bulk devolatilization. In the latter case, heat must then be supplied to the system to avoid quenching of the devolatilization process.

In a rather crude analysis, the macroscopic heat balance on a non-adiabatic devolatilization unit involves the heat supplied to the system and the devolatilization latent heat; hence, the solution temperature may increase or decrease according to the relative magnitude of those two terms. Let ΔT_w be the imposed driving force for heat transfer. The heat supply is easily estimated by considering that the polymer flow is usually viscous, so that the heat transfer coefficient can be assumed to be of order k/l, where k represents the solution thermal conductivity and l is the characteristic dimension across which the heat transfer takes place. It is always possible to define an aspect ratio Θ such that the cross-sectional area S is given by Θl^2. Thus, the heat supplied per unit time over the devolatilization length Z is of order $\Theta Zk\,\Delta T_w$, and the overall heat balance is

$$\tilde{Q}c\,\Delta T_{int} = \lambda\,\Delta W_1\tilde{Q} - \Theta Zk\,\Delta T_w \tag{1}$$

By defining the Graetz number as $Gz = \tilde{Q}c/kZ$, and the Stefan number as $St = \lambda/(c\,\Delta T_w)$, Eq. (1) reads

$$St\,\Delta W_2 - \frac{\Delta T_{int}}{\Delta T_w} = \Theta\,\frac{1}{Gz} \tag{2}$$

To avoid the quenching of the devolatilization process, the operating conditions must guarantee that the second term on the left-hand side of Eq. (2) is negligible with respect to $St\,\Delta W_1$. This is certainly true if Gz is of order $\Theta/(St\,\Delta W_1)$. This conclusion gives a design criterion for a bulk devolatilization device with an optimal heat transfer.

A process designed to optimize the heat transfer phenomenon is the so-called multislit devolatilizer (EP, 1990; Maffettone et al., 1991; Ianniruberto et al., 1993), which is derived from the multislit cooler of Lynn and Oldershaw (1984). Actually, in this process the polymeric solution is fed to a huge number of heated slits, which discharge into a chamber where vacuum is imposed. The slit device can achieve efficient separations (Maffettone et al., 1991), and since the heat exchange here is very effective, the solution quickly reaches optimal devolatilization conditions. Thus, the polymer resides at high temperature for a short time, thus minimizing the thermal degradation of the polymer.

The residence time in a generic unit is given by $\tau = SZ/Q^*$, where Q^* is the volumetric flow rate. In terms of the Graetz number, τ can be written as $S/(\alpha\,Gz)$, where α is the thermal diffusivity. If the device is *properly* designed, then $Gz = \Theta/(St\,\Delta W_1)$, and the optimal residence time, τ_{opt}, is given by

$$\tau_{opt} = \frac{l^2}{\alpha}\,St\,\Delta W_1 \tag{3}$$

The slit geometry has an aspect ratio, Θ, much larger than any other conceivable geometry. This property implies that, for any given cross-sectional area S, the slit is characterized by the smallest l value, and therefore the residence time can be reduced to a very small value. The pressure drop in the slit, however, is proportional to $1/l^2$, and thus reducing l results in progressively larger pressure drops.

Actually, the pressure drop along the slit has an upper limit dictated by practical arguments. Now, to perform an *optimal* heat transfer, the volumetric flow rate has to be

$$Q^*_{opt} = \frac{wlZ}{\tau_{opt}} = \frac{\alpha w}{St\,\Delta W_1}\frac{Z}{l} \tag{4}$$

where w is the slit width. Let Δp_{max} be the maximum pressure drop allowed for the operation. The momentum balance over the slit for the optimal operating conditions roughly gives

$$\Delta p_{max} \propto \frac{ZQ^*_{opt}}{l^3} \tag{5}$$

Combining Eq. (4) with Eq. (5) gives the two geometrical parameters Z and l:

$$l \propto \frac{Q^*_{opt}}{\sqrt{\Delta p_{max}}}, \qquad Z \propto \frac{Q^{*2}_{opt}}{\sqrt{\Delta p_{max}}} \tag{6}$$

Equation (6) is the design equation for a single-slit devolatilizer. It can be seen that the slit length increases quadratically with the optimal flow rate, and thus a practical limit may easily occur for high values of Q^*_{opt}. Moreover, the slit height is proportional to Q^*_{opt}, and thus the residence time, Eq. (3), can become too large even for moderate values of the flow rate. Therefore, from a practical point of view, it is necessary to feed a small flow rate to the slit, and as a consequence, the real process needs to be carried out with a multislit unit, where a large number of slits work in parallel.

In this chapter, the modeling of the slit devolatilization process (Maffettone et al., 1991; Ianniruberto et al., 1993) is presented. Some experimental evidence, obtained in a single-slit unit, proves the effectiveness of the slit devolatilization process. These results motivate our modeling approach, which is based on an analysis of the three transport phenomena involved in any devolatilization process: momentum transfer, heat transfer, and mass transfer. The predictions of the model are then compared with the experimental results. Finally, the multislit device is considered; in particular, we focus on an instability phenomenon that typically occurs in a multislit unit.

II. MODELING OF THE SINGLE-SLIT UNIT

The behavior of a single slit has been experimentally studied in the typical process conditions at the Centro Ricerche Enimont Anic Bollate. The experimental setup is sketched in Fig. 1: a polymeric solution of ethylbenzene in polystyrene is fed from a stirred steel tank to an instrumented slit, which discharges into a chamber kept under vacuum. The slit walls are maintained at constant temperature by means of an oil bath. The temperature profile inside the slit is measured with five thermocouples distributed along the flow direction. The pressure is measured at the slit entrance and at two different positions along the slit. The residual ethylbenzene mass fraction is measured with a chromatographic technique by sampling the stream exiting from the slit.

No devolatilization occurs before the slit entrance because the pressure in that region is sufficiently high so that no driving force for devolatilization exists. For this reason, the initial part of the slit is used to heat the polymeric solution. This region is called *heating section*, while the rest of the slit, where devolatilization takes place, is referred to as *main section*. Needless to say, in optimal operating conditions the heating section should be a small fraction of the slit.

The experimental results (shown in detail in Section III) indicate that the slit device is able to perform a significant bulk devolatilization. Indeed, the ethylbenzene mass fraction can be reduced from 50% to a few thousand ppm for a solution flow rate of order 1 kg/hour (corresponding to a residence time of about 1 min).

Since the volumetric gas flow rate is orders of magnitude larger than the liquid flow rate over most of the slit, the removed gas phase cannot be thought to flow just as bubbles contained in the polymeric solution. Rather, the stripped volatile component forms a continuous gas region leading to the slit exit. For the sake of clarity, note that this consideration does not exclude the presence of bubbles inside the liquid phase; rather, it implies that the gas flow rate is only slightly affected by the gas flowing toward the slit exit as bubbles. Actually, since in the main section a driving force for boiling does exist, it is likely that gas bubbles form in the liquid phase, migrate, and eventually feed the gas region that has been progressively developing.

The modeling of the slit devolatilizer is developed in the classical

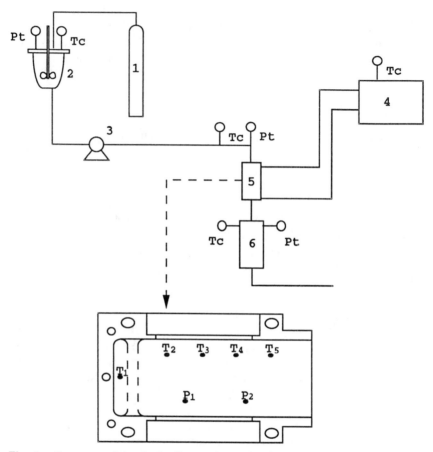

Fig. 1 Geometry of the single-slit experimental unit. 1. Nitrogen cylinder; 2. Tank; 3. Pump; 4. Thermostat; 5. Slit; 6. Vacuum chamber. In the lower part, the slit is sketched with indication of the location of pressure transducers and thermocouples. (From Maffettone et al., 1991, reproduced with permission of the American Institute of Chemical Engineers. © 1991 AIChE. All rights reserved.)

framework of transport phenomena. Devolatilization involves three inter-acting phenomena: mass transport, heat transport, and momentum transport. For the sake of simplicity, from now on we assume steady-state conditions.

Since the gas flows as a continuous phase, we assume that the fluid dynamics in the main section can be modeled as a stratified two-phase flow. Though a multitude of different configurations are, in principle, conceivable, we focus on a simple, two-dimensional, symmetric flow configuration. We in fact assume that a gas layer forms at the slit midplane, while two liquid layers flow touching the walls. This flow configuration is unstable to gravity, but since the experiments were conducted with a vertical slit, such an argument plays no role in this case. Also an asymmetric configuration, with a gas layer touching one of the walls and the liquid layer touching the other, could be considered as a simple idealization of the two-phase flow. However, the predictions of the model based on this flow configuration prove to be worse than those obtained with the assumption of a symmetric configuration (Maffettone et al., 1991). In view of the better agreement with experimental results, in the following we only present the symmetric case.

In the heating section no mass transfer occurs, and thus, only momentum and heat balance equations have to be considered. In this region of the slit, the solution temperature progressively increases and the pressure gradually decreases. When the pressure attains the local value of the volatile vapor pressure, a driving force for the mass transfer exists and devolatilization starts taking place. This point marks the transition from the heating to the main section. In the latter region, all transport phenomena occur, and three balance equations have to be written.

To simplify the analysis, the mass fraction is assumed to be uniform over the cross section; in view of our ignorance about the modeling of the mass transport phenomenon (compare Section II.C), there is no reason to assume a variable mass fraction over the cross section. Moreover, all the physical properties are considered uniform over the cross section, and they are evaluated at the cross-sectional mean temperature. This rather crude simplification is motivated by the large uncertainties regarding most of these quantities; any refinement would add only useless information.

A. Momentum Transfer

In the heating section, the polymer solution fills the whole slit cross section. The slit geometry and the coordinate system are illustrated in Fig. 2. The solution is fed at $z = 0$, and the discharge section is located at $z = L$. The x coordinate gives the vertical distance from the upper wall, with H being the slit thickness. The slit width is indicated as w. Since the slit aspect ratio w/H is much larger than unity, we approximate the slit with two parallel infinite plates. As a consequence, the two coordinates x, z are sufficient to describe the system.

For typical operating conditions the Reynolds number (Re $= Q/\mu_P$, where Q is the solution mass flow rate per unit width, and μ_P is the characteristic viscosity of the polymeric solution) is much smaller than unity, and thus, it is reasonable to assume viscous flow conditions. The flow can be considered hydrodynamically well developed since the Prandtl number

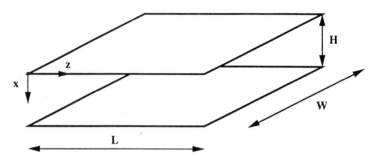

Fig. 2 Slit geometry and coordinate system. Flow takes place in the z direction. See text for details. (From Maffettone et al., 1991, reproduced with permission of the American Institute of Chemical Engineers. © 1991 AIChE. All rights reserved.)

($\text{Pr} = \mu_P c_P / k_P$, c_P and k_P being the specific heat per unit mass and the thermal conductivity of the polymeric solution, respectively) is significantly larger than unity for the polymeric solution used in the experiments; in other words, the velocity profile develops much faster than the temperature profile. With these assumptions, the momentum balance in the heating section is given by the well-known Hagen–Poiseuille equation:

$$-\frac{dp}{dz} = \frac{12\mu_P Q}{\rho_P H^3} \tag{7}$$

where p is the local pressure, and ρ_P is the density of the polymeric solution.

In the main section, the two-phase flow is modeled as a stratified flow. This flow condition has been widely studied (Bird et al., 1960; Yu and Sparrow, 1967; Charles and Redberger, 1962; Charles and Lilleleth, 1965; Navratil DiAndreth, 1984). The existing literature mainly refers to situations where nothing changes along the flow direction. On the contrary, in the case of slit devolatilization, the gas flow rate, zero at the beginning of the main section, progressively increases proceeding toward the slit exit. Moreover, the viscosity of the liquid phase may increase of some orders of magnitude as the mass fraction of the volatile component decreases due to devolatilization.

A significant simplification is obtained by making use of the lubrication approximation (Batchelor, 1967): the flow conditions at any axial position z are the same one would have in a steady flow through a channel of uniform geometry with the local values of flow rate, cross section, and physicochemical parameters. While of this approximation is clearly acceptable as far as the liquid flow is considered (since the Reynolds number of the polymeric phase is very small), its applicability to the gas flow has to be ascertained.

The Reynolds number relative to the gas layer is $Re_G = G/\mu_G$ where G is the gas mass flow rate per unit width and μ_G is the gas viscosity. Re_G attains its highest value at the slit exit, where it is still significantly less than 1000. Now, defining h as the gas layer thickness, it appears that the derivative dh/dz is invariably small, and thus, the condition for the lubrication approximation (i.e., $Re_G\, dh/dz \ll 1$) is satisfied in the gas phase as well.

Within the framework of the lubrication theory, the steady-state momentum balance in the main section reduces to

$$\frac{d\tau}{dx} = -J \tag{8}$$

where τ is the shear stress, and J is the axial pressure gradient. Equation (8) is readily integrated to give the triangular profile for the tangential stress τ:

$$\tau(x) = -J\left(x - \frac{H}{2}\right) \tag{9}$$

The integration constant has been obtained by imposing symmetry at the midplane position $x = H/2$.

Equation (9) holds true whatever the rheological behavior of the fluid might be. Now, to obtain the velocity profiles in both the polymer and gas phase, a constitutive assumption has to be made for both fluids. We regard both the polymer and the gas phase as Newtonian fluids. This constitutive choice has to be carefully motivated, at least for the polymeric phase. Since the single-slit flow rates are invariably small in typical process conditions, the shear rates in the slit never exceed the shear-thinning onset value for the system here considered (Mendelson, 1979, 1980); furthermore, viscosity data available in the literature do not encompass the whole volatile concentration range of interest. Thus, the Newtonian assumption seems to be the best we can do. In the evaluation of the transport properties of the polymeric solution, we neglect any possible influence of gas bubbles.

Within the assumption of Newtonian behavior, Eq. (9) reduces to the following set of two balance equations

$$\mu_P \frac{dv_P}{dx} = -J\left(x - \frac{H}{2}\right), \qquad 0 \leq x \leq \frac{H-h}{2} \tag{10}$$

$$\mu_G \frac{dv_G}{dx} = -J\left(x - \frac{H}{2}\right), \qquad \frac{H-h}{2} \leq x \leq \frac{H}{2} \tag{11}$$

where v_P and v_G are the velocities of the liquid and gas phase, respectively. Needless to say, symmetry with respect to the midplane is implied in deriving

Eqs. (10) and (11). By imposing the no-slip conditions at the wall, and at the interface between the two fluids, located at $x = (H - h)/2$, the foregoing equations readily give the polymer and gas velocity profiles

$$v_P = \frac{J}{2\mu_P} (Hx - x^2) \tag{12}$$

$$v_G = V + \frac{J}{2\mu_G} \left(Hx - x^2 - \frac{H^2 - h^2}{4} \right) \tag{13}$$

where V is the velocity of the gas–polymer interface, given by

$$V = \frac{J}{8\mu_P} (H^2 - h^2) \tag{14}$$

Integration of Eq. (12) between 0 and $(H - h)/2$, and of Eq. (13) between $(H - h)/2$ and $H/2$, gives the polymer and gas mass flow rates per unit width, P and G, respectively:

$$P = \frac{\rho_P J H^3}{12\mu_P} \left(1 + \frac{\delta^3}{2} - \frac{3}{2}\delta \right) \tag{15}$$

$$G = \frac{\rho_G J H^3}{12\mu_G} \left(1 + \frac{3}{2}\frac{\varepsilon}{\delta^2} - \frac{3}{2}\varepsilon \right) \tag{16}$$

where $\delta = h/H$, $\varepsilon = \mu_G/\mu_P$, and ρ_G is the gas density. As will be seen, δ is significantly smaller than unity, and ε is much smaller than δ^2. Hence, Eqns. (15) and (16) become

$$P \approx \frac{\rho_P J H^3}{12\mu_P} \tag{17}$$

$$G \approx \frac{\rho_G J h^3}{12\mu_G} \tag{18}$$

Dividing Eq. (18) by Eq. (17) side by side gives an order of magnitude estimate of the nondimensional gas layer thickness:

$$\delta^3 \approx \frac{\rho_P}{\rho_G} \frac{G}{P} \varepsilon \tag{19}$$

The gas layer thickness attains its largest value at the slit exit. There, in the typical operating conditions, the ratio ρ_P/ρ_G may be as large as 10^5, G/P is of order unity, and ε is of order 10^{-9}. Thus, Eq. (19) yields an estimate for δ as of order 0.1. On the other hand, close to the beginning of the main section, ρ_P/ρ_G is of order 10^3, and ε is of order 10^{-6}. One may thus conclude that, throughout the main section, $\varepsilon \ll \delta^2$ and $\delta \ll 1$, provided that G/P is larger than 10^{-6}. This latter condition is obviously violated in a very small region at the beginning of the main section.

In conclusion, the momentum balances in the polymer and gas phase, at the first order in δ, are, respectively

$$-\frac{dp}{dz} = \frac{12\mu_P P}{\rho_P H^3} \tag{20}$$

$$-\frac{dp}{dz} = \frac{12\mu_G G}{\rho_G h^3} \tag{21}$$

Equations (20) and (21) contain four physicochemical parameters: ρ_P, ρ_G, μ_P, μ_G. The polymer density is assumed to be constant and equal to 9.5×10^2 kg/m^3; the gas density is evaluated from the ideal gas law by assuming that the gas temperature is uniform over the gas layer thickness h, and equal to the temperature of the polymer at the gas interface. The polymer viscosity strongly depends on both temperature and concentration of volatile components; viscosity measurements carried out at the Centro Ricerche Enimont Anic Bollate gave results in good agreement with those presented by Mendelson (1979, 1980) for the same polymer–solvent system (Maffettone et al., 1991). Therefore, we have used the same interpolation law proposed by Mendelson, which reads:

$$\mu_P(T, W_2) = \mu_P^* M_P^{3.4} W_2^{4.2} \exp\left[\frac{E_{att}}{R}\left(\frac{1}{T} - \frac{1}{T_{rif}}\right)\right] \tag{22}$$

In Eqn. (22), T is the absolute temperature, T_{rif} is a reference temperature chosen to optimally fit experimental data, μ_P^* is the viscosity of the polymer at T_{rif}. M_P is the molecular weight of the polymer molecules, R is the gas constant, and W_2 is the polymer mass fraction. E_{att} is the activation energy, which depends on concentration. This function is evaluated by interpolating experimental data as explained by Maffettone et al. (1991). Finally, the viscosity of the gas phase, μ_G, is assumed to be a constant quantity, equal to 8×10^{-4} Pa sec.

B. Heat Transfer

The heat transfer is analyzed first considering the main section. In the following, we neglect the frictional heating term and heat conduction in the axial direction (as usual in a Graetz-like problem). With these simplifications, the heat transfer equation in the main section involves only three terms: one accounting for convection in the axial direction, one accounting for conduction in the transverse direction, and finally, a term accounting for the effect of devolatilization. The conduction term is $k_P \, \partial^2 T/\partial x^2$. The other two terms need some discussion.

First, we consider the effect of the devolatilization phenomenon. Let T^0 be the temperature at which the liquid-phase enthalpy is assigned the value zero. The enthalpy per unit mass of the polymeric phase at temperature T is $c_P(T - T^0)$; similarly, with c_G being the gas specific heat per unit mass, the enthalpy of the gas at temperature T_G is $\lambda + c_G(T_G - T^0)$, where λ is the enthalpy of vaporization per unit mass at the reference temperature, T^0. As a consequence, at temperature T the equilibrium enthalpy of vaporization is $\lambda + (c_G - c_P)(T - T^0)$. It should be remarked that the choice of the reference temperature T^0 is arbitrary, and we choose it to be the feed temperature T_0. Now, let $M(x,z)$ be the mass of volatile component that devolatilizes per unit solution volume and time. This quantity is equal to the local change of the solution flow rate along the axial direction, and it is then given by $-\rho_P \, \partial v_P/\partial z$. Hence, devolatilization determines a local rate of heat removal in the polymer phase equal to

$$-\rho_P(\partial v_P/\partial z)[\lambda + c_G(T_G - T_0) - c_P(T - T_0)]$$

The term representing convection in the axial direction can be written in the usual form: $\rho_P c_P v_P \, \partial T/\partial z$. Therefore, in the main section, the heat balance equation in the polymer phase takes the following form:

$$\rho_P c_P v_P \frac{\partial T}{\partial z} = k_P \frac{\partial^2 T}{\partial x^2} + [\lambda + c_G(T_G - T_0) - c_P(T - T_0)]\rho_P \frac{\partial v_P}{\partial z} \qquad (23)$$

Now, since the thickness of the gas phase is very small, it is reasonable to assume that no transverse temperature gradient develops in the gas phase. Therefore, the local average gas temperature is simply the interface temperature of the polymeric phase: $T_G(z) = T(H/2,z)$, where the interface has been considered as located at $x = H/2$ rather than at $x = (H - h)/2$, since $h/H \ll 1$.

The heat transfer equation in the heating section is obtained by slightly modifying Eq. (23). Obviously, in that region the devolatilization term cancels out, and v_P is determined by Eq. (12) with $x \in [0,H/2]$; thus, the heat balance equation in the heating section reduces to

$$\rho_P c_P v_P \frac{\partial T}{\partial z} = k_P \frac{\partial^2 T}{\partial x^2} \qquad (24)$$

As previously mentioned, all the parameters entering the model have been evaluated at the cross-sectional average temperature T^*, defined as the mixing-cup temperature of the polymeric phase:

$$T^*(z) = \frac{2\rho_P}{P} \int_0^{(H-h)/2} v_P(x) T(x,z) \, dx \tag{25}$$

The mixing-cup temperature in the heating section can be calculated with Eq. (25) by replacing P with Q, and by changing the upper limit of integration to $H/2$.

In the heat balance equation, the frictional heating contribution has been neglected. At low flow rates, the devolatilization process is very efficient, and it can be seen that the frictional heating is at most 0.1% of the latent heat of devolatilization. At larger flow rates, devolatilization is not so important, and the frictional heating has to be compared with the heat conduction term. In these conditions, however, the solution viscosity is relatively low since the polymeric solution is still rich in volatile components, and hence, frictional heating is again negligible.

Parameters appearing in the heat balance have to be evaluated. The gas specific heat is 1.841×10^3 m^2/sec^2,K (TPM, 1976); the polymer specific heat is 2.092×10^3 m^2/sec^2,K and its conductivity is 1.255×10^{-1} W/m,K (EPSE, 1985). Finally, the latent heat at the inlet temperature is determined by using the correlation proposed by Watson (1943).

C. Mass Transfer

Devolatilization is a very complex process that involves nucleation and growth of gas bubbles, their migration to the interface, and their release to the continuous gas phase. Modeling of such a complex phenomenon would be a formidable task. We therefore simply assume that, whatever the actual mechanism of the devolatilization process is, the rate of devolatilization (i.e., of release of volatile components to the gas phase) is proportional to the local driving force, which is the difference between the volatile vapor pressure, $P_1(T^*, W_2)$, and the actual gas-phase pressure p. With A being the proportionality parameter, the mass transfer equation is written as

$$-\frac{dP}{dz} = A[P_1(T^*, W_2) - p] \tag{26}$$

A is the only adjustable parameter of the model, and it is assumed to be constant. Its value is chosen to fit the experimental data as follows. The residual volatile mass fraction at the slit exit depends on the imposed flow rate: if one carries out experiments at very low flow rates, the residual mass fraction in the discharged stream is very low, and, in fact, close to the equilibrium value corresponding to the temperature and pressure conditions at the slit exit. In this situation, the slit is obviously overdesigned, and the value of the parameter A cannot be extracted from the data. As the flow rate is increased, one reaches a point where the exit stream residual mass fraction significantly exceeds the equilibrium value, that is, where the volatile component leaks out. In these conditions, the mass transfer becomes the rate-determining step, and the value of the parameter A has a large influence on the calculated exit mass fraction. The parameter A is then chosen to fit the measured residual mass fraction at the flow rate of incipient leak. The value of A is determined once and for all, and it is not changed with experimental conditions. Of course, much better data fits could have been obtained by allowing A to depend on operating conditions.

The volatile vapor pressure is evaluated as a function of temperature and polymer volumetric fraction, ϕ_2, with the Flory–Huggins equation (Flory, 1942)

$$\ln\left[\frac{P_1(T,\phi_2)}{P_1^\phi(T)}\right] = \ln(1 - \phi_2) + \left(1 - \frac{1}{DP}\right)\phi_2 + \chi\phi_2^2 \qquad (27)$$

where P_1^ϕ is the vapor pressure of the pure ethylbenzene, which depends on temperature, DP is the chain polymerization index, and χ is an interaction parameter. The function $P_1^\phi(T)$ has been determined by using the Antoine law (Reid et al., 1988). The predictions of the Flory–Huggins equation with $\chi = 0.4$ are in good agreement with those obtained by using the approach suggested by Holten-Anderson et al. (1986, 1987) based on experimental data for the activity coefficients at infinite dilution.

D. Boundary Conditions and Nondimensional Formulation

The overall and polymer mass balances are needed to close the model:

$$P + G = Q \qquad (28)$$

$$PW_2 = QW_{2,0} \qquad (29)$$

where $W_{2,0}$ is the polymeric mass fraction at the slit entrance.

The boundary conditions required to solve the differential equations are the following:

$$\begin{aligned} z = 0, \quad & W_2 = W_{2,0}, \quad && \forall x \\ z = L, \quad & p = p_0, \quad && \forall x \\ z = 0, \quad & T = T_0, \quad && \forall x \end{aligned} \qquad (30)$$

$$\begin{aligned} x = 0, \quad & T = T_w, \quad && \forall z \\ x = \frac{H}{2}, \quad & -2k_P \frac{\partial T}{\partial x} = c_G G \frac{\partial T}{\partial z}, \quad && \forall z \end{aligned} \qquad (31)$$

where p_0 is the pressure imposed at the slit discharge section. The last boundary condition says that the heat flux toward the gas phase must balance the local rate of heating of the gas phase itself.

The relevant model equations, reduced to nondimensional form, are reported in Table 1. In the equations given there, β is the nondimensional pressure p/p_0, ζ is the axial position z/L, ξ is the transversal position x/H, θ is the temperature $(T - T_0)/(T_w - T_0)$, and ω is the ratio $W_2/W_{2,0}$. The nondimensional gas temperature θ_G is of course $\theta(1/2,\zeta)$. The dimensionless vapor pressure function β^0 and the dimensionless viscosity function μ are defined as

$$\beta^0(\theta,\omega) = \frac{P_1(T,W_2)}{p_0} \tag{32}$$

$$\mu(\theta,\omega) = \frac{\mu_P(T,W_2)}{\mu_P(T_0,W_{2,0})} = \frac{\mu_P(T,W_2)}{\mu_{P0}} \tag{33}$$

The model equations contain five dimensionless groups: a dimensionless mass transfer coefficient $E = Ap_0L/Q$, a dimensionless exit pressure $U = \rho_P p_0 H^3/L\mu_{P0}Q$, a modified Graetz number $Gz = Hc_PQ/k_PL$, a Stefan number $St = \lambda/c_P(T_w - T_0)$, and a specific heat ratio $\gamma = c_G/c_P$. All the physical properties are evaluated at the dimensionless cross-sectional

Table 1 Model Equations, Nondimensional Form

	Heating section	Main section
Momentum balance	$-\dfrac{d\beta}{d\zeta} = \dfrac{12\mu}{U}$	$-\dfrac{d\beta}{d\zeta} = \dfrac{12\mu}{U\omega}$
Heat transfer	$(\xi - \xi^2)\dfrac{\partial\theta}{\partial\zeta} = \dfrac{1}{6\,Gz}\dfrac{\partial^2\theta}{\partial\xi^2}$	$(\xi - \xi^2)\dfrac{\partial\theta}{\partial\zeta} = \dfrac{1}{6\,Gz\,\omega}\dfrac{\partial^2\theta}{\partial\xi^2}$ $+ \dfrac{1}{\omega}(St + \gamma\theta_G - \theta)(\xi - \xi^2)\dfrac{d\omega}{d\zeta}$
Mass transfer	$\omega = 1$	$E(\beta^0 - \beta) = \dfrac{d(1/\omega)}{d\zeta}$

temperature θ^*, actually given by

$$\theta^*(\zeta) = 12 \int_0^{1/2} (\xi - \xi^2)\theta(\xi, \zeta) \, d\xi \tag{34}$$

The boundary conditions in nondimensional form are

$$\omega(0) = 1, \qquad \beta(1) = 1, \qquad \theta(\xi,0) = 0, \qquad \theta(0,\zeta) = 1 \tag{35}$$

$$\xi = \frac{1}{2}, \qquad -2\frac{\partial\theta}{\partial\xi} = \gamma\left(1 - \frac{1}{\omega}\right)\mathrm{Gz}\frac{\partial\theta}{\partial\zeta} \tag{36}$$

These boundary conditions make the problem a two-point boundary value problem. We have integrated the model by using a shooting scheme. Starting from a guessed value of the pressure at the slit entrance, the heating section equations are integrated forward until $\beta = \beta^0$; from that point on, the main section equations are integrated forward. Thus, one obtains a value for the pressure at the discharge section, which must be compared with that imposed by the boundary condition. If this condition is not satisfied, the pressure at the slit entrance is changed, and a new value of the pressure at the discharge is calculated. A simple bisection scheme was then used to converge to the right value of $\beta(0)$ (for more details on the numerical algorithms, refer to Ianniruberto, 1992; Maffettone, 1993). All the results presented in the following have been obtained by considering $A = 10^{-4}$ sec/m, $T_w = 260°C$, and $p_0 = 4$ kPa.

III. SMALL FLOW RATE LIMIT

From a practical point of view, it is worthwhile to investigate in detail the limit of small flow rates, since these are the typical operating conditions of the industrial process. At small flow rates, the heat balance can be significantly simplified. The Graetz number, in the range of flow rates typical of the process (<2 kg/hour), is of order 10. This implies that the slit is *long* in the sense of Graetz, and that the Nusselt number, Nu, does not change with Gz (the slit is long as far as heat transfer is concerned, since most of it is used for devolatilization). Under these conditions, Nu attains its asymptotic value for the flow geometry at hand, i.e., Nu = 3.8 (Drew, 1931; Norris and Streid, 1940; Jacob, 1949). It is then reasonable to rephrase the heat balance in an one-dimensional formulation by integrating Eq. (23) over the cross section.

By considering the gas temperature equal to the polymer mean temperature, the global heat balance is

$$\Phi(T_w - T^*) = (c_P P + c_G G)\frac{dT^*}{dz} - [\lambda + (c_G - c_P)(T^* - T_0)]\frac{dP}{dz} \tag{37}$$

where Φ is the heat transfer coefficient, given by

$$\Phi = \frac{2k_{\mathrm{P}}\mathrm{Nu}}{H} \tag{38}$$

Equation (37) holds true in the main section, whereas in the heating section it is written as

$$\Phi(T_{\mathrm{w}} - T^*) = c_{\mathrm{P}}Q\frac{dT^*}{dz} \tag{39}$$

In nondimensional terms Eqs. (37) and (39) become, respectively,

$$\frac{2\,\mathrm{Nu}}{\mathrm{Gz}}(1 - \theta^*) = \left(\gamma + \frac{1-\gamma}{\omega}\right)\frac{d\theta^*}{d\zeta} - [\mathrm{St} + (\gamma - 1)\theta^*]\frac{d(1/\omega)}{dz} \tag{40}$$

$$\frac{2\,\mathrm{Nu}}{\mathrm{Gz}}(1 - \theta^*) = \frac{d\theta^*}{d\zeta} \tag{41}$$

In the limit of small flow rates, these equations replace the heat transfer equations given in Table 1. Once the problem has become one dimensional, the boundary conditions reduce to

$$\omega(0) = 1, \qquad \beta(1) = 1, \qquad \theta^*(0) = 0 \tag{42}$$

We now present the model predictions, and compare them with experimental results. Figure 3 shows the calculated profiles of volatile mass fraction along the slit axis. At the lowest flow rate (0.5 kg/hour), the devolatilization process starts early in the slit—that is, the heating section is very small. At this flow rate the slit is largely overdesigned, and the volatile mass fraction is everywhere very close to the equilibrium value at the local temperature and pressure. The volatile mass fraction decreases over the whole slit, since the local pressure decreases approaching the slit exit. For flow rates below 2 kg/hour the mass fraction at the discharge section is close to its equilibrium value, whereas for larger flow rates the leak rate is not negligible. The comparison between the calculated and the measured volatile mass fraction at the discharge section is also reported in Fig. 3. We believe that the agreement between theory and experiments is acceptable since the uncertainty of measured values is quite large because of sampling problems. It should be noted, in this regard, that the inlet mass fraction is of order of 500,000 ppm, so that the calculated total removal agrees with the measured value extremely well.

Figure 4 reports the discharge mass fraction of volatile versus flow rate. The dashed line represents the equilibrium mass fraction at the wall temperature and at the discharge pressure. At low flow rates, the mass

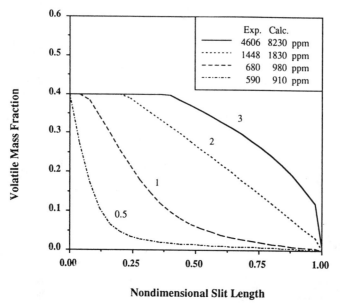

Nondimensional Slit Length

Fig. 3 Calculated profiles of volatile mass fraction. Parameter on curves is feed flow rate in kg/hour. The inset shows measured and calculated mass fractions in the exit stream. (From Maffettone et al., 1991, reproduced with permission of the American Institute of Chemical Engineers. © 1991 AIChE. All rights reserved.)

fraction is very close to the equilibrium value. Since the feed mass fraction is 40% by weight, it appears that the slit devolatilization can perform very effective separations.

Figure 5 presents calculated temperature profiles, which should be compared with those experimentally measured reported in Fig. 6. The initial conditions for the simulations have been chosen to exactly match the experimental conditions. For all the flow rates, an initial heating ramp is visible in Fig. 5. The axial extent of the ramp corresponds to the length of the heating section. As one might expect, the length of the heating section increases with increasing flow rate. At the lowest flow rate, the temperature quickly increases since most of the devolatilization takes place in a very short initial region. In this case, the temperature already reaches the wall temperature at the middle of the slit length. As the flow rate increases, the heating ramp is followed by a rather flat temperature profile, where most of the devolatilization takes place: in this region, the heat supplied from the walls is balanced by the latent heat. It is worth noting that for the highest value of the flow rate, a temperature reduction is predicted close to the discharge

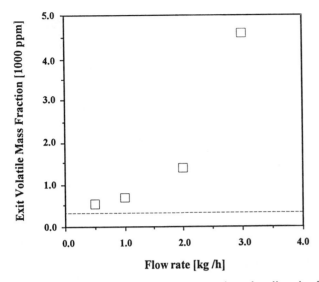

Fig. 4 Volatile mass fractions measured at the slit exit plotted versus flow rate. Dashed line is the equilibrium value at the wall temperature and exit pressure.

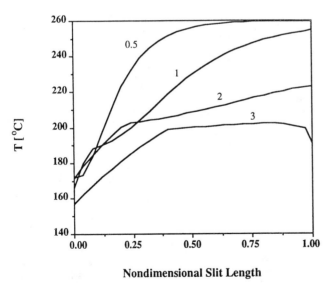

Nondimensional Slit Length

Fig. 5 Calculated temperature profiles. Parameters as in Fig. 3. (From Maffettone et al., 1991, reproduced with permission of the American Institute of Chemical Engineers. © 1991. AIChE. All rights reserved.)

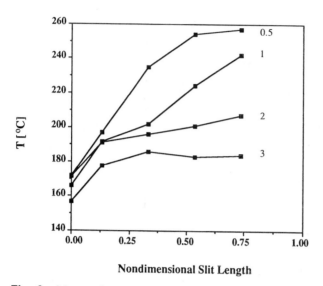

Fig. 6 Measured temperature profiles. Parameters as in Fig. 3. (From Maffettone et al., 1991, reproduced with permission of the American Institute of Chemical Engineers. © 1991 AIChE. All rights reserved.)

section. Actually, for these values of the flow rate, the solution reaches the discharge region still rich in volatile; on the other hand, as will be shown in Fig. 7, the pressure abruptly reduces to the discharge value. Thus, in this region a significant separation takes place, and the latent heat of devolatilization exceeds the amount of heat supplied from the walls since there is still a significant fraction of dissolved volatile to strip. The comparison between predicted and measured temperature profiles is good, although they do not superimpose. The qualitative trends, however, appear to be well predicted by the model.

Figure 7 shows the calculated pressure profiles, and the measured values of the pressure. As a matter of fact, only two measured points are significant, since the discharge value is intrinsically well predicted by the model (it is the pressure boundary condition). The comparison may then be just qualitative. Only the curve for the flow rate of 0.5 kg/hour shows three different regions. At the slit entrance, the pressure profile shows a mild slope since the fluid has a low viscosity due to the presence of a large amount of volatile. In the next region, most devolatilization takes place, and a rapid pressure decrease is predicted for the higher viscosity of the concentrated solution. Finally, the mass fraction does not change anymore, whereas the temperature increases (compare Fig. 5). Thus the average viscosity lessens,

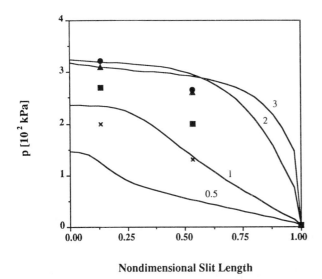

Nondimensional Slit Length

Fig. 7 Calculated pressure profiles compared with measured values of pressure. Parameters as in Fig. 3. Symbols correspond to the four different flow rates (kg/hour): ×, 0.5; ■, 1.0; ●, 2.0; ▲, 3.0. (From Maffettone et al., 1991, reproduced with permission of the American Institute of Chemical Engineers. © 1991 AIChE. All rights reserved.)

and the pressure profile becomes smoother. At larger flow rates only the first two trends are predicted.

The model underpredicts systematically the actual pressure drop at the lower flow rates. This is probably because we have modeled the thermal conductivity of the solution as that of the pure polymer without bubbles. This approximation is obviously worse for the case of small flow rates because, in that case, devolatilization occurs earlier in the slit.

Figure 8 presents the comparison between predicted and measured pressure drops between the first pressure gauge in the slit and the exit. It can be seen that the agreement is good only at high flow rates. Figure 8 clearly shows that the diagram of pressure drop versus flow rate exhibits a maximum, which is correctly predicted by the model. Moreover, the single slit works optimally at flow rates corresponding to this maximum. Indeed, in those conditions, a satisfactory compromise between the processed flow rate and the separation is attained (compare Fig. 4). The existence of such a maximum is not without consequences. In the next section, in fact, instabilities arising for the presence of the maximum in the multislit process will be discussed.

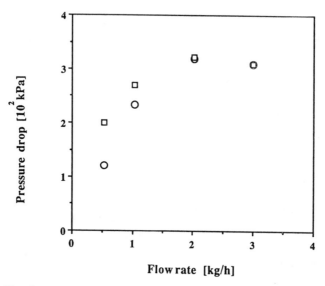

Fig. 8 Calculated and measured pressure drops from first pressure gauge to exit. Symbols: ○, theory; □, experiments. (From Maffettone et al., 1991, reproduced with permission of the American Institute of Chemical Engineers. © 1991 AIChE. All rights reserved.)

IV.   MULTISLIT DEVOLATILIZATION: THE STABILITY PROBLEM

The flow rate that can be actually processed in a single slit under acceptable devolatilization conditions is quite low for any industrial process, which, in fact, is performed by means of a multislit configuration. Let Q_T be the total flow rate imposed to the system made of N slits working in parallel. The major difficulty arising when modeling the multislit device is that we do not know, a priori, at which flow rate each slit effectively works. If all the slits work at a flow rate equal to Q_T/N, we say that the operation is stable. Depending on the operating conditions, however, there may exist a window of total flow rates in which such a stable operation cannot be attained; that is, it may happen that some slits work at a flow rate smaller than Q_T/N, whereas the others treat flow rates larger than Q_T/N. When this situation occurs, we say that the multislit operation is unstable. The devolatilization process cannot be carried out in these conditions since, while the slits working at low flow rates are largely overdesigned and give a good separation, those working at high flow rates perform a very inefficient removal. In view of this unsatisfactory behavior, it is worthwhile to analyze when unstable operations may be likely to occur in typical industrial conditions.

In a multislit device all the slits discharge into a chamber where vacuum is imposed, and are all fed at the same pressure; thus, the pressure drop Δp is bound to be the same for all the slits. Unstable conditions could occur when the function $Q(\Delta p)$ is multivalued. Figure 9 shows a typical steady-state $Q(\Delta p)$ curve that can give rise to an instability phenomenon. Indeed, there exists a range of Δp values (the shaded area in Fig. 9) in which three different flow rates are possible: Q_I, Q_{II}, Q_{III}, say. A simple linear stability analysis (Ianniruberto et al., 1993) shows that Q_{II} is a locally unstable solution, whereas Q_I and Q_{III} are locally stable. As a consequence, at steady state some slits may work in a low-flow-rate mode (Q_I) performing a significant separation, while the remaining slits work in a high-flow rate mode (Q_{III}), where devolatilization is not satisfactory.

In the case of a single slit, we have shown that the optimum flow rate is close to that corresponding to the maximum of the $\Delta p(Q)$ curve. As just shown above, that flow rate certainly falls in the region of instability. As a matter of fact, the largest flow rate per slit that guarantees a stable multislit operation may well be significantly smaller than the optimal one for the single-slit process (as will be shown in the following). The physical origin of the fluid-dynamic instability occurring in a multislit apparatus is strictly

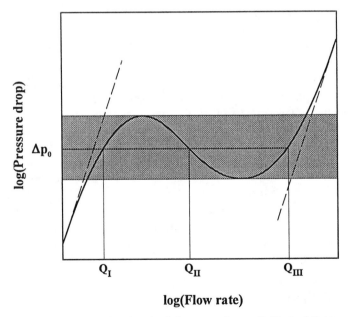

Fig. 9 A qualitative sketch of the curve Δp vs. Q. Dashed lines are the asymptotes at low and high flow rates. The shaded area is the unstable region.

linked to the strong dependence of the solution viscosity on temperature and volatile mass fraction. Useful information is obtained by analyzing the asymptotic behavior of a single slit in both the low flow rate and the high flow rate limits.

At extremely low flow rates the slit is largely overdesigned, and therefore we may assume that the liquid is everywhere at the wall temperature, T_W, and with a polymer mass fraction everywhere equal to $W_{2,R}$, the (large) equilibrium mass fraction at T_W and p_0. It follows that the viscosity of the liquid is, almost everywhere, $\mu_P(T_W, W_{2,R})$; even though the temperature is the highest attainable, such a viscosity is quite large because of the very low volatile mass fraction. Thus, the low flow rate asymptote predicts a pressure drop proportional to $\mu_P(T_W, W_{2,R})Q$. In the plane $(Q, \Delta p)$, this asymptote is a straight line with a very large slope.

Conversely, at very high flow rates, the heat transfer is completely inefficient and, therefore, so is the mass transfer. Hence, the temperature and the volatile mass fraction remain in fact those of the inlet conditions. Thus, the viscosity of the polymeric phase is everywhere $\mu_P(T_0, W_{2,0})$, and the high flow rate asymptotic pressure drop is proportional to $\mu_P(T_0, W_{2,0})Q$. In conclusion, this asymptote is again a straight line in the plane $(Q, \Delta p)$, but with a very low slope since the viscosity is that of a solution rich in volatile components. It is worth noting that Fig. 9 reports Δp vs. Q in logarithmic scales, so that the two asymptotes, indicated with dashed lines, are two parallel straight lines.

That the low flow rate asymptote lies above the high flow rate one implies that the Δp vs. Q curve may exhibit both a maximum and a minimum (see Fig. 9). This behavior can be explained with a simple physical argument. Starting from the low flow rate limit, as Q increases, the devolatilization progressively becomes less efficient (but still significant), thus determining a lower and lower average viscosity of the polymeric solution flowing in the slit. This effect reduces the frictional dissipation, and as a consequence, the pressure drop starts leaving the asymptotic line. It then may happen that, with further loss of efficiency of the separation, the viscosity reduction more than balances the effect due to increasing flow rate. Thus, a decreasing branch of the Δp vs. Q curve is predicted. Needless to say, since the high flow rate asymptotic behavior has to be recovered, the pressure drop must pass through a minimum and then start increasing again with a slope close to $\mu_P(T_0, W_{2,0})$.

A. Results and Discussion

The stability analysis requires the knowledge of the Δp vs. Q curve over the whole range of Q values, well beyond the range of flow rates of practical interest. As a consequence, the Graetz number cannot be considered *low*; hence, the Nusselt number does not attain its asymptotic value anymore. The heat balance equation has then to be solved in its original two-dimensional form.

Figure 10 shows the diagram of the function $\Delta p(Q)$ for a typical operating condition. The pressure drop goes through a maximum and a minimum, and thus, a region of unstable behavior is predicted. The optimal single-slit flow rate, that is, the largest Q value at which the separation is still efficient, is close to Q_{max}. Unfortunately, such a Q value falls well within the instability window $[Q',Q'']$. As a consequence, to carry out an efficient devolatilization process in stable conditions, one must feed a flow rate per slit smaller than Q', a value an order of magnitude less than Q_{max}, as can be seen in Fig. 10. From a practical point of view, this implies that to devolatilize a total flow rate Q_T one has to employ a multislit device made of N slits, with N limited below by the following relationship:

$$N \geq \frac{Q_T}{Q'} \tag{43}$$

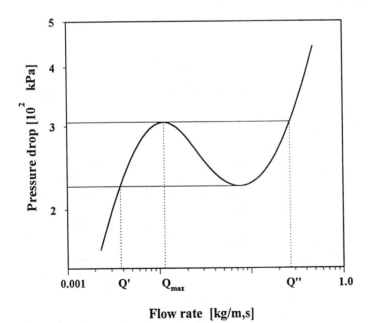

Flow rate [kg/m,s]

Fig. 10 Pressure drop versus flow rate calculated with $W_{2,0} = 0.5$, $p_0 = 4\,\text{kPa}$, $T_W = 260°C$, and $T_0 = 158°C$. Q' and Q'' limit the instability window.

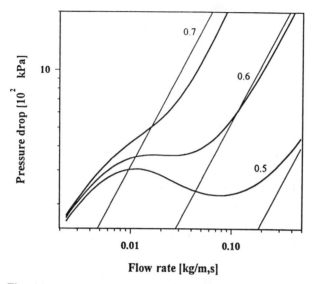

Fig. 11 Pressure drop versus flow rate at three different inlet mass fractions. High flow rate asymptotes are also reported ($p_0 = 4\,\text{kPa}$, $T_w = 260°C$, and $T_0 = 158°C$). (From Ianniruberto et al., 1993, reproduced with permission of the American Institute of Chemical Engineers. © 1993 AIChE. All rights reserved.)

Figure 11 shows the effect of the feed mass fraction on the instability phenomenon. It should be noted that the value of $W_{2,0}$ affects the position of the high flow rate asymptote, whereas the lower one remains unaltered. Furthermore, it clearly appears that for $W_{2,0} = 0.7$ the instability window has disappeared. Also note that for $W_{2,0} = 0.6$ and 0.7, the large flow rate asymptote is reached from below.

Figure 12 shows the effect of the exit pressure p_0. This parameter alters only the low flow rate asymptote, since it affects the exit mass fraction by changing the equilibrium conditions. Notice that, even for exit pressures unrealistically high, the instability window does not in fact disappear.

Figure 13 presents the influence of the inlet temperature on the instability phenomenon. As one might expect, the low flow rate asymptote is not affected by T_0. It should be mentioned that as T_0 increases, the instability window tends to disappear. This fact, however, cannot be used as an actual design criterion since, as the temperature increases, polymer thermal degradation becomes more and more important.

A significant internal check for testing the reliability of the numerical results is obtained by comparing them with analytical predictions derived from an asymptotic analysis. Indeed, that analysis (Ianniniberto et al., 1993)

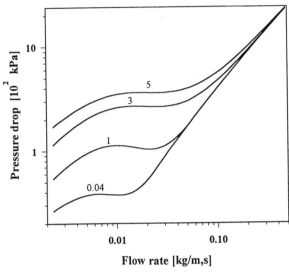

Fig. 12 Dependence of the pressure drop in the slit on the flow rate for four different values of the pressure at the slit discharge ($W_{2,0} = 0.5$, $T_W = 260°C$, and $T_0 = 158°C$). (From Ianniruberto et al., 1993, reproduced with permission of the American Institute of Chemical Engineers. © 1993 AIChE. All rights reserved.)

predicts a quite anomalous behavior of the dependence of the heating section axial extent, say z^*, on the flow rate. In fact, z^* is predicted to be a nonmonotone function of the flow rate; that is, the asymptotic analysis predicts that, in a range of Q values, the extent of the heating section can decrease as the flow rate increases. The possibility of such an anomalous behavior depends on the value of the Stefan number. Figure 14 shows the dependence of z^* on the flow rate for two different values of St. The upper curve, calculated at a St value corresponding to the system for which experiments were performed, does not show this anomalous behavior. On the other hand, the curve for a St value three times as large (the lower curve) exhibits the predicted anomalous phenomenon.

It should be remarked that the modeling has been carried out by considering the polymer phase as a Newtonian fluid (Ianniruberto et al., 1993). This assumption, while acceptable when dealing with the small flow rate limit discussed before, is not realistic when high flow rates are considered, since in that case high shear rates develop in the slit. Although the qualitative essence of the instability problem is not affected by the non-Newtonian character of the polymeric liquid, some simple considerations may show possible quantitative effects of the polymeric solution rheology.

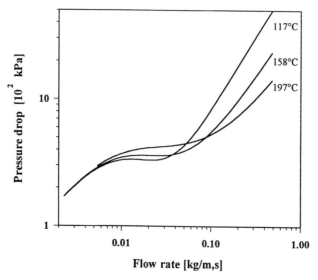

Fig. 13 Dependence of the pressure drop in the slit on the flow rate for three different values of the inlet temperature ($W_{2,0} = 0.5$, $T_W = 260°C$, and $p_0 = 4\ kPa$). (From Ianniruberto et al., 1993, reproduced with permission of the American Institute of Chemical Engineers. © 1993 AIChE. All rights reserved.)

Minima of the Δp vs. Q curve are predicted to occur at relatively high flow rates, where the viscosity is significantly smaller than the Newtonian value because of the shear thinning. As a consequence, the non-Newtonian behavior gives rise to a smaller pressure drop over the slit. Thus, the actual minimum value of Δp is likely to be less than that calculated on the basis of the Newtonian assumption. Hence, the instability window is enlarged since Q' (see Fig. 10) decreases as the minimum becomes smaller.

V. CONCLUSIONS

Slit devolatilization appears to be a very efficient separation process. Indeed, experiments have shown that the volatile contents can be reduced from 50% by weight down to a few thousand ppm. A rather simple model developed in the framework of transport phenomena gives predictions in good agreement with the experimental observations. The model contains just one adjustable parameter, which can be estimated on the basis of a well-defined criterion.

The single-slit unit is in fact capable of performing a bulk separation, but only working at relatively small flow rates. As a consequence, actual

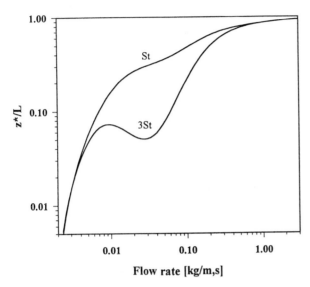

Fig. 14 Axial position where devolatilization starts for two different values of the Stefan number ($W_{2,0} = 0.5$, $p_0 = 4\,\text{kPa}$, $T_\text{w} = 260°C$, and $T_0 = 158°C$). (From Ianniruberto et al., 1993, reproduced with permission of the American Institute of Chemical Engineers. © 1993 AIChE. All rights reserved.)

industrial processes are carried out with multislit devices. The modeling has shown that multislit devolatilization is an inherently unstable operation, with flow rates per slit being quite different from each other in most situations. A criterion for estimating the maximum flow rate per slit treatable under stable conditions has been presented. Such a (maximum) flow rate appears to be significantly smaller than the optimal one for a single-slit operation.

The modeling of the slit devolatilization has shown that this process is essentially driven by heat transport. Indeed, in typical industrial operating conditions, the model predicts that the mass fraction of the volatiles dissolved in the polymeric phase is everywhere close to the equilibrium value. Hence, the resistance to mass transport is bound to be small. This behavior could in principle be attributed to the A value chosen, which in fact could determine a very effective mass transport. On the other hand, the predicted temperature profiles are in good agreement with the experimentally measured ones, thus implying that the A value has been chosen reasonably. Moreover, temperatures inside the slit are always significantly less than the wall temperature, thus confirming that the rate-determining step is indeed the heat transfer phenomenon. The importance of heat transport has often been neglected in the modeling of many devolatilization processes. Though this modeling approach appears to be largely justified for refinement stages, it may turn out not to be reasonable when dealing with bulk devolatilization, like the one considered here.

NOMENCLATURE

A mass transfer coefficient
c specific heat per unit mass
c_G specific heat per unit mass of the gas phase
c_p specific heat per unit mass of the polymeric phase
DP   polymerization index
E dimensionless mass transfer coefficient
E_{att} activation energy
G mass flow rate per unit width of the gas phase
Gz   Graetz number
H slit height
h thickness of the gas layer
k thermal conductivity
k_G thermal conductivity of the gas phase
k_p thermal conductivity of the polymeric phase
L slit length
l characteristic thermal length
M mass of volatile that devolatilizes per unit solution volume and time
M_p molecular weight of polymer molecules
N number of slits
Nu   Nusselt number
P mass flow rate per unit width of the polymeric phase
p local pressure
p_0 exit pressure
P_1 vapor pressure of volatiles in the polymeric solution
P_1^ϕ vapor pressure of pure ethylbenzene
Pr   Prandtl number
$\underset{\sim}{Q}$ slit mass flow rate per unit width
$\bar{Q}$ mass flow rate
Q^* volumetric flow rate
Q^*_{opt} volumetric flow rate for an optimal heat transfer
Q_T total mass flow rate per unit width
R ideal gas constant
S cross-sectional area
St   Stefan number
T Solution temperature
T_G gas temperature

T_0 inlet temperature
T^0 temperature at which the liquid phase enthalpy is zero
T^* cross-sectional average temperature
T_{rif} reference temperature in the correlation for the solution viscosity
U dimensionless exit pressure
V velocity of the polymer–gas interface
v_G gas velocity
v_P polymer velocity
x distance from the upper wall of the slit
w slit width
W_2 polymeric mass fraction
$W_{2,0}$ inlet polymeric mass fraction
$W_{2,R}$ equilibrium mass fraction
W_1 volatile mass fraction
$W_{1,0}$ inlet volatile mass fraction
Z slit length
z axial distance from the slit entrance
z^* axial extent of the heating section

Greek Letters

α thermal diffusivity of the polymeric solution
β dimensionless pressure
β^0 dimensionless vapor pressure
γ ratio between the gas and polymer specific heats
Δp pressure drop over the slit
Δp_{max} maximum pressure drop allowed in the devolatilizer
ΔT_w temperature difference imposed on the system
ΔT_{int} temperature change in the devolatilizer
ΔW_1 volatile mass fraction reduction in the devolatilizer
δ dimensionless gas layer thickness
ε ratio between the gas and polymer viscosities
ζ dimensionless axial position
Θ aspect ratio of the devolatilization unit
θ dimensionless solution temperature
θ_G dimensionless gas temperature
θ^* dimensionless average temperature
λ latent heat of vaporization
μ dimensionless solution viscosity
μ_G gas viscosity
μ_P polymeric solution viscosity
μ_P^* polymer viscosity at a reference temperature

ζ dimensionless transversal coordinate
ρ_G gas density
ρ_P polymeric solution density
τ residence time
τ_{opt} residence time at optimal heat transfer conditions
ϕ_2 polymer volume fraction
Φ heat transfer coefficient
χ interaction parameter in the Flory–Huggins equation
ω dimensionless polymer mass fraction

REFERENCES

Batchelor, G. K. (1967). *An Introduction to Fluid Dynamics*, University Press, Cambridge.

Bird, R. B., Stewart, W. E., and Lightfoot, E. N. (1960). *Transport Phenomena*, John Wiley and Sons, New York.

Charles, M. E., and Lilleleth, L. U. (1965). Co-current stratified laminar flow of two immiscible liquid in a rectangular conduit, *Canad. J. Chem. Eng.*, *43*(6): 110.

Charles, M. E., and Redberger, P. J. (1962). The reduction of pressure gradients in oil pipeline by addition of water: Numerical analysis of stratified flow, *Canad. J. Chem. Eng.*, *40*(4); 70.

Denson, C. D. (1983). Stripping operations in polymer processing, *Adv. Chem. Eng.*, *12*: 61.

Drew, T. B. (1931). Mathematical attacks on forced convection problems: A review, *Trans. AIChE*, *26*: 26.

EP (1990). European Patent 352,727.

EPSE (1985). *Encyclopedia of Polymer Science and Engineering*, 2nd ed., Wiley Interscience, New York.

Flory, P. J. (1942). Thermodynamics of high polymer solutions, *J. Chem. Phys.*, *10*: 51.

Holten-Andersen, J., Frederslund, A., Rasmussen, P., and Carvoli, G. (1986). Phase equilibria in polymer solutions by group contribution, *Fluid Phase Equil.*, *29*: 357.

Holten-Andersen, J., Rasmussen, P., and Frederslund, A. (1987). Phase equilibria of polymer solutions by group contribution. 1. Vapor–liquid equilibria, *Ind. Eng. Chem. Res.*, *26*: 1382.

Ianniruberto, G. (1992). Stabilità della Devolatilizzazione in Sistemi a Multifessura, Senior thesis in Chemical Engineering, University of Naples.

Ianniruberto, G., Maffettone, P. L., and Astarita, G. (1992). "Polymer Devolatilization: How Important Is Rheology," Proc. XIth Int. Cong. on Rheology (R. Keunings and P. Moldenaers, eds.), Elsevier, Amsterdam, p. 366.

Ianniruberto, G., Maffettone, P. L., and Astarita, G. (1993). Stability of multislit devolatilization of polymers, *AIChE J.*, *39*: 140.

Jacob, M. (1949). *Heat Transfer*, Wiley, New York.

Lynn, S., and Oldershaw, C. F. (1984). Analysis and design of a viscous-flow cooler, *Heat Tr. Eng.*, *5*: 86.

Maffettone, P. L. (1993). Devolatilizzazione di Polimeri, Ph.D. thesis in Chemical Engineering, University of Naples.

Maffettone, P. L., Astarita, G., Cori, B., Carnelli, R., and Balestri, G. (1991). Slit devolatilization of polymers, *AIChE J.*, *37*: 724.

Mendelson, R. A. (1979). A method for viscosity measurements of concentrated polymer solutions in volatile solvents at elevated temperatures, *J. Rheol.*, *23*: 545.

Mendelson, R. A. (1980). Concentrated solution viscosity behavior at elevated temperatures—polystyrene in ethyl-benzene, *J. Rheol.*, *24*: 765.

Navratil DiAndreth, A. (1984). Two Phase Flow in Screw Extruders, Ph.D. thesis, University of Delaware.

Norris, R. H., and Streid, D. D. (1940). Laminar flow heat transfer coefficients for ducts, *Trans. Am. Soc. Mech. Eng.*, *62*: 525.

Reid, R. C., Prausnitz, J. M., and Poling, B. E. (1988). *The Properties of Gases and Liquids*, McGraw-Hill, New York.

TPM (1976). *Thermophysical Properties of Matter*, Vol. 6, Suppl., Plenum Press, New York.

Yu, H. S., and Sparrow, E. M. (1967). Stratified laminar flow in ducts of arbitrary shape, *AIChE J.*, *13*: 10.

Watson, K. M. (1943). Thermodynamics of the liquid state. Generalized predictions of properties, *I.E.C.*, *35*: 398.

11

Devolatilization in Single-Screw Extruders

Martin H. Mack and Armin Pfeiffer

Berstorff Corporation, Charlotte, North Carolina

I. INTRODUCTION

Today, even tighter emission controls and better emission monitoring techniques are required from resin manufacturers by the latest regulations of the EPA under the Clean Air Act (La Lumonider, 1992). Accordingly, the devolatilization step has become a high priority for all major polymer production lines. The single-screw extruder has proved to be a competitive solution.

Due to its lower investment costs, excellent pressure buildup capacity and effective melt cooling characteristics, the use of single-screw extruders up to 600 mm in screw diameter is reported. In Table 1, a variety of polymers is listed with the target residual volatile levels for environmentally acceptable conditions. Most applications include a melt coming from the reaction process, which then enters the single-screw extruder for the finishing step, followed by a pelletizer.

In the manufacture of polyethylene resins more recent developments have focused on the removal of hydrocarbons with higher molecular weight, and therefore higher boiling points. These substances (oligomers) are responsible for unpleasant odors and for releasing of waxy deposits in subsequent extrusion processes, such as sheets or films.

Table 1 Objectives for Polymer Devolatilizations

Polymer	Solvents	Low-molecular-weight volatiles	Residuals in ppm
LLDPE	C2, C8		100
LDPE	C2, C6		30–500
PC	MCB		500
PC	MC		10
ABS		styrene	100
ABS		acrylonitrile	5
PMMA		MMA	3000
Nylon 6		caprolactam	1500
		oligomers	1500
		cycl. trimers	1500
PS		styrene	300
EVA		VA	50

Abbreviations: C2 = Ethylene; C6 = Hexane; C8 = Octane; MCB = Mono chlorbenzene; MC = Methyl chloride; MMA = Methyl-meth-acrylate; VA = Vinyl acetate.

II. HISTORICAL BACKGROUND

The first publications and patents on polymer degassing in single-screw extruders focused on the screw geometries and physical layout of the vent openings of the barrel (Price, 1915). The degassing tasks were rather simple and concentrated on the removal of moisture or trace monomers. It was well understood that the screw channels should be partially filled, resulting in a free surface area of the melt pool where the degassing process may occur. The effect of the conveying capacity on the free surface area was later calculated (Schenkel, 1963; Squires, 1981). After gaining more experience from production lines and after calculating the metering zone downstream of the degassing zone, the actual degassing process could be evaluated in greater detail (Biesenberger, 1983).

The "efficiency" of the degassing has been studied in single-screw extruders by Guionneau and Schenkel (1972). The surface area of the melt pool and its renewal rate were reported to be functions of screw geometry and degree of fill. A characteristic number for degassing was defined, based on the average shear stress in the vent zone, and supported with results from measuring the capacity in a model fluid. Although the shear stress is a function of screw speed, channel depth, and degassing length, the model did not allow accurate characterization of the degassing process, because rate and surface area for the mass transfer were not considered.

Latinen (1962) presented a model for the devolatilizing process in a single-screw extruder that takes into account the melt layer at the barrel wall. This layer is created by the leak flow over the flights and is reintroduced into the rotating melt pool by the wiping action of the following flight. The limitations of Latinen's model, which assumed devolatilization to be a simple process of molecular diffussion, are discussed elsewhere in this book.

Coughlin and Canevari (1969) calculated the degassing effect through the free surface area of the rotating melt pool, but they did not consider the melt film at the wall. Roberts (1970) based his results on the work of Coughlin and Canevari; however, he took into consideration the degassing through the melt film at the barrel. His results showed that up to 50% of the degassing effect is contributed by this melt layer.

Martin and Schuler (1991) presented a degassing constant that describes the load on the melt surface in kg/m^2 of the free degassing surface in single- and twin-screw extruders. The surface load is presented as the quotient of throughput and the surface renewal rate of the free degassing surfaces. The surface area of the melt pool in partially filled screw channels is given by the number of screw flights, the screw diameter, and the length of the zone under vacuum. The screw speed is a measure of the surface renewal rate. The devolatilization conditions are good, if the surface load is low. This means, for example, that the best degassing results are obtained at low throughput and high screw speed, since the value of the surface load is at a minimum. This model allows one to compare different sizes of extruders and even to compare process conditions from single-screw extruders with twin-screw extruders. However, the model also has its restrictions when used for scaling up: it considers the newly generated melt surface in relation to the throughput but does not consider the residence time of that surface. The agreement between measured and calculated results is not satisfactory, especially when the screw speed and throughput rates are, for example, doubled—which should result in the same value for the surface load. In this case, the residence time of the melt surface is shortened to half, and therefore the efficiency of the devolatilization is reduced.

III. THEORIES OF DEVOLATILIZATION APPLIED TO SINGLE-SCREW EXTRUDERS

The general layout of the degassing section is based on the physical principles that apply to devolatilization processes: Henry's law describes the solubility boundaries of low-molecular-weight substances in polymer melts. The first and second laws of Fick describe the mass flow from the polymer melt into the gas space. According to Fick's first law, the rate of mass transfer is given by

$$\dot{m} = DA\left(\frac{dC}{dx}\right) \tag{1}$$

where D is the diffusion coefficient, dC/dx is the concentration gradient, and A is the boundary surface area. High mass transfer and, consequently, high degassing efficiency can be obtained when all terms of Eq. (1) are maximized.

Fick's second law describes the change in equilibrium concentration (C) over time (t):

$$\frac{dC}{dt} = \frac{d}{dx}\left(D\frac{dC}{dx}\right) \tag{2}$$

To maximize the concentration gradient dC/dt, the layer thickness x of the polymer melt should be as small as possible.

Since the mass transfer is a function of time, the melt surface should be renewed frequently. The concentration gradient has its highest value each time a new layer of melt is formed, and it decreases shortly afterward. A rapid layer renewal leads to a high average concentration gradient and, therefore, according to Eq. (1), to better degassing results.

In the following, Eq. (1) will be applied to a single-screw extruder with the goal of establishing a characteristic number for the degassing process. This number may be used for scaling up from pilot plant trials to production-size extruders.

For single-screw extruders, the melt surface renewal occurs at the rotating melt pool and at the barrel surface exposed to the gas space (Fig. 1a). The trailing edge of the flights cannot be taken into account since single-screws are not self-wiping systems.

To define the amount of volatiles that will be "devolatilized" during the life span of the newly created layer, Eq. (1) is integrated over time:

$$\dot{m}_{vol} = \int_0^t \dot{m}_{vol}(t)\,dt = 2A(C_0 - C_e)\frac{\sqrt{D}}{\pi}\sqrt{t} \tag{3}$$

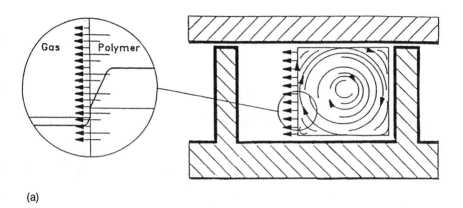

(a)

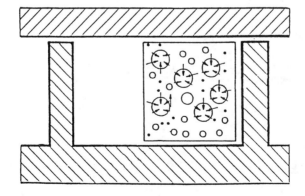

(b)

Fig. 1 Mass transfer principles in degassing extruders: (a) surface renewal of the rotating melt pool; (b) bubble formation inside the melt pool.

where C_0 is the inlet concentration and C_e the concentration at equilibrium. The residence time of the layers at the rotating melt pool is

$$t_{MP} = \frac{h}{D\pi n \sin \phi} \qquad (4)$$

and the residence time of each layer at the barrel surface is

$$t_B = \frac{1}{in} \qquad (5)$$

By using t_{MP} and t_B as the boundary conditions, the mass flow rate of volatiles can be described as

$$\dot{m}_{vol} = \frac{\int_0^{t_{MP}} \dot{m}_{vol, MP}(t)\, dt}{t_{MP}} + \frac{\int_0^{t_B} \dot{m}_{vol, B}(t)\, dt}{t_B} \qquad (6)$$

with Eq. (3) as follows:

$$\dot{m}_{vol} = 2(C_0 - C_e) \frac{\sqrt{D}}{\pi} \left(\frac{A_{MP}}{\sqrt{t_{MP}}} + \frac{A_B}{\sqrt{t_B}} \right) \qquad (7)$$

The mass flow of volatiles, $\dot{m}_{vol}$, can be described in a different way using the volume flow of polymer for the inlet concentration and the residual concentration:

$$\dot{m}_{vol} = V(C_0 - C_f) = \frac{\dot{m}}{\rho}(C_0 - C_f) \qquad (8)$$

where ρ is the melt density.

By combining Eq. (7) with Eq. (8), one obtains the term for the performance characteristic of the degassing zone:

$$C_f = C_0 - 2\rho(C_0 - C_e) \frac{\sqrt{D}}{\pi} \frac{A_{MP}/\sqrt{t_{MP}} + A_B/\sqrt{t_B}}{\dot{m}} \qquad (9)$$

IV. THE CHARACTERISTIC DEGASSING NUMBER FOR SCALE-UP

The goal of the scale-up calculation is to predict the residuals content in large production extruders or to give the extruder size when residual content and rates are given. Equation (9) can be rearranged to show the ratio of the actual reduction of monomers to the theoretical possible reduction of

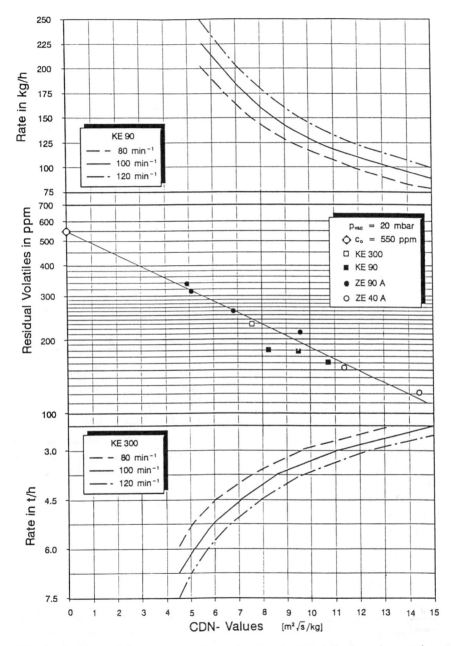

Fig. 2 Scale up of degassing extruders using the characteristic degassing number (CDN).

monomers on one side and the operating conditions in the vent zone on the other side of the equation:

$$\frac{c_0 - c_f}{c_0 - c_e} = 2\rho \frac{\sqrt{D}}{\pi} \frac{A_{MP}/\sqrt{t_{MP}} + A_B/\sqrt{t_B}}{\dot{m}} \tag{10}$$

The left-hand side of Eq. (10) is also referred to as the efficiency of the devolatilization process. For scale-up, the types of volatiles and polymers stay the same, and therefore, the coefficient of diffusion D and the melt density ρ may be assumed to remain constant.

The characteristic number of the degassing process for various extruder sizes is given by

$$CDN = \frac{A_{MP}/\sqrt{t_{MP}} + A_B/\sqrt{t_B}}{\dot{m}} \tag{11}$$

The residual monomer content for various extruder sizes can be determined through the degassing number CDN, when polymer temperatures, inlet concentration of volatiles, and vacuum level are kept constant during scaling up from pilot plant results to production lines. A typical case study of a scale-up procedure is shown in Fig. 2.

In the diagram for the pilot plant extruder (a 90-mm single-screw extruder, KE90), the rate versus the CDN value is calculated for various screw speeds. The middle diagram shows the actual residual values measured during the trials. With the target residual volatile content of the production line known, the CDN value of the production extruder can be calculated.

In the bottom diagram, the rate and screw speed combinations are shown for a 300-mm single-screw extruder. The CDN value also allows one to compare single- and twin-screw extruders after adjusting for the additional surface renewal in the twin-screw extruder's trailing side of the flights. The CDN value is limited to degassing through the surface of the melt pool and does not apply when high amounts of volatiles are present or when foaming is initiated by using stripping agents, because the screw channels are typically completely filled with foamy melt.

V. FOAM DEVOLATILIZATION WITH INJECTION OF STRIPPING AGENT

The use of water as a stripping agent not only contributes to reducing the partial pressure, it also delivers an additional devolatilization effect compared with the described surface process. Bubbles are generated inside the melt pool. The same diffusion process that takes place at the surface of the melt also takes place on the surfaces of these bubbles (Fig. 1b). The volatiles also diffuse from the melt into the bubbles produced by the stripping agent, causing the bubble to grow and finally burst. The effectiveness of this volume process depends on the distribution of the stripping agent, the formation of bubbles, their growth, and the production of melt membranes. A fine distribution of the stripping agent has a positive effect on degassing, as reported by Mack and Pfeiffer (1993). A finer distribution of the stripping agent produces a higher number of bubbles and, as a consequence, a larger bubble surface, thus reducing the thickness of the melt membrane.

The basic processes of bubble growth can be illustrated by

$$R = \left(\frac{1}{R_i^3} - \frac{\pi \Delta P t}{\eta V} \right)^{-1/3} \tag{12}$$

which describes the growth of an ideally round bubble formed in a viscous liquid as a function of the initial bubble radius, R, the melt viscosity, η, and the pressure difference, ΔP, between the system pressure and the bubble pressure (Roberts, 1970).

Low initial radii and high viscosities slow down the growth or prevent premature burst of bubbles. Furthermore, the residence time determined by the length of the devolatilization zone must be sufficient for the bubbles to grow and burst.

The vacuum level is one of the factors that determine the rate of bubble growth. For the scale-up of the devolatilization process, the differentiation between surface renewal rate and bubble growth is important, because in the first case, a scale-up from laboratory machines to production machines will follow the CDN term in Eq. (11). As a first approximation, this implies that the throughput can be raised by the square of the screw diameter. With the use of a stripping agent, however, the throughput is limited by the channel volume required for the foam formation. In this case the throughput can be increased by the third power of the screw diameter. It also suggests that excellent degassing results might be expected if mixing of a stripping agent is accomplished at a low screw speed.

Figure 3 shows the devolatilization results for low-density polyethylene (LDPE) obtained without a stripping agent (A) and with a stripping agent (B) as a function of the surface load as described by Roberts (1970). It becomes obvious in case A that the devolatilization results improve at low surface load. In case B, this correlation is less dominant, as the surface load model does not cover volume-controlled processes such as the bubble growth in the melt and volatile diffusion into the bubbles. If the stripping agent is distributed as finely as possible and, thus, devolatilization inside the melt pool is dominant, the screw speed and the surface load have no effect on the final degassing results!

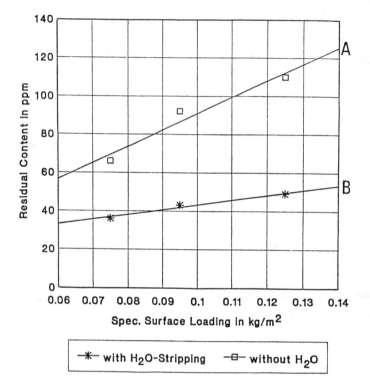

Fig. 3 Residual hydrocarbons in LDPE at various surface renewal rates. See text for discussion.

VI. DESIGN AND FUNCTION OF SINGLE-SCREW DEVOLATILIZATION EXTRUDERS

Typically, single-screw extruders are located next to the polymerization reactor, where polymer melt can be delivered at a constant rate to the feed section. Volatiles can be removed through a rear vent or a forward vent. The use of stripping agents is common.

A. Extruders with Rear Vents

Figure 4 shows the design of a rearward devolatilization extruder. The figure shows a hydraulically actuated slide valve is arranged under the low-pressure separator of an LDPE reactor. It controls the level in the separator and ensures that the feed zone of the extruder is supplied with melt as evenly as possible. The cross section of the slide valve is designed to achieve a linear

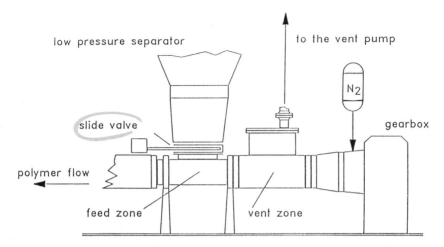

Fig. 4   Schematic of a rearward devolatilization system.

relation between the position of the slide valve and the cross section of the opening. This ensures excellent flow control. In the feed zone, partially filled channels are maintained by measuring the pressure in the compression zone located downstream. The degree of fill is changed by a change in screw speed or in the position of the slide valve.

In Fig. 5, the devolatilizing results for a melt-fed single-screw extruder (KE450 × 26D) are shown as a function of the degree of fill. The degree of fill was adjusted by increasing the screw speed at a constant feed rate of 22,000 lb/hour. The rear vent was operated at atmospheric pressure and 500 mbar vacuum. The equilibrium lines of ethylene in LDPE are marked for the high and low pressure range, to evaluate the efficiency of the vent. As expected, the residual ethylene content drops when the screw speed is increased. At atmospheric pressure, a 10% starvation was sufficient to reach the equilibrium line of 550 ppm. At 500 mbars, a 30% starvation was required to approach the equilibrium of 220 ppm.

To ensure that the residence time is adequate, the pressureless feed zone is longer than usual. The volatile substances separated from the melt are removed through the rear vent zone. Toward the gear box, the screw shaft is sealed against atmosphere by mechanical seal. It can be operated safely and economically with a vacuum of 500–1000 mbars. However, a slight overpressure of 10–30 mbars is recommended. This prevents leakage of oxygen to the extruder.

The advantage of rearward vents over siloventing is that the removed

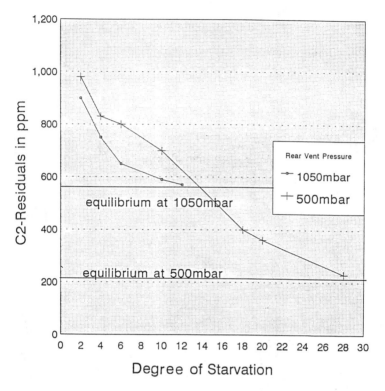

Degree of Starvation

Fig. 5 Results of C2-residuals versus degree of starvation from a rear vent system on a Berstorff 450-mm single-screw extruder. KE450 single screw; 60 rpm, 22,000 lb/hour LDPE.

≈ 83.~~t/yr ?.

volatiles are free of oxygen when operating at a low overpressure and can thus be returned to the process. This becomes quite important if large quantities are to be removed. Furthermore, the machine and space require-ments are lower than with forward devolatilization. In contrast to con-ventional discharge extruders, which are normally flood-fed, the starve-fed situation allows one to adjust the screw speed for optimization of the degassing results, without change in rate. Rear venting is recommended when gas bubbles are entrained in the feed stream.

B. Extruders with Forward Venting Systems

Figure 6a shows the basic design of a single-screw extruder with forward devolatilization. Its characteristic features are the feed zone with the subsequent melt seal, the devolatilization zone with the deeply cut screw profile, the barrel opening, and the vent dome, as well as the pumping zone following further downstream for generating the head pressure. The melt seal technically separates the devolatilization zone from the feed zone. It guarantees absolute gas tightness between the feed zone and the devolatilization zone under vacuum conditions. The melt seal is obtained by restricting the free-flow cross section. A pressure drop toward the devolatilization zone of 10–60 bars is quite common. In some cases, a drop of up to 120 bars may be achieved. Figure 6b shows a single-screw extruder, model KE450 × 18D by Berstorff, with a forward vent for devolatilizing LDPE melt.

The screw design in the devolatilization zone is tailored to the process conditions, which determine the length of the devolatilization zone required, the channel depth, and the number of flights. The following criteria must be met to ensure satisfactory operation:

Sufficiently large devolatilization surface
Surface renewal
Sufficiently long residence time
No foam rising up into the devolatilization dome

For the use of a *stripping agent*, an additional mixing element for the homogeneous distribution of the stripping agent in the melt is needed. This mixing element is arranged upstream of the devolatilizing zone and the retaining blister. The design of such a system is illustrated in Fig. 7.

Forward venting is more efficient than rear venting due to the options of better vacuum levels and the use of stripping agents. The devolatilization efficiency can be defined as the ratio of residual volatiles to incoming volatiles, in percent. Figure 8 summarizes results from several single-screw extruders, operating with forward vents for the removal of various hydrocarbons from LDPE melt. The initial volatile levels are 500–1500 ppm. It is obvious that for hydrocarbons with higher molecular weights (higher boiling points) the devolatilizing step becomes more difficult. In any case, stripping agents such as water improve the process.

Ethylene (C2), for example, can be reduced by 95% in one devolatilization step without a stripping agent and by 98% when using water injection. Starting with the initial content of 1000 ppm, the residual content obtained is 50 ppm without and 20 ppm with a stripping agent. In two-stage devolatilization, C2 can virtually be 100% removed. For C8, a reduction to approximately 35% without stripping agents and 20% with stripping agents can be expected. Compared with C8, the conditions are much worse for C12, where devolatilization without a stripping agent should not be attempted.

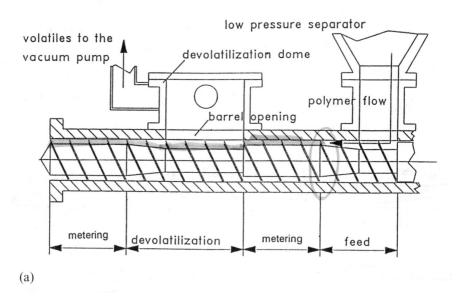

(a)

Fig. 6 (a) Schematic of an extruder with a forward devolatilization system. (b) A melt-fed single screw extruder, model KE450 × 18D by Berstorff with a forward vent for devolatilization of LDPE melt.

(b)

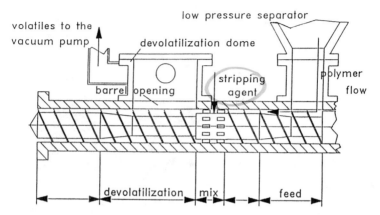

Fig. 7 Schematic of an extruder with a forward devolatilization system using a stripping agent.

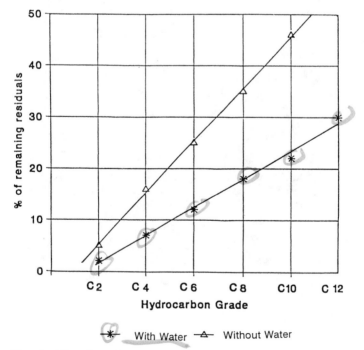

Fig. 8 The effect of water on a single-step forward venting process.

C. Dual-Stage Forward Vents with Stripping Agents

Multistage vacuum vents are used in cases where large amounts of volatiles are to be removed. The vacuum levels are adjusted in steps to keep the vapor velocities at a constant value in each vent dome.

In the following production situation, dual venting was necessary to achieve a C8 reduction to 10% for linear low density polyethylene (LLDPE) resins. Based on Fig. 8, it is expected that for each vent stage, a C8 reduction by 80% is feasible. For a dual stage, this will mean a reduction of 96%, or 4% residuals.

Especially for the high-viscosity resins, the melt temperature increases due to the extended barrel length. High shear rates in mixing sections and blister elements also contribute to higher processing temperatures and cross-linking reactions due to degradation, and they are therefore monitored closely for design improvements.

The residual C8 concentration has been measured between 1% and 5% for this application (Fig. 9). The lowest C8 residual levels are seen for the products with a low Melt Flow Index (MFI). The specific rate has no effect on the results. For the 0.7 MFI resin, the devolatilizing process is volume driven, as seen by the flat slope of the curve.

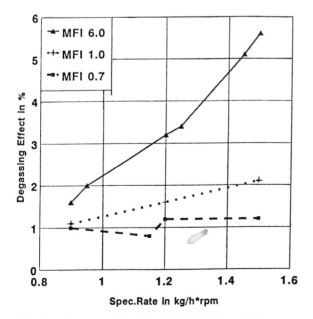

Fig. 9 Octene residuals in LLDPE after a dual-stage venting process using a KE 90 single-screw extruder; 1% water, 30 mbar.

VII. CONCLUSIONS

Most melt-fed single-screw extruders are equipped with forward vents and special mixing sections for stripping agents. In special cases, where high amounts of volatiles have to be removed, a two-stage degassing process is used. In future projects, higher-molecular-weight substances such as ocetene, trimers, or benzene will be removed. The growth in this market is partially dictated by new emission laws. As a result, more polymers will become odorless and available for food packaging. It is economical to retrofit existing extruders with devolatilizing sections to reduce overall plant emission levels.

NOMENCLATURE

A	free surface area of melt to vapor space
A_{MP}	free surface area of melt pool
A_B	free surface area of melt skin at barrel
C	volume concentration of volatiles in the polymer
C_e	equilibrium concentration of volatiles
C_f	residual concentration at extruder discharge
C_0	inlet concentration of volatiles
CDN	characteristic devolatilization number
D	screw diameter
h	channel depth
i	number of screw flights
$\dot{m}$	mass flow rate
$\dot{m}_{vol}$	mass flow rate of volatiles
n	screw speed
R	radius of bubble
t	time
t_B	renewal time for the phase boundary at the barrel
t_{MP}	renewal time for the phase boundary layer at the melt pool
V	volume flow rate
x	thickness of melt layer
ΔP	pressure difference
ϕ	angle of pitch
η	viscosity
ρ	melt density

REFERENCES

Biesenberger J. A., ed. (1983). *Devolatilization of Polymers*, Hanser, Munich, and Macmillan, New York.

Coughlin, R. W., and Canevari, G. P. (1969). Drying polymers during screw extrusion, *AIChE J.*, *15*: 560.

Guionneau, H., and Schenkel, G. (1972). Experimentelle untersuchung am modell eines entgasung extruders, *Kunststoffe*, *62*: 245–264.

La Lumondier, R. (1992). The Clean Air Act Federal Overview, Conference Manual, SPI—Technology Workshop, Houston, Chapter 4.

Latinen, G. A. (1962). Devolatilization of viscous polymer systems, *Adv. Chem. Ser.*, *34*: 235.

Mack, M. H., and Pfeiffer, A. (1993). "Effect of Stripping Agents for the Devolatilization of Highly Viscous Polymer Melts," 51st SPE ANTEC, New Orleans.

Martin, G., and Schuler, W. (1991). "Degassing During the Polymer Production and Processing," Various Aspects of Ethylene Propylene Based Polymers, Conference Proc., Leuben.

Price, R. B. (1915). U.S. Patent 1,156,096.

Roberts, G. W. (1970). A Surface renewal model for the drying of polymers during screw extrusion, *AIChE J.*, *16*: 878.

Schenkel, G. (1963). *Kunststoff-Extrudertechnik*, G. Hanser Verlag.

Squires, P. (1981). "Design of Two-Stage Extractor Screws," 39th SPE ANTEC, Boston.

12

Devolatilization in Co-Rotating Twin-Screw Extruders

John Curry and Faivus Brauer

Werner & Pfleiderer Corporation, Ramsey, New Jersey

I. INTRODUCTION

The ultimate goal of any polymerization or compounding process is production of a high-performance material in a convenient form. Contaminant volatiles are powerful degradative agents of properties. Consider the following issues:

A. Hazards of Volatile Contaminants

1. Health hazards of solvent-laden products in packaging, household, and health care applications
2. Optical deterioration of glazing, optical, or film products
3. Dielectric strength loss in electrical connector materials
4. Surface blistering on molded and sheet-extruded products
5. Mechanical stress failure of composites with voided interfaces
6. Ballistic inconsistency of voided composite propellants
7. Molecular weight loss of condensation polymers that have not been optimally devolatilized

B. Sources of Volatiles

Common volatile sources in polymer processing include the following:

1. Residual solvents of rubbers, olefins, or high-temperature amorphous polymers that have been polymerized as cements, emulsions, or suspensions to high molecular weight in an excess of a suitable solvent
2. Residual monomers from bulk polymerized materials produced through suspension polymerization in an excess of their own monomer
3. Condensation by-products such as water, alcohols, or acids formed when heat and vacuum promote desiccation and polymerization of polymers, or during alloying of polymers

4. Similarly, condensation of by-products from a reacting interface of a polymer composite
5. Surface contaminants on feedstocks due to contaminant diffusion to the surface, vapor condensation on the surface during storage, or upstream washing operations
6. Ingested air (or other gases) accompanying feedstocks into the process

To remove the undesirable volatiles, the polymer processor needs to supply these criteria:

1. A thermodynamic potential
2. A vapor–solution interface
3. Enough time to accommodate the rate of removal

Contaminants will coexist with their polymer hosts at an equilibrium level depending on the environment (temperature, pressure, strain) and the nature of the system (interaction strength between polymer and volatile, synergy or antagonism towards solubility in a multiple-component system, proximity of environmental condition to phase transitions of the polymer). To provide a thermodynamic driving force, modify the vent environment so that the equilibrium volatile level is lower than the volatile level of the processing material. The greater the difference, the stronger the thermodynamic potential.

C. Volatile Transport Processes

Volatile contaminants are transported from the mass to a convenient interface by diffusion, a process that, like heat transfer, has a rate dependency upon local gradient (in this case, a volatile concentration gradient). This is a painfully slow process (for example, at a typical 10^{-6} cm^2/sec diffusion coefficient, a molecule of volatile could take 2.8 hours to diffuse 1 mm!). Polymer processing equipment operates with deeper (tens of cm) flow channels, and material degradation kinetics are far too fast to depend on diffusion as the sole devolatilization mechanism. Successful devolatilization processes depend optimally on successively presenting volatile-rich layers of polymer solution to an interface maintained at high concentration gradient for an adequate time. The preceding statement has two aspects: first, a flow profile conducive to exposure of new, volatile-rich surfaces; second, an environmental condition and geometry that maintain maximum thermodynamic potential and interfacial area.

D. Extruders as Devolatilizers

How does one best devolatilize a polymer solution? As *safely* and *price effectively* as possible. Safety concerns include vapor containment, effluent control, and plant operability. Economic issues are utility costs, capital equipment costs, plant throughput, and product quality. A characteristic of polymer solution devolatilization is that the processing fluid becomes more viscous and the thermodynamic potential drops as the process proceeds. Fluid viscosity is the determinant for the shaft power required to drive a process. Fluid viscosity and product contaminant tolerance dictate the preferred machine dynamics (gravity-driven, wiped, self-wiping). High-powered, self-wiping devices are expensive but operable over a wide range of fluid characteristics and throughputs. Combined with an appropriate preconcentrator and vent management system, a robustly powered self-wiped twin-screw extruder is a popular system for polymer devolatilization.

II. TWIN-SCREW CO-ROTATING EXTRUDERS

A hydrodynamic hybrid between nonintermeshing devices and the positive conveying screw pump is the co-rotating intermeshing and self-wiping extruder. Self-wiping mechanically removes screw residue that would other-wise retard fluid flow.

The channel geometry for these machines is defined by sliding disks around one another. The screw cross section is then lenticular in shape. At the crest the flight walls are still convergent, so that the crests are narrower and the free volume is larger than for the closely intermeshing counter-rotating devices. The intrusion of one flight into the channel of the opposite screw is a barrier over which material must flow, causing a twist restraint that forwards material directly down the machine when overcome. The same restraint causes a restriction to pressure flow from the die or high-pressure areas (Fig. 1).

A. Geometry and Kinematics

The co-rotating geometry is open to cross-flow at the intermesh except for the wiping flight, and therefore the screws have an open path from feed to discharge. The screws are generally cut with two or three starts. Fewer flights (or starts) result in wider crests (and lower free volume) at a fixed diameter. Although the screws are open down-channel, interchannel flow across the flights is restricted by the flights.

Collections of profiled disks with stagger angle, called kneading blocks, allow cross-channel contact, and their wide crests and flow-restricting dynamics cause the stresses that plasticate and mix polymers. Kneading blocks maintain the self-wiping profile (Fig. 2a).

Screw elements are available in reverse pitch for maintaining back pressure dynamically against mixing elements.

The transport efficiency of the self-wiping geometry is the nemesis of certain processes, especially diffusion-limited types like solution forming, thermal equilibration, or plasticizing. A series of non-self-wiping elements has been developed for these processes. Slotted screw elements, turbine-shaped elements, and turbine-shaped elements cut on a helix have all proved successful. In the compromise between adequate machine volume and adequate strain history, novel mixing elements have been designed (Fig. 2b). Eccentric lobed disks and multilobed disks are effective tools for reliable mixing in high-volume machines.

B. Co-Rotating Intermeshing History

Co-rotating intermeshing designs were developed and applied in the 1930s for mixing and pumping of ceramic pastes. The first commercially produced machine was invented by Colombo of LMP (Turin, Italy) in 1939 and was used for thermoset and polyvinyl chloride (PVC) processing.

In a separate development Erdmenger and colleagues at I. G. Farben developed co-rotating machines for processing of plastic masses. In 1949 Erdmenger (with Bayer) developed co-rotating extruders with modular elements and kneading disks for improved mixing. In 1953 Bayer licensed Werner and Pfleiderer (Stuttgart), who expanded the Erdmenger development to a commercially produced series of so-called ZSK machines. Bayer and the other engineering giants of the German chemical industry provided a stream of applications in polymerization, devolatilization, plastification, and mixing of thermoplastics.

In a third independent development, Loomans and Brennan of Readco developed an intermeshing, self-wiping extruder with screws and paddles. The patent was transferred to Baker Perkins, who developed commercial equipment originally applied to condensation reactions. Since the expiration of the original patents, more than 40 manufacturers have produced or marketed a version of the twin-screw co-rotating and self-wiping extruder.

C. The Extruder Devolatilizer

Screw extrusion is a highly developed technology for melting polymers, mixing process aids, additives, or other polymers, and pressurizing the mixed melt to downstream transport, pelletizing, or profiling devices. The extruder environment has many desirable features for devolatilization processes:

1. Continuous operation that matches the characteristic of modern upstream reactors
2. Transport of relatively thin bands of material, which are easily homogenized, heated, or cooled

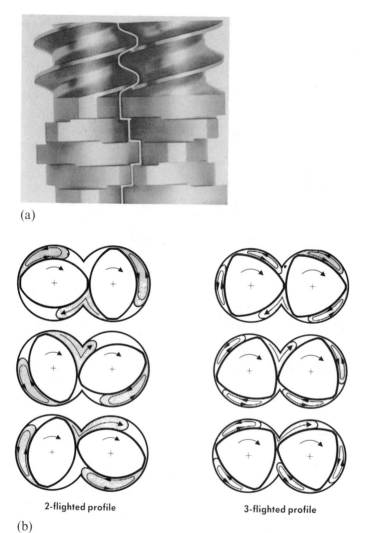

(a)

(b)

2-flighted profile   3-flighted profile

Fig. 1 Fully intermeshing profile and material transport in screw elements of twin screw co-rotating extruders. (a) Intermeshing self-cleaning profile. (b) Cross-sectional shapes. (c) Material movement. (d) Flow dislocation in intermesh zone.

(c)

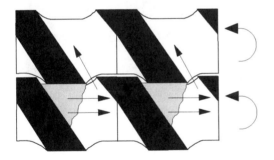

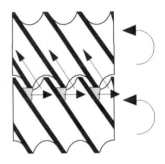

single-flighted elements   triple-flighted elements

(d)

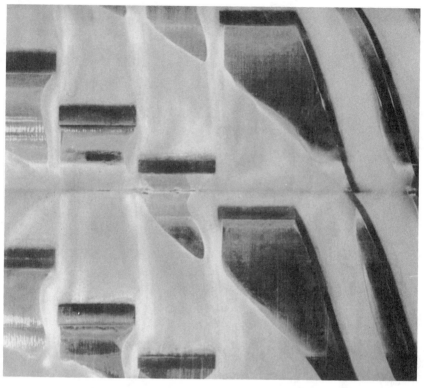

(a)

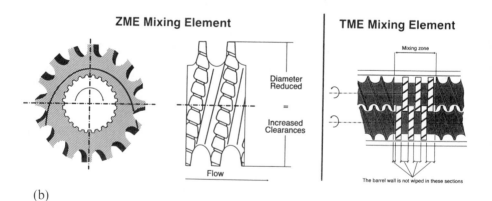

ZME Mixing Element

Diameter
Reduced

=

Increased
Clearances

Flow

TME Mixing Element

Mixing zone

The barrel wall is not wiped in these sections

(b)

Fig. 2 Kneading blocks and other mixing elements. (a) Flow distribution around kneading blocks. (b) Toothed and turbine elements for flow distribution.

3. Capability of successive treatments in, for example, successively higher vacuum environments (staging)
4. Capability for injection and mixing of a stripping agent
5. In the case of multiscrew, intermeshed extruders, the capability to automatically load and drain the devolatilizer during startup and shutdown

To operate successfully, the extruder devolatilizer must

1. Be mechanically designed and operated to sustain the required vacuum environments
2. Control the natural vent climbing behavior of elastic melts
3. Accommodate the volume of polymer foam in the extruder (not in the vent lines)
4. Control the vapor velocity to reduce incidence of entrained particles or apply deentrainment devices
5. Resupply the latent heat lost to volatile vaporization so that the processing melt stays molten and at a desirable temperature
6. Mix polymer melt with stripping aids and present the mixture to the vented section

III. CONTINUOUS DEVOLATILIZATION THEORY

Much material data is required for responsible process design. Prediction of volatile transport presupposes a knowledge of the

1. Size and shape of the vapor–solution interface, which comes from continuity and momentum balances
2. Capability of the system to supply energy to the solution, which comes from energy balances

The successful process design needs to provide an ultimate vacuum and thermal environment that at least drives equilibrium volatile concentration to target levels. This obliges the designer to become familiar with vapor pressure data of the polymer solution in the applied range of pressure (vacuum), temperature, and concentration. He or she will also need to know how much energy to resupply the process as volatiles vaporize, or for that matter to simply control solution temperature. Solution viscosity data allows estimation of heat made available through viscous dissipation and provides a clue to the interstitial free volume available for volatile transport. Solution viscosity also guides the selection of interstage sealing and mixing elements. The rate of volatile transport is described by the temperature- and concentration-sensitive diffusion coefficient, or a mass transport partitioning coefficient in the event of nondiffusion mechanisms.

Use of stripping agents to facilitate devolatilization is common. Usually water vapor or inert gas, the stripping agent should ideally

1. Be environmentally acceptable and easily separated from the contaminant
2. Not catalyze or participate in degradation or other undesirable reactions
3. Have a relatively high vapor pressure and low molecular weight so that it is effective at low concentration and easily removed
4. Have a low heat of vaporization to reduce thermal loads on the solution

Stripping agents are effective for two reasons: partial pressure reduction in the vapor space, and bulk volatile transport in the solution. The stripping agent reduces the partial pressure of the volatile in the vapor space contiguous to the melt. Since the equilibrium volatile concentration in the melt corresponds to the partial pressure of that volatile in the vapor, the presence of a second vapor effectively improves the thermodynamic driving force at the expense of generated vapor volume.

If the stripping agent (or for that matter, the contaminant) concentration is high enough, a large pressure difference (customarily termed a superheat) is present between the total vapor pressure in the melt and the vacuum at which the vent operates. Dust and dissolved gas in the melt provide sites for nucleation of a vapor bubble. Some bubbles nucleate spontaneously, but most are initiated by the void-forming cavitation action during screw rotation. The bubble grows while the volatile concentration difference from the melt feeds it. The bubble is transported along flow streamlines and under the power of its own buoyancy. An equilibrium bubble size develops, depending on the balance of local disruption forces (pressure $P_B - P$ and shear deformation $\mu\dot{\gamma}$) to stabilizing force ($2\sigma/R$), where P_B is the gas pressure in bubble; P is the vent pressure; μ is the local melt viscosity; $\dot{\gamma}$ is the local shear deformation rate; σ is the bubble surface tension; R is the bubble radius.

The vapor bubbles burst especially at the melt–vapor interface, where the suddenly free surface is allowed to stretch until a critical film thickness is violated. The bubbles also burst, coalesce, and regenerate within the melt under the action of local extensional and shearing strains. The system properties of interest for predicting bubble-driven devolatilization include

1. Critical superheat for bubble nucleation
2. Temperature- and composition-dependent surface tension of the polymer solution
3. Critical thickness of the bubble film

The stripping agent, then, increases the interfacial area (by the net surface of all submerged bubbles and irregularities in the solution surface) available for exchange of volatile contaminant from the polymer solution.

There is yet another reason for using stripping agents. Some fluid mixtures form azeotropes; that is, the mixture boils at a lower temperature than either pure component. Water and styrene monomer is a notable example.

A. An Axial Diffusion Model with Transverse Evaporation

Once the local devolatilization mechanism, the values for local equilibrium volatile concentrations, and diffusion coefficient are known, a differential mass balance on volatile component (1) can be performed so that the process mechanics can be understood and optimized. If the separation is controlled by volatile diffusion,

$$\frac{\partial \rho_1}{\partial t} + \nabla \cdot \rho_1 V - \nabla \cdot \rho D \nabla W_1 = R_1 \tag{1}$$

where, for a binary system:

W_1 = weight fraction of volatile
V = local velocity of polymer solution (m/sec)
ρ_1 = local density of volatile (kg/m^3)
ρ = local density for polymer solution (kg/m^3)
D = local mass diffusion coefficient (m^2/sec)
R_1 = rate of mass formation of volatile (kg/m^3/sec)
∇ = del operator

Following Biesenberger (1987), for the special one-dimensional case of constant density and constant diffusion coefficient at steady state without reaction, and using an alternative expression for convected mass flux of the volatile:

$$\rho_1 V = \frac{W_1 \dot{m}}{A} \tag{2}$$

where $\dot{m}$ is the local polymer solution mass flow rate (kg/sec), and A is the flow cross section (m^2). Adding boundary conditions to the homogeneous equation, a boundary flux equation analogous to a heat flux expression results:

$$D_S \frac{\partial W_1}{\partial z} = \rho k_{\text{loc}} S (W_1 - W_\text{e}) \tag{3}$$

where

k_{loc} = a local mass transfer coefficient (m/sec) whose value depends on the nature and sum of mass exchange mechanisms happening at the melt–vapor interface

S = solution-vapor interfacial area

W_e = equilibrium weight fraction of the volatile

The final equation is

$$0 = \dot{m}\frac{\partial W_1}{\partial z} - \rho AD_S\frac{\partial^2 W_1}{\partial z^2} + \rho k_{loc}S(W - W_e) \tag{4}$$

where z is the down-channel direction, and D_S is the effective axial dispersion coefficient in the z direction.

B. Geometric, Operational, and Mechanism Sensitivity of the Solution–Vapor Partitioning Behavior

Devolatilizer design affects many of the equation's coefficients:

A: The cross-sectional flow area is a consequence of channel shape and occupied channel volume.

D_S: The axial dispersion coefficient is influenced by backmixing caused by configuration and operating condition.

k_{loc}: The interface partitioning coefficient depends on the average rate of volatile depletion from the vapor–solution interface, which is different, for instance, for bursting bubbles as opposed to a diffusing species. The ability to create randomly placed bubbles and contain their volume depends on contaminant distribution and vent hardware.

S: The total vapor–solution interfacial area clearly depends on presence or absence of bubbles as well as the geometry and operating conditions in the vent area.

Equation (4) in dimensionless form is

$$0 = -\frac{\partial W_1}{\partial \hat{z}} + \text{Pe}\frac{\partial^2 W_1}{\partial \hat{z}^2} - \text{Ex}(W - W_e) \tag{5}$$

where

$$\text{Pe} = \frac{LV}{D_S}; \qquad \text{Ex} = \frac{\rho k_{loc}S}{\dot{m}}$$

Pe = dimensionless Peclet number, an indication of backmixing in the devolatilization stage (high value = low backmixing)
L = a length of a devolatilization stage
$\hat{z}$ = dimensionless length, z/L
Ex = extraction number, an indication of the capability of achieving the desired separation

Biesenberger (1987) and Werner (1981), among others, solved Eq. (5) for a range of Peclet and extraction numbers (Fig. 3). The general result is expressed as separation efficiency, E_f, versus extraction number, Ex, where

$$E_f = \frac{W_0 - W_f}{W_0 - W_e} \qquad (6)$$

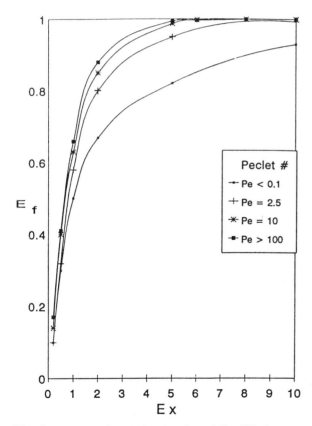

Fig. 3 Parametric solution for the axially diffusing transversely evaporating continuous devolatilizer. Pe < 0.1, well backmixed; Pe > 100, plug flow.

W_0, W_f, W_e are volatile concentration entering, leaving, and at equilibrium in the devolatilization stage. Separation efficiency versus extraction number shows the intuitive result that backmixing (low Pe) during devolatilization dilutes the volatile gradient and degrades devolatilization performance. Small Pe_L values are not available to extruders where flow direction in a down-channel direction dominates the flow dynamics.

Backmixing is an important type of extrusion mixing and the subject of many papers (Herman, 1966; Todd, 1975; Curry et al., 1991). Classical experiments track the residence time of tracers and usually report a composite system residence time distribution, which is useful, for instance, for specifying acceptable feeder variance, establishing controllability of a process, or investigating the feasibility of a reaction. In the devolatilization sense, we are interested only in the backmixing behavior of a polymer solution flowing in a partially full channel in a vent zone. Tools for this type of experimentation are just now being developed (Zeuner, 1993), and analytical methods to predict flow details in partially full channels are not yet available.

The boundary condition flux expression deserves special discussion. There are several mechanisms acting in parallel to bring the volatile contaminant to the vapor–solution interface.

The interface is conceptually divided between a pool surface and a film surface. If the devolatilization mechanism is diffusion, an expression is available from shallow penetration theory for the partitioning coefficient:

$$k_{loc} = 2\left(\frac{D}{\pi\lambda}\right)^{1/2} \tag{7}$$

where λ is the characteristic time of exposure. The exposure time depends on the renewal rates of pool and film, which in turn depend on rpm, throughput, and flow behavior.

The exposure time of films in co-rotating closely intermeshing devolatilizers was analyzed by Collins et al. (1985). They recognized films in the screw intermesh, on the barrel wall, and on the screw surfaces. The exposure time is related to geometric features, rpm, and degree of fill. For example, when only films exist in a devolatilizing zone and no polymer solution fills the channel volume, but the films are regenerated by screw rotation and self-wiping:

1. The intermesh volume is negligible.
2. The screw film exposure time

$$\lambda_s = \frac{2\pi}{N}$$

3. The barrel film exposure time

$$\lambda_b = \frac{2\pi}{ZU} \left[\frac{1}{2} + \frac{Z}{\pi} \cos^{-1} \frac{\rho_c}{2} \right]$$

where Z is the number of screw flights, U is the screw RPM, and ρ_c is the ratio of centerline to barrel diameter.

Foam devolatilization is different. A strict accounting of foam devolatilization requires expressions for nucleation, transport, and rupture of the foam. Furthermore, since polymer solutions rapidly expand then usually quickly collapse during transit under the vent, the volatile flux to the vapor space:

1. Varies with position under the vent;
2. Is erratic since bubbles each contain significant volatile volume and burst suddenly;
3. Is not easily calculated, since bubble population and size vary with available nucleation sites, solution superheat, and exposure time.

A sort of partitioning coefficient can be calculated from an appropriate model for bubble count (nucleation), size (growth), and volatile release (rupture). Observation of foam volume growth and collapse in vent stages verifies the models. Foster (1989) developed and fit foam regime models for a counter-rotating tangential devolatilizer and found that k_{loc} drops with increasing foam volume in the vent and rises with screw rpm (hypothetically because surface renewal, cavitation effects, and bubble-convected transport to the surface are improved with rpm).

Estimations of total vapor–solution interface area, S, or even indeed, the operating vent stage length, L, are not trivial exercises. The flow dynamics of a fluid with changing composition and foam structure is complicated. While it is generally agreed that the vented length extends beyond the vent opening, how the disengagement and consolidation of polymer solution from and against the interstage mixing element behaves is not well understood. Furthermore, how fluid obstructions interfere with the free passage of vapor and contribute to vapor pressure drop and vapor composition gradient in the vent stage are unknown. For these reasons, experimental observations of solution expansion, collapse, and free surface shape and movement are valuable indicators of the local separation mechanism.

C. Design Efficiency

The macroscopic, steady-state component balance for the volatile component is

$$\dot{m}(W_0 - W_f) = \rho k_m S_m (W_0 - W_e) \tag{8}$$

where

> W_0 = inlet mass fraction of volatile
> W_f = outlet mass fraction of volatile
> k_m = mass transfer coefficient for volatile
> S_m = composite solution–vapor interfacial area

By comparison with the definition of separation efficiency (Eq. 6),

$$k_m = \frac{\dot{m}}{S_m} E_f \qquad (9)$$

The most efficient equipment therefore maintains the highest partitioning coefficient at highest rate and lowest interfacial area. The interfacial area here relates only to machine size and specifically not to surface renewal and stripping tricks that improve the interphase transport without increasing the equipment size.

For the special case of dilute solutions, the solution volume decreases and solution viscosity increases dramatically as devolatilization proceeds. In extreme cases, the solution can contain regions of free solvent or solvent that can be precipitated into two phases. Mechanical filters are always preferred as phase separators for energy efficiency, safety, and product quality reasons. Low-viscosity solutions can be most economically processed on gravity-driven flash chambers. The thermodynamical flash potential is economically developed in heat exchangers and pumps. The intermediate concentrated solutions can be preconcentrated in light-duty, high-volume twin-screw devolatilizing extruders before final processing in a high-power, staged devolatilizer.

Vacuum staging (Figs. 4 and 5) is universally practiced for solutions with high solvent content (ca. 10%), for many reasons:

1.   High separation efficiencies are more easily maintained for smaller thermodynamic driving forces, and therefore the process will be more load tolerant and the product more uniform.
2.   Foam growth will be more modest at higher applied pressure (lower vacuum), and therefore rates can be improved or stripping agent content can be increased.
3.   Volumetric vacuum pump loads can be reduced, resulting in less expensive systems and easier effluent reclaim.

On the other hand, an eventual environment consistent with the target residual contaminant level must be developed in a practical machine length, and therefore the number of vent stages is limited.

Fig. 4 Staged vented devolatilizer type VDS-V83. (Courtesy Werner & Pfleiderer Corp.)

Following Biesenberger (1987), and using Eq. (9), a difference expression for volatile concentrations in staged systems is

$$W_j - (1 - E_{fj})W_{j-1} = E_{fj}W_{ej} \tag{10}$$

where

E_{fj} = separation efficiency of jth stage
W_j = volatile content of solution entering jth stage
W_{j-1} = volatile content of solution entering $(j-1)$-th stage

Following Todd (1974), a minimum total vapor flow rate is realized in N stages, when

$$P_j = \left(\frac{W_{N,e}}{W_0}\right)^{1/N} P_{j-1} \tag{11}$$

where

$W_{N,e}$ = equilibrium volatile content in last stage
W_0 = volatile content entering apparatus
N = number of staged vents

However, this minimum is realized for ideal stages that are realized with oversized equipment (allowing for efficiency E_f).

(a)

(b)

Fig. 5 Looking upsteam along a melt-fed staged devolatilizer. (a) 57-mm test extruder. Upsteam view to melt feed connection. (b) 300-mm production extruder in the assembly hall. (Courtesy Werner & Pfleiderer-KSVT.)

The corresponding machine fractional separation using the minimum vapor pressure profile is

$$F_s = \frac{a(1 - a^N E_f) - (1 - E_f)[1 + (a - 1)(1 - E_f)^N]}{a - (1 - E_f)} \tag{12}$$

where

$$F_s = W_0 - \frac{W_N}{W_0} = (1 - E)E_f, \qquad \text{where } E = \frac{W_{N,e}}{W_0} \tag{13}$$

$$a = \left(\frac{W_{N,e}}{W_0}\right)^{1/N} \tag{14}$$

$$E_f = W_0 - \frac{W_N}{W_0} - W_{N,e} \tag{15}$$

Both the vacuum requirements and machine size can be minimized by

1. Estimating the ultimate vacuum and maximum number of stages
2. Calculating the staged pressure profile
3. Estimation, experiment, or computation to establish the overall process thermodynamic separation efficiency at the desired rate for a single-stage process
4. Calculating the efficiency improvement for the staged process

Adjust parameters for desired result and repeat.

D. Vapor–Solution Interface in ZSK Extruders

Some insight into the dependency of available interfacial area on operation and configuration of a twin-screw devolatilizer can be realized by studying Fig. 6. The figures were developed for a ZSK-30 using the following assumptions:

1. Surface areas include films on barrel surfaces and exposed pool area. Screw surfaces are assumed to be wiped clean.
2. The polymer solution fills the flight from the pushing flight in an orderly fashion with increased fill degree. The solution does not roll during transport. Films are formed by the troweling effect of a passing flight.
3. The analysis reflects steady-state fill without effect of disengagement or consolidation effects of neighboring elements.

The analysis in the graphs in Fig. 6 yields a sort of normalized extraction number (S/Q) versus degree of fill for a particular rpm or screw pitch. The highest extraction potential is available for the lowest rpm (or pitch) *and* lowest fill degree. The same surface area is available for many

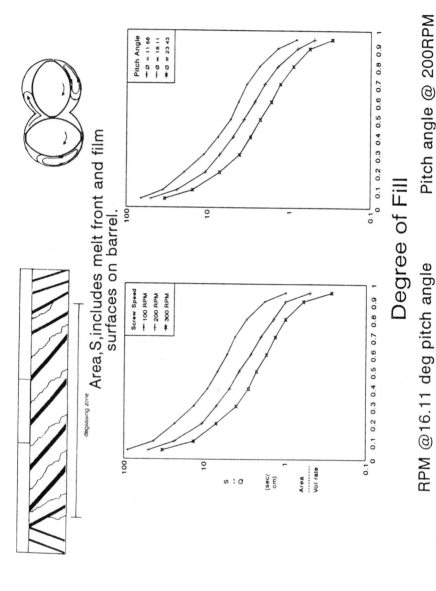

Fig. 6 Normalized solution–vapor interface area as a function of degree of fill for ZSK-30 devolatilizer.

appropriate combinations of pitch, rpm, and fill degree. However, the combination of low rpm (or pitch) and low degree of fill improves the value of the specific area. The meaning of the graphs is clear. For maximum efficiency of a diffusion-controlled separation, use low screw pitch, rpm, and throughput. This recommendation is in practice neither economical or feasible. It is more practicable to investigate process possibilities such as viscous heating at high rpm, exposure of fresh surface with screw design or operating condition, or use of stripping agents to produce quality product at competitive rates.

IV. DEVOLATILIZATION EQUIPMENT

A generalized view of a devolatilizer is presented in Fig. 7. The extruder consists of solution feed connection, back vent, primary vent, and several successive staged vents with interspersed stripping agent injection and mixing barrels. The devolatilized solution may be finally mixed with additives and discharged.

A. Preconcentrators

Molten solutions can easily be superheated and flashed into finishing devolatilizers, releasing high volumes of vapor. Depending on the polymer solution and volume of vapor removed, snowing (freezing of free polymer particles), popcorning (a kind of rapid localized foaming), foaming of the bulk solution, and film draining can occur. The flash chamber must:

1. Provide adequate volume—maintain a clear path to vent line for the separation
2. Collapse foam into the finishing extruder
3. Wipe adhering particles from the vent hardware
4. Present a large surface area to the vent space
5. Maintain pressure and temperature conditions required to sustain the separation

Flash chambers, falling-strand and strip chambers, and wiped chambers are available. A configuration for a wiped flash chamber and the connection to a finishing devolatilizer are shown in Fig. 8.

For feed solutions with modest flash potential, the first barrels of the finishing devolatilizer can be used to collect entrained polymer and release vapor in a technique called back venting. The technique is especially useful for twin-screw devolatilizers, which have a tortuous flow path and a self-cleaning mechanism. The number of polymer-collecting screw flights and vapor pressure gradient between solution feed and back vent can be

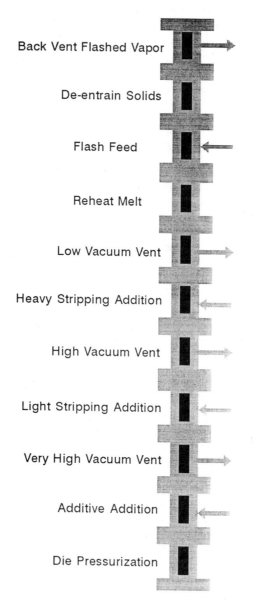

Back Vent Flashed Vapor

De-entrain Solids

Flash Feed

Reheat Melt

Low Vacuum Vent

Heavy Stripping Addition

High Vacuum Vent

Light Stripping Addition

Very High Vacuum Vent

Additive Addition

Die Pressurization

Fig. 7 Conceptual diagram of a devolatilizing extruder.

optimized by judicious choice of screw elements. Large-diameter twin-screw solution concentrators consisting of a short, back vented extruder discharging into a smaller, close-coupled finishing devolatilizer are used when large solution volume charges are effected by the flash devolatilization (Fig. 9).

B. Screw Designs for Extruder Devolatilizers

Proper screw design is critical for

1. Controlling flow behavior in the rear vent
2. Resupplying heat via shear-intensive elements to the polymer solution
3. Separating the differential pressures of adjacent vacuum stages
4. Mixing in a stripping agent
5. Foam disengagement from the melt seal to the vent, and foam collapse in and after the vent
6. Optimizing surface exposure and exposure time in the vent stage

Several candidate designs are shown in Fig. 10.

A priori designs are subject to optimization for operability and performance during trials.

C. Vent Hardware

Polymer solution behavior in a typical devolatilization stage is shown in Fig. 11. Mildly superheated solution is presented to the vent opening, where it foams. The foam builds volume in the screw channel and rolls against the downstream vent wall. Some melt climb is also evident in the intermesh. The foam collapses under the action of vent pressure and the moving screws and leaves behind a translucent melt populated with frozen particles of polymer. The filled channel volume and vent roll dissipate near the discharge end of the vent.

A number of vent adapters have been developed (Fig. 12) to improve foam and melt management. Pocketting the down-channel vent wall and plugging the intermesh area reduces melt climb. Of course covering vent area to improve melt management subtracts from the open vent area, which increases vapor velocity (and entrainment) and degrades the pressure differential to vapor space from melt that is not directly under the opening. A recent vent adaptor development is the side-mounted vent for twin-screw devolatilizers (Fig. 13). In this vent location, contaminants and entrained polymer are drained away from the process stream, avoiding incidental contamination.

In an extreme vent management situation, a small twin-screw device called a vent stuffer is used to compress solution into the devolatilizer

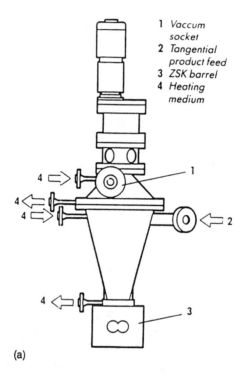

1 *Vaccum socket*
2 *Tangential product feed*
3 *ZSK barrel*
4 *Heating medium*

(a)

Fig. 8 A wiped surface flash evaporation chamber, type VS30. (a) Process connections; (b) internal details. (Courtesy Werner & Pfleiderer Corp.)

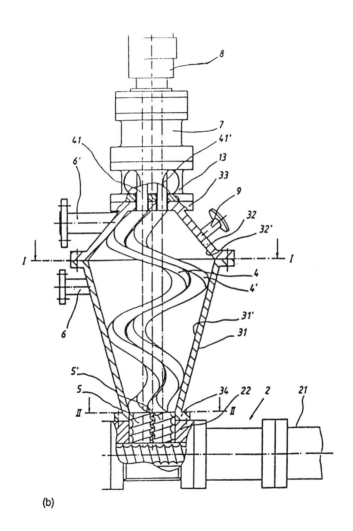

(b)

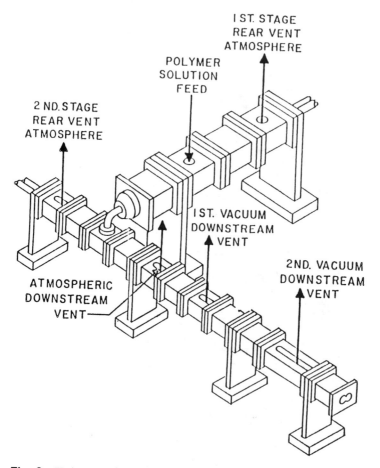

Fig. 9 Twin-screw flash evaporation extruder feeding a devolatilizer. (Courtesy Werner & Pfleiderer Corp.)

extruder while back venting releases vapor to the vent line. The device, shown in Fig. 14, is also useful as a straining filter for two-phase separations.

D. Injection and Mixing of Stripping Agents

Stripping agents should be heated to reduce quenching of the polymer solution unless coolant action is desirable. The agents are infinitely more easily metered and mixed as liquids rather than as vaporized gas. An injection nozzle (Fig. 15) with adjustable restriction and sprayed discharge facilitates metering and initial distribution. A pressurized mixing region in the extruder

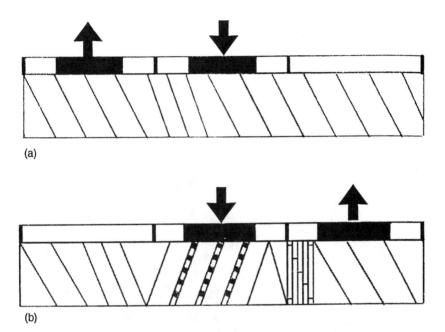

(a)

(b)

Fig. 10 (a) Screw and barrel arrangement for back venting technique. (b) Screw and barrel arrangement in a stripped vent stage.

devolatilizer (Fig. 10b) improves stripping agent distribution and restricts stripping agent vapor escape away from the targeted vent stage.

V. APPLICATIONS

A. Bulk Polymerization Finishing Devolatilizer

Bulk polymerization of polystyrene is terminated at 10–30% monomer content and further concentrated in a devolatilizer suitably designed to handle the high-viscosity syrup. Residual volatile content should be minimized to reduce flammability and odor from parts made from the product.

At impact grade with MI = 2 and containing 10.2% monomer is fed at 220°C and 6 bar through a slot in the side of a vented barrel of a ZSK-90 devolatilizer. The solution is flashed in the first vent stage and stripped and degassed in three successive staged vents. Water is the stripping agent. The residual volatile content was measured in the melt between vent stages, and the product contained 123 ppm residual styrene. The configuration, vent operating conditions, and residual styrene in solution are plotted

(a)

Fig. 11 (a) View in the vent port of a devolatilizing stage under operation. (b) Melt behavior transitions from foam to diffusing melt.

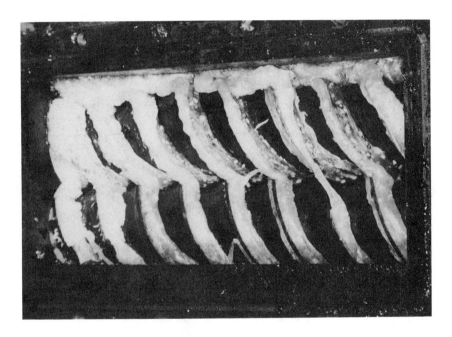

(b)

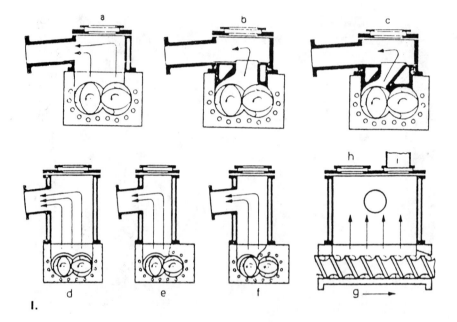

Fig. 12 Vent adaptor type for solution control in the vent: (I) mounted vent domes; (II) vent adaptors.

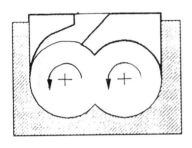

design A

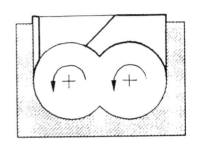

design B

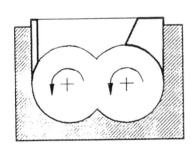

design C

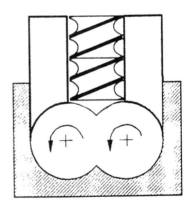

twin-screw vent stuffer

II.

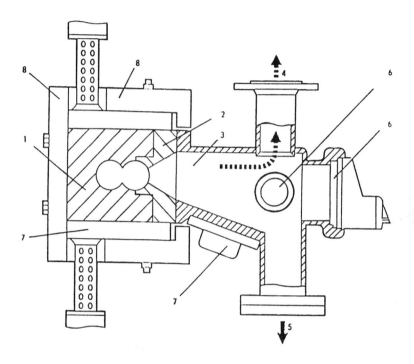

1.ZSK Barrel Section
2.Vent Insert
3.Side Vent Port
4.Volatiles
5.Degraded Product
6.Sight Glass
7.Heater Shell
8.Insulating Cover

Fig. 13   The side vent. (Courtesy Werner & Pfleiderer Corp.)

in Fig. 16. The process is analyzed for efficiency and apparent mass transfer coefficient, k_m, in Table 1. Equilibrium volatile concentrations are calculated on the basis of partial styrene pressure in the vent. The separation efficiency of the last vent is degraded because of monomer production through polystyrene degradation, which is appreciable above 235°C. Degree of fill and effective vent L/D are calculated using the program SULIZ (Burkhardt, 1974) for a polymer with flow properties of the base resin. Melt–vapor interface area is developed from an analysis like Fig. 6; that is, the total interface is represented as the sum of film and pool surfaces for a nonfoaming

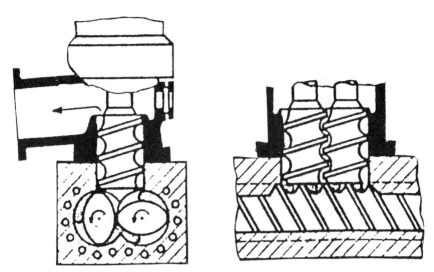

Fig. 14   Twin-screw vent stuffer. (Courtesy Werner & Pfleiderer Corp.)

melt. Foaming was observed in stage 1 and 2 vents. As expected, the stage efficiencies fall off with decreasing volatile content, and the mass transfer coefficient is lower in the diffusion zone (stages 3 and 4) where monomer superheat is low. Higher vacuum strengths or higher stripping rates (providing melt temperature could be maintained) would improve performance.

The grades of polymer and types of additives have effects on devolatilization performance even if equilibrium volatile contents are indistinguishable. A high viscosity grade may generate excessive heat by shear dissipation as the polymer concentrates and regenerate monomer as a degradation mechanism. Internal low-volatility lubricants help moderate shear overheating. Impact modifiers can act as stress concentrators and help collapse an otherwise resilient foam.

B. Solution Polymerization Finishing Devolatilizer

Polyethylene, polymerized in solution, has some unique properties and can contain various comonomers to produce products over a wide density range. The reaction is routinely carried out in ca. 85% solvent. The solution is concentrated to ca. 90% solids in flash devolatilizers and finished in a devolatilizing extruder. High residual volatile content causes dangerous environments in storage and void defects in films and moldings. Care must be taken during devolatilizing to avoid degradation, cross-linking, and gel and speck formation in the melt.

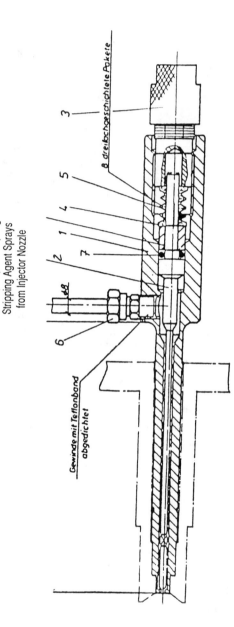

Line Pressure Lifts
Injector Body against
Manually Adjusted Springs
Stripping Agent Sprays
from Injector Nozzle

8 dreibchgeschichtete Pakete

Gewinde mit Teflonband
abgedichtet

Fig. 15 Stripping agent injection nozzle. (Courtesy Werner & Pfleiderer Corp.)

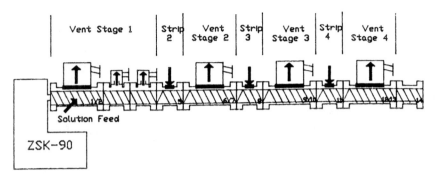

Fig. 16 Flow design for staged separation of polystyrene from its monomer.

Table 1 Analysis of PS–Styrene Devolatilization Process of Fig. 16[a]

Stage j	1	2	3	4	Product
Applied vacuum pressure (torr)	83	26	26	9	
Stripping agent injected (mass fraction)	0	0.019	0.038	0.038	
Inlet volatile concentration to stage (mass fraction)	0.102	$2.09e^{-3}$	$4.07e^{-4}$	$1.75e^{-4}$	$1.22e^{-4}$
Mole fraction styrene in vapor	1.0	$1.5e^{-2}$	$1.1e^{-3}$	$2.5e^{-4}$	
Partial vacuum on styrene (torr)	83	0.4	$2.9e^{-2}$	$2.3e^{-3}$	
Stage entrance melt temp. (°C)	230	278	278	265	
Equilibrium volatile level in melt, W_{ej}	$3.6e^{-3}$	$1.8e^{-5}$	1.3^{-6}	$1e^{-7}$	
Available superheat on styrene, SH_j (atm)	3.0	$2.8e^{-2}$	$2.9e^{-2}$	$5.2e^{-3}$	
Poststage volatile conc. (mass frac.), W_{aj}	$2.09e^{-3}$	$4.07e^{-4}$	$1.75e^{-4}$	$1.22e^{-4}$	
Stage efficiency, E_{fj}	1.00	0.81	0.57	0.30	
Average fill degree, E_j	0.131	0.178	0.178	0.178	
Vent, L/D	8.0	5.4	5.7	6.7	
Melt–vapor interface area, (cm²)	$3.11e^3$	$2.80e^3$	$2.95e^3$	$3.47e^3$	
Overall stage partitioning coefficient, k_{mj}[b] (mm/sec)	0.38	0.33	0.22	0.10	
Overall efficiency, E_f	0.999				

[a] Impact PS, 2MFR at 415 kg/hour; use Henry's constant, 30 atm/mass fraction.
[b] Interface calculated for nonfoaming process.

A solution containing 10% of cyclohexane and high-density poly-ethylene (HDPE) (10 MI) was flashed into a ZSK-57 devolatilizer at 200°C and 20 bar. The devolatilizer had a rear vent held at atmospheric pressure, a downstream vent at reduced pressure, and two stripped and staged vacuum vents. Water was the stripping agent. Mass balance was done in stage 1 vent condensate, and product volatiles were measured. The machine design is shown in Fig. 17, and the process analysis in Table 2. Stage efficiency in the back vent stage is low because the vent volume was inadequate for foam expansion and rupture of a large number of foam cells. Stage mass transfer coefficient are higher than for the styrene process, reflecting the higher superheats achieved and the relative ease of the solvent separation for the polyethylene process.

Indiscriminate devolatilization of polyethylene solutions achieves very low residual volatile levels, but at the expense of cross-link degradation to the polymer. High viscosity (low MI) grades are especially sensitive in this regard. Successful designs distribute stripping agent in an optimal fashion to derive maximum benefit from both the heat sinking and bulk volatile transport potential.

Depending on the molecular weight distribution of the polymer, low end fractions could lubricate an efficient, high-rpm separation without unacceptable damage to the polymer. The solution behavior in the vent zone, recognized by relative foam expansion, and banding or chunky flows, is affected by screw pitch within the vent and mixing elements upstream of the vent.

C. Engineering Polymers and Composites

A popular, efficient method, for free radical polymerization is direct poly-merization of a syrup or monomer–host polymer mixture in the twin-screw

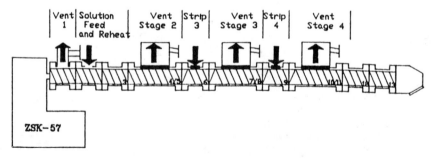

Fig. 17 Flow diagram for staged separation of polyethylene from its solvent.

Table 2 HDPE–Cyclohexane Process of Fig. 17[a]

Stage j	1	2	3	4	Product
Vent pressure (torr)	760	115	35	25	
Stripping agent (mass fraction)	—	—	0.01	0.01	
Inlet volatile (mass fraction)	0.10	$3e^{-2}$	$2.3e^{-3}$	$8e^{-4}$	$3.3e^{-4}$
Mole fraction cyclohexane in vapor	1	1	$3.1e^{-2}$	$1e^{-2}$	
Cyclohexane partial pressure (torr)	760	115	1.1	0.25	
Equilibrium volatile concentration, W_{ej}	$2e^{-2}$	$2.3e^{-3}$	$1.9e^{-5}$	$3.9e^{-6}$	
Available superheat (atm)	4.1	1.8	0.17	$7e^{-2}$	
Post-stage concentration, W_{aj}	$3e^{-2}$	$2.3e^{-3}$	$8e^{-8}$	$3.3e^{-4}$	
Stage efficiency, E_{fj}	0.875	1.0	0.658	0.590	
Average fill degree, E_j	0.172	0.246	0.246	0.246	
Vent L/D	4.2	7.4	6.3	6.3	
Melt–vapor interfacial area[b] (cm^2)	$7.19e^2$	$1.22e^3$	$1.04e^3$	$1.04e^3$	
Overall stage partitioning coefficient, k_{mj}[b] (mm/sec)	0.76	0.51	0.40	0.35	
Overall efficiency, E_f	0.997				
Product MI	7.8				

[a] HDPE at 225 kg/hour; feed MI = 8; discharge temp. = 240°C.
[b] Interface calculated for nonfoaming process.

extruder followed by vent extraction of the unpolymerized monomer. High clarity grades of acrylic polymers, graft rubber impact modifiers, and custom formulated vinyl acetate polymers are manufactured in this way.

High-performance engineering plastics become intractable in reactors unless polymerized in solvent. Polycarbonate–methylene chloride is a rare example of a high-volatility solvent system for an engineering plastic. Polysulfone–sulfonated solvent and ether–imide–diphenyl solvent separations, where the low solvent vapor pressure and high polymer viscosities require process temperatures above 250°C and applied vacuum below 10 torr, place high performance demands on process equipment. Furthermore, a suitable, safe, economical, easily separated, and easily mixed stripping agent may not be available for these systems, so that final devolatilization takes place by the slow diffusion process. Process development is often delayed by slow polymer solubility in expensive, dangerous solvents and nondetectable limit analyses.

Incidental volatiles borne by feedstocks, or volatiles created by in-process reactions can wreak havoc with appearances and properties. Stripping volatiles from the primary feed in the counter-current gas stream venting from the feed hopper is usually an unproblematic process. Interstitial air and water vapor are usually displaced through the consolidating and heating action of the plasticating region of the extruder screw. If, however, the feed contains particulates with poor coating behavior (high-energy surfaces), rough surfaces, hydrated moisture, or other residual volatiles, design of the feed and plasticating screws and effective heat transfer from barrel to the developing melt are key considerations for high-quality product. Reinforcements fed into the matrix melt are usually hydrophobic (glass or silica) or calcined and packed to prevent moisture absorption. Moisture would not only interfere with the interface but also could degrade the molecular weight of polymer matrices. Glass fiber reinforcements are sized with low-molecular-weight vinyl alcohol adhesives to improve feedability and reduce fiber damage during handling. The vinyl alcohol is vaporized and removed. Particulate fillers often are treated with vinyl or anhydride surfactants, which improve adhesion to the matrix via condensation reaction or polar interaction. These residuals or by-products generally do not interact with the matrix and are easily removed if the vent can be designed and operated to stay clear. Applied vacuum improves the intimacy of the matrix–filler interface and, therefore, the stress transfer across it. Well-degassed composites are better dispersed, more reliably stranded, and less prone to transverse swell.

VI. FINAL COMMENTS

Polymer devolatilization is a process common to polymer isolation, reactive extrusion, composite compounding, and alloying. Extrusion devolatilization has been widely developed in the last 40 years. Specialized equipment has been developed for isolation of bulk and solvent polymerized materials. The axial mass transport equation with traverse boundary conditions has been applied to unify thermodynamic, bulk transport, and surface transport data into a model with sensitivities of axial backmixing, available surface area for mass exchange, throughput, thermodynamic driving force, and mass transport mechanism. Developed surface area, backmixing, and available mass transport mechanisms are responsive to machine type and screw design.

The long aspect ratios of devolatilization extruders and limited back-mixing in screw-type elements substantiate the application of one-dimensional mass transport models. The two-dimensional recirculation flows are accommodated in the mass transfer coefficient at the solution–vapor interface by making the coefficient a function of pool and film surface regeneration times. A mass transfer coefficient that is sensitive to the three-dimensional flows characteristic of the intermesh in twin-screw machines has yet to be developed. Until two- and three-dimensional free surface modeling in twin-screw extruders is developed, axial models with manipulated boundary conditions will probably survive.

Economical designs allow the highest mass transport with the smallest machine diameter and length, lowest vacuum level and volume requirement, and lowest externally supplied heat source to perform the desired separation. In addition, the design must be operable and accommodating to the quality requirement of the product. Stripping aids, when available, must be used to increase the solution–vapor interface. Staged vacuum improves process load tolerance and minimizes the devolatilizer diameter (in favor of length).

NOMENCLATURE

A interfacial areas across which the volatile is flowing (m^2)

D local mass diffusion coefficient (m^2/sec)

D_S axial dispersion coefficient along extruder channel (m^2/sec)

E_f separation efficiency

E_{fj} separation efficiency in jth stage

Ex   extraction number

k_{loc} local mass transfer coefficient (m/sec)

k_m an average mass transfer coefficient for the volatile over an entire vent stage (m/sec)

L length of a devolatilizing stage measured in a down-channel direction (m)

$\dot{m}$ local polymer solution mass flow rate (kg/sec)

N number of vent stages

P vent operating pressure ($Pa/absolute$)

P_B gas pressure in a devolatilizing bubble ($Pa/absolute$)

Pe   Peclet number for axial backmixing over the vent stage length

P_j vent pressure in a jth stage ($Pa/absolute$)

Q volumetric flow rate (m^3/sec)

R bubble radius (m)

R_1 rate of mass formation of volatile ($kg/m^3/sec$)

S solution–vapor interfacial area (m^2)

S_m total solution vapor interfacial area in a vent stage (m^2)

U screw speed (rev/sec)

V local velocity of the polymer solution (m/sec)

W_1 weight fraction of volatile in solution

W_e local equilibrium weight fraction of volatile in polymer solution

$W_{N,e}$ equilibrium volatile mass fraction in last vent stage

W_f outlet mass fraction of volatile from last vent stage
W_j outlet mass fraction of volatile from jth vent stage
W_{j-1} inlet mass fraction of volatile to jth vent stage
W_0 inlet mass fraction of volatile to vent stage or the entire devolatilizer
Z number of screw flights
z down-channel distance in a screw helix
$\hat{z}$ dimensionless down-channel distance $= z/L$
$\dot{\gamma}$ local shear deformation rate $(\sec^{-1})$
λ exposure time of solution to a vent (sec)
λ_b exposure time of solution film on barrel (sec)
λ_s exposure time of solution film on screw (sec)
μ solution viscosity (Pa·s)
ρ local density of polymer solution (kg/m^3)
ρ_1 local density of volatile (kg/m^3)
ρ_c ratio of centerline distance to barrel diameter for a twin-screw extruder
σ devolatilizing bubble surface tension (N/m)

REFERENCES

Biesenberger, J. (1987). Polymer melt devolatilization: On equipment design considerations, *Adv. Polym. Tech.*, *7*(3): 267–278.

Burkhardt, U. (1974). SULIZ, Werner and Pfleiderer. (A program for computing the drag capability of ZSK screw elements according to a two-dimensional analytical analysis for power low fluids. The program is not publicly available.)

Collins, G., Denson, C., and Astarita, G. (1985). *AIChE. J.*, *31*: 8.

Curry, J., Kiani, A., and Dreiblatt, A. (1991). *Intern. Polym. Process.*, 6(2): 148.

Foster, R. (1989). Mass Transfer Processes in the Extrusion Devolatilization of Polymer Solutions, Ph.D. Dissertation, Univ. of Pittsburg.

Hermann, H. (1966). *Chem. Ing. Tech.*, *38*: 25.

Todd, D. (1974). 32nd SPE Antec Preprints.

Todd, D. (1975). *Polym. Engin. Sci.*, *15*: 437.

Werner, H. Devolatilization of polymer in multi-screw devolatilizers, (1981). *Kunstoffe*, *71*: 18.

Zeuner, A. (1993). "Local In-Line Detection of Melt Flow (RT, RTD) Along Extruders by Radio Tracers," Proceedings of Int'l Conf. on Advanced Polym. Materials, IUPAC Conference, Dresden, p. 64.

13

Devolatilization in Counter-Rotating Nonintermeshing, Twin-Screw Extruders

Russell J. Nichols

Farrel Corporation, Ansonia, Connecticut

J. Thomas Lindt

University of Pittsburgh, Pittsburgh, Pennsylvania

I. INTRODUCTION

A. Polymer Isolation

Counter-rotating, nonintermeshing (CRNI) twin-screw extruders have found wide acceptance in the isolation of polymers from solutions, the isolation of polymers from emulsions or suspensions, and in devolatilization-driven reactive extrusion processes. While the equipment for these different processes might appear to be mechanically similar, the underlying physics determining the performance can be quite different, as solutions are single-phase systems, emulsions and suspensions are two-phase systems, and reactive extrusion systems add the complexity of time-dependent composition.

B. Single-Phase Systems

Devolatilization is the thermodynamically driven separation of a volatile species from a less volatile material. When devolatilizing relatively dilute low-viscosity solutions, the dominant factors are the thermodynamic driving force (temperature and partial pressure) and the free volume available for separation of vapor from the boiling solution. As solvent is removed, and viscosity increases to the intermediate viscosity range, bubble nucleation and growth dominate the mass transfer process. Finally, in high-viscosity concentrated solutions, the process of volatile removal is primarily diffusion controlled. A complete extrusion process must provide an appropriate environment to satisfy the substantially different requirements of each stage of the process.

Many polymerization reactions are either carried out in dilute solutions, or as a result of the polymerization process create a substantial volume of reaction by-products that must be removed. Systems of commercial interest often involve complex multicomponent polymers and/or multicomponent solvents. The physical principles governing separation of such polymer–solvent systems include multicomponent solvent–solvent interactions, multicomponent polymer–solvent interactions, and physical constraints introduced by the geometry of the machine in which separation takes place. In addition, the vast changes in viscosity that take place as a dilute solution is processed into a solvent-free polymer must be accommodated.

C. Two-Phase Systems

In order for a two-phase system to be efficiently processed in an extruder, the phases must separate fairly well and the high-viscosity (or solid) phase must form a reasonably continuous mass. Common examples of such systems are numerous commercial elastomers (e.g., SBR, NBR, EPDM, butyl) recovered as wet crumb or in slurry from emulsions or suspensions. Technology for extrusion dewatering and drying such wet rubber crumb has been employed for many years. These methods typically squeeze water out of the product by mechanical compression or raise the product temperature by controlled shear so that when the product is expelled through a die, efficient flash drying occurs.

Today it is possible to continuously coagulate polymer latex, and wash, dewater, dry, and pelletize finished polymer in a continuous twin-screw extrusion process. This process is described in detail in Section IV.B.

D. Reactive Systems

Many kinds of reactions are carried out successfully in CRNI twin-screw extruders. Most reactive systems require both excellent mixing and volatile removal. Other design considerations of particular importance to reactive extrusion are reactor volume and residence time distribution. Devolatilization requirements for reactive systems are discussed briefly Section IV.C.

II. PROCESS PRINCIPLES OF COUNTER-ROTATING, NONINTERMESHING TWIN-SCREW EXTRUDERS

A. Types of Twin-Screw Extruders

There are three basic types of twin-screw extruders: (1) counter-rotating, nonintermeshing (CRNI) twin-screw extruders, which are used for compounding, polymer isolation (including devolatilization, coagulation, and drying of latex emulsions), and for a wide variety of reactive extrusion processes; (2) counter-rotating, fully intermeshing twin-screw extruders, which are principally used for extrusion of polyvinyl chloride (PVC) pipe and profiles; (3) co-rotating, fully intermeshing twin-screw extruders, which are also widely used for compounding and devolatilization. Figure 1 is a

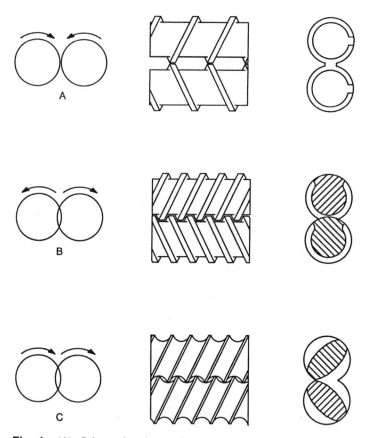

Fig. 1 (A) Schematic views of nonintermeshing, counter-rotating screws. (B) Schematic views of intermeshing, counter-rotating screws. (C) Schematic views of intermeshing, co-rotating screws.

schematic representation of these three basic machine types. Not included is the theoretically possible fourth case of co-rotating, nonintermeshing twin screws. Although nonintermeshing mixing sections are employed in some co-rotating twin-screw extruders, no commercial machines are manufactured in which either solid or melt transport through the entire machine depend solely on non-intermeshing co-rotating screw elements.

B. Screw Design Principles

The full range of screw design technology applicable to single-screw extruders can be adapted to the CRNI twin-screw extruder. However, many of the mixing devices introduced in single-screw extruders to overcome limitations in distributive mixing are not needed in CRNI twin-screw extruders because of the excellent mixing characteristics arising from the natural interaction of polymer flows between the screws. As a result, the most commonly utilized screw elements are usually simple square pitch, single flight conveying screws; longer pitch, multiflight screws for vent zones; and cylindrical or reverse flight compounders or seal sections. Typical screw elements are illustrated in Fig. 2.

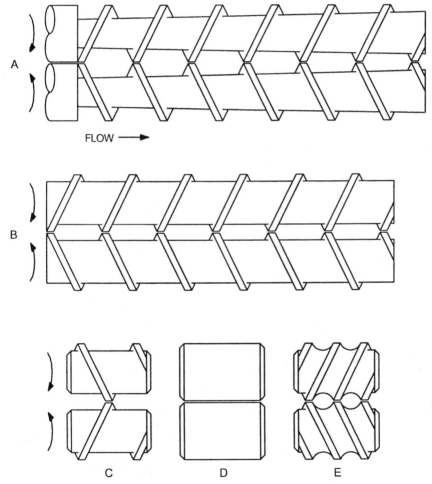

Fig. 2 Typical screw elements: (A) deep flighted feed screws; (b) shallow flighted melt conveying screws; (C) forward flighted compounders; (D) cylindrical compounders; (E) reverse flighted compounders.

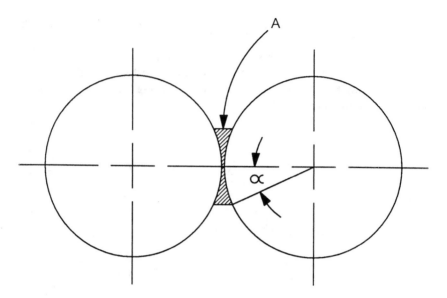

Fig. 3   Apex flow region (includes 2 × flight clearance).

Transport of molten polymer in CRNI twin-screw extruders is by drag flow, and it is quite similar to that in single-screw extruders except that there is an additional component of pressure flow resulting from the open area at the apex of the extruder barrel, as shown in Fig. 3. Additionally, there are two principal orientations in which the screw flights can be positioned between the main and auxiliary screw. As illustrated in Fig. 4, the screw flights can either be staggered, in which the flight tips of one screw oppose the root of the other screw midway between the flights, or they can be matched, in which the flight tips of one screw are directly opposed to the flight tips of the other screw. The matched flight configuration optimizes pressure-building capability, and the staggered flight configuration optimizes distributive mixing. An analytical solution for the matched flight configuration (developed by Lindt) as presented in Nichols (1984) is shown in the following section. As yet, flow performance of the staggered flight configuration has only been solved by using numerical methods. The flow equations that follow are derived in the same fashion as those by Kaplan and Tadmor (1974), but they include provision for the apex flow, which had been previously overlooked.

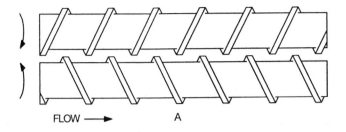

FLOW ⟶   A

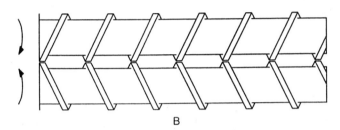

B

Fig. 4 (A) Staggered screw flight opposition. (B) Matched screw flight opposition.

C. Flow Equations (Matched Flights)

$$q = \frac{1}{2} V_z W H F_{DCRT} - \frac{W H^3}{12\mu} \frac{\Delta P_T}{\Delta z_T} F_{PCRT} \tag{1}$$

where F_{DCRT} and F_{PCRT} are given by

$$F_{DCRT} = f F_d \frac{(2 + a/H)^3}{(2 + a/H)^3 (f/F_p) + 2(1 - f)} \tag{2}$$

$$F_{PCRT} = \frac{(2 + a/H)^3}{(2 + a/H)^3 (f/F_p) + 2(1 - f)} + \frac{L}{B} M_0 \frac{a}{W} \left(\frac{\lambda}{H}\right)^3 \tag{3}$$

and

$$V_z = \pi D_b N \cos \theta \tag{4}$$

D. Mixing

Mixing can be subdivided into two major areas of consideration: dispersive and distributive. Dispersive mixing, or intensive mixing, is the process of particle size reduction, which is induced by shear deformation and/or elongational stress. In most polymer isolation processes, such as devolatilization and latex emulsion coagulation, dispersive mixing is not a critical process requirement unless a secondary polymer or additive stream is added to the process. In contrast, distributive mixing is fundamental to the success of nearly every aspect of these processes.

Distributive, or extensive, mixing is the rearrangement of particles. The dominant flow regime in extruders processing viscous polymers is laminar flow. In early studies of mixing in laminar flow, Spencer and Wiley (1951) studied the growth of interfacial area of an arbitrarily placed element in a fluid subjected to simple shear. They found that the growth of interfacial area depended on both the magnitude of the shear and the initial orientation of the element. The relationship for uniform shear deformation shows that the ratio of interfacial area grows linearly with applied shear strain:

$$\frac{A}{A_0} = \frac{1}{2}s \tag{5}$$

This theoretical framework remained unchanged for more than 25 years, until the late Lewis Erwin and his coworkers at MIT recognized and quantified the effect of reorientation (Gailus and Erwin, 1981).

It has long been recognized, on a practical technology level, that if a simple strain is applied (as, for example, with a two-roll mill) and no change in orientation is introduced, then the mixing will be quite nonuniform (i.e., well-mixed bands can co-exist side by side with poorly mixed bands of material). However, if the material is cut periodically, turned 90°, and then reintroduced to the simple strain, the mixing will be very effective. In early studies of melting in CRNI twin screws, Nichols and Kheradi (1984) determined that the staggered screw arrangement gave rise to a unique pressure distribution that promoted flow of material back and forth from one screw channel into the channel of the opposing screw. Later results by Howland and Erwin (1983) concluded that the interaction of flow between the two screws in the CRNI twin-screw extruder introduced reorientation of layers, which repeated as the material moved down channel and was exposed to each new interaction with the adjacent screw.

In contrast to uniform shear deformation, interrupted shear mixing with optimal reorientation leads to an exponential relationship between interfacial area growth and applied shear strain:

$$\frac{A_f}{A_0} = \left(\frac{s}{N}\right)^N \tag{6}$$

Although Erwin's kinematic explanation (of exponential interfacial area growth by interrupted shear mixing with optimal reorientation) provides an

Table 1 What Kinematics Can Explain

	Simple fluids	Complex fluids
Simple devices or Simple flow paths	X	
Complex devices or Complex flow paths		

attractive explanation of the phenomena observed in CRNI twin-screw extruder mixing, it is not the only possible explanation.

Table 1 illustrates the shortcoming of a purely kinematic explanation. Alone, kinematics is limited to the study of simple fluids in simple devices. Erwin (1991) warns of the limitations of decoupling flow and mixing, explaining that a decoupled mixing problem was one in which the fluid mechanics is unaffected by component distribution, and acknowledging that most mixing problems of practical interest involve systems in which flow and mixing are coupled. In fact, in a system where flow and mixing are coupled, rheological phenomena could account for similar results.

The lubrication approximation has been for many years a cornerstone for modeling polymer processing machinery. Fully developed flow is an assumption implicit to the use of that approximation. In the case of single-screw extruders, particularly those without sophisticated mixing sections or other flow channel interruptions, this assumption is reasonable, as the channel is long and end effects are minimal. However, in the case of the CRNI twin-screw geometry, due to the interaction of converging and diverging flow streams between the screws, it is unlikely that fully developed flow is ever attained. In viscoelastic liquids, in the environment of thermally and, frequently, compositionally developing flows, the nonlinear effects (Eq. 6) due to kinematics are bound to be a subset of the nonlinear mixing effects associated with the process.

Regardless of the underlying mechanism, the CRNI twin-screw extruder exhibits outstanding distributive mixing performance, and the role of distributive mixing will be a consistent theme throughout the discussion of devolatilization mechanisms.

E. Stage Separation

Polymer isolation and recovery, whether from solution or from emulsion, involves multiple stages, which may have considerably different process

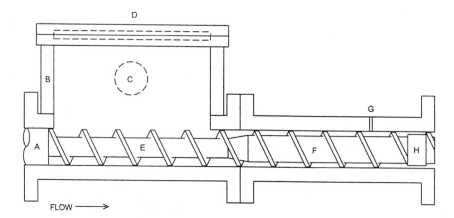

Fig. 5   Typical functional stage with (A) upstream seal, (B) vent stack, (C) vent effluent connection, (D) vent cover with sight glass, (E) deep flighted vent screw, (F) shallow flighted metering screw, (G) stripping injection provision, and (H) downstream seal.

requirements. However, one requirement common to all processes is the need to separate the stages with effective polymer seals. The polymer seals utilized most frequently by CRNI twin screws are cylindrical compounders. Figure 5 shows a typical stage arrangement, which provides a deep flighted screw section suitable either for downstream ingredient addition or as a vent zone; followed by a transition zone to a shallower conveying and mixing zone suitable for building pressure in conjunction with a compounder or sealing section. Most stages are made up of these common building blocks, with adjustments as required to optimize the function being performed.

Seal requirements must be defined by the combination of process objectives of the stage ending with the seal, and the requirements of the stage downstream of the seal. For example, the primary concern upsteam may be to assure sufficient preheat in the polymer prior to a vacuum vent zone, while the concern downstream may be to maintain a high vacuum. The characteristics of cylindrical compounder seals were studied experimentally by Nichols and Yao (1982), who found that that temperature rise and pressure drop increased exponentially with increasing cylinder diameter (or decreasing annulus gap), as shown in Figs. 6 and 7.

Jerman (1986) developed a finite difference model for pressure drop and temperature rise across cylindrical compounders and showed good agreement to the data of Nichols and Yao, as well as including experimental data collected on a 20-mm-diameter CRNI twin-screw extruder.

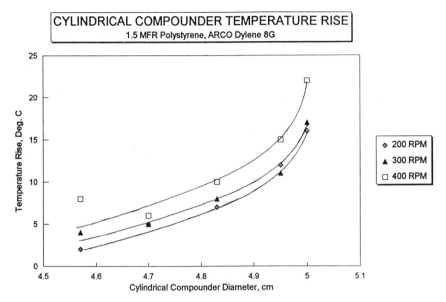

Fig. 6 Cylindrical compounder temperature rise as a function of compounder diameter.

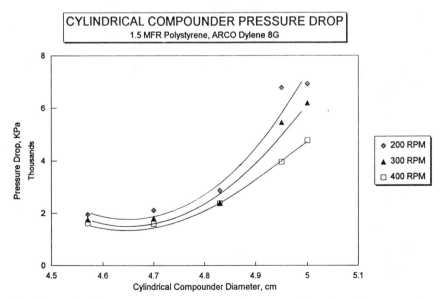

Fig. 7 Cylindrical compounder pressure drop as a function of compounder diameter.

F. Discharge Pumping

One of the design characteristics of the nonintermeshing twin-screw extruders mentioned earlier is that the screws do not have to be of equal length. In fact, one standard configuration of this type of extruder includes the extension of the main screw to create the pressure to pump the product through the die. However, there are several alternative methods of pressurizing the die. Quite often, the optimum screw speed for process steps like mixing, devolatilization, and latex coagulation is quite high (usually in the range of 400–500 rpm). Discharge pumping at high screw speed may result in excessive shear heating of the polymer, resulting in unacceptable product temperature. In such cases, melt pressurization can be accomplished with gear pumps or a separate, slower speed, single-screw crosshead extruder.

G. Residence Time Distribution

Measuring the residence time distribution (RTD) allows characterization of the behavior of a mixing machine in comparison with idealized flow models. At one extreme, plug flow is a rigorous first in, first out process. At the opposite extreme, the model of a continuously stirred tank reactor (CSTR) has the chance of a particle leaving the vessel as completely random. The basic residence time measurements are the exit age distribution and the cumulative distribution. The exit age distribution is obtained by monitoring exiting samples from a process after introducing a pulse of tracer. The cumulative distribution is the integral of the exit age distribution, and it is physically analogous to the results obtained by measuring the exiting samples from a process after introducing a step change as in white to black material of identical rheological properties.

Exit age distribution:

$$E(t) = \frac{C}{\Sigma_0^\infty \, C \, \Delta t} \tag{7}$$

Cumulative distribution:

$$F(t) = \int_0^t E \, dt = \sum_0^t E(t) \, \Delta t = \frac{\Sigma_0^t \, C \, \Delta t}{\Sigma_0^\infty \, C \, \Delta t} \tag{8}$$

Mean residence time:

$$\bar{t} = \frac{\Sigma_0^\infty \, tC}{\Sigma_0^\infty \, C} \tag{9}$$

The mean residence time is also defined by the physical relationship:

$$\bar{t} = \frac{V}{Q} = \frac{\text{volume}}{\text{flow rate}} \tag{10}$$

After extensive measurements of CRNI residence time distributions for conventional screw orientations (i.e., staggered and matched flights), Nichols et al. (1983) compared data against eight RTD model system equations and concluded that the closest model fit was the Pinto and Tadmor (1970) model derived for single-screw extruders. This result, which shows that extruders perform somewhere between the lower boundary limit of perfect mixing (modeled as a CSTR) and the upper boundary limit of plug flow is not particularly surprising. However, Lu et al. (1993) recently showed that if the helix on one of the screws was reversed, so that one screw was attempting to pump from inlet to discharge and the other screw was attempting to pump from discharge to inlet, a dramatic change in the RTD occurred. The machine performed like a CSTR. This is a radical breakthrough, as conventional CSTRs cannot handle highly viscous materials.

H. Scale-Up Considerations

Scale-up criteria for various processes using nonintermeshing twin-screw extruders have been presented previously by Nichols and Lindt (1988). In that study, several scale-up methods were compared and evaluated to determine if they were applicable to CRNI twin screws. The comparison included the following:

Carley–McKelvey (1953) single-screw model	$Q_2 = Q_1(D_{b2}/D_{b1})^3$
Nichols–Lindt (1988) CRNI historical method	$Q_2 = Q_1(D_{b2}/D_{b1})^{2.5}$
Maddock (1959) single-screw model	$Q_2 = Q_1(D_{b2}/D_{b1})^2$
Pearson (1976) single-screw model	$Q_2 = Q_1(D_{b2}/D_{b1})^{(1 + 5n)/(1 + 3n)}$

Of these models, the Pearson approach is the most conservative, with output rate increasing only as the 1.5 power of the diameter ratio, and represents the lower bound of all scale-up theories. In contrast, with the output rate increasing as the cube of the diameter ratio, the Carley–McKelvey model represents the upper bound, which is fixed by geometric limitations disregarding heat transfer. In practice, the Maddock model is the most widely used to scale up single-screw extruder processes. Because it is based on the square of the diameter ratio, the output rate is growing proportionally with heat transfer area while shear rate is held constant.

It is apparent that when scaling up one cannot scale simultaneously in a consistent fashion all parameters. The Carley–McKelvey model, which scales product rate proportional to the volume, and the Maddock model, which scales product rate proportional to the barrel surface area, exemplify the opportunity to optimize the scale-up of one parameter in preference to another. Each of the scaling parameters can be varied in such a manner to optimize performance (e.g., based on throughput, heat transfer, or specific energy). The actual choice made depends on the process being scaled up, with a clear understanding of both the parameters that will be compromised as well as the parameters that will be optimized.

The method used historically for the nonintermeshing twin-screw extruder is a hybrid system in which the screw channel depth is scaled up directly by the diameter ratio and the screw speed is held constant. However, the product rate is decoupled from the actual volumetric displacement (which grows as the cube of the diameter ratio), while the throughput rate is independently controlled to the 2.5 power of the diameter ratio. As a result, although the screw speed (and hence shear rate) is constant, the residence time, total shear, and specific energy decrease as the square root of the diameter ratio. This combination has worked quite well for scale-up of devolatilizers. A plausible argument for the relative degree of success in scale-up of devolatilizers is that mass transfer limitions imposed by lack of free volume often arbitrarily limit devolatilization performance. This scale-up approach is conservative in the sense that the free volume grows upon scale-up at a greater rate than the filled channel width.

III. THEORY OF DEVOLATILIZATION

A. Introduction

The removal rate of volatiles, Φ, from a flowing solution can be calculated from

$$\Phi = k_m A (C - C_e) \tag{11}$$

provided that the mass transfer coefficient k_m, the vapor–liquid interfacial area, A, and the concentration driving force (the difference between the bulk concentration of the solvent, C, and its equilibrium concentration, C_e, at a given devolatilization pressure) are available. Conversely, by measuring the devolatilization rate Φ, the devolatilization performance of the extruder can be characterized by an experimental value of $k_L A$, the ratio between the measured devolatilization rate and the known concentration driving force, which thus forms the bridge between the experiment and theory.

In reality, none of the three quantities required for theoretical predictions of Φ is necessarily known a priori. The question of thermodynamic equilibrium at the vapor–liquid interface is obscured by the presence of multiple species, typical of industrial devolatilization, and the fundamental uncertainty of whether the interface is at equilibrium at all. Even when no foaming occurs, the true interfacial area is not known with any degree of precision due the geometric complexity of the dynamically interacting nonintermeshing screws, breaking up the polymer ribbon and affecting its swelling, and the compounder and vent designs, all affecting the extruder's ability to pump and thus vacate a part of the extruder volume in which devolatilization is allowed to occur. In foam-enhanced devolatilization, the task of determining the collective surface area of the entire bubble population appears insurmountable. Finally, the mass transfer coefficient depends on the diffusivity coefficient, which in turn depends on the composition and temperature, and on the rheological conditions at the vapor–liquid interface, making it necessary to include accounts of flow and heat transfer in the already difficult problem.

Assuming that the vapor–liquid interface is at known equilibrium, the averaged product of the mass transfer coefficient and the overall mass transfer area, $k_m A$, can be determined experimentally from the measurements of the amount of volatiles drawn from the vents. The mass transfer coefficient—the volatile removal rate per unit interfacial area per unit of concentration driving force—is to be extracted from $k_m A$ using an independently determined interfacial area, A, experimentally or theoretically.

Any attempt to predict theoretically the performance of a devolatilizing extruder includes the separate, yet intimately linked, tasks of determining the interfacial area that the solvent molecules cross and the mass transfer coefficient. When choosing to do so from first principles, one is left with a coupled momentum, heat, and mass transfer problem invoking Flory–Huggins thermodynamics. This approach is common to all extruder types. Leaving aside the design of the vapor flow conduits and of the vent opening itself, the individual extruder devolatilizers, single versus twin, intermeshing versus nonintermeshing, co-rotating versus counter-rotating, differ merely in their hydrodynamic working, namely, their ability to create free surface and to mix, both connected with pressure fields specific to a given machine. The same mass transfer mechanisms operate in all cases, differing merely in intensity, affected by the various flow environments. In this light, our treatment of mass transfer in the nonintermeshing, counter-rotating twin-screw extruder is quite similar to those for other types of extruder devolatilizers while dealing with the removal of small amounts of volatiles. Our analysis of devolatilization of relatively dilute systems contains new elements not because the high-mass-transfer-rate mechanisms are inherent to the nonintermeshing, counter-rotating twin-screw extruder, but because only incomplete information on this regime had existed. Our analysis of devolatilization of the dilute systems rests on integration of the physicochemical elements involved, and thus it can be extended to other devolatilizers, if needed. To reduce duplication with other chapters, the sources quoted in this section refer specifically to the nonintermeshing, counter-rotating twin-screw extruder where possible, despite the commonality of most of the underlying ideas with other extruder types.

B. Flow Environment for Devolatilization

Equation (1) and the subsequent finite element modeling by Nguyen and Lindt (1989), illustrated in Fig. 8, offer means of determining the lengths of the fully filled portions of the extruder in which the pressure necessary to transport the liquid past the pressure-drop elements is generated, for example, compounders and die. The balance of the extruder volume is available for devolatilization. In the partly filled channels, free surfaces exist as the exposed surface of the melt pool and the barrel surface film. The barrel surface film is assumed stagnant until removed by the advancing flight and added to the melt pool. At a given flow rate, q, the lateral screw fillage, f, can be estimated from Squires (1958):

$$F_D = \frac{32f^2}{\pi^3(H/W)} \sum_{i=1,3,\ldots}^{\infty} \frac{1}{i^3} \tanh \frac{i\pi(H/W)}{4f} \tag{12}$$

upon substitution for F_{DCRT} in Eq. (1), remembering that in the partly filled zone only drag flow exists ($F_{PCRT} = 0$). The cross-channel flow can be assessed numerically (as by Chen and Lindt, 1995) to determine the circulation pattern and cross-channel pressure distribution (Fig. 9), relevant to the determination of the exposure time for diffusion and to considerations of bubble nucleation and motion (Foster and Lindt, 1990b).

C. Mass Transfer Mechanisms

In addition to being a prerequisite for any form of mass transfer across a given interfacial area (Eq. 11), the driving force for devolatilization, $\Delta C = C - C_e$, also significantly contributes to whether foaming occurs or not. If ΔC, and thus the pressure differential between the operating pressure and the vapor equilibrium pressure (ΔP), is large enough, the nuclei present in the solution will grow and form bubbles. At small ΔC, the nuclei will remain inactive, and the diffusion will take place in a homogeneous medium. As an approximation, neglecting the possible effects of hydrodynamic stress

(Han and Han, 1988), the condition dividing the high- from the low-mass-transfer-rate regime, foaming from nonfoaming devolatilization, follows from the force balance on the nucleus surface: if $\Delta P > 2\sigma/R$ the nuclei will grow; if the reverse holds, the surface tension, σ, does not allow the nuclei to grow. The critical nucleus size has been measured by Han and Han (1988) to be of the order of 10^{-7} m.

During devolatilization, predominantly heterogeneous nucleation occurs at the screw and barrel surfaces as well as in the bulk of the liquid, at rates specific to the actual polymer–solvent system, to the amount and nature of solid impurities and microbubbles, and the metallurgy of the material of construction of the extruder. The effective nucleation density is not known a priori and introduces considerable uncertainty into all known attempts to analyze the problem of bubble-enhanced devolatilization, regardless of the equipment used.

We will give separate treatments of the high- and the low-mass-transfer-rate regimes. In the intermediate regime both mechanisms appear to operate simultaneously, and the combined theories have been used successfully to interpret the experimental results obtained on the first downstream vent (Foster and Lindt, 1990b).

D. The High-Mass-Transfer-Rate Regime ($\Delta P \gg 2\sigma/R$)

1. Macroscopic Interactions Between Flow and Mass Transfer (Competition for Volume Between Liquid Conveying and Devolatilization)

If a large amount of the low-molecular-weight component is involved, say 50% by weight, the first stage of separation is a flashing process carried out in an atmospheric pressure vent.

Foster and Lindt (1989) measured the devolatilization rate, Φ, of polystyrene–ethylbenzene solutions and expressed their results in terms of the devolatilization efficiency, E:

$$E = \frac{\Phi - \Phi_a}{\Phi_i - \Phi_a} \tag{13}$$

where Φ_a and Φ_i refer to the separation rates that would have occurred if the adiabatic or isothermal thermodynamic equilibrium, respectively, had been obtained. Φ_i follows from the mass balance and the Flory–Huggins thermodynamics. On the other hand, Φ_a requires the inclusion of a heat balance (Padlyia and Tebbens, 1981). As seen in Fig. 10A, the high rate of devolatilization does not necessarily mean that a thermodynamic equilibrium is achieved. Except for small feed flow rates, that is, long residence times, the solution leaves the high-mass-transfer-rate regime, present in the rear vent of the extruder, far from equilibrium, suggesting kinetic control.

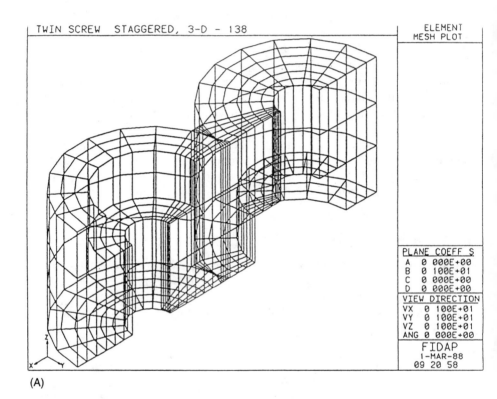

(A)

Fig. 8 Finite element model of CRNI twin-screw extruder: (A) mesh for the staggered screws (example); (B) leakage flow equations; (C) down-channel flow equations. (From Nguyen and Lindt, 1989.)

$$\frac{\partial u_x}{\partial x} + \frac{\partial u_y}{\partial y} + \frac{\partial u_z}{\partial z} = 0,$$

$$-\frac{\partial p}{\partial x} + \frac{\partial^2 u_x}{\partial x^2} + \frac{\partial^2 u_x}{\partial y^2} + \frac{\partial^2 u_z}{\partial z^2} = 0,$$

$$-\frac{\partial p}{\partial y} + \frac{\partial^2 u_y}{\partial x^2} + \frac{\partial^2 u_y}{\partial y^2} + \frac{\partial^2 u_y}{\partial z^2} = 0,$$

$$-\frac{\partial p}{\partial z} + \frac{\partial^2 u_z}{\partial x^2} + \frac{\partial^2 u_z}{\partial y^2} + \frac{\partial^2 u_z}{\partial z^2} = 0,$$

$\underline{u} = 0$ on screws and flights,

$$\underline{u} = \underline{e}_z \text{ on barrel,}$$

(B)

$$\frac{\partial u_x}{\partial x} + \frac{\partial u_y}{\partial y} = 0,$$

$$-\frac{\partial p}{\partial x} + \frac{\partial^2 u_x}{\partial x^2} + \frac{\partial^2 u_x}{\partial y^2} = 0,$$

$$-\frac{\partial p}{\partial y} + \frac{\partial^2 u_y}{\partial x^2} + \frac{\partial^2 u_y}{\partial y^2} = 0,$$

$u_x = u_y = 0$ on the barrel,

$\underline{u} - \underline{e}_t$ on the screws,

$$\frac{\partial p}{\partial t} = \text{constant,}$$

$\underline{e}_t$ is the local tangential unit vector,

(C)

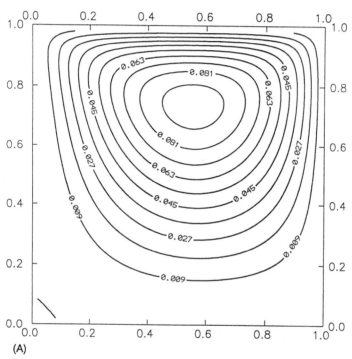

(A)

Fig. 9   Finite difference analysis of the cross-channel flow in a partly filled channel: (A) stream lines; (B) pressure distribution. (From Chen and Lindt, 1995.)

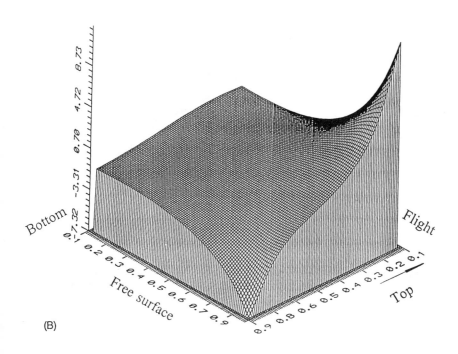

(B)

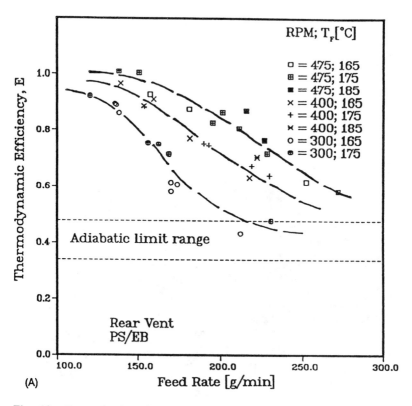

Fig. 10 Determination of volume available for devolatilization: (A) Apparent mass transfer coefficient depends on screw speed and flow rate. (B) Compounder/screw characteristics obtained from FEM allow the liquid volume to be determined; it correlates with mass flux, regardless of screw speed and flow rate. (C) Mass transfer takes place while competition for volume occurs between liquid and vapor. (From Foster and Lindt, 1989.)

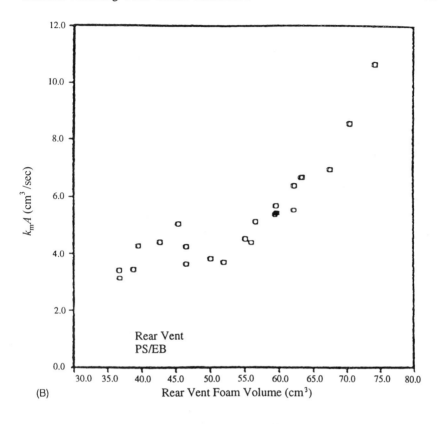

(B)

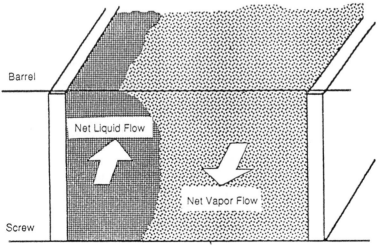

(C)

The apparent drop in E with decreasing screw speed and increasing feed rate is a result of limited screw channel volume. The devolatilization in the high-mass-transfer-rate regime is associated with a very rapid growth of a large bubble population. Discounting the cavitation and stress effects, the bubbles are able to exist in the upstream fully filled zone, if the local hydrodynamic pressure is relatively low, $\Delta P > 2\sigma/R$. Furthermore, bubbles carried by the flow at velocity v_{zP} into the filled zone are pushed back by the down-channel pressure gradient, $v_{zP} \approx R^2(\partial P/\partial z)/\mu$. By increasing the feed rate or decreasing the screw speed, the filled length grows and the volume for bubble formation and expansion diminishes. This is confirmed in Fig. 10B, where the volumetric mass transfer coefficient $k_L A$, closely related to the thermodynamic efficiency, is shown as a function of volume available for devolatilization, determined by finite element modeling of liquid conveying (Nguyen and Lindt, 1989). In this representation, the widely stratified data of Fig. 10A collapse within a narrow band, within the experimental error of the measurement of the devolatilization rate. This indicates that the channel is fully packed (Fig. 10C). The competition for volume between the liquid and bubbly flows limits the devolatilization capacity in the high-mass-transfer-rate regime. Improvements might be sought by deepening the screw channels in the rear vent where this tends to occur, mechanical limitations of the design permitting.

2. Microscopic Interactions Between Flow and Mass Transfer (Bubble Growth, Motion, and Rupture)

In the partly filled channels, spontaneous bubble growth occurs, tending to dominate both the fluid flow and mass transfer patterns. Although, in principle, stress deformations of the foam occur, they can be assumed secondary as long as the bubble growth is sufficiently fast: $(dR/dt)/R \gg \pi DN/H$.

Beyond the surface tension–controlled regime, the initial stages of the bubble growth tend to be controlled by diffusion of volatiles into the bubble if $R^2/D \gg \mu/\Delta P$. On the other hand, the hydrodynamic resistance to bubble expansion may control if $R^2/D \ll \mu/\Delta P$. Given enough time, all bubble growth will be controlled by hydrodynamic forces, and their growth may eventually cease, due to an appreciable increase in viscosity, or as ΔP diminishes, both a result of solvent depletion. Although these extremes are of practical and theoretical interest, a sufficiently general model covering all such mechanisms, and their combinations, is desirable for understanding continuous foam-enhanced devolatilization (Chella and Lindt, 1986).

In a population of closely packed growing bubbles, significant hydrodynamic and diffusion mutual interactions exist. Amon and Denson (1984), Chella and Lindt (1986), and others have used an ordered lattice of identical cells, each containing a single bubble, to represent the bubble-to-bubble interactions. Chen et al. (1993) introduced into their cell model the simultaneous multispecies diffusion at the bubble surface. All these models involve simultaneous solutions of the Rayleigh equation (see, e.g., Middleman, 1977) and the diffusion equation(s), retaining the radial diffusion and transient terms. The results shown in Fig. 11 have been obtained from Chella's model accounting for an adiabatic temperature drop due to vaporization, and for significant concentration and temperature dependencies of diffusivity and viscosity from the free-volume theory (Duda et. al., 1982) utilizing the viscosity data of Foster and Lindt (1987).

The spatial distributions of concentration and temperature within the cells have been calculated by finite differences after the time-dependent location of the bubble surface has been fixed in the transformed space by a convenient combination of the spatial and time coordinates. The net result of the numerical cell model is the kinetics of bubble growth, $R(t)$, to be used in conjunction with the residence time distribution in the partly filled zone

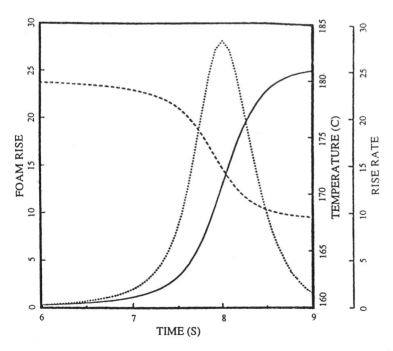

Fig. 11 Polymer devolatilization—theory for batch experiment. —— = foam rise; --- = temperature; ···· = rise rate. (From Chella and Lindt 1986.)

to theoretically predict the devolatilization rate. Fortunately, the mean residence time of the bubbles in any given vent zone is typically less than a second, preventing the appreciable bubble growth, and thus bubbles remain spheroidal, as assumed. This has been confirmed by the visual observations of Amon and Denson (1984) and ourselves. The devolatilization rate is an external manifestation of the vapor release from the bubble population at rupture. Mass conservation dictates that the rate of vapor removal equals the rate of foam generation, expressed in terms of nucleation density and the kinetics of bubble growth (Fig. 11).

3. Simulation of the Devolatilization Performance

In the fully filled rear vent (Fig. 12A) the extruder bubbles nucleate continuously within the liquid and grow explosively, providing intense mixing within the liquid volume available for devolatilization. The assumption of equal probability for any bubble, regardless of age, reaching the disengagement zone in the rear vent and releasing its content into the vapor stream is equivalent to assuming ideal mixing in the rear vent.

A transformation of the well-known exponential RTD for a CSTR into the bubble-size distribution (BSD) function can be achieved once the bubble growth kinetics, obtained from the cell model, is introduced into it. As expected and seen in Fig. 12B, the BSD tends to narrow with decreasing mean residence time in the vent, assuming equality between the number of bubbles nucleated in the vent and the number of bubbles removed from it at any time. For the mean residence time in the rear vent of some 0.5 sec, the mean radius is calculated to be of the order of millimeters. In the rear vent, we have indeed observed visually a population of spheroidal bubbles about 1–2 mm in diameter. Combining the BSD with a population balance and the overall balances of mass and heat produces a prediction of the mass transfer efficiency. For the apparent nucleation density of approximately 200 nuclei/sec/cm^3 a good agreement with the experiment is observed (Fig. 12C).

In the absence of direct in situ measurements, the nucleation density remains a parameter into which much of the theoretical, and experimental, uncertainty is lumped. The collapse of all data into a narrow continuous band, falling within the experimental scatter, suggests some internal consistency of this concept. The validity of these ideas has further been confirmed by independent Monte Carlo simulations of the bubble phenomena (Foster and Lindt, 1990b) that lead to similar results, using similar nucleation densities.

E.   The Low-Mass-Transfer-Rate Regime ($\Delta P \ll 2\sigma/R$)

1.   Macroscopic Interactions Between Flow and Mass Transfer (Drag Flow–Controlled Diffusion Environment)

In solutions containing sufficiently low amounts of volatiles, for example, during degassing or in the downstream vacuum vents, the thermodynamic driving force may be insufficient for bubbles to form. The residual volatiles diffuse toward the free surface formed in the partly filled channels in a fashion similar to conventional drag flow, provided no excessive concentration effects on viscosity occur. For the sake of simplicity, we will assume that the flow will affect both the interfacial area and the mass transfer coefficient (cf. Eq. 1) but that the base drag flow in the partially filled screw, discussed in the preceding section, remains unchanged.

In addition to fully filled portion of the screw, two distinct drag flow situations coexist along the screw that constitute the flow environment of the low-mass-transfer-rate regime (Foster and Lindt, 1990a):

1.   In the partially filled channels, sufficiently far away from the vent opening, the flow is driven by the relative motion between the screw and barrel. The exposed surface is divided between the barrel and liquid pool films. As suggested in Fig. 13A, the respective surface areas of the barrel and pool films per unit down-channel length are taken as $W(1 - f)$ and $H - \delta/2$. The surfaces are continuously created and again merged into the bulk of the rotating melt pool, available for diffusion over relatively short exposure times: $[W(1 - f)/\pi]D_b N \sin \theta$ for the barrel film and $W(1 - \delta/2)D_b N \sin \theta$ for the pool.

2.   Under the vent opening, in the upper part of the exposed screws, the drag flow becomes inoperable as the barrel surface is not available to support the necessary flow stress. The melt emerges from the lower part of the channel driven by drag flow; its top surface becomes exposed while the flow stresses cease, until the liquid is merged back into the lower half of the closed barrel, where shear flow resumes (Fig. 13B). The time spent in the barrel opening is proportional to the circumference fraction occupied by the vent opening and inversely proportional to the screw speed, F/N. The exposed surface per screw, per unit down-channel length is $\pi D_b F/\cos \theta$.

2.   Simulation of the Devolatilization Performance

The low-rate-mass-transfer regime in a CRNI twin-screw extruder lends itself to a rational analysis without the uncertainties plaguing the analysis of the foam-enhanced, high-mass-transfer-rate devolatilization. In the absence of nucleation and bubble growth, both the mass transfer coefficient, k_m, and the mass transfer area, A, necessary for a determination

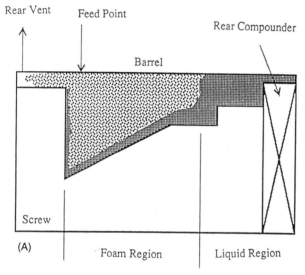

Fig. 12 Determination of mass transfer coefficient (high-mass-transfer regime): (A) flow and mass transfer pattern; (B) bubble size distribution from cell model and population balances; (C) theory versus experiment (nucleation density as a parameter). (From Foster and Lindt, 1990.)

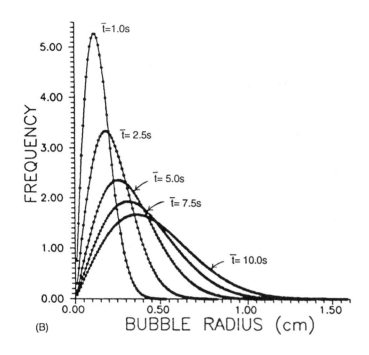

(B)

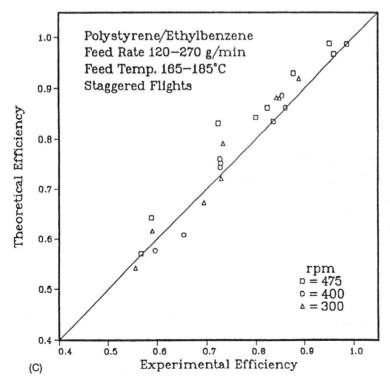

Polystyrene/Ethylbenzene
Feed Rate 120–270 g/min
Feed Temp. 165–185°C
Staggered Flights

rpm
□ = 475
○ = 400
△ = 300

(C)

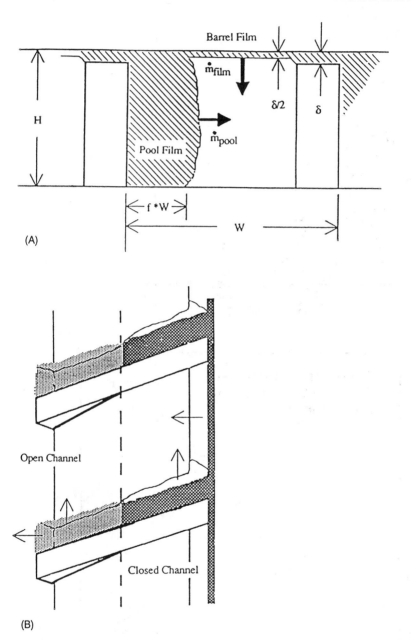

Fig. 13 Extrusion devolatilization (low-rate-mass-transfer regime): (A) mass fluxes in partly filled screw channel; (B) open versus closed channel. (From Foster and Lindt, 1990.)

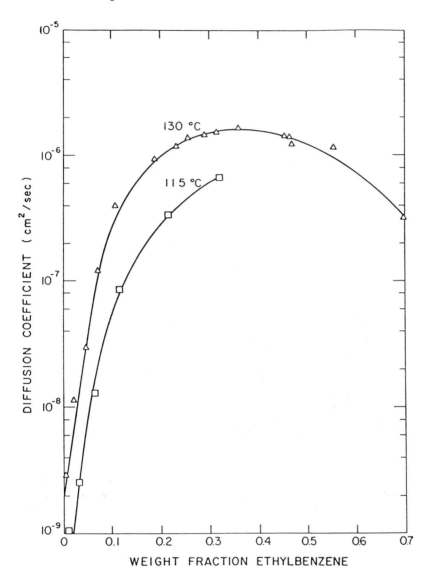

Fig. 14 Diffusivity data for the ethylbenzene–polystyrene system over a range of concentration and temperature. (From Duda et al., 1982.)

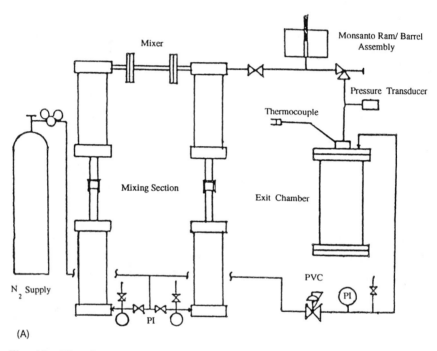

(A)

Fig. 15 Viscosity measurements: (A) schematic of solution viscometer; (B) summary of data—the ranges of shear rate, temperature, and polymer weight fraction, respectively, are $5 < \gamma < 2000\ \text{sec}^{-1}$, $130 < T < 240°C$, $0.6 < X_P < 1$. (From Foster and Lindt, 1987.)

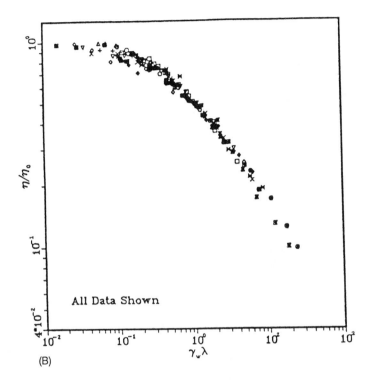

(B)

of the devolatilization rate (Eq. 11) can be readily approximated from the flow and diffusion equations (Foster and Lindt, 1990a).

The exposure of all free surfaces, present in the low-rate-mass-transfer regime, is short, $t_e \ll \delta^2/D$, and therefore the established approximations for diffusion into an infinite medium (Higbie, 1935; Latinen, 1955) apply. The mass transfer coefficient can be estimated from the exposure time, t_e, and the mass diffusivity, D; the penetration theory suggests that $k_m = \sqrt{D/\pi t_e}$, as long as the dependence of diffusivity on concentration is weak, $(\partial D/\partial C)/D \ll 1$. While the exposure times and surface areas available for devolatilization are known with reasonable certainty, the diffusivity coefficient is not always known to the same degree. In Foster's work (Foster and Lindt, 1990a) dealing with polystyrene–ethylbenzene solutions, the diffusivity data of Duda et al. (1978) were used along with their free-volume theory

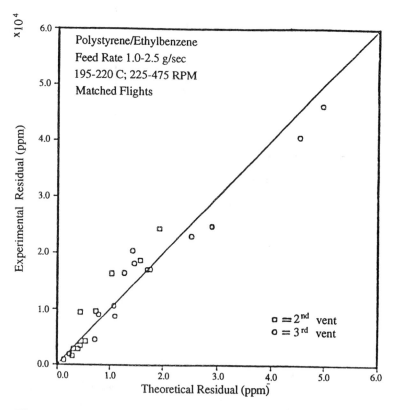

Fig. 16 Extrusion devolatilization (low-rate-mass-transfer regime)—experimental versus theoretical residual. (From Foster and Lindt, 1990a.)

representation of the concentration and temperature dependence (Duda et al., 1982), illustrated in Fig. 14, complemented by the viscosity data of Foster and Lindt (1987), shown in Fig. 15.

The rate of evaporation from the vent zone is determined by summation over all areas where the devolatilization occurs. Overall mass balances allow the residual concentration of the volatiles to be assessed. The agreement between the measurements of the residual concentrations and theory is good; see Fig. 16. For scaling purposes, it is interesting to note that the mass transfer coefficient is inversely proportional to the square root of screw speed as expected and shown in Fig. 17. In conclusion, the good agreement with experiment reconfirms the utility of this well-established approach and its applicability to devolatilization in the CRNI twin-screw extruder.

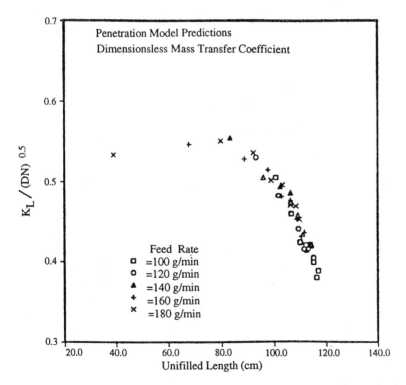

Fig. 17 Normalized mass transfer coefficient (low-mass-transfer regime). (From Foster and Lindt, 1990a.)

IV. APPLICATIONS

A. Solution Devolatilization Tutorial

Every experimental program starts by setting objectives, assessing the information previously available, and defining the experimental program. After the experiments are completed, of course, it is then necessary to analyze the data, determine if the results adequately meet the objectives set forth, and formalize the conclusion with a report documenting the activities and results obtained. Typical objectives of solution devolatilization experiments are to define a suitable machine geometry to devolatilize a specific solution; or to determine the best operating conditions for a polymer system. In complex, highly coupled systems, great care must be exercised to assure that upsteam stages are performing at or near optimum conditions before drawing conclusions about the performance of subsequent process stages. The following tutorial covers preparation of equipment for solution devolatilization (with details of each major subsystem) and then discusses the operation of a solution devolatilizer in process stage sequence. An addendum on general laboratory safety practices relating to solution devolatilization is included at the end of this chapter.

1. Solution Preparation

To conduct a solution devolatilizing test, a solution of known composition must be prepared. In a binary polymer–solvent system, this may be as simple as preheating a predetermined amount of diluent in an appropriately heated and stirred reactor vessel and slowly adding polymer pellets until the right ratio is achieved. Then the reactor vessel can be sealed, and inert gas pressurized. It greatly facilitates both preparing the solution and obtaining material balance data on the process later if such a vessel is mounted on load cells so that the net weight of the contents can be readily determined.

SAFETY NOTE: The reactor vessel must be equipped with an appropriately rated relief valve. The relief valve rating must never exceed the design ratings found on the manufacturer's nameplate for the reactor vessel; lower temperature and pressure ratings may be appropriate for the polymer–solvent system under study. The relief valve must be installed in such a manner that it will discharge safely if excessive pressure or temperature is reached. Figure 18 shows a water-quenching system that is appropriate for many polymer–solvent systems. Before installation of such a system, make certain that the quenching medium is suitable for the chemistry of the materials to be tested.

Multicomponent systems may require some study to prepare large batches of solution efficiently. While it may be sufficient to preblend the

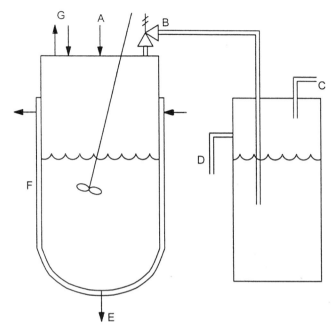

Fig. 18 Details of a water-quenched vent from a pressurized solution feed system: (A) solution feed port; (B) pressure relief valve; (C) atmospheric vent to suitable exhaust; (D) safely contained water overflow; (E) solution discharge to process; (F) heating jacket for vessel; (G) inert gas purge.

diluents and then to introduce the polymeric materials, you may find that some ingredients may dissolve more rapidly in one solvent than in another and that preparing two solutions that are later combined may be easier. In any event, understand that preparation of materials can be quite time consuming.

2. Solution Feed Systems

Adequate means of transport of the solution to the process extruder from the reactor vessel is necessary. Both gear pumps and piston pumps have proven suitable for this application. In either case, both the pump and its feed piping must have appropriate heating and/or cooling capability to maintain a constant temperature of the solution delivered to the process. The feed valve assembly should be designed so that the solution delivery pump can be started and the solution recirculated until steady-state conditions are established.

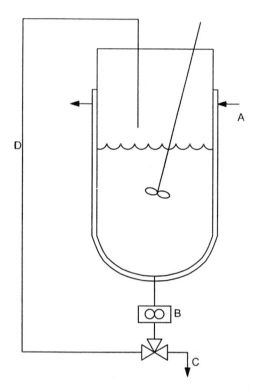

Fig. 19 Overview of a solution feed system: (A) temperature-controlled, pressure-rated vessel; (B) solution discharge/transfer pump; (C) solution discharge to process; (D) solution recycle to feed vessel.

Safe means of obtaining feed samples should be provided. Figure 19 is a schematic diagram of a typical solution feeding system. Figure 20 illustrates a feed valve arrangement with provision for feed sampling.

3. Solution Viscosity

It is particularly useful to the extrusion process engineer to know the viscosity of the solution as a function of composition, temperature, and shear rate. It is not unusual for the viscosity to change four or five orders of magnitude from a dilute solution containing, for example, 50% solids to a fully devolatilized polymer melt containing only trace residual solvent. Combining an understanding of the viscosity with an estimated material balance allows reasonable selections for screw elements for the initial tests. In the absence of any other data, a reasonable assumption for starting material balance is that about 50% of the solvent can be extracted using

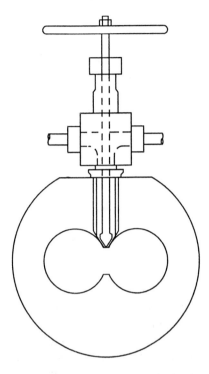

Fig. 20 Detail of solution feed valve assembly.

the rear vent and that each of the downstream vents can remove approximately 90% of the remaining solvent in each stage.

4. System Overview

CRNI twin-screw extruders have been used with good success as devolatilizers since the 1950s as found in U.S. Patents to Street and to Skidmore. Current technology differs in a number of ways (e.g., more modular design, more sophisticated instrumentation, optimization of many parts of the process as a result of more complete phenomenological understanding of devolatilization).

Figure 21 is a schematic representation of a fairly common extruder arrangement for solution devolatilization including a rear vent and three downstream vents, with a total twin-screw L/D ratio of 60:1 plus a 6:1 L/D single-screw discharge pumping zone. In the first stage, dilute polymer solution is introduced through a feed control valve. As the pressurized feed material exits the feed control valve into the first stage of the CRNI

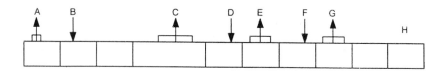

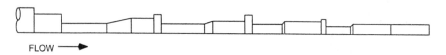

FLOW ⟶

Fig. 21 Schematic representation of multistage solution devolatilizer: (A) rear vent; (B) solution feed; (C) first downstream vent; (D) stripping agent injection port; (E) second downstream vent; (F) stripping agent injection port; (G) third downstream vent; (H) polymer discharge pumping zone.

twin-screw extruder, a substantial fraction of the solvent flashes to vapor and escapes through the rear vent. In dilute solution devolatilization, the rear vent is typically operated at or slightly above atmospheric pressure, but if the equilibrium requirements of the polymer–solvent system necessitate it, operation under vacuum can be sustained. Typically from 30% to 60% of the solvent can be removed via the rear vent. As the enriched polymer solution is conveyed forward, it is reheated through a combination of conductive heat transfer through the barrel wall and by viscous dissipation of energy from the rotating extruder screws.

5. Vent Stack Assemblies

Vent assemblies provide space for vapor to disengage from the polymer at a velocity that will avoid entrainment of particulate materials. Unlike co-rotating twin-screw extruders in which the combined rotation of the screws tends to push material out of the vent, in CRNI twin screws the screws rotate inward at the top to constantly draw material into the vent. However, in all extruders, it is possible for operational imbalance to create a condition where material may be forced out of the vent. Therefore, whenever feasible, vents are provided with ports that can be opened with reasonable speed so that material fouling the vent line can be removed. Similarly, it is necessary to see into the vent system to determine how the vent is operating and to make adjustments to equilibrium conditions to change the vent stage performance. There are three basic types of vents frequently employed on CRNI twin-screw extruders, as shown in Fig. 22.

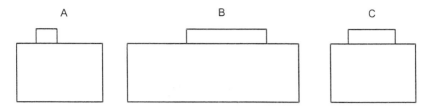

Fig. 22 Typical vent zone barrel modules: (A) 6 L/D barrel module with 1 L/D round vent opening; (B) 12 L/D barrel module with 6 L/D rectangular vent opening; (C) 6 L/D barrel module with 4 L/D rectangular vent opening.

Vent stacks can be customized to accommodate various materials (i.e., materials of construction and means of heating and/or cooling).

6. Solvent Recovery Systems

Solvent recovery systems usually comprise three elements: a means of cooling the exiting vapor stream, and appropriate means of collecting the exiting vapor stream, and a source of vacuum. Care should be taken in preparing the recovery system piping so that the operator can quickly and easily transfer the recoverable stream from one container to another without interrupting the process operation. For small-scale devolatilizers (e.g., 20 mm) where only small quantities of vapor are involved, glass traps immersed in dry ice–alcohol slurry are commonly used in series with small shell- and-tube heat exchangers. Figure 23 illustrates a typical solvent recovery system for a rear vent (note that the expectation is that the rear vent will be operated at or slightly above atmospheric pressure), and Fig. 24 illustrates a typical solvent recovery system for a downstream (vacuum) vent.

Due to the higher volume of effluent, it is very difficult to use glassware for solvent recovery in larger systems. Also, larger-volume containers are difficult to cool in dry ice–alcohol baths. Slightly larger-scale systems (30–40 mm) may be accommodated with jacketed stainless steel traps cooled by chilled water, but it is likely that some quantity of vapor will not be condensed and will pass through the vacuum pump. As the volume of effluent grows, the most practical means of condensing the vapor is with spray contact cooling. This system, illustrated in Fig. 25, uses a recirculating flow of chilled solvent spraying into a receiving vessel where the vapor is introduced. The spray of chilled solvent condenses the vapor stream on contact so long as the heat load of the absorbed vapor does not exceed the heat removal rate of the system heat exchanger. As with any engineered system, one must have a reasonable estimate of the vapor rate to be

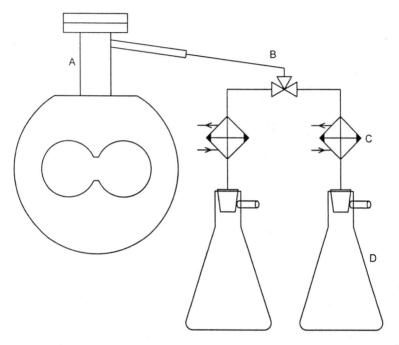

Fig. 23   Rear vent effluent collection arrangement: (A) vent stack assembly; (B) three-way valve assembly; (C) heat exchangers; (D) collection vessels.

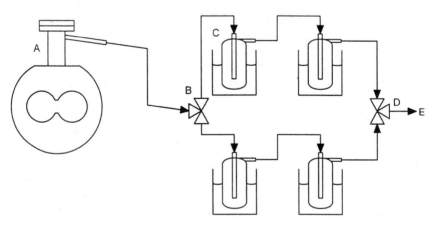

Fig. 24   Downstream vent effluent collection arrangement: (A) vent stack assembly; (B) three-way valve assembly; (C) condenser vessels immersed in alcohol–dry ice coolant; (D) three-way valve assembly; (E) connection to vacuum system.

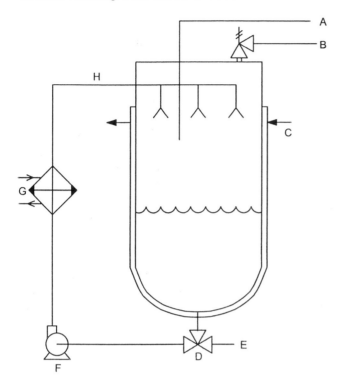

Fig. 25 Spray contacting condenser arrangement: (A) vapor inlet; (B) pressure relief system; (C) vessel jacketed for cooling; (D) three-way valve; (E) drain port; (F) recycle port—inlet to recirculation pump; (G) heat exchanger; (H) recycle cold solvent to spray heads.

handled to design the system in terms of heat exchange capacity, recycle flow rate, and receiver volume.

7. *Vacuum Pump Selection*

Single-stage, rotary vane water-sealed pumps are usually adequate for most first-stage vacuum vents, as the vacuum level required is usually in the range of 50–100 mm Hg. Single-stage, rotary oil-sealed pumps are better choices for second-stage vacuum vents, as they need to operate in the range of 20–50 mm Hg. Final-stage vacuum vents, which operate in the range of 5–25 mm Hg, require two-stage pump systems. One typical combination is to use a single-stage, rotary oil-sealed pump backed up by a single-stage, rotary vane water-sealed pump. There are many other types of vacuum pumps that can provide similar performance capability. Vacuum pump selection should be based on discussion with qualified vendors and careful consideration of the vapor load requirements, run duration, and overall system capability (such as leakage rate).

8. First-Stage Design

The rear vent system is illustrated in Fig. 26. From this sectional view, one can visualize the deep flighted portion of the screws into which the boiling polymer solution is pumped. As the boiling solution enters the extruder screws, the foam instantaneously expands to fill the available space. Vapor separates and is driven out through the vent stack by the differential pressure. The resulting more concentrated solution is carried downstream by the drag flow induced by rotation of the screws and is reheated to provide driving force for the second-stage separation. The first and second stages are isolated by cylindrical compounders that function as seals.

9. First-Stage Operation

Yield from the rear vent is a function of the temperature of the feed material, back pressure on the vent, the product rate, and the volume available for separation. Obviously superheat and back pressure are closely related, as are product rate and free volume in the extruder screws. The usual practice is to begin feeding material at a constant rate and gradually increase preheat temperature, monitoring rear vent recovery rate as a function of feed temperature. After the rear vent performance is adequately characterized, select an operating point that is very reliable (both steady and reproducible) and shift your attention to the second stage.

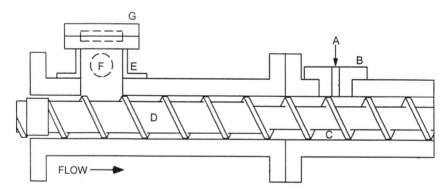

Fig. 26 Typical rear vent zone arrangement: (A) solution feed; (B) adapter for solution feed valve; (C) deep flighted screws for initial polymer solvent separation; (D) shallow flighted screws to prevent entrainment of polymer solids; (E) rear vent stack assembly; (F) vent effluent connection; (G) vent lid with sight glass.

10. Second-Stage Design

Entering the second stage, the polymer is usually exposed to a large rectangular vent (up to six diameters long and one diameter wide), where about 90% of the remaining volatile is separated under vacuum (typically 25–100 mm Hg). Depending on the nature of the polymer–solvent system, a stripping agent may be introduced in this stage to enhance separation. Skidmore (1974) showed that counter-current stripping was more effective in reducing final volatiles content than co-current stripping. Figure 27 shows how stripping agents introduced on a pressure gradient in the reheating zone will be forced to flow counter-current to the upstream vent. Obviously, a small amount of stripping agent may become entrained with the polymer and carried forward from the second stage to the third stage.

11. Second-Stage Operation

Gradually apply vacuum while watching the polymer through the vent sight glass. Increase vacuum to the system maximum unless excessive foaming occurs. Check the solvent recovery system to assure that it is functioning properly. Experimental variables that may be evaluated include

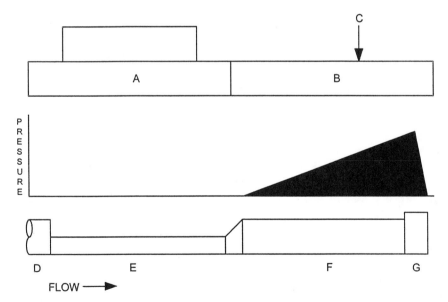

Fig. 27 Schematic arrangement of typical vent stage: (A) barrel module with rectangular vent; (B) closed barrel module; (C) stripping injection; (D) upstream seal (compounder); (E) deep flighted vent screws; (F) shallow flighted reheating/conveying screws; (G) downstream seal (compounder).

product temperature entering the zone (if sufficient reheat is provided following the rear vent zone for the first downstream stage to operate efficiently); feed rate and/or screw speed (volume effects: is there sufficient disengagement space for optimum bubble growth?); equilibrium conditions (temperature, pressure, or vacuum level).

12. Third-Stage Design

Proceeding downsteam to the third stage, the polymer encounters the second vacuum vent. Once again, about 90% (or more) of the remaining volatiles are removed in this stage, and stripping is almost always recommended in this stage. Vacuum levels are usually somewhat higher than in the first vent (typically 20–50 mm). Once again the polymer-enriched solution is pumped over a cylindrical sealing compounder to the fourth stage. In this stage, care usually has to be taken to prevent excessive temperature rise in the polymer, as the viscosity is approaching that of the neat polymer.

13. Third-Stage Operation

With a well-defined operating stage for the rear vent and first downstream vent, the volatiles level entering the third stage should be less than 5%. It is at this level that stripping agents become very useful. Stripping agents effectively lower the partial pressure in the separation zone and hence reduce the residual volatile content without exposing the polymer to excessive temperature or forcing the machine designer to resort to excessively expensive vacuum levels. As noted earlier, counter-current stripping has been shown to be more efficient than co-current stripping. Experimental variables to be evaluated include stripping addition rate, selection of stripping agent (although water is the most common stripping agent, it is not the only possible choice), operating feed rate and/or screw speed (although these are likely to be limited by the upstream stage optimization process), and equilibrium conditions (temperature, pressure, or vacuum level). There will probably be a trade-off between vacuum level and rate of stripping agent addition. The optimum condition must be determined experimentally for each polymer system.

14. Final-Stage Design

In the fourth stage, the polymer is exposed to the third vacuum vent. This final vent is usually operated at high vacuum (typically in the range of 5–25 mm Hg), but for special processes sustained vacuum levels of 1 mm Hg have been attained through the use of special seals and operating procedures. The final fraction of volatiles is removed at this point. Stripping is generally not introduced in this stage unless an inert gas is used as the stripping agent, because of the risk of entrainment of some small amount of the stripping agent in the final product. The fourth stage terminates in whatever means is selected for discharge pumping of the product through the die, as previously described.

15. Final-Stage Operation

In this stage, most volatiles are already removed, the concentrated polymer usually exhibits a high viscosity, and there are few remaining unconstrained variables that can be adjusted independent of the previous stages. In this zone bubble formation and growth will be limited; however, it often can be enhanced by the use of multiflighted vent screws that subdivide the polymer melt into multiple small streams, thereby creating more surface area. The highest (economically) practical vacuum levels can be applied at this stage, as the amount of contaminant to be removed will be quite small. It is possible to apply stripping agents in this zone, but usually at reduced addition rate compared with the next upstream zone. Sometimes, the preferred stripping agent in the final zone is an inert gas rather than a liquid, although liquids (especially water) have the usually beneficial side effect of slightly cooling the polymer.

16. Data Acquisition and Analysis

The range of operating data that should be obtained during devolatilization experiments includes the following:

1. Product feed rate and condition (temperature and pressure)
2. Machine operating parameters (screw speed, power, and temperatures)
3. Stage-wise process parameters (material temperature and vacuum levels)
4. Exiting product state (product rate, temperature, pressure, and quality)
5. Vent yields (rate and equilibrium conditions)

From this data, material balances around the process can be calculated. Although there are usually some minor losses, such material balance data gives valuable insight into the performance of the process. Automate as much of the data collection process as is practical. With automated data collection tools, the amount of data collected quickly becomes unmanageable unless you also use statistical analysis methods to reduce the data to representative data sets for study.

B. Latex Coagulation and Drying Tutorial

1. Two-Phase Systems

Although similar to devolatilization, substantially different techniques are required to process two-phase systems. Many two-phase systems use water as the diluent. Dewatering and drying of elastomeric materials is carried out predominantly by mechanical means rather than by heat transfer. Such systems function analogously to wringing water out of a sponge, with one key difference: wringing out a sponge is a batch process, while commerical dewatering and drying of elastomeric materials requires continuous processes.

A key concept in the continuous process is that the forces imposed for conveying the elastomer or polymeric material are separable from the forces imposed for conveying the water. Therefore, the polymer can be conveyed by drag flow in one direction while the water is forced in another direction by pressure differential, or even just gravity. Such two-phase flow exists because of the very large viscosity differences (about six or seven orders of magnitude) between the polymeric phase and the water. One such commercial system consists of a conical single-screw extruder tilted on an incline, so that as the elastomer is conveyed upward the water simply drains out the back end. Another relatively common commercial method is based on conveying the elastomeric material through a screw extruder with a slitted barrel, so that the water is squeezed out through the slits.

2. System Overview

In the continuous system for coagulation, dewatering, and drying of polymer latex emulsions of Skidmore (1973, 1976), a multistage, CRNI twin-screw extruder is used in which the first stage mixes the latex with coagulants (usually acids or salt solutions) and flocculation takes place. The flocculated particles are massed together by compression, and the massed polymer and free water are conveyed into a second stage where separation of approximately 90% of the free water takes place. Usually, toward the downstream end of this stage, a counter-current flow of water is introduced to wash contaminants from the product. Again, the massed polymer and a substantially smaller quantity of free water are conveyed into a third stage, where a second counter-current washing takes place and further separation of water from the process is carried out. Finally, the product (now containing less than 5% moisture) is passed into a third dewatering zone, which is operated as a conventional vacuum vent stage. After devolatilization of the remaining water, the final product is discharged through a die and is pelletized. Figure 28 is a schematic representation of a typical coagulation, dewatering, and drying system for ABS polymer.

3. Other Multiphase Systems

This technology has also been applied to polymer solutions to which an antisolvent has been added to precipitate the polymer, followed by

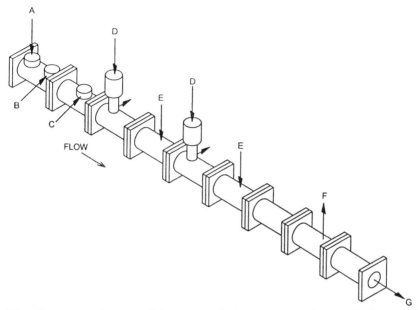

Fig. 28 Schematic view of latex coagulation process: (A) latex feed port; (B) coagulant feed port; (C) steam injection port; (D) mechanical filter(s); (E) wash water injection port(s); (F) vacuum vent port; (G) polymer melt discharge.

deliquefaction, washing, and devolatilization steps. See also Powell (1993) and Golba (U.S. Patent 5,043,421).

4. Handling of Polymer Latex Emulsion

The preparation of polymer latex emulsions is beyond the scope of this chapter. However, bringing a latex emulsion to the process safely and without compromise to its quality is pertinent. Although the flammability hazard of solvent solutions is avoided with water-based emulsions, there are other hazards. Of these, the likeliest personnel hazard is associated with breathing fumes or making skin contact with the material. Follow the guidance of the material safety data sheets (MSDS) for appropriate personnel safety and handling precautions. The second concern is possibility of an unconfined spill. Such materials are generally considered hazardous wastes and are not permitted to simply be dumped into the environment. Appropriate spill containment, cleanup, and disposal methods must be followed.

Finally, latex emulsions are electronchemically stable only within limited ranges of temperature, shear, hydration, and age. The greatest risk is probably shipping and handling latex emulsions in the winter, with the attendant risk of freezing (which in turn coagulates the latex). Although it is unlikely that excessive shearing will occur during transport, excessively vigorous handling can cause similar problems. In well-sealed containers, changes in hydration should not become a concern. Latex emulsions should always be used as soon as possible, as excessive time in storage may compromise product quality. Storage should always be in a cool area protected from sunlight.

5. Emulsion Feed Systems

The first choice for transfer of latex from drums to other containers should always be gravity flow whenever possible. It is simply the easiest, most reliable means of transferring latex with no risk to the latex itself. However, transfer of latex from drums to other containers can be facilitated with inexpensive pneumatic drum pumps—but they will need to be field stripped, thoroughly cleaned, and reassembled after each use. Some seals may require frequent changing. Latex emulsions vary widely, and so it is difficult to generalize without first testing your latex material with each specific pump. In any case, none of the foregoing examples involved pumping against a significant restriction.

To feed the continuous extrusion latex coagulation process described earlier, a pump was needed that would (1) treat the latex gently enough to avoid coagulation within the pump, (2) continue to pump the coagulated material anyway, should coagulation take place, (3) sustain normal operation in the range of 20–35 bar discharge pressure, and (4) have a sufficiently positive displacement characteristic to deliver uniform rate in spite of variation in discharge pressure. The only pump type found to fulfill all these criteria has been the progressive cavity pump.

6. Wash Water Systems

One of the unique functions of the latex coagulation process is that the product that has been coagulated is contaminated by residual soaps and other chemicals, which must be removed in order for the final product to develop optimal physical properties. In prior technology, large vats of water were used in cascade sequence to wash the product. Unlike stripping in devolatilization, which uses very small quantities of water to enhance separation efficiency, in this process the amount of water needed to contact the polymer is on the order of magnitude of 100% (i.e., the wash water rate needed to at least equal the product rate) and higher wash water rates are preferred. The washing system requires only a source of clean water, an appropriately sized pump, a flow meter, and an injection valve.

7. *Mechanical Filter Design*

Unlike the vent stuffers used on co-rotating, intermeshing twin-screw extruders, mechanical filters are not primarily used to force a continuing flow of polymer back into the screws; they are designed to capture moderate-size entrained solids particles and to reject them back into the process. Figure 29 is a schematic view of a mechanical filter (Skidmore, 1978, 1979). The design clearances in the mechanical filters are a careful balance

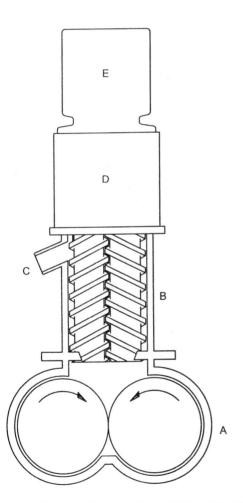

Fig. 29 Details of mechanical filter: (A) CRNI twin-screw extruder–coagulator; (B) intermeshing twin-screw "mechanical filter" screws; (C) effluent discharge port; (D) gear reducer; (E) motor.

between hydrodynamic (water flow rate through the device) and filtration requirements. Commercial mechanical filters have been designed to handle flow rates in excess of 0.38 m³/min. Total process solids loss through the mechanical filters is usually on the order of about 0.5%, which usually represents an improvement over prior commerical latex conversion technology. This potential loss stream is both clean and recoverable through conventional filtration technology.

As mechanical filters are relatively short L/D devices, their filtration efficiency is improved if the pressure drop across the mechanical filter is limited. Therefore, rather than allowing the effluent stream from the mechanical filter to blast out in a plume of steam, a heat exchanger and back pressure valve combination are provided to cool the effluent below the boiling point and to maintain a system pressure that assures only a modest pressure drop across the mechanical filter.

8. Minor Ingredient Addition

Polymer finishing of both dilute solution polymers and latex emulsion polymers may require the addition of minor ingredients to bring the final product to specification. Downstream solids addition would require providing an opening into the melt stream for the solids to be introduced. Quite often the resulting oxygen entrainment in a high-shear environment would compromise product quality. Direct injection of low-viscosity liquids is readily accomplished. Alternatively, if the additives are higher in viscosity, then addition of a molten master batch by side arm extruder is the preferred approach. Addition of thermodynamically compatible materials is of course the easiest to handle, but addition of reasonable quantities of immiscible materials has been accomplished.

9. Coagulation Zone Design

In the first process zone, a stable queous dispersion, or latex, is subjected to a combination of chemicals, dilution, heating, and shear to almost instantaneously coagulate into particulate polymer and free water. The working of the screws tends to knead the polymer into a continuous mass with only a small content of fines in the water phase. The latex can be introduced concurrently with a molten polymer stream from an upstream process to form a polymer blend. In cases where an upstream molten polymer stream is introduced, there must be appropriate polymer seals at both ends of the coagulation zone to contain the pressurized water and steam in that zone. The seal upstream of the coagulation zone is limited by both the rheology of polymer and the typically low rate of that first feed stream. This can lead to the need for augmenting tight cylindrical compounders with a barrel with a special small apex opening. The seal downstream of the coagulation zone usually provides no special difficulty, as there is a generous length of pumping zone immediately upstream and the volume is satisfied by the polymer solids added by the latex.

Performance of the coagulation zone is influenced strongly by the selection of coagulant chemistry. The other significant variable in this zone is the rate of steam injection. Steam is quite useful, as it helps to satisfy the energy requirements imposed by introducing a large volume of cold water into a polymer melt process.

10. Coagulation Zone Operation

Startup of a latex coagulation process is fairly easy; if there is an upstream polymer stream, it is always started first and stopped last. If there is no upstream polymer stream, then first the latex is introduced at a moderate rate, with immediate addition of coagulant flow. Initially, the effluent from the first dewatering zone will include some "white water," or uncoagulated latex. As soon as the white water has disappeared, the operation is usually sufficiently stable to introduce steam and then to gradually increase the rate to the target value.

11. Dewatering and Washing Design

Washing is normally accomplished in two sequential stages, with approximately the same quantity of wash water rate per stage. Figure 30

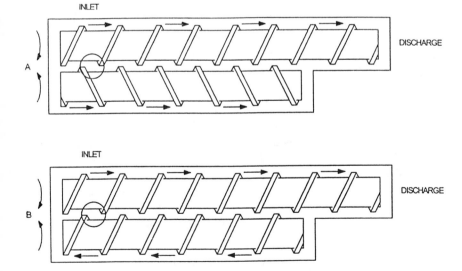

Fig. 30 Comparison of schematic views of CRNI twin-screw extruder: (A) conventional "plug flow" screw arrangement; (B) unique "backmixing" screw arrangement.

illustrates how a counter-current stripping stage functions. The principles of the washing stage are the same, except the water flow rate of 100–150% of product rate is vastly greater than the 1–2% of product rate usually used for stripping. Briefly, the coagulated and now massed polymer is being conveyed downstream by drag flow induced through the rotation of the extruder screws. A cylindrical compounder at the discharge end of the zone causes a pressure rise over 4 or 5 diameters of filled screw length. A high volume of water is injected in the filled zone, and the pressure gradient imposed by the polymer forces the low-viscosity water to flow counter-current to the polymer flow to a controlled exit through a mechanical filter.

12. Dewatering and Washing Operation

When a stable process is established in the coagulation zone, and the effluent from the first mechanical filter is clear, then wash water can be added (initially at about half the target rate) and gradually raised to the full rate desired. So long as the effluent has remained clear, external back pressure can be applied to the mechanical filter discharge line to raise the pressure to about 3–4 bar. Care should be taken periodically to check in-line filters in the mechanical filter discharge as during startup (or any significant process upset) the process is likely to discharge some excess fines.

13. Second-Stage Dewatering and Washing Design

This stage functionally duplicates the first dewatering and washing stage with the exception that the water content entering the stage from upstream is considerably lower than in the first wash stage. Also note that as the contaminants removed from the second stage are usually much less than removed from the first stage, it is possible to recycle the wash water by using the second-stage effluent as a feed water supply to the first wash stage. This results in substantial savings in water with only minor loss of contaminant removal efficiency in the first stage.

14. Second-Stage Dewatering and Washing Operation

When a stable process is established through the first dewatering and washing zone, the same process startup steps can be continued within the second-stage dewatering and washing zone (i.e., start the wash water system, gradually raise the wash water rate to the target value, and apply external back pressure). Again, care should be taken periodically to check in-line filters, although the fines loss from the second wash stage is usually much less than that from the first stage.

15. Vacuum Vent Zone—Final Dewatering—Design

In the final stage, the polymer is exposed to a vacuum vent. This vent is usually operated at moderate to high vacuum (typically in the range

25–100 mm Hg). The final fraction of volatiles is removed at this point. Stripping is generally not used in this process, because the volatile content is predominantly water. The fourth stage terminates in whatever means is selected for discharge pumping of the product through the die.

16. *Vacuum Vent Zone—Final Dewatering—Operation*

Always visually check any vent to be sure that it is clear prior to applying vacuum. With the high-volume wash water systems applied upstream, there is always the possibility that excess water migrated downstream during startup. Apply vacuum gradually so that the excess water can be flashed off and then, when the process permits, apply full vacuum. For this system "full vacuum" usually is the application of a single-stage water ring vacuum pump (i.e., about 25–100 mm Hg).

C. Reactive Extrusion (Examples)

1. *Polymer Degradation*

Visbreaking of polypropylene is one of the most widely practiced examples of reactive extrusion. Controlled-rheology polypropylene yields narrower molecular weight distribution and corresponding improved physical properties of particular importance in film products. Although limited visbreaking can be accomplished by thermal and shear degradation, the most widely used commercial practice is to introduce organic peroxide. The product viscosity can then be easily adjusted by controlling process temperature and residence time. A typical peroxide in commercial use a 2,5-dimethyl-2,5-di(t-butyl-peroxy)hexane. Thermal decomposition of this compound yields *tert*-butyl alcohol (TBA) and acetone along with several trace by-products. The resulting devolatilization requirement is to reduce the TBA and acetone content to low ppm levels.

2. *Graft Reactions*

Visbreaking may also be used as a first step in a multistage polymer modification process in which a graft reaction with a comonomer takes place. Nichols and Kheradi (1988) describe such a process in which the first zone includes solids conveying, melting, and peroxide-induced degradation, followed by a zone in which an acrylic acid comonomer is introduced, followed by addition of glass fibers and vacuum venting.

Another common type of graft reaction is the production of two-phase dispersions of a rubbery polymer in a thermoplastic matrix. Grafting of the rubbery dispersion to the thermoplastic matrix promotes adhesion between the phases and thus improves mechanical properties compared with mechanical blending without reaction. Devolatilization requirements vary with the actual chemistry involved, but systems of this sort are generally well served with a single vacuum vent stage.

A unique example of a grafting process using a CRNI twin-screw extruder is the halogenation of butyl rubber, which has been described by Kowalski (1992). In this process, there are competing reactions, and the preferential reaction required to obtain quality product is at a relatively low process temperature. A successful process was obtained by separating the rubber into multiple flow paths under low shear conditions and exposing it to chlorine (or bromine) gas. Due to the hazardous nature of the reactant, the process required good stable dynamic seal formation and very good devolatilization, as excess reactant could not be permitted to remain in the product.

3. Functionalization

Functionalization of end groups in polymers usually provides involved chemical and/or thermal stability. Other functionalization reactions improve mechanical or optical properties and may alter the morphology of the material. One example of a commercial functionalization reaction using a CRNI twin-screw extruder is the modification of an ethlene–acrylic acid ionomer with an amine to improve mechanical and optical properties in thick extruded sheet. The process begins with plastification of the base polymer. Following addition and mixing of the amine, the reaction is completed and the devolatilized melt is pumped through a sheet die. The resulting thick extruded sheet is used in the construction of security glazing laminates. In the absence of effective devolatilization, the resulting product lacks the clarity required for optical applications.

4. In Situ Polymerization

In situ polymerization is defined here as a process in which polymerization is carried out from monomers (or a low-molecular-weight pre-polymer) to a finished high polymer in the extruder and may include both addition and polycondensation reactions. Residence time considerations make these processes economically viable for reaction times of less than approximately 5 min, unless the value of the product is extremely high. Examples of commercially viable in situ processes include the condensation reaction of polyether imides and the anionic polymerization of Nylon 6 from caprolactam (see Nichols et al., 1986). These reactions require excellent distributive mixing and devolatilization. In the case of poly-condensation reactions, the reaction rate is usually driven by the rate of removal of the volatile by-product. Applebaum and Nichols (1987) described the statistical optimization of a polycondensation extrusion process in which a pelletized pre-polymer was melted in one extruder and pumped into a CRNI reactor/devolatilizer. Variables such as screw geometry, barrel temperature, feed rate, and vent pressure were evaluated. The resulting statistical model was able to predict intrinsic viscosity as a result of process condition with reasonable accuracy.

5. New Developments in Reactive Extrusion

A recent discovery by Biesenberger and associates at the Polymer Processing Institute, Stevens Institute of Technology, is that the CRNI twin-screw extruder can be configured to perform with a residence time distribution (RTD) similar to a continuous stirred tank reactor (CSTR). In a normal configuration, the CRNI exhibits a fairly narrow RTD similar to that of other twin-screw and single-screw extruders, and hence reactions that are favored are those which are optimal in a plug flow environment. The discovery is that when the screw flights of one screw are reversed, as illustrated in Fig. 30, a recirculatory flow pattern can be established in which the chance that an entering particle gets mixed with any other particle at random within the system volume is equal. Thus, for the first time, a backmix reactor exists that can handle viscous fluids (Lu et al., 1993). CSTRs are very well known and understood in the chemical engineering field, but they are limited to low-viscosity fluids because mixing is induced by the combination of the mechanical action of the impeller and the gravity (or pressure) flow of the entering and exiting fluid streams.

In the case of the CRNI twin-screw extruder functioning as a backmix reactor, the driving force for fluid transport is drag flow and the mixing is induced by a combination of drag and pressure flows within the extruder. This discovery opens the door to the application of viscous polymer systems in which the reaction is not favored by plug flow. For example, CSTRs can produce chain addition polymers with narrower molecular weight distribution than plug flow reactors, and copolymers with narrower composition distributions. Experimental confirmation of this concept was produced using a urethane polymerization in which a higher average molecular weight and a broader molecular weight distribution were observed (Lu et al., 1994). This discovery provides opportunities for many new reactive extrusion processes to be created.

D. Addendum—Safety Issues

Safety is the primary concern when preparing to work with polymer solutions or emulsions. Before any materials are actually handled, all personnel involved in the project should be thoroughly briefed on the nature of the materials to be studied, and material safety data sheets (MSDS) should be read and discussed. Any special hazards of using the materials in conjunction with the equipment or physical area should be identified, and appropriate preventative actions taken to assure safe operation. These should include, but are not limited to the following:

1. Posting of relevant MSDS and instructions to all personnel regarding the materials to be handled.
2. Issuance of any appropriate special safety equipment (gloves, goggles, full-face masks, respirators, etc.).
3. Appropriate isolation of the test area from nearby activities.
4. Assure that the area ventilation is sufficient to handle anticipated vapor load resulting from normal activities and the possibility of accidental spills.
5. Immediate access to adequate and appropriate spill containment materials.
6. Placement of appropriate fire extinguishers, personnel safety showers, and eyewash stands within ready access of the experimental equipment.
7. Appropriate monitoring equipment for detecting trace levels of hazardous materials.
8. Plan ahead—have a list of emergency numbers ready in advance for all contingencies, listing both intracompany and community resources for

   a. Fire safety
   b. Medical assistance
   c Environmental authorities
   d. Hazardous waste disposal

If you adopt and rigorously follow good safety practices, it is unlikely that you will ever need to call on such resources, but good safety planning is essential for safe operation of these complex processes.

NOMENCLATURE

a apex land length
b screw flight width, axial direction
A Vapor–liquid interfacial area
A interfacial area element
A_0 initial interfacial area
A_f final interfacial area
B axial flight width
C concentration (of tracer re: RTD)
C bulk concentration of solvent
C_e equilibrium concentration of solvent

ΔC concentration driving force, $= C - C^*$ at a given pressure
D diffusion coefficient
D_b inside diameter of extruder barrel, nominal screw outside diameter
E Devolatilization efficiency, $= \Phi - \Phi_a / \Phi_i - \Phi_a$
f fraction of uninterrupted barrel circumference
f fraction fillage of liquid in the screw channel
F_d shape factor for drag flow in screw channel
F_p shape factor for pressure flow in screw channel
F_{DCRT} Overall drag flow factor for nonintermeshing twin-screw extruder
F_{PCRT} Overall pressure flow factor for nonintermeshing twin-screw extruder
H channel depth
k_m mass transfer coefficient
$k_m A$ overall mass transfer area
L helical screw channel length
M_0 apex flow resistance coefficient (dimensionless)
n power law exponent
N screw speed
N number of optimal reorientations in laminar mixing
P pressure
ΔP differential between operating pressure and vapor equilibrium pressure
ΔP_T total pressure gradient over filled screw length
q volumetric flow rate (per screw)
Q flow rate
R bubble radius
s applied shear strain
t mean residence time $= V/Q =$ volume/flow rate
Δt time interval
V volume
V_z Down-channel velocity, $= \pi D_b N \cos \theta$
W width of flow channels
W^* $(1 - f)\pi D$
Δz_T total screw length, filled screw length
α apex angle
χ Flory–Huggins interaction parameter
δ radial screw flight clearance, melt film thickness
Φ removal rate of volatiles from a flowing solution
Φ_a adiabatic separation rate
Φ_i isothermal separation rate
λ length of apex opening
μ viscosity of a Newtonian fluid
θ screw helix angle

ρ_s solution density
σ surface tension

REFERENCES

Amon, M., and Denson, C. D. (1984). *Polym. Engin. Sci.*, *24*: 1026.

Applebaum, M. D., and Nichols, R. J. (1987). Presentation at AIChE, New York, November.

Carley, J. F., and McKelvey, J. M. (1953), *Ind. Eng. Chem.*, *45*: 985.

Chella, R., and Lindt, J. T. (1986). Technical Papers, 44th Annual Technical Conference, SPE, *XXXII*: 851.

Chen, L., and Lindt, J. T. (1995). *AIChE J.* In Press.

Chen, L., Hu, G. H., and Lindt, J. T. (1993). *AIChE J.*, *39*: 653.

Duda, J. L., Ni, Y. C., and Vrentas, J. S. (1978). *J. Appl. Polym. Sci.*, *22*: 689.

Duda, J. L., Vrentas, J. S., Yu, S. T., and Liu, H. T. (1982). *AIChE J.*, *28*: 279.

Erwin, L. (1991). in *Mixing in Polymer Processing* (C. Rauwendaal, ed.) Marcel Dekker, New York, p. 3.

Foster, R. W., and Lindt, J. T. (1987). *Polym. Engin. Sci.*, *27*: 1292.

Foster, R. W., and Lindt, J. T. (1989). *Polym. Engin. Sci.*, *29*: 178.

Foster, R. W., and Lindt, J. T. (1990a). *Polym. Engin. Sci.*, *30*: 424.

Foster, R. W., and Lindt, J. T. (1990b). *Polym. Engin. Sci.*, *30*: 621.

Gailus, D. W., and Erwin, L. (1981). Technical Papers, 39th Annual Technical Conference, SPE, *XXVII*: 639.

Golba, J. C., (1991). U.S. Patent 5,043,421.

Han, J. H., and Han, C. D. (1988). *Polym. Engin. Sci.*, *28*: 1616.

Higbie, R. (1935). *Trans. AIChE*, *31*: 365.

Howland, C., and Erwin L. (1983). Technical Papers, 41st Annual Technical Conference, SPE, *XXXI*: 113.

Jerman, R. E. (1986). Technical Papers, 42nd Annual Technical Conference, SPE, *XXXII*: 996.

Kaplan, A., and Tadmor, Z. *Polym. Engin. Sci.*, *14*(1): 58.

Kowalski, R. (1992). in *Reactive Extrusion—Principles and Practice* (M. Xanthos, ed.), Hanser Publishers, Munich, p. 29.

Latinen, G. A. (1955). *Adv. Chem. Ser.*, *34*: 235.

Lu, Y., Biesenberger, J. A. and Todd, D. B. (1993). Technical Papers, 51st Annual Technical Conference, SPE, *XXXIX*: 27.

Lu, Y., Biesenberger, J. H., Todd, D. B. (1994). Technical Papers, 52nd Annual Technical Conference, SPE, *XXXX*: 113.

Maddock, B. H. (1959). *SPE J.* *14*(12): 853.

Middleman, S. (1977). *Fundamentals of Polymer Processing*, McGraw-Hill, New York, p. 40.

Nichols, R. J. (1984). Technical Papers, 42nd Annual Technical Conference, SPE, *XXX*: 6.

Nichols, R. J., and Kheradi, F. (1984). *Mod. Plast.*, *61*(2): 70.

Nichols, R. J., and Kheradi, F. (1988). Presentation, ACS, 3rd Chemical Congress of North America, Toronto, June.

Nichols, R. J., and Lindt, J. T. (1988). Technical Papers, 46th Annual Technical Conference, SPE, *XXXIV*: 80.

Nichols, R. J., and Yao, J. (1982). Technical Papers, 38th Annual Technical Conference, SPE, *XXVIII*: 416.

Nichols, R. J., Golba, J. C., and Shete, P. K. (1983). Paper presented at AIChE Diamond Jubilee Meeting, Washington, D.C.

Nichols, R. J., Golba, J. C., and Johnson, B. C. (1986). Presentation, Polymer Processing Society, 2nd Annual Meeting, Montreal, April.

Nguyen, K. T., and Lindt, J. T. (1989). *Polym. Engin. Sci.*, *29*: 709.

Padlyia, D., and Tebbens, K. (1981). Paper presented at the 2nd World Congress of Chemical Engineering, Montreal.

Pearson, J. R. A. (1976). *Plas. and Rubber Proc.* Sept.: 113.

Pinto, G., and Tadmor, Z. (1970). *Polym. Engin. Sci.*, *10*(5): 279.

Powell, K. (1993). Technical Papers, 51st Annual Technical Conference, SPE, *XXXIX*: 1046.

Skidmore, R. H., (1963). U.S. Patent 3,082,306.

Skidmore, R. H., (1963). U.S. Patent 3,085,288.

Skidmore, R. H., (1973). U.S. Patent 3,742,093.

Skidmore, R. H., (1974). U.S. Patent 3,799,234.

Skidmore, R. H., (1976). U.S. Patent 3,993,292.

Skidmore, R. H., (1978). U.S. Patent 4,110,843.

Skidmore, R. H., (1979). U.S. Patent 4,148,991.

Spencer, T. S., and Wiley, R. M. (1951). *J. Colloid. Sci.*, *6*: 133.

Squires, P. H. (1958). *SPE J.*, *14* (May): 24.

Street, L. F., (1956). U.S. Patent 2,733,051.

14

Devolatilization of Polymers in Industrial Applications

Colin Anolick

E. I. du Pont de Nemours & Co., Wilmington, Delaware

I. INTRODUCTION

In the production of polymers, the removal of solvents or residual volatile components is often a critical step, one that can control manufacturing costs as much as does polymerization kinetics or even monomer costs. It is the

devolatilization step that often requires expensive equipment, causes down-
time due to fouling or mechanical failure, and is as likely as any to
be the cause of product quality problems. Although the devolatilization
step is much better understood than it was only a few years ago, more
work is needed before we can feel comfortable with our ability to choose
the right equipment and to scale it up well. The results depend greatly
on the specific equipment used, and since this equipment is large and
expensive and is not usually available in university laboratories, thorough
academic studies are rare. This chapter gives examples of how the devolatilization process is handled in industrial applications, some commercial,
some only experimental, to try to classify the devolatilization process into
areas where research is likely to be helpful, and to show that there are useful
simplifying principles.

In the typical industrial situation, a polymer is made in a solution
or in the presence of its monomers, and after removal from the reactor,
the polymer retains some level of these organics still dissolved. In cases
of condensation polymerization, volatile by-products must be removed
before the reaction can be completed. In all cases, devolatilization of
the retained solute is critical. The most frequent problems with which
the engineer has to cope is how to predict efficiencies and how to
scale up the process. We are either faced with designing a commercial
process based on a small laboratory experiment, or we have some
limited tests at a vendor's laboratory. Seldom do we have the luxury of
lengthy full-scale tests. There is no obvious choice of equipment, of
scale-up criteria, or even of ways to solve process problems when they do
occur.

Before describing some useful principles, let me state what should
be the first rule of polymer devolatilization. Devolatilization is like driving
a car; it is not how well you drive most of the time that is important, but
how badly you do it occasionally. A devolatilizer that plugs with polymer
only once a week will give unacceptable commercial problems. Similarly,
only small amounts of polymer caught for too long in a crevice and
degraded before being ejected into the main stream of product can give
your customer serious problems and send them to your competitors. So
materials handling is critical. For example, the twin-screw extruder is
gaining in popularity for devolatilization operations. It is certainly not the
most effective devolatilizer; it does not renew surface as well as other,
more complex devices, and it has little effective heat transfer surface.
However, it is a reliable machine for pumping viscous melts and this
reliability is critical and the reason for its increasing popularity.

What are the principles that should guide the engineer? How does
he, or she, choose between drum driers, wiped-film evaporators, steam

strippers, vacuum extruders, or even driers that handle solid pellets. This is just a short list of the devolatilization equipment available. Each scales up differently and handles the resin differently.

The following four questions are the major considerations:

1. Will the machine handle the resin—before, during, and after devolatilization?
2. What is the thermodynamic equilibrium? No machine can do magic, yet too often the engineer has only scant equilibrium data. The effort to get good equilibrium data is always worthwhile since the equilibrium is what often limits evaporation.
3. How close to equilibrium can we expect the devolatilization process to get? This will be termed throughout this chapter as the *approach to equilibrium.* Surprisingly, this can be estimated and scaled, if not with confidence at least with the knowledge of where scaling problems will most likely occur. The examples that follow will help with this. For example, it is crucial to know whether foaming occurs or not.
4. Is heat transfer limiting? In some cases this is critical; in others, it is of little importance. It is usually clear when this will be an issue.

II. EFFICIENCY AND APPROACH TO EQUILIBRIUM

A few words are needed about the thermodynamic equilibrium. Like any other mass transfer operation, devolatilization cannot progress further than thermodynamic equilibrium, and so it is critical to evaluate an operation against the equilibrium state. Usually the devolatilization process is characterized by the percent of equilibrium attained: $W_f/W_0 \times 100$. However, the logarithmic form of the equation is preferable:

$$\frac{k_m A}{V} = -\ln \frac{W_f - W_e}{W_0 - W_e} \tag{1}$$

in which W_f, W_0, and W_e are the respective weight fractions of solute in the product, in the feed, and at equilibrium if the product had sufficient time and surface area to come to equilibrium with the exit vapors. The quantity $k_m A/V$ contains the mass transfer coefficient, k_m, the surface area available for evaporation, A, and the volumetric flow rate of resin, V. It is a dimensionless number akin to "number of transfer units." This equation can be derived from the usual "log mean" driving force equations of mass transfer, with the assumption that the driving forces at the inlet and outlet are $W_0 - W_e$ and $W_f - W_e$, respectively. Unfortunately, we seldom can separate the terms k_m and A, although we usually know V quite accurately.

We too often have to be content with using this number of transfer units and scaling using this alone. Until we gain a better understanding of k_m and A, and what controls them, we are forced to use dimensionless quantities like this one. The most useful aspect of this dimensionless quantity is that it makes scale-up easier. If we double the length of the equipment, we should double the value of $k_m A/V$. For example, if $k_m A/V$ changes from 1.0 to 2.0, the fractional attainment of equilibrium will change from 63% to 86%. Since the equilibrium is approached logarithmically, it is easier to think in terms of number of transfer units than in terms of approach to equilibrium.

It is unwise to measure the amount of material devolatilized without reference to the equilibrium state, W_e. Frequently, trade publications and even publications in scientific journals show data in terms of "percent devolatilized" since this is indeed the bottom line in commercial applications. At best, this is because it is assumed that the equilibrium concentration of volatiles is so low that it can be ignored. However, serious errors in scaling can occur if we make no reference to the equilibrium state. It is far better to use an equation, such as Eq. (1), that normalizes with reference to equilibrium.

Equilibrium between W_e, the weight fraction of solute in the polymer, and partial pressure of the solute at melt temperature can be measured by various techniques, including the chromatographic technique (for example, see Newman and Prausnitz, 1972; Young, 1968). Flory–Huggins equations and interaction parameters can be used to estimate the vapor–liquid equilibrium relationship or a Henry's law coefficient. However, these estimates are seldom as accurate as the measured data. Engineers should only feel as confident of their scale-up as their confidence in the equilibrium data; particularly since many industrial devolatilization processes get relatively close to equilibrium, and small errors in W_e can make a large difference to the final level of organic volatiles remaining in the product.

Where Henry's law data are available, they usually take the form

$$\ln K_W = a - \frac{b}{T} \tag{2}$$

where K_W is the Henry's law coefficient, usually in units such as atmospheres per weight fraction, a and b are constants, and T is the absolute temperature. This is similar to Antoine's equation for vapor pressure of pure components. The relationship between the equilibrium concentration of solute in the polymer, c_e, and the partial pressure of the solute in the gas phase, P_1, also at equilibrium, is then linear:

$$W_e K_W = P_1 \tag{3}$$

Antoine's equation is an equation developed to correlate vapor pressures of pure components as functions of temperature, and sometimes the Henry's law coefficient will be found in terms of a function of the pure component vapor pressure, P_1^0 (values of P_1^0 are listed in Appendix A for various solvents and monomers).

Ironically, often the error in measuring polymer temperature is large enough to lead to serious doubts about how close the system may get to thermodynamic equilibrium. Many types of mechanical devices work the polymer significantly and thus increase its temperature. This can lead to significant scale-up problems; often the large-scale process runs at higher temperatures, because heat loss through the equipment walls is less. This is an advantage in devolatilization, but it can lead to serious product degradation problems.

In this chapter we do not concentrate on the theoretical aspects of devolatilization. However, that is not to imply that shortcuts are warranted. Not only is it essential to know the thermodynamic equilibrium, but the heat balance must be described as well. Many industrial problems, which occur years after startup, are caused by the heat balance being neglected and by production rates being increased too far for the heat input capability of the equipment.

III. EXAMPLES

A summary of many of the practical aspects of devolatilization in industry will be instructive. Obviously many of the practical solutions and design criteria are proprietary to the industry involved, but there is a remarkable degree of consistency.

A. Steam Stripping

In the manufacture of polymers such as styrene-butadiene rubber (SBR), acrylic, or neoprene latices, the polymer is made as a dispersion or latex in water. The latex usually contains residual unconverted monomers, which need to be removed. Since the latex is a relatively stable suspension of tiny polymer particles in water, retained monomers can be removed by steam stripping without agglomeration, and the process can be run for long periods without polymer fouling the equipment. Stability is important. The resulting product is a latex free from volatiles and that might be used as is, or forwarded to a separate operation to remove the water. Figure 1a is a schematic representation of a typical process. Here two stages of stripping are employed to reduce the dissolved monomer to acceptable levels. Performance of strippers in this configuration is usually close to theoretical equilibrium if a sufficient length of tube is used.

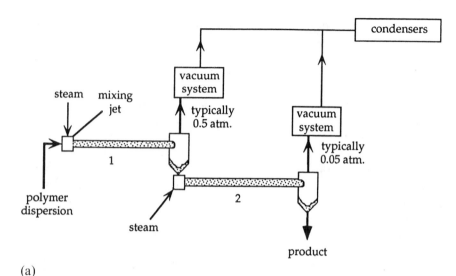

(a)

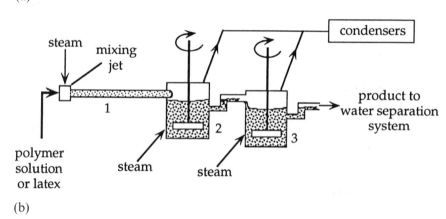

(b)

Fig. 1   (a) Two-stage stripping system for a polymer dispersion. (1) and (2) are turbannular strippers. (b) Three-stage steam strippers. (1) is a co-current tube stripper. (2) and (3) are agitated vessels.

The theoretical limits can be calculated from the partial pressure of water and volatiles, from the mass balance, and from the Henry's law coefficient. Typically, volatiles will be removed from about 5 to 15% based on the weight of polymer, down to 0.1% or lower. Design of the tube must consider gas velocities since many strippers operate with velocities approaching sonic velocity.

Many other polymers, such as high-density polyethylene (HDPE), polypropylene (PP), or ethylene/propylene/diene rubber (EPDM), are made not as water-based latices, but as solutions in an organic solvent, often at 10 to 15% solids only. They can have the residual solvent and monomers removed to produce a solid with acceptable levels of volatiles by a series of steam stripping stages. This is a good case of how materials handling is critical. As the solution loses solvent, it usually passes through a stage where it is a sticky mass. Sometimes handling this without fouling the equipment is an insurmountable problem. Other times good, mechanical agitation and addition of excess water is all that is needed. Similarly, if portions of the drying polymer agglomorate and trap regions containing solvent that has not yet evaporated, the process might be unreliable and the resulting polymer will have too much residual solvent and might be a flammability and environmental hazard. However, where the polymer can be kept as a separate crumb that does not stick to the walls, the process will operate smoothly. Again materials handling is critical, Unfortunately it is difficult to predict how a solution will handle without experimental trials.

Figure 1b shows a typical steam stripping operation for removing large volumes of solvent. A number of stages can be installed in series, although in practice two or three stages are usually employed. The first is a mixing jet and tube, in which steam and solution are mixed and move co-currently in highly turbulent flow. This first stage requires the design of an efficient contacting jet where the steam and latex or polymer solution are first mixed. Solution contactors are particularly difficult to design for sufficient contact between steam and solution, and the scale-up criterion is usually the surface area of contact between the steam and solution streams. A well-designed mixing jet with a moderate-length tube will take the polymer devolatilized to approximately the thermodynamic equilibrium at the end of the tube. The flow in the tube is intentionally kept very turbulent to enhance mass transfer and decrease the chance of plugging. Since there is normally still too much dissolved volatile with the polymer (typically about 50%), additional stages are added. However, usually the polymer is now too viscous to handle in a tube stripper without the polymer adhering to the walls. For this reason it is usual to suspend the polymer in excess water, to agitate the water vigorously while adding more steam. However, the second and third stage will not approach equilibrium as closely, since the polymer is now in a lump or crumb form and solvent diffusion is slower. The scaling criterion for this type of stripper is the agitation intensity, since the particles must be kept well suspended and in contact with the steam–water mixture. If agitation is effective, the approach to thermodynamic equilibrium can then be scaled from relatively small laboratory experiments.

The major disadvantage of this type of devolatilization technique is that although the polymer crumb is now separated from the solvent, the polymer is wet and dispersed in a large quantity of water. Therefore dewatering steps are needed, and design of these is also a complex art. Often dewatering screens followed by dewatering extruders are used to remove the water. It is surprising that such a multistep process can be run economically, but there are many examples where these processes are.

B.   Hot Surface Evaporators

Polymer solutions can often be devolatilized by contact with a hot surface. Good examples are drum driers and wiped-film evaporators. (See Figs. 2a and b.) In practical terms, the time the solution must stay on the surface is a function of the film's thickness and the heat transfer coefficient. This operation is best analyzed as a heat transfer process rather than a mass transfer operation, although the goal is to transfer mass. As in many chemical engineering operations, where both heat and mass are transferred simultaneously, it is dangerous to consider only one of the two. In these cases the temperature of wall surface is usually increased, producing an increase in production rate, until a limit is found. At this point the surface is boiling too vigorously, causing a barrier of vapor that insulates the drying film from the wall. One notably clever patent is an old one issued to R. A. Bernard, W. T. Dove, and J. C. Smith (1969). They added an electrostatic field across the boiling film on a drum drier, which held the film on the surface at higher boiling rates and increased capacity.

The drying polymer film does not know if it is on the outside of a drum or inside a wiped-film evaporator. In both cases heat is transferred to the film, which boils, or at least dries by diffusion, and the resulting vapors are carried away to a vapor port, usually by an inert gas. The gas quantity needs to be known since it defines the approach to thermodynamic equilibrium. However, these processes are usually heat-transfer-limited and in the final stages limited by the diffusion of the solute to the film surface. The design engineer will be required to choose an evaporator type and to specify a scaling technique. Scaling is based on film area and time on the surface between lay down and removal. Scale-up is thus unusually accurate and reliable. Unfortunately, most industrial processes of this type require large equipment sizes since the surface area of drier controls the rates.

Choice of evaporator type (drum versus wiped film) is not so easy. At first glance a wiped-film evaporator looks like a better choice for flammable fluids, since the gases are contained. However, the engineer must still assure against air leakage, especially during startup and upsets. Also the frequent wiping of a thin-film evaporator renews the surface and improves mass transfer. However, most often the choice becomes one of how the equipment

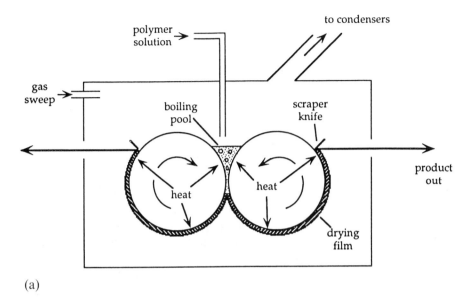

(a)

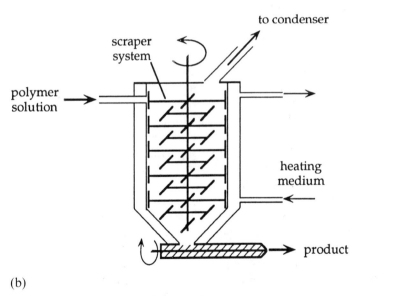

(b)

Fig. 2 (a) Rotating drum driers. (b) Wiped-film evaporator.

works in terms of product quality. No matter how well a blade or scraper is designed, some polymer remains on the wall or on the blade for longer than average. If this degrades, it will cause a quality problem. One advantage of a drum drier is that the operator can see this material accumulate and deal with it. So once again materials handling becomes the rationale for equipment choice, not devolatilization efficiency.

The difficulty of designing reliable hot surface evaporators is most easily appreciated when one considers the change in rheology of the polymer solution as it changes from a thin mobile solution to a solvent-free melt. The equipment must handle this complete range of viscosity.

There are other types of polymer driers (e.g., the List drier and the Buss Kneader) where the polymers solution is essentially dried by heat transfer through a wall and through paddles or pins, combined with mechanical working of the polymer. The principle of operation is essentially equivalent for both, but they might handle an individual polymer better or worse and there are few criteria to use to make a choice without testing. Sometimes most of the heat is generated by mechanical work. This can be an important benefit since the polymer always goes through a viscous stage where heat transfer is difficult in any form except thin films.

C. Phase Decantation

If the polymer can be precipitated from solution in the solvent in which it is produced, we have an advantage in isolating it to a solvent-free form. In the laboratory this is often easy; one can add a nonsolvent for the polymer that extracts the solvent. This causes a phase to separate that is concentrated with polymer. This can be decanted and washed a few times with nonsolvent and then oven dried to a clean dry sample of polymer. However, scaling this process to commercial facilities is not so easy, since the engineer now has to deal with two solvents and the auxiliary equipment needed to purify and separate solvents yields a separate set of problems. Few such processes exist at commercial scale where solvent–nonsolvent pairs are used.

On the other hand low-density polyethylene (LDPE) is phase decanted at gigantic scales, and EPDMs (ethylene, propylene diene rubbers) can be isolated from their reaction solvent by similar processes (Anolick and Goffinet, 1971; Anolick and Slocum, 1973). Both of these use the large changes in solvent power associated with pressure changes above the solvent critical temperature. At high pressure the polymer remains in solution, but decreasing the pressure precipitates a phase containing an increased concentration of polymer. The lower the pressure, the higher the polymer concentration. In thermodynamic terms, this uses the lower critical solution temperature/pressure. Figure 3 shows the process in schematic terms. The

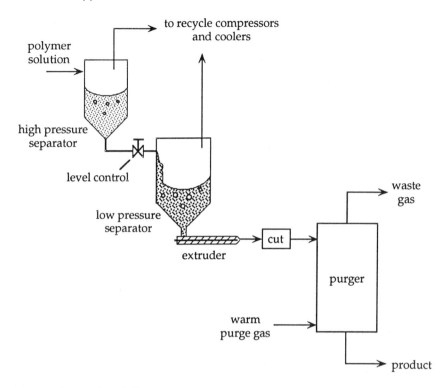

Fig. 3 Phase decantation.

polymer is polymerized in solution, either in a solvent (EPDMs) or in its monomer mixture (LDPE and ethylene copolymers). The simplest case to describe is LDPE, where the pressure and temperature of the reaction are both high. To keep the polymer soluble in the reactor, and to keep reaction rates practical, the ethylene is polymerized by free radical initiation at pressures from 10,000 to 50,000 psi. Since the reactor is adiabatic, the heat of reaction takes the solution to above 200°C. At the high-pressure conditions, the polymer is soluble in the reaction medium. However, when pressure is released to below about 5000 psi a phase separates richer in polymer than the reaction solution. The choice of separator pressure and reactor temperature and concentration are the art highly valued by the individual manufacturing companies involved. The practicality of the process hinges on these choices. Higher polymer concentrations in the reactor will reduce the cost of monomer recycle, but since the reaction is adiabatic, the reactor temperature will increase and the chance of polymer degradation or, worse, unexpected explosions will increase. So, in practice the reactor temperature is very carefully controlled.

Also carefully controlled are the pressures of the separators. Usually two stages of separation are used. In the first, a concentrated phase, containing some 30–60% polymer, will precipitate. Too low a pressure will produce a phase that is too highly concentrated and too viscous to separate cleanly, and this will result in excessive recompression costs. Too high a pressure, on the other hand, will result in high concentrations of monomer being retained in the polymer-rich phase, and higher concentrations of low-molecular-weight fractions being recycled through the high-pressure compression loop. A well-designed separator is simple and reliable and the separated phase concentrated in polymer. Another difficult aspect is how to control the level of polymer-rich phase in the separator.

The separated polymer-rich phase flows directly to a low-pressure separator, where almost all the retained monomer separates. Actually, both first- and second-stage separators can be thought of as flash devolatilization steps rather than simple separations. This is an important distinction. Since there are such large pressure changes between separators and a resulting large difference in solvent density, physical driving forces are large. The solvent dissolved at reactor pressure "erupts" from the polymer phase in the high-pressure separator, and the remaining dissolved solvent once again expands rapidly and also "erupts" from the polymer phase surface in the low-pressure separator.

Where there are large density changes, which is true for both LDPE when separated from ethylene, and EPDMs when separated from hexane, the phases separate close to equilibrium. In practical terms the real limit to the approach to equilibrium appears to be the usual materials handling problems. Where phases are very viscous, the polymer-rich phase might entrain bubbles of solvent-rich phase. Entrainment of solvent in the polymer-rich phase is minimized if there is a large density difference between phases. However, if the volume of solvent leaving the polymer-rich phase is too large, entrainment of small polymer particles in the solvent-rich phase will result.

In this, like many other devolatilization processes where dissolved solute is being removed in a lower-pressure zone, if spontaneous phase changes occur the process seems to rapidly approach the thermodynamic equilibrium. Diffusion is not controlling. These phase changes can be either foaming or precipitation, but they are always pressure induced.

What is the thermodynamic equilibrium? This is not well defined. In these phase decantation processes, sometimes we have a measured phase diagram, but we possess little fundamental knowledge of the equilibria involved. The phase diagrams giving only where precipitation occurs are certainly available for LDPE within the manufacturing organizations and in the literature (e.g., Hasch et al., 1993), but the relationships between phase equilibrium concentrations at various temperatures and pressures are less often available. However, if a Henry's law coefficient is measured at the appropriate temperature, or an equivalent relationship between solute partial pressure and concentration at low pressure is available, linear equations can be used to estimate the phase relationships even at the high-pressure separator conditions. It is better to run through these calculations than to simply guess the relationship. It is surprising how well these estimates fit the real large-scale operation. The most accurate method would be measurement of the equilibrium at high pressure and temperature, but this is not easy to do. Figure 4 depicts the phase relationships in schematic form. There is little reliable theory on which to base phase relationships, nor methods to predict the phase concentrations at any particular temperature and pressure. Experience shows that concentrations are best correlated to solvent density, and possibly solubility parameters, but it is hard to get good predictions.

Let us concentrate on the final steps of the devolatilization process in

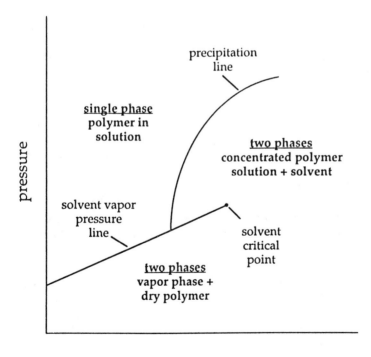

Fig. 4 Schematic phase diagram for EPDM in hexane.

the case of LDPE. The low-pressure separator is usually operated at some slight positive pressure to keep air and moisture out of the system and reduce explosion hazards and purification costs. Thus the concentration of unreacted monomer in equilibrium with the polymer is sufficient to give more than 1 atm of vapor pressure of monomers above the polymer under the low-pressure separator conditions. Usually, the polymer is removed from the low-pressure separator by an extruder and then cooled. Even after cooling this polymer to normal temperatures, the polymer will have too much organic volatile to ship safely. If transported in a closed container on a hot day, the concentration of ethylene in the air in the container could accumulate, presenting a serious explosion hazard. Also, any solvent left with the polymer is likely to find its way into our environment. Therefore another step of devolatilization and recovery is required.

Here there are at least two techniques worth considering. In the first, the polymer is extruded and cut into pellets, then purged with large volumes of warm air. This is the technique shown in Fig. 3. Here the remaining monomers slowly diffuse from the pellets until their level is low enough for safe shipment and to keep subsequent environmental pollution to a minimum. This is a slow process and can be expensive. Another alternative is to add a vacuum port to the extruder and remove more monomers this way. Since extruder devolatilization is such an important technique in many applications, it will be described separately.

D. Extruder Devolatilization

Many equipment manufacturers currently recommend the use of extruders equipped with vacuum ports for the devolatilization of polymer melts. It is even possible to start with a relatively dilute solution of polymer and do most of the devolatilization in a twin-screw extruder. Both counter-rotating and intermeshing co-rotating extruders are recommended. Results are so encouraging, particularly with twin screw extruders, that more and more commercial processes are likely to use such equipment in the future. We will present one example to show the versatility of the equipment. It is not that extruders have any special ability to generate mass transfer surface nor that they have large amounts of heat transfer surface, but they do have a useful combination of features. They are exceptionally good pumps of the viscous polymer melts that often have to be devolatilized, they can generate heat by the application of mechanical work on the melt, and they can handle the foaming action provided the ports are designed well.

In our example, a solution of polymer in organic solvent (see Fig. 5) is first heated in a traditional heat exchanger. The heater is normally held under pressure to eliminate evaporation of the solvent therein. On exiting

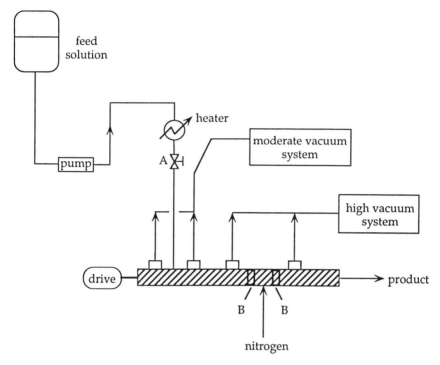

Fig. 5 Multistep extruder system with four vacuum ports for drying of polymer solutions. A: pressure release valve; B: seal sections.

the heater, pressure is released, and the solvent is flashed in a separator or even into the extruder entrance port. Most extruder manufacturers will recommend the arrangement of ports shown in Fig. 5. Here both an upstream and downstream port are used. The approach to equilibrium will be quite good, and if the Henry's law coefficient is known, the efficiency of this flash step can be calculated quite accurately. Normally a solution of about 15% polymer will increase to as high as 95% in this step. In this flash step the polymer must be increased in viscosity sufficiently to allow the extruder to function efficiently and to allow the maintenance of different pressures on the vent ports. Vapor handling can sometimes be a problem if foaming occurs. The concentration of polymer in solution might be as low as 60%, but usually better operation will occur if the flash step generates a more concentrated melt. More heat is now generated both by heat conduction through the extruder walls and through mechanical action. The pressure in the next port is usually lowered, and more solvent is evaporated. Thus the concentration of polymer is increased further.

The number of stages or ports used depends on how low a solvent content is needed. However, it is not really practical to have more than two or three vacuum ports at decreasing pressure. If the solvent content is not low enough, the last port can be assisted by adding an inert gas, usually nitrogen, but even water or steam have been used. The best approach is to have a section of the extruder in which the inert gas is added, usually with a special screw design so that the sweeping gas will be incorporated under pressure (Werner, 1980). Thus there are two sealing sections that isolate the region where inert gas is added from both upstream and downstream sections. The polymer melt is then pumped into a lower-pressure part of the extruder, where foaming occurs. The approach to thermodynamic equilibrium is usually remarkably good, with values of the dimensionless relationship $(k_m A/V)$ between 3 and 4 (see Eq. 1).

Using this approach, the solvent can usually be reduced to levels that satisfy environmental, safety, and economic needs, usually fractions of a percent. Unfortunately, sometimes the last small fraction is very difficult to remove, even though the thermodynamic equilibria are correct. Why this is so remains unclear.

Extruders are also useful where a small amount of dissolved material must be removed. Often only one or two vacuum ports are then needed, or one port with an inert gas addition. Usually the gas sweep helps considerably but there are examples where the effect is disappointing, and at present there seems to be no way to predict how a system will behave.

There are other ironic effects that occur in extruder devolatilization. Normally if a solvent, which might be present at a 1 or 2% level, is evaporated from a melt using an extruder with a vacuum port, approximately 50–70% of the solvent will be removed by the application of vacuum ports. This might be at least as much as would be predicted from the Henry's law coefficient and the temperature and pressure in the ports. In fact, many examples exist where more than the theoretical maximum is evaporated this way. This might be because there is a certain amount of air dissolved in the polymer or sucked into the extruder, which will reduce the partial pressure of the evaporating species. Theoretical efficiencies are therefore more than 100%. When an inert gas is added, significantly more solute is evaporated (up to about 95–98%), and one can expect a theoretical efficiency of about 98%. Now the theoretical efficiencies are lower and closer to 100%, even though more solvent is devolatilized.

What controls the approach to equilibrium in these operations, and why is the use of a gas sweep usually so efficient? These questions are not well understood, and a good academic answer to this question would certainly help in designing more reliable equipment. For example, much of the literature that discusses these operations focuses on the importance of surface renewal by the screw flights. This includes work by Denson (1983), Biesenberger (1983), Amon and Denson (1984), Vrentas and Duda (1985), Powell (1987), Ravindranath and Mashelkar (1988), Mack and Pfeiffer (1993), and Coelho Pinheiro and Guedes De Carvalho (1994). Yet many operations seem almost independent of screw rpm. Our present theories fail to address why it seems to be important to incorporate an inert gas at some moderate or high pressure, allowing it to expand and to foam. It is easy to state that this forms a good foam with a large surface area, but how do we quantify this foaming effect? Research is under way (Bigio et al., 1995) to improve our understanding of these questions.

IV. CONCLUSIONS

There are a few important principles that seem to control all these examples and many other devolatilization processes. There seem to be two general cases. In the first, the process is relatively slow and diffusion controlled; in the second, the process is very rapid and the product is close to equilibrium. Of course both can be effective. Let us summarize each case.

A. Diffusion-Controlled Processes

Examples of these types of processes are drum driers, wiped-film evaporators, and many other types of equipment designs that generate surface by mechanical means. In these cases the process has little superheat. In fact, to devolatilize the polymer we have to add heat during the evaporation process. The principles are then mainly those of heat transfer, and the design of commercial equipment based on laboratory tests is usually one of scaling up the heat transfer surface. Thus, drum driers and wiped-film evaporators scale well. In the cases where polymer melts are subjected to vacuum under non-foaming conditions, in extruders or other surface generation equipment, but where the surface area is not well understood, scaling is more difficult since surface area is probably proportional to the square of the equipment dimensions, while the capacity might be otherwise proportional to the cube of the dimensions. So large-scale equipment is invariably mass-transfer-limited.

B. Phase Change Processes

In those processes where there are large changes of density caused by changes in pressure and a phase change, a large thermodynamic driving force separates the solvent from a polymer-rich phase. The solvent might be a supercritical fluid, but in any case, a large density change exists. Under these circumstances, the phases will separate without the need for mechanical surface renewal. Conventional mass transfer theory, as described in the traditional chemical engineering literature, is not applicable. We do not know the mass transfer coefficients, the surface area, nor the correct diffusion coefficient. In fact it can be argued that a high diffusion coefficient may hinder foaming since the molecule will diffuse to the surface instead of generating surface area through bubble generation and growth. Although this argument is probably not valid, the effect of diffusion coefficients is not well understood. It would appear that the relevant driving force is the difference between the total vapor pressure of volatile components at the inlet conditions, which might be tens or even hundreds of atmospheres, and the absolute pressure in the flasher, which might be fractions of an atmosphere. Figure 6 depicts this process. Under these conditions we are forced to make an assumption regarding the approach to equilibrium. The following expression is remarkably easy to apply and gives reliable results in many cases:

$$\frac{k_m A}{V} = -\ln \frac{W_f - W_e}{W_0 - W_e} = 3 \text{ to } 4 \qquad (4)$$

This does not work when only small volumes of foam are produced (the pressure driving force is too small), or where the polymer melt is very viscous and is not given time to disengage from the volume of gas. Then significant gas volumes might be redissolved in the melt. In fact, recent work indicates that the limit to the process is the separation of the last few gas bubbles. The value of 3 to 4 for $k_m A / V$ can be developed by a very reasonable assumption of the volume of vapor phase redissolved in the polymer melt, but this is conjecture based on work under way (Bigio et al., 1995).

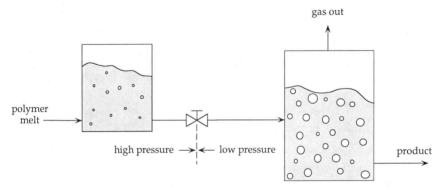

Fig. 6 A schematic representation of foaming devolatilization. A stripping agent may be added to the polymer.

Even though this expression is currently empirical, it seems to work reasonably well as a scale-up tool in foaming devolatilization, and the engineer in industry does not have the luxury of waiting for better scale-up tools. We are therefore in the uncomfortable position of having little that is fundamentally sound and reliable.

C. General

From these examples we can make some general conclusions. The best rules therefore seem to be as follows:

1. Where diffusion is likely to be controlling and little or no foam is produced, scale-up is based on the area of mass transfer and the effective heat transfer process. Scale-up is usually proportional to the heat transfer surface area.
2. Where large pressure drops are involved, and where there is a large volume of foam produced, scale-up can be based on Eq. (1) with a value based on small-scale tests.
3. Efficiency must be measured relative to thermodynamic equilibrium. Unfortunately there is no easy way of estimating the Henry's law coefficient, on which the thermodynamic efficiency can be based. Here is one more place where sound academic work is desperately needed.
4. Be aware (and beware) of the equipment's ability to handle the polymer. If small amounts stick in crevices or surfaces and slowly degrade, quality problems will be severe. Can the ports handle the vapor volumes, especially if there is a process upset? It is more important that the equipment can handle the polymer throughout its drying sequence than whether it is slightly more or less efficient in approach to equilibrium.

Although these rules look essentially empirical, they are the best we have. Hopefully, academic and industrial research will soon make improvements and bring a set of fundamental relationships into our science. Many questions remain unanswered. How do we estimate Henry's law coefficients? What controls the approach to equilibrium? How is foaming or phase change devolatilization really controlled? When are mechanical techniques for increasing surface area useful? Even important are such mundane questions as, what is the real temperature in an extruder under high-shear conditions when the melt is being worked by the screws? In addition, we need to understand what happens to these devolatilization process if chemical reactions are occuring simultaneously, such as in condensation polymerization. The answer to these questions would make life easier for the engineer in industry.

NOMENCLATURE

a constant
A surface area across which devolatilization will occur
b constant
k_m mass transfer coefficient
K_W Henry's law coefficient
P_1 vapor pressure of the volatile component over the polymer solution
P_1^0 vapor pressure of the pure volatile component
T temperature
V volume flow rate of polymer
W weight fraction

Subscripts

0   initial
e   equilibrium
f   final

REFERENCES

Amon, M., and Denson, C. D. (1984). A study of the dynamics of foam growth, *Polym. Engin. Sci.*, *24*: 1026.

Anolick, C., and Goffinet, E. P. (1971). U.S. Patent 3,553,156.

Anolick, C., and Slocum, E. W. (1973). U.S. Patent 3,726,843.

Bernard, R. A., Dove, W. T., and Smith, J. C. (1969). U.S. Patent 3,442,026.

Biesenberger, J. A. (1983). *Devolatilization of Polymers*, Hanser, Munich; Macmillan, New York.

Bigio, D., Smith, T., and Yang, C. T. (1995). Ph.D. thesis, University of Maryland.

Coelho Pinheiro, M. N., and Guedes De Carvalho, J. R. F. (1994). Stripping in a bubbling pool under vacuum, *Chem. Eng. Sci.*, *49*: 2689.

Denson, C. D., (1983). Stripping operations in polymer processing, *Adv. Chem. Eng.*, *12*: 61.

Hasch, B. M., Meilchen, M. A., and McHugh, M. A. (1993). Cosolvent effects on copolymer solutions at high pressure, *J. Polym. Sci., Polym. Phys. Ed.*, *31*: 429.

Mack, M. H., and Pfeiffer, A. (1993). "Effect of Stripping Agents for the Devolatilization of Highly Viscous Polymer Melts," 51st SPE ANTEC.

Newman, R. D., and Prausnitz, J. M. (1972). Polymer solvent interactions from gas–liquid partition chromatography, *J. Phys. Chem.*, *76*: 1492.

Powell, K. G. (1987). The Thinning and Growth of Gas Bubbles on Viscous Liquid/Gas Interfaces, Ph.D. thesis, University of Delaware.

Ravindranath, K., and Mashelkar, R. (1988). Analysis of the role of stripping agents in polymer devolatilization, *Chem. Eng. Sci.*, *43*: 429.

Vrentas, J. S., Duda, J. L., Ling, H.-C. (1985). Enhancement of impurities removal from polymer films, *J. Appl. Polym. Sci.*, *30*: 449.

Werner, H. (1980). *Devolatilization of Plastics*, Verein Deutscher Ingeniere VDI-GmbH, Düsseldorf.

Young, C. L. (1968). The use of gas–liquid chromatography for the determination of thermodynamic properties, *Chromatog. Rev.*, *10*: 129.

15

Single-Rotor Wiped-Film Devolatilization of Styrenic Polymers

Leo F. Carter

University of Puerto Rico, Mayagüez, Puerto Rico

I. INTRODUCTION

Interest in wiped-film devolatilization (WFD) of polystyrene (PS) at Monsanto became intense in the 1950's with a focus on the removal of residual styrene (ST) from PS made in batch polymerization processes (Latinen, 1962a, 1962b). This work sought to improve those polymer properties adversely influenced by residual low-molecular-weight species. It was found to be more effective to remove the last 1 to 2 wt % of unpolymerized styrene by vacuum vented extrusion than it was to complete the reaction. Efforts to polymerize the last monomer led to excessive reaction times, and to the production of low-molecular-weight species. Low helix angle vented single-screw extruders were shown to be effective for this need. The present chapter focuses on the developments leading from this point to the achievement of commercial-scale WFDs used in a continuous PS process (Latinen, 1974b).

It was evident that two stages of devolatilization (DV) would be required. First, a high energy input "primary" WFD would operate at pressure high enough to permit efficient removal and condensation of the majority of the volatiles at conditions consistent with economical condensing system design and operation. Then, "final" WFD would focus on attaining required product purity via low-pressure operation requiring more expensive vapor handling.

Proof was needed of the feasibility of using WFD to remove the larger quantities of monomer consistent with desirable reactor operation. Latinen's observations on ethyl benzene/polystyrene (EB/PS) extraction using a "plow-blade" rotor WFD with counter-current vapor/melt flow showed that total residuals of less than 0.1 wt % were attainable from polymer feed streams with greater than 40 wt % volatiles. The plow-blade design consisted of four axial rows of angled tabs wiping the circumference. Later EB/PS laboratory work using a new co-current WFD design was consistent with his earlier results, and it showed improved processing capabilities. Four-start 55° helices were shown to be effective as primary WFDs. Final purification capability using various rotor geometries confirmed the superiority of the low helix angle, single flighted rotor. A ST polymerization process with staged WFD was demonstrated at this point. These studies combined with extrusion experience were the basis for scaling to a pilot plant.

In spite of what had been learned in the laboratory, an unsuccessful effort was made to use a commercially available (Marco) counter-current, single flighted deep-channeled helix rotor as a primary WFD in the pilot plant. Significantly better results were obtained with co-current, high-energy input four-start helix rotor designs. The use of a notched version of the four-start helices was motivated by the idea that the notched version would give a better balance of holdup, conveying, area generation, and energy input. Final WFD was attempted with the multistart rotor designs, but the low helix angle, single helix again provided the best results.

The commercial primary WFD rotor was a tabbed rotor that incorporated features similar to the notched helix and the early plow blade. Not surprisingly, the low helix angle rotor became the commercial final WFD. Results for these and selected small-scale units are presented here.

II. DISCUSSION

The work done in the development of WFD for the continuous PS process was guided by underlying concepts concerning mass transfer and viscous flow. Extraction and power data supplemented by visual observations of flow provided the basis for developing commercial PS WFDs and carrying out their successful use. We will briefly review the basis for the relationships used in assessing results and in scaling.

A. Theory

The key elements of successful WFD design and operation are (1) ensuring adequate mass transfer from the melt to the vapor phase; (2) achieving vapor release without polymer flow into vent pipes; (3) attaining desired power input with minimal fluctuation; and (4) maintaining stable flow rate and holdup.

1. Extraction

We have chosen the extraction efficiency E_f,

$$E_f = \frac{C_o - C_e}{C_i - C_e} \tag{1}$$

where

C_o = outlet volatile concentration
C_i = concentration of volatile in the inlet
C_e = outlet equilibrium volatile concentration

as the measure of mass transfer effectiveness to be reported in addition to the actual concentrations. Equilibrium concentration, C_e, has been estimated from the well-known Flory–Huggins (Flory, 1953) and Antoine equations (Dean, 1979; Biesenberger and Sebastian, 1983) using the parameters listed in Table 1. The equations for mass transfer from a film regenerated by a WFD flight (Latinen, 1962a) led to the extraction number Ex:

$$Ex = \frac{K_N D L N_s \sqrt{N}}{F_p} \tag{2}$$

where K_N is a constant and

N_s = Number of rows or starts
N = rotor speed
D = WFD diameter
L = WFD length
F_p = product flow rate

Table 1 Parameters Used in Estimating C_e

	Density (g/cm^3)	χ	Antoine constants		
			a	b	c
ST	0.905	0.33	6.924	1420	206
EB	0.867	0.40	6.957	1424	213
PS	1.05				

Another scaling basis is A_r, the area generated per unit of product:

$$A_r = \frac{\pi D L N_s N}{F_p} \qquad (3)$$

Equations (2) and (3) fit our data comparably well so that A_r, which is conceptually simpler, will be used in our discussion.

Axial vapor velocities V_a in the WFD have been calculated to assess the potential for foam entrainment. This is defined as

$$V_a = \frac{Q_v}{A_c} \qquad (4)$$

where Q_v is the volumetric vapor flow rate and A_c is the WFD free cross-sectional area.

2. Power

A relation for power, P_c, was developed using a power law fluid. The equation used is

$$P_c = K_p L (ND)^{(3n+1)/2} F_p^{(1-n)/2} N_s^{1+n} \qquad (5)$$

where K_p is a constant, and n is the constant from the power law expression

$$\tau = K_v G^n \qquad (6)$$

relating shear stress, τ, and shear rate, G. The value of n used in all estimates of P_c shown here is 0.34. Equation (5) proved useful in scale-up in spite of the substantial simplifications involved in averaging over conditions that varied substantially from inlet to outlet and with differing polymers. A more detailed quantitative model of a final WFD energy balance confirmed the use of this approach. One must supply a significant fraction of the process heat load Q_L through viscous dissipation in a commercial primary WFD. Q_L has been approximated by

$$Q_L = F_i C_p (T_o - T_i) + F_v \Delta H_v \qquad (7)$$

with

$F_i = F_p/(1 - C_i/100)$ = flow rate into the WFD
$F_v = F_p/(1 - C_o/100)$ = flow of vapor out of the WFD
F_p = flow rate of polymer through the WFD
ΔH_v = heat of vaporization
C_p = heat capacity

We also present the observed power per unit volume of WFD in our tables as P_s/V, where

$$V = \frac{\pi D^2 L}{4} \qquad (8)$$

is the volume of the DV.

3. Other Considerations

In addition to those items quantitatively described above, there were several other factors that influenced WFD design, operation, and scale-up. These included concern about avoiding the vent flow of polymer via wiping action of the rotor, and the desire to use centrifugal force to concentrate material at the periphery of the unit.

One idea that guided design and operation was that the strength of the material stretching across a vent opening should be minimized by minimizing rotor–barrel clearance in the vent region. Vent opening dimensions needed to be maximized in the tangential direction.

It was also thought that centrifugal forces were important in avoiding erratic flow by forcing melt to the periphery, so that we have reported the g forces.

B. Experiment

Experimentation leading to a commercial ST/PS WFD happened in three distinct steps. First, a laboratory WFD was used to prove the feasibility of primary WFD using EB/PS. Second, a new laboratory WFD setup confirmed more efficient co-current devices and demonstrated the intended process. Finally, a pilot plant–scale operation (1) gave assurance that the needed energy input for primary WFD was achievable, (2) demonstrated acceptable final product purity, and (3) provided the basis for selecting the commercial rotors described here.

1. Equipment

The WFDs used at Monsanto for PS are are described in patents (Latinen, 1973a, b, 1974a), and they illustrated here in Figs. 1–6. Key geometrical details are shown in Tables 2 and 3. Table 2 takes the point of polymer entry as "zero" and locates other feature from this point. Rotor–barrel clearances were maintained small, ranging from about 0.2 to 3% of the diameter. Clearances at vents were minimized.

The laboratory WFD AA (barrel A and rotor A) is the only Monsanto counter-current device used in this development. A commercial counter-current WPD (not detailed here) was tried unsuccessfully at the pilot plant scale. Counter-current vapor–melt flow causes foamy polymer of minimum density and viscosity to encounter maximum vapor velocity close to the vapor exit. This requires a design for low internal vapor velocity. Hence, AA had a low L/D and a high H/D. The design also required a seal around the drive shaft to keep air out of the vapor during vacuum operation. All further Monsanto designs were co-current and used the superior "viscoseal" technology (Butler et al., 1972; Carter and Latinen, 1974), which was applied to the polymer exit end of the devices. The rotor design of the AA laboratory unit had one of the features ultimately included in the commercial-scale primary WFD. It is referred to here as a "plow-blade" or "tab" rotor. The numerous tabs were intended to provide good surface generation and were angled at 45° for good drag flow conveying.

Laboratory barrel B was used for single- and two-stage evaluations. Rotors B and C with an interstage seal provided a two-stage operation. This

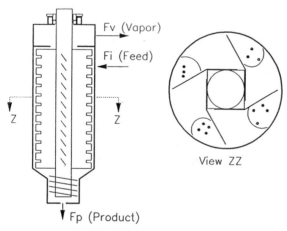

Fig. 1 WFD AA.

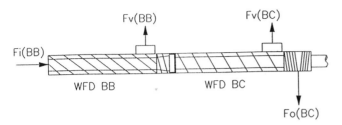

Fig. 2 WFDs BB and BC.

Table 2 WFD Barrel Geometry

Barrel	A	B	C	D
Type of DV[a]	1	2	1/2	1/2
Diameter, cm	8.9	3.7	14.9	71.2
Lenght, cm	19.1	21.6	61.0	366.0
Area, m²	0.05	0.02	0.27	7.9
L/D	2.1	5.8	4.1	5.1
Vapor exit, cm	0.0	21.6	61.0	366.0
Vent diameter, cm	4.4	2.5	12.7	35.6
Vent area, cm²	60	20	492	3861
Feed direction[b]	Rd	Ax	Ax	Ax
Flow direction[c]	Dn	Hz	Dn/Hz	Dn/Up
Vent direction[c]	Up	Up	Hz/Up	Hz/Hz

[a] 1 = primary, 2 = final.
[b] Rd = radial, Ax = axial.
[c] Dn = down, Hz = horizontal.

WFD ran with its shaft horizontal and its vents on top, so that gravity tended to return any escaping polymer from the vent back to the rotor. Rotor B was generally used for the primary WFD, but it was also tested as a final rotor. Rotor C was our first melt-fed final PS devolatilizer.

Separate, but essentially identical, DV barrels C were used for the pilot plant work. This permitted independent setting of rotor speed. Rotor D had notches that created roughly square teeth less than half the effective channel depth, and it was otherwise geometrically similar to laboratory-scale rotor B. The effect of notches is believed to be a reduction in conveying capacity with an increase in surface generation. Vents were generally oriented and vent inserts were used to encourage PS return to the rotor.

Table 3   WFD Rotor Geometry

Rotor	A	B	C	D	E	F	G
Diameter, cm	8.9	3.7	3.7	14.9	14.9	71.1	71.1
Length, cm	19	22	22	61	61	366	366
Channel depth, cm	2.0	0.6	0.6	2.2	2.3	10.7	10.2
L/D	2.1	5.8	5.8	4.1	4.1	5.1	5.1
H/D	0.22	0.16	0.16	0.15	0.15	0.15	0.14
Surface area, m^2	0.05	0.03	0.03	0.28	0.28	8.17	8.17
Vapor flow area, cm^2	44	6	6	88	92	2026	1946
Type[a]	Tb	Hx	Hx	NH	Hx	Tb	Hx
Rows/starts	4	4	1	4	1	4	1
Tb or Hx angle	45	55	15	55	15	55	15

[a] Tb = tabs, Hx = helix, NH = notched helix.

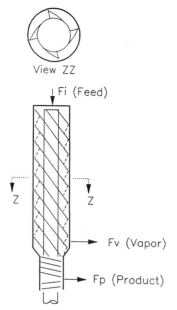

Fig. 3   WFD CD.

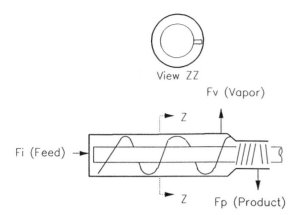

View ZZ

Fig. 4 WFD CE.

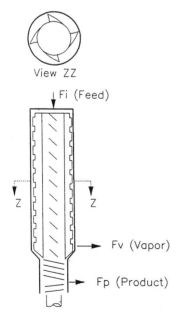

Fig. 5 WFD DF.

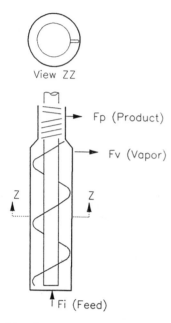

View ZZ

Fp (Product)

Fv (Vapor)

Z   Z

Fi (Feed)

Fig. 6   WFD DG.

Barrel C and rotors D and E were the pilot plant choices that were used to define/select final geometry and operating conditions for the commercial PS process WFDs. Rotor D was tested for primary and final DV. Rotor 9 was the preferred final rotor.

The commercial rotor choices G and H were used in separate D barrels. Rotor clearance was maintained near 1.5 mm. The vent openings were located as close to the exit screw pumps as practical. These large-scale units were vertically oriented to minimize bearing loads and shaft flexing. Avoidance of unwanted gravity flow of melt to the trailing side of flights dictated the upward flow in the final commercial stage.

All Monsanto co-current WFDs were axially fed. Care was taken to provide adequate end clearance in small units, and special effort was made to ensure flow to the periphery, in the commercial ones. Sealing was accomplished in all co-current WFDs at the product exit end of the device by a "viscoseal."

2.   Results

Extraction and rotor power information is presented for all stages of the WFD development for PS in the following sections. This information represents the quantitative basis for the commercial WFDs as well as commercial operating results. In addition, a description of other qualitative matters is included.

    a. Extraction. Primary DV results were first obtained in system AA using solutions of 40/60 Wt. EB/Wt. PS. These mixtures reasonably represented ST/PS DV since vapor pressure and solubility of the two systems are very similar. Table 4 shows the DV results obtained. The extraction efficiency E_f was essentially 100% within the accuracy of the available information. This suggested that the values of A_r (> 488 m²/kg) was greater than necessary. Channel velocities were maintained very low except for one case at 10 km/hour. Exit residuals more than met the expected process requirements. The relative unimportance of C_e for primary DV is seen in case at 760 torr, where more than 94% of the incoming EB was still removed.

    The small co-current WFDs BB and BC were tested with ST/PS in the polymerization process laboratory and provided the data shown in Table 5. Again the EB results for E_f were nearly 100% in spite of the reduced level of A_r. Importantly, much higher internal vapor velocities were seen to be acceptable, with values up to 60.7 km/hour reported. Extraction efficiencies for this unit of 95 and 96% for ST/PS were very encouraging in view of the values of A_r and channel velocity. The data for PS/ST (Table 6) guided the design of the pilot plant units, and showed the four-start 55° helix rotor (BB) to be effective as a primary DV. The single-start 18° helix rotor (BC) chosen from extrusion experience was proved satisfactory. Rotor B was ineffective as a final stage (residuals double those attained with rotor C), possibly because it collapsed bubbles without breaking them.

Table 4 Primary Extraction for Laboratory WFD AA (360 rpm, 40 wt % EB Feed)

F_p (kg/hour)	P (torr)	T_i (°C)	T_o (°C)	C_o (wt %)	C_e (wt %)	E_f (%)	A_r (m²/kg)	V_a (km/hour)	Ex
3.1	760	119	234	2.2	2.7	101	1463	0.2	4092
3.4	22	97	231	0.3	0.1	100	1346	7	3765
5.2	66	104	221	0.7	0.3	99	886	3	2477
6.0	27	107	232	0.4	0.1	99	759	10	2123
8.0	82	107	227	0.3	0.3	100	571	4	1595
8.0	82	107	227	0.3	0.3	100	571	4	1595
8.0	86	104	213	0.4	0.5	100	571	4	1595
9.4	92	109	227	0.5	0.4	100	488	4	1364

Table 5 Primary Extraction for Laboratory WFD BB (360 rpm, 51 wt % EB)

F_p (kg/hour)	P (torr)	T_i (°C)	T_o (°C)	C_o (wt %)	C_e (wt %)	E_f (%)	A_r (m²/kg)	V_a (km/hour)	Ex
0.8	55	140	232	0.1	0.2	100	2700	6	7551
2.3	55	140	232	0.3	0.2	100	972	18	2718
4.0	55	140	232	0.3	0.2	100	546	32	1527
7.7	55	140	232	0.5	0.2	99	286	61	799

Table 6 Two-Stage Laboratory ST/PS Extraction at 200 rpm

F_p (kg/hour)	P (torr)	T_i (°C)	T_o (°C)	C_i (wt %)	C_o (wt %)	C_e (wt %)	E_f (%)	A_r (m²/kg)	V_a (km/hour)	Ex
3.5	53	148	246	32	1.7	0.20	95	355	18	1333
3.5	8	246	274	1.7	0.1	0.03	97	89	7	333

Pilot plant data are shown in Table 7. Extraction efficiencies approaching 100% were attained with A_r as low as 98 m²/kg in primary DV (CD). Somewhat lower values of E_f are seen in the final DV (CE) for A_r going down to 60 m²/kg (Table 8). The ability to manipulate outlet melt temperature with jacket oil temperature was used to permit the study of the influence of stage pressure (or C_e) on residuals. The results defined the needed levels of P and T_o to ensure commercial success ($C_o < 0.1\%$). During this work it was observed that lowering C_o of the primary DV could lead to higher C_o out of the final DV. This and visual observation of the final WFD suggested bubble growth as a key factor in area generation, and supported the selection of a rotor less prone to collapse bubbles.

The commercial unit extraction results are shown in Table 9. Both stages were evaluated at two rotor speeds. The primary unit showed some loss of effectiveness at higher throughput rates corresponding to A_r less than 100 m²/kg. The final WFD performed satisfactorily for A_r of 14 m²/kg and above (Table 10). Flow surging at lower rotor speed favored the higher speeds for these units. Although the data included here do not show it, it is also possible to attain higher values of T_o (lower C_e at higher speeds for better control of C_o.

b. Power. Power and other mechanical information obtained from laboratory WFD AA are shown in Table 11. Power per unit volume of WFD,

Table 7 ST/PS Primary Extraction Data for Pilot Plant WFD CD

F_p (kg/hour)	N (rmp)	P (torr)	T_i (°C)	T_o (°C)	C_i (wt %)	C_o (wt %)	C_e (wt %)	E_f (%)	A_r (m²/kg)	V_a (km/hour)	Ex
23	72	60	148	238	30	1.3	0.3	97	219	6	1371
28	220	98	148	227	30	0.5	0.5	100	544	5	1945
28	88	148	148	227	30	0.8	0.8	100	544	3	1945
28	44	190	148	227	30	0.8	1.0	101	544	2	1945
28	220	60	148	227	48	0.6	0.3	99	544	13	1945
31	220	58	148	227	30	0.4	0.3	100	491	9	1758
31	220	58	148	227	30	0.9	0.3	98	197	9	1112
31	220	58	148	227	30	1.1	0.3	97	98	9	786

Table 8 ST/PS Final Extraction Data for Pilot Plant WFD CE

F_p (kg/hour)	N (rmp)	P (torr)	T_i (°C)	T_o (°C)	C_i (wt %)	C_o (wt %)	C_e (wt %)	E_f (%)	A_r (m²/kg)	V_a (km/hour)	Ex
79	23	8	238	274	1.3	0.10	0.02	94	60	2	358
176	23	11	227	274	1.0	0.04	0.03	99	134	1	536
176	23	17	227	274	1.0	0.07	0.04	97	134	1	536
176	34	12	227	274	2.0	0.08	0.03	98	88	3	354
176	34	17	227	274	2.0	0.09	0.04	98	88	2	354
176	34	30	227	274	2.0	0.15	0.07	96	88	1	354

Table 9 ST/PS Primary Extraction Data for Commercial WFD DF

F_p (kg/hour)	N (rmp)	P (torr)	T_i (°C)	T_o (°C)	C_i (wt %)	C_o (wt %)	C_e (wt %)	E_f (%)	A_r (m²/kg)	V_a (km/hour)	Ex
1341	105	70	147	230	35	1.0	0.49	99	154	16	795
1500	105	70	147	225	35	2.7	0.52	94	137	17	711
1500	105	70	147	235	33	1.1	0.45	98	101	17	609
1773	77	70	147	225	35	1.0	0.52	99	116	21	602
2182	105	70	142	200	45	3.7	0.85	94	94	31	489

Table 10 ST/PS Final Extraction Data for Commercial WFD DG

F_p (kg/hour)	N (rmp)	P (torr)	T_i (°C)	T_o (°C)	C_i (wt %)	C_o (wt %)	C_e (wt %)	E_f (%)	A_r (m²/kg)	V_a (km/hour)	Ex
1341	62	5	230	275	1.0	0.10	0.01	91	16	7	129
1500	62	5	225	265	2.7	0.10	0.01	97	20	22	137
1500	44	5	235	285	1.1	0.10	0.01	92	14	9	115
1773	44	5	225	282	1.0	0.08	0.01	93	17	9	116
2182	44	5	200	270	3.7	0.12	0.01	97	10	44	79

Table 11 Power Data for Primary WFD AA with EB/PS (C_i = 40 wt %)

F_p (kg/hour)	N (rpm)	T_i (°C)	T_o (°C)	C_o (wt %)	Visc (kpoise)	FF (vol %)	P_s (W)	P_s/V (km/m³)	g	P_s/Q_L
6.3	49	104	213	0.4	14	18	30	2.5	0.02	0.02
6.3	76	107	227	0.3	13	12	36	3.0	0.04	0.03
6.3	320	97	231	0.3	4	3	47	4.0	0.15	0.03
7.0	76	119	234	2.2	2	14	25	2.1	0.04	0.02
7.0	114	104	221	0.7	2	4	16	1.4	0.05	0.01
13.4	320	107	232	0.4	4	8	77	6.5	0.15	0.03
16.2	168	109	227	0.5	1	23	86	7.3	0.08	0.03
16.2	247	107	227	0.3	1	6	53	4.5	0.12	0.02

P_s/V, fell in the range of $1-7\,\mathrm{kW/m^3}$, and this indicated the need for more power-intensive operations to supply the commercial-scale units with a larger fraction of the process heat load Q_L through viscous dissipation. It was believed that centrifugal forces were needed to keep material from accumulating on the shaft and erratically releasing. Successful operation was achieved at levels as low as 0.02 g. Observed levels of relative holdup FF in the device were under 23 vol % of the available space on a nonfoamed basis (approximately less than half full, including retained gas). This avoided entrainment, as did the low vapor velocities mentioned previously. This work, although done with an EB/PS solution, gave an estimate of shaft power input capability, P_s. It also showed the dependence of power on speed, fractional fillage, and viscosity, which led to Eq. (5). The low ratio of shaft power, P_s, to Q_L pointed to the need for higher energy input rotor design. This was achieved in subsequent designs primarily via channel depth reduction.

Laboratory-scale co-current combinations BB and BC were tested both as single stages on EB/PS mixtures and as a combined two-stage device on ST/PS. Tables 12 and 13 present information obtained without DV. These

Table 12 WFD BB Power Data for EB/PS (no DV,
$T_i = T_o = 25°C$, $C_i = C_o = 32$ wt %, Visc. = 1.1 kpoise)

F_p (kg/hour)	N (rpm)	P_s (W)	P_c (W)	P_s/V (km/m³)	g
2.7	40	3	21	1.1	0.01
5.5	78	7	53	3.0	0.02
7.7	120	12	91	5.0	0.02
12.3	200	25	178	10.6	0.04
16.4	300	46	295	19.6	0.06
20.0	400	63	422	26.7	0.08

Table 13 WFD BC Power Data for EB/PS (no DV,
$T_i = T_o = 25°C$, $C_i = C_o = 32$ wt %, Visc. = 1.1 kpoise)

F_p (kg/hour)	N (rpm)	P_s (W)	P_c (W)	P_s/V (km/m³)	g
2.7	50	1	11	1	0.01
5.5	82	3	22	1	0.02
7.7	110	5	33	2	0.02

Table 14 WFD BB Final DV Power Data for EB/PS (no DV, $T_i = T_o = 25°C$, $C_i = C_o = 32$ wt %)

F_p (kg/hour)	N (rpm)	T_i (°C)	T_o (°C)	P_s (W)	P_c (W)	P_s/V (km/m^3)	g	P_s/Q_L
4.5	75	238	246	52	48	22	0.01	1.5
7.8	130	238	252	104	99	44	0.03	1.2
7.8	170	238	257	134	130	57	0.03	1.2
9.3	170	238	249	112	138	47	0.03	1.3
10.6	170	238	252	112	144	47	0.03	1.0
13.6	210	238	254	156	194	66	0.04	0.9
15.7	170	238	252	142	164	60	0.03	0.8
15.7	240	238	252	186	232	79	0.05	1.1
15.7	270	238	252	253	262	107	0.05	1.5

data were used to confirm the speed and throughput dependencies predicted by Eq. (5). The higher power input capability of the B rotor was demonstrated here, with values up to 26.7 kW/m^3 attained. Absolute agreement with Eq. (5) is not expected, since we have used a pilot plant primary WFD case as the reference condition for power estimation. Data from single-stage operations helped define the percentage of power input going into each stage of the combined rotor, and they have been used here to provide the data shown in Tables 15 and 16. The information in Table 14 suggested that rotor B would lead to excessive stock temperatures if used in a commercial final stage. Information obtained with no DV also helped confirm the power input capabilities of the different rotor designs.

Table 15 shows the primary WFD results of a ST/PS operation closely simulating the intended commercial process. It showed that the primary rotor B was capable of supplying all the required heat for the process stream ($P_s/Q_L = 1$). The final WFD rotor C gave acceptable stock temperatures in spite of the high rotor speed required by the primary stage. The data in

Table 15 WFD BB Power Data for ST/PS

F_p (kg/hour)	N (rpm)	T_i (°C)	T_p (°C)	C_i (wt %)	C_o (wt %)	P_s (W)	P_c (kw)	P_s/V (km/m^3)	g	P_s/Q_L
1.9	200	149	238	30	2	237	96	100	0.04	1.0

Tables 12–16 formed the basis for selecting rotor designs, motor sizes, and speed ranges for the pilot plant phase of the development.

Independently driven pilot plant units provided the data presented in Tables 17 and 18 for rotors CD and CE, respectively. This information was obtained during typical continuous process operation. Although g forces increased substantially over the laboratory levels, there was no significant change in WFD performance with respect to flow and power variations.

Table 16 WFD BC Power Data for ST/PS

F_p (kg/hour)	N (rpm)	T_i (°C)	T_o (°C)	C_i (wt %)	C_o (wt %)	P_s (W)	P_c (W)	$P_s V$ (km/m³)	g	P_s/Q_L
1.9	200	238	254	2	0.1	88	38	37	0.04	3.7

Table 17 WFD CD Power Data (F_p = 31 kg/hour, T_i = 160°C, T_o = 254°C, C_i = 30 wt %, C_o = 1 wt %)

F_p (kg/hour)	N (rpm)	T_i (°C)	T_o (°C)	C_i (wt %)	C_o (wt %)	P_s (kw)	P_c (kw)	P_s/V (km/m³)	g	P_s/Q_L
30.7	42	160	254	30	1	0.7	0.6	7	0.03	0.2
30.7	88	160	254	30	1	1.3	1.2	12	0.07	0.3
30.7	180	160	254	30	1	2.5	2.5	23	0.14	0.6
30.7	220	160	254	30	1	2.8	3.0	27	0.18	0.7
30.7	285	160	254	30	1	4.1	3.9	39	0.23	1.0

Table 18 WFD CE Power Data, Mst/Mst_0 = 0.8 (F_p = 23 kg/hour, T_i = 238°C, T_o = 252°C, C_i = 2 wt %, C_o = 0.1 wt %)

N (rpm)	P_s (kw)	P_c (kw)	P_s/V (km/m³)	g	P_s/Q_L
42	0.2	0.2	2	0.03	0.9
75	0.3	0.4	3	0.06	1.4
135	0.7	0.7	6	0.11	2.8
180	1.3	0.9	12	0.14	5.4
260	2.4	1.3	22	0.21	9.9

Table 19 WFD DF Power Data

F_p (kg/hour)	N (rpm)	T_i (°C)	T_o (°C)	C_i (wt %)	C_o (wt %)	P_s (kw)	P_c (kw)	P_s/V (km/m³)	g	P_s/Q_L
1341	105	147	230	35	1.0	106	144	7.3	0.4	0.6
1500	77	147	235	33	1.1	126	109	8.7	0.3	0.6
1500	105	147	225	35	2.7	118	150	8.2	0.4	0.6
1773	105	147	225	35	1.0	122	158	8.4	0.4	0.5
2182	105	142	200	45	3.7	108	169	7.4	0.4	0.3

Table 20 WFD DG Power Data

F_p (kg/hour)	N (rpm)	T_i (°C)	T_o (°C)	C_i (wt %)	C_o (wt %)	P_s (kw)	P_c (kw)	P_s/V (km/m³)	g	P_s/Q_L
1341	44	230	275	1.0	0.10	22	24	1.5	0.2	0.6
1500	44	235	285	1.1	0.10	37	25	2.6	0.2	0.8
1500	62	225	265	2.7	0.10	64	35	4.4	0.2	1.6
1773	62	225	282	1.0	0.08	25	37	1.7	0.2	0.4
2182	44	200	270	3.7	0.12	33	28	2.3	0.2	0.3

The agreement of the calculated and actual values for the commercial-scale WFDs may be seen in Tables 19 and 20. It can be seen in Table 19 that the fraction of the needed power input attained was more than 0.5 except for the case at maximum throughput. The centrifugal forces have increased over those attained in the smaller units, again without large changes in operation. P_s/V has given acceptable DV at 7–9 kW/m³. Power predictions were in reasonable agreement with the plant data.

The power data presented in Table 20 have been adjusted by removing the estimated pump and viscoseal power (19–23 kW) from the totals observed. Here the need to minimize power input has been quite well met, with P_s/V typically 2 kW/m³.

Flow variations initiated in the primary WFD were damped sufficiently by the intervening screw pump and the final WFD for normal downstream product handling to be like that of normal extrusion operations.

c. Other Observations. Visual observations with model fluids and geometry similar to WFD AA led Latinen to conclude that minimizing holdup and maximizing centrifugal effects was the best way to avoid surging in the WFD. He reported that polymer accumulated on the trailing side of flights until it suddenly released, causing disruptive power and exit flow surges. A second visual observation related to gravity flow in the final WFD. Operation at plant-scale Froude numbers using a C rotor in a glass tube showed that the commercial final stage unit must pump upward to avoid buildup of fluid on the trailing side of the flights.

Vent wipe-off was most severe with the small units and was solved partly by use of vent inserts. Larger-scale units showed little difficulty with this phenomenon as a result of the larger ratio of vent dimensions to bridging film thickness or rotor–barrel clearance.

III. SUMMARY AND CONCLUSIONS

Highlights of the successful development of a two-stage commercial WFD system for PS have been presented.

High energy input and gross monomer removal have been achieved at pressures needed for convenient condensation in a primary WFD using a rotor with four rows of 45° tabs. Extraction efficiencies of 94–99% were obtained for A_r of 100–150 m^2/kg. Absolute exit ST concentrations were 1–4 wt %. Satisfactory commercial operation was attained with power per unit WFD volume of about 8 kW/m^3.

The lower power input single-start 18° helix performed final-stage DV well at all scales, and it was shown capable of maintaining residual ST below 0.1 wt % in commercial operation. Extraction efficiencies were 91–97% for A_r of 10–20 m^2/kg for the commercial final WFD. Power per unit volume of less than 4.7 kW/m^3 was sufficiently low to avoid excess melt temperatures. Commercial WFDs DF and DG were shown to have capacities of at least 1770 kg/hour.

This project clearly demonstrated the practicality of using WFDs in the purification of PS. In general, this type of equipment would be useful for the purification of many viscous nonvolatile substances able to tolerate the conditions needed for vaporization and conveying during the short time in the device.

Since the completion of the work described in this case study, a substantial effort has been made to improve the understanding of DV in general and WFD of PS in particular. Biesenberger (1980, 1983), Collins (1982, 1985), Denson (1985), and Tukachinsky et al. (1994) have investigated mechanically enhanced DV. This work combined with the mechanistic description of vapor release into bubble swarms of Newman and Simon (1980) and the subsequent observations of actual PS foam structure of Albalak (1987, 1990) have greatly increased our understanding. These observations of foam structure show the presence of more complex geometry (e.g., bubbles with internal "blisters"), thus providing important clues to the underlying mechanisms of nucleation and bubble growth. The connection between stresses and nucleation is described by Han and Han (1988), Lee (1991), and Tukachinsky et al. (1994). The latter relates the observation of blisters inside larger bubbles to elongational stresses, and it shows the vapor removal rate of mechanically enhanced DV to be considerably more rapid than for unagitated systems under comparable conditions. This suggests the use of WFD for thermally sensitive materials.

Much of what has been reported in the literature relates to rather dilute (with respect to volatiles) systems, and while complete rigorous design equations may not exist, one is encouraged by the progress seen. More concentrated systems, however, need to be investigated to aid in the design of primary-stage DV. Further work is needed to establish the criteria for stable flow in WFDs. What are the conditions for uniform steady vapor release and melt flow? Bubble rupture, vapor escape, and foam volume prediction methods are certainly valid areas for future research. The details of bubble rupture and vapor escape should be determined.

NOMENCLATURE

a, b, c	Antoine constants
A_r	area generated per unit of product
C_e	equilibrium volatile concentration
C_i	inlet polymer solution volatile concentration
C_o	outlet polymer solution volatile concentration
C_p	heat capacity
D	WFD diameter
E_f	mass transfer efficiency, Eq. (1)
Ex	extraction number, Eq. (2)
F_i	$F_p/(1 - C_i/100)$ = flow rate into the WFD
F_p	polymer output rate
F_v	$F_p/(1 - C_o/100)$ = vapor flow rate from the WFD
g	gravity constant, 980 cm^2/sec
g	centrifugal acceleration/in terms of acceleration of gravity
H	WFD channel depth
ΔH_v	heat of vaporization
K_N	constant in Eq. (2)
K_p	constant in Eq. (5)
K_v	power law constant, Eq. (6)
L	WFD length
Mst	Staudinger molecular weight
Mst$_0$	reference Staudinger molecular weight

N rotor speed
N_s number of rotor flights or tab rows
P WFD stage pressure
P_c calculated rotor power
P_s rotor (shaft) power
Q_L process heat load, Eq. (7)
T_i inlet polymer solution temperature
T_o outlet polymer solution temperature
V WFD volume, Eq. (8)
V_a velocity of volatiles in the WFD
χ polymer–volatile interaction parameter

REFERENCES

Albalak, R. J., Tadmor, Z., and Talmon, Y. (1987). Scanning electron microscopy studies of polymer melt devolatilization, *AIChE J.*, *33*: 808.

Albalak, R. J., Tadmor, Z., and Talmon, Y. (1990). Polymer melt devolatilization mechanisms, *AIChE J.*, *36*: 1313.

Amon, M., and Denson, C. D. (1984). A study of the dynamics of foam growth: analysis of the growth of closely spaced spherical bubbles, *Polym. Engin. Sci.*, *24*: 1026.

Biesenberger, J. A., and Kessidis, G. (1980). Devolatilization of polymer melts in single-screw extruders, *Polym. Eng. Sci.*, *22*: 832.

Biesenberger, J. A., and Sebastian, D. H. (1983). *Principles of Polymerization Engineering*, John Wiley and Sons, New York.

Blander, M., and Katz, I. L. (1975). Bubble nucleation in liquids, *AIChE J.*, *21*: 833.

Butler, R. G., Carter, D. E., Latinen, G. A., (1972). U.S. Patent 3,700,247.

Carter, D. E., and Latinen, G. A. (1974). U.S. Patent 3,795,386.

Collins, G. P. (1982). Devolatilization of polymeric solutions in an intermeshing co-rotating twin-screw extruder, M. S. Thesis, University of Delaware.

Collins, G. P., Denson, C. D., and Astarita G. (1985). Determination of mass transfer coefficients for bubble-free devolatilization of polymeric solutions in twin-screw struders, *AIChE J.*, *31*: 1288.

Dean, J. A., ed. (1979). *Lange's Handbook of Chemistry*, 12th ed., McGraw-Hill.

Denson, C. D. (1985). Stripping operations in polymer processing, *Advances in Chemical Engineering* (J. Wei, ed.), vol. 12, Academic Press, p. 61.

Flory P. J. (1953). *Principles of Polymer Chemistry*, Cornell University Press, Ithaca.

Foster, R. W., and Lindt, J. T. (1990). Twin screw extrusion devolatilization: From foam to bubble free mass transfer, *Polym. Engin. Sci.*, *30*: 621.

Han, I. H., and Han, C. D. (1988). A study of bubble nucleation in a mixture of molten polymer and volatile liquid in a shear flow field, *Polym. Engin. Sci.*, *28*: 1616.

Latinen, G. A. (1962a). *Devolatilization of Viscous Polymers Systems*, ACS Advances in Chemistry series, vol. 34, 235.

Latinen, G. A. (1973a). U.S. Patent 3,720,479.
Latinen, G. A. (1973b). U.S. Patent 3,781,132.
Latinen, G. A. (1974a). U.S. Patent 3,797,550.
Latinen, G. A. (1974b). U.S. Patent 3,812,897.
Latinen, G. A., and Simon, R. H. M., (1962b). U.S. Patent 3,067,812.
Lee, S. T. (1991). Shear Effects on Thermoplastic Foam Nucleation," 49th SPE ANTEC Preps.
Lee, S. T., and Biesenberger, J. A. (1989). A fundamental study of polymer melt devolatilization. IV. Some theories and models for foam-enhanced devolatilization, *Polym. Engin. Sci.*, *29*: 782.
Mehta, P. S., Valsamis, L. N., and Tadmor, Z. (1984). Foam devolatilization in multichannel corotating disk processors, *Polym. Process. Eng.*, *2*: 103.
Mendelson, R. A. (1980). Concentrated solution viscosity behavior at elevated temperature—polystyrene in ethylbenzene, *J. Rheol.*, *24*: 765.
Newman, R. E., and Simon, R. H. M. (1980). "A Mathematical Model of Devolatilization Promoted by Bubble Formation", AIChE 73rd Ann. Meet., Chicago.
Petel, R. D. (1980). Bubble growth in a viscous Newtonian liquid, *Chem. Eng. Sci.*, *35*: 2352.
Tukachinsky, A., Talmon, Y., and Tadmor, Z. (1994). Foam enhanced devolatilization in a vented extruder, *AIChE J.*, *400*: 670.

16

Devolatilization of Polyvinyl Chloride

Helmut M. Joseph

Israel Plastics and Rubber Center, Technion City, Haifa, Israel

I. INTRODUCTION

The suspension and emulsion polymerization of polyvinyl chloride (PVC) from vinyl chloride monomer (VCM) accounts for about 90% of the total production of PVC. The remaining 10% is produced mainly by the mass polymerization process.

During the 1970s, serious health hazards connected with VCM became known, including hepatic angiosarcoma (a liver cancer) and acro-osteolysis (decalcification of the finger tips). In 1974 it was reported that several employees of the Goodrich Co. died from angiosarcoma and that their deaths were linked to exposure to VCM. More deaths and health problems caused by VCM were known in Europe and Canada.

However, VCM and PVC were established as important raw materials and could not easily be given up. The emissions of VCM, though, had to be reduced drastically, and the extent of the problem had to be investigated. The biggest health hazard was cleaning of the PVC reactor, which was originally done manually after every polymerization.

The VCM emissions were now measured in the workplace of the PVC and VCM plants, and new severely restrictive standards were set up. Investigations were then extended to the users of PVC, to examine whether VCM was present in the PVC compound or possibly even in articles like toys and food or drinks packaged in PVC bottles or containers.

II. CONTROL OF VINYL CHLORIDE IN PVC PRODUCTION

A. PVC Production

Vinyl chloride boils at $-13°C$ and is polymerized by dispersing the liquid monomer under pressure in water. The volume ratios of water to VCM are usually 1.25–1.50. Essential ingredients are suspending agents (usually partially hydrolyzed polyvinyl acetates and/or cellulose derivatives) and initiators soluble in VCM. The latter are organic peroxides, peroxydicarbonates, or any other compound that easily decomposes into radicals. The reaction temperature is about 50–70°C; the higher the temperature, the shorter the polymer chain formed. The choice and amounts of the suspending agents influence particle size distribution and morphology.

Suspension PVC (SPVC) reactors vary now in size between 15, 76 and 113 m^3 (respectively 4000, 20,000, and 30,000 gallons). Economics of design, use of reflux condensers, and other considerations for choosing reactor sizes have been reported (Albright and Soni, 1982). Figure 1 shows a typical flow sheet of SPVC polymerization.

PVC is insoluble in VCM, precipitates as very fine particles, and is swollen by large amounts of VCM. At about 70% conversion, all the free

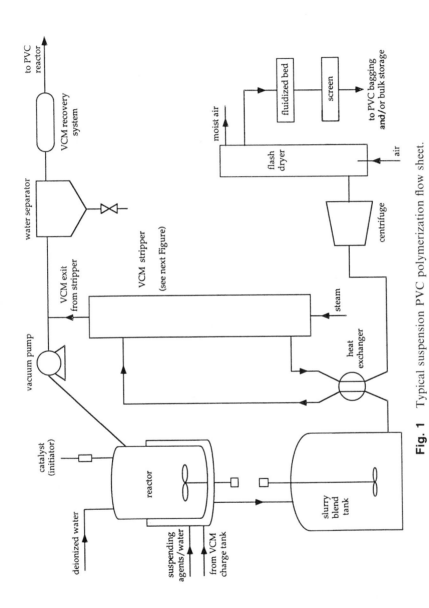

Fig. 1 Typical suspension PVC polymerization flow sheet.

VCM has been used up and the pressure in the reactor starts to drop. The polymerization is usually continued up to conversion of 75–90%.

B.   Reduction of Residual VCM

Prior to 1974 and the revelations of the VCM toxicity, PVC was polymerized in relatively small reactors, and some VCM was recovered by vacuum and subsequently reliquified. After venting, the slurry still contained large amounts of VCM—up to about 3%. The reactors were usually cleaned manually after every reaction. Emissions of VCM during all stages of transport from the reactor up to the PVC bag into the surrounding atmosphere were considerable. The PVC contained residual VCM (RVCM) in excess of 100 up to 1000 ppm (mg/kg) even after drying (Lobo, 1974).

Immediate steps were taken by the PVC industry to eliminate VCM emissions from the manufacturing areas of the VCM and PVC plants, in addition to reduction of the RVCM content. Procedures were developed to clean the reactors by chemical and mechanical means without opening them. This "clean reactor technology" eliminated the need for manual cleaning of the reactors and allowed introduction of the "jumbo" reactors. At the same time, sophisticated computerized on-line gas chromatographic (GC) or other analytical systems were installed to determine VCM levels in the work areas of the plants. Analytical methods were set up to determine VCM levels in PVC, water, slurries, and PVC compounds based on GC head-space analysis. Control of emissions from pumps, seals, and plant systems were performed daily using a variety of portable VCM analyzers.

Determinations of VCM in the ppb level were run on-line in areas near the PVC plants. Worker exposure standards were set up by OSHA (Occupational Safety and Health Administration) and environmental standards by EPA. Similar regulations were issued in Europe and other areas. (See Sections VII and VIII.) When operating with the "clean reactor technology" and the VCM stripper column, the PVC plants conformed to the official standards, and the RVCM was reduced in all PVC grades to below 1 ppm. As to diffusion of VCM from the PVC packaging material and bottles into food, drinks, or liquids like orange juice or cooking oil, a "no-diffusion" level could be shown for the RVCM (Joseph, 1979).

C.   Diffusion of VCM from PVC

VCM is present during the polymerization, in particular inside the PVC particles, but also in the water phase of the slurry and in the gas phase inside the reactor. Knowledge of the distribution of the VCM in the various phases allows estimation of the polymerization rate and development of procedures for removal and recovery of the VCM (Chan et al., 1982a). The VCM concentration in the solid phase (PVC) increases during the polymerization, up to about 0.225 mole fraction at 70–75% conversion, at the start of the pressure drop; it then drops to about 0.105 mole fraction at 90% conversion; the VCM in the vapor phase increases steadily from 0.011 mole fraction at the start of the polymerization up to about 0.017 mole fraction at 80% conversion. The VCM concentration in the water phase is constant at about 0.015 mole fraction, meaning saturation of the water with VCM. The solubility of VCM in water is known to be about 2000 ppm at 20°C and about 1000 ppm at 60°C and atmospheric pressure. About 20% of the total free VCM is present in the water and gas phases in the reactor. It should be understood that the reported mole fractions of VCM refer neither to the liquid VCM present at the start of the pressure drop at 75% conversion, nor to the VCM that has been converted to solid PVC during the polymerization.

To remove the RVCM from the PVC by vacuum, it has to undergo:

1. Diffusion from the PVC into the water, which is saturated with VCM
2. Transfer of the VCM from the water into the vapor phase
3. Evaporation from the vapor into the recovery system

A detailed investigation of the vacuum batch stripping of VCM has been described (Chan et al., 1982b). A typical example of batch stripping of a PVC slurry by vacuum at 80°C in a 38-m^3 reactor resulted in reduction of the VCM concentration in the solid phase (PVC) from about 15,000 ppm at time 0 to about 200 ppm after 30 min, and to about 80 ppm after 60 min. At that time, the VCM concentration in the liquid phase (water) was still above 10 ppm, compared with about 1000 ppm at time 0. This type of batch stripping cannot produce PVC slurry with the required low RVCM content (< 10 ppm) even after uneconomical long evaporation times. On the other hand, these vacuum evaporations in the reactor reduced the VCM concentrations in the reactors considerably and allowed specially equipped operators to enter the reactors for cleaning purposes. Subsequent modifications of the procedure in some SPVC plants included sparging of the PVC with steam during the evaporation and adding a separate monomer recovery tank.

Desorption and solubility data for VCM in PVC have been described (Berens, 1981). Emulsion PVC powders with particle size in the submicron range desorbed VCM at 90°C to the low ppm range in less than a minute. The desorption of suspension PVC was sometimes as fast as predicted for the 2–3 μm size, indicating that the primary PVC particles govern the process and not the 100–200 μm agglomerates. In practice, however, SPVC varied widely in VCM desorption times and RVCM. It is believed that the presence of undesired "glassy particles" causes these deviations from the fast desorption times. The morphology of the PVC particle certainly has a decisive influence on the ease of VCM removal. High-molecular-weight SPVC is usually more porous, and desorption of VCM is relatively easy; the opposite is true for low-molecular-weight SPVC produced at higher reaction temperatures. In addition to the foregoing, desorption depends on particle size and, in particular, particle size distribution. Presence of small concentrations of large PVC particles (> 250 μm) increases the diffusion times and the RVCM obtained.

III. VCM STRIPPING OF SUSPENSION PVC

Major innovations in the SPVC production since relevation of the VCM toxicity were the "clean reactor technology," eliminating the need of manual cleaning of the reactors, and introduction of the "jumbo" reactors and the VCM stripper.

The PVC slurry (about 30 wt % solid content) is transferred to a closed slurry blend tank and from there to the stripper. On the way, the slurry is usually preheated by the stripped slurry in a heat exchanger. Use of the VCM stripper (Fig. 2) allows reduction of recovery times by transferring the slurry at relatively high RVCM content. In the stripper, RVCM in the PVC is reduced to below 10 ppm (mg/kg PVC). The RVCM in the dried PVC is below 1 ppm.

Ease of removal of RVCM depends on PVC particle size, particle size distribution, presence of "glassy" particles, and its porosity. Thus proper stripping conditions have to be determined for the various PVC grades.

The stripper is equipped with perforated plates. Steam enters the column from the bottom, while the PVC slurry enters the column either near the top or at other available entrances further down. The RVCM is stripped from the slurry counter-currently on the various plates of the stripper. The various entrance levels allow adjustment of the residence time of the PVC slurry according to the treatment needed for the various PVC grades.

The desorption of VCM from the PVC has been described in the previous paragraph. The success of the stripper is due, in particular, to the continuous process ensuring near-zero VCM concentration around the PVC particle.

Performance of the stripper also can be adjusted by changing the ratio of steam–PVC slurry. Higher steam ratios and temperatures will speed up the diffusion from the liquid phase to the gas phase. The minimum time required for reaching low RVCM concentrations in the PVC depends on particle size and morphology. Higher temperatures in the column will thus speed up devolatilization.

Typical steam consumption of the described stripper operation is about 0.25 kg/kg PVC. Between 1 and 3% of the total VCM charge is recovered

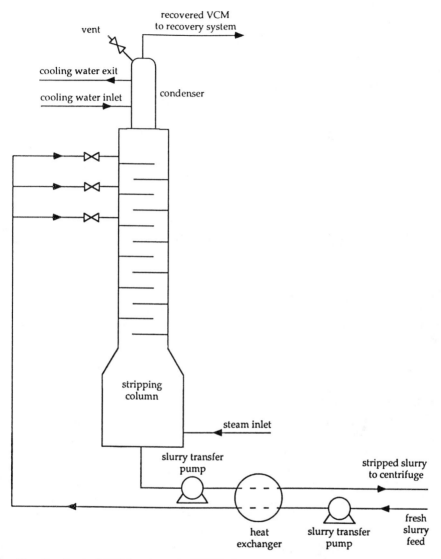

Fig. 2 Typical VCM stripper in SPVC plant.

by the stripper. Thus the operation is not only indispensable for ecological reasons but has also an economic importance. Residence times of the slurry in the stripper should be up to about 10 min.

The steam stripping subjects the PVC to relatively high temperatures, up to about 110°C, in particular at the bottom of the stripper. Most of the steam will condense by the time it reaches the top of the stripper. The excess steam is removed by condensation, and the VCM is liquified and returned to the recovery system (Smallwood, 1989).

The operating conditions of the stripper have to be adjusted to minimize the deleterious effects on the properties of the PVC. Dehydro-chlorination of the PVC is known to occur at temperatures around 100°C, and a certain reduction of the whiteness is usually noted. Influence on heat stability is negligible at worst.

IV. DEVOLATILIZATION OF EMULSION PVC

Various technologies are available for the production of emulsion PVC (EPVC), to be used in PVC plastisols. Usually, EPVC is prepared from VCM in the presence of water, an emulsifying agent (one or a mixture of water-soluble surfactants), and a water-soluble initiator. Most polymeriza-tions use small-particle-size seed latex and result in bimodal latices with particles of 0.3–3 μm.

A significant portion of the RVCM is removed from the latex in the reactor to the recovery system by vacuum. RVCM will still be about 3%. All the water is removed in a spray dryer, and the PVC is ground mechanically in a mill.

The PVC obtained in this system conforms to RVCM standards (less than 1 ppm); apparently, the small particle size of the EPVC facilitates removal of the RVCM still present after recovery in the reactor. This procedure will, however, not conform to emission standards; additional devolatilization is required.

Removal of the RVCM in a continuous stripper as described for SPVC is impractical. The large amount of foam from the emulsifier would fill the whole column, and continuous operation would be impossible. Thus batch removal is a preferred solution. In this case, devolatilization is performed using vacuum and at low temperatures. Such a process is possible for EPVC because of its small particle size and its relatively high porosity.

V. RECOVERED VCM

As discussed, VCM is recovered from various points in the SPVC and EPVC systems. Large amounts of residual VCM (RVCM) are thus obtained, which have to be returned to the polymerization. Impurities in the RVCM may be quite considerable. There are two sources of contamination: internal contaminants due to concentration of original traces in the VCM, and external contaminants coming from the polymerization systems. In particular, inert components present in the original VCM in trace amounts only are concentrated in the RVCM. These may be methyl chloride, ethyl chloride, and 1,2-dichloroethane (EDC), which are all easily checked by GC analysis. However, the RVCM may also be contaminated with traces of impurities originating in additives to the PVC polymerizations. In addition to the required suspending agents, emulsifiers, and initiators, buffers, chelators, and other products may have been added. Thus an additional procedure for purification of the RVCM is required. The RVCM quality is monitored by GC and infrared (IR) analyses (see Section VII). Selective addition of purified RVCM to the various grades is then carried out.

VI. DRYING OF SUSPENSION PVC

Devolatilization of PVC also includes removal of the large amounts of water and of the still remaining traces of RVCM. The stripped slurry is transferred to the centrifuge, where the bulk of the water is separated from the slurry. However, the slurry from the centrifuge still contains about 20% water, mainly inside the pores of the PVC and on the particle surface.

A typical drying system consists of a flash dryer followed by a heated fluidized bed. In the flash dryer, the slurry is suspended in a stream of hot air; this is a short heat treatment lasting less than a minute; on passing from the flash dryer to the fluidized bed, the water content of the PVC is less than 3%. The last traces of water and of RVCM are removed in this drying system. The final PVC has a moisture of less than 0.3% and RVCM of less than 1 ppm.

VII. ANALYTICAL DETERMINATIONS

Various analytical procedures have been set up by the PVC industry. In particular, the RVCM content has to be controlled in various stages of PVC production from the polymerization to the dried resin. A complete analysis procedure using head-space gas chromatography (GC) has been published (EPA, 1976); this procedure refers to analyses of PVC resins, process water, slurry, cake, and latex. This head-space GC is based on the vapor equilibrium established between RVCM, the sample, and air at temperatures above the glass transition temperature of the PVC. The procedure has a low detection limit, well below 0.1 ppm (mg/kg PVC). The current procedure for determination of RVCM in PVC homo-/copolymer resins (ASTM-D 3749) is based on the aforementioned EPA method. Note that RVCM in PVC compounds and articles should not be determined by this method. These samples have to be dissolved in a suitable solvent like dimethylacetamide (DMAC) before subjecting them to head-space GC (ASTM-D 4443). Various calibration and calculation alternatives can be applied in this method. The sensitivity limit is up to 5 ppb.

The determination of VCM in air vented from VCM, EDC (1,2-dichloroethane), or PVC manufacturing processes can be performed by sampling the vent gases into Tedlar bags of 100-liter capacity and connecting the latter to a GC equipped with a gas sampling valve (EPA, 1976). Alternative methods involve sample collection with charcoal tubes rather than the Tedlar bags; the sample can then be thermally desorbed from the charcoal into the GC (State of California, 1978). A simpler method—without use of special desorption equipment—is transfer of the charcoal into a closed vial, addition of dichloromethane (DCM), and injection of the obtained VCM–DCM solution into a GC.

All these procedures involve use of suitable GC columns to eliminate possibilities of errors due to elution of acetaldehyde or other components at the elution time of VCM. Calibration gases of VCM in nitrogen can be purchased or prepared from permeation tubes.

Analysis of RVCM is done as for fresh VCM. This analysis may be done separately for "light" and for "heavy" impurities by GC procedures, or in one run using a suitable column (ASTM-D 3834); additional procedures may involve concentration of impurities by passing the RVCM through suitable solvents like carbon tetrachloride and subsequent analysis by IR and/or high-performance liquid chromatography (HPLC).

Ambient air determination for VCM and leak detection from plant systems can be performed at the low ppm level by portable GC or IR analyzers.

VIII. OFFICIAL PUBLICATIONS

The Environmental Protection Agency (EPA) is responsible for setting standards of vinyl chloride monomer emissions from VCM and PVC plants in the United States. These standards have originally been listed in the aforementioned EPA publications and have subsequently been confirmed (EPA, 1979). The concentration of VCM in exhaust gases discharged to the atmosphere from reactors, vents, and strippers during operation is not to exceed 10 ppm; the maximum allowable emission of VCM from latex (emulsion resins) is 400 ppm.

The Occupational Safety and Health Administration (OSHA) is responsible for standards relating to workers exposure in PVC plants in the United

States. These standards have been reviewed (Hart, 1992) and vary somewhat in various countries:

Workers Exposure Limits

United States, old plants	5 ppm VCM/8 hours TWA*
United States, new plants	1 ppm VCM/8 hours TWA
Sweden	1 ppm VCM/8 hours TWA
United Kingdom	10 ppm VCM/8 hours TWA
Italy	10 ppm VCM/8 hours TWA
EEC	3 ppm VCM/1 year TWA

* TWA: time-weighted average.

In Germany, the emission standards are not identical in the various States. The German Federal emission standards specify the permissible VCM concentration in exhausts or vents in units of mg/m^3 (wt/vol) up to VCM weights/hour. At the present level of low RVCM/PVC, these emission standards do not present any limiting problem.

As stated in Section III, the RVCM concentration in PVC resins is now less than 1 ppm, which conforms to official limitations. Obviously, the RVCM concentration in compounds and articles like packaging and bottles will be significantly lower (Joseph, 1979) and can be determined according to procedures listed in Section VII.

NOMENCLATURE

DCM	dichloromethane
DMAC	dimethylacetamide
EDC	1,2-dichloroethane
EPA	Enviromental Protection Agency
EPVC	emulsion polyvinyl chloride
GC	gas chromatography
HPLC	High-performance liquid chromatography
IR	infrared
OSHA	Occupational Safety and Health Admininstration
PVC	polyvinyl chloride
RVCM	residual vinyl chloride monomer
SPVC	suspension polyvinyl chloride
VCM	vinyl chloride monomer

REFERENCES

Albright, L. F., and Soni, Y. (1982). Design and operation of reactors for suspension polymerizations of vinyl chloride, *J. Macromol. Sci.—Chem.*, *A17*(7): 1065.

ASTM-D-3749-87 (08.03), Residual VCM in poly(vinyl chloride) homo/copolymer resins by headspace technique. Published January 1988

ASTM-D-3834-80 (08.03), Standard test method for purity of vinyl chloride monomer by gas chromatography. Published January 1980

ASTM-D-4443-84 (08.03), Residual VCM in ppb range in homo/copolymers by headspace gas chromatography. Published January 1985

Berens, A. R. (1981). Vinyl chloride monomer in PVC, *Pure App. Chem.*, *53*: 365.

Chan, R. K. S., Langsam, M., and Hamielic, A. E. (1982a). Calculation and applications of VCM distribution in vapor/water/solid phases during VCM polymerizations, *J. Macromol. Sci—Chem.*, *A17*(6): 969.

Chan, R. K. S., Patel, C. B., Gupta, R., Worman, C. H., and Grandin, R. E. (1982b). Batch stripping of vinyl chloride, *J. Macromol. Sci.—Chem.*, *A17*(7): 1045.

Environmental Protection Agency (EPA) (1976). *Federal Register*, *41*(205), October.

Environmental Protection Agency (EPA) (1979). *340/1–79-006*, April.

Hart, D. G. (1992). "European Regulatory Update," Vinyl Chloride Safety Association (VCSA) Meeting, Baltimore, October 1–2.

Joseph, H. M. (1979). "Migration of Vinyl Chloride from PVC into Food," International Conference on Critical Current Issues in Environmental Health Hazards, Tel Aviv, Israel.

Lobo, P. A. (1974). "VCM and PVC Manufacture," Regional Technical Conference of the Society of Plastics Engineering, Inc., New York.

Smallwood. (1989). Suspension and mass polymerization. *Encyclopedia of Polymer Science and Engineering*, Vol. 17. John Wiley & Sons, New York.

State of California, 1978. Public hearing to consider establishment of state ambient air quality standards, Air Resources Board, 78-8-3, April.

17

Devolatilization of Fibers, Latices, and Particles

Eric A. Grulke

University of Kentucky, Lexington, Kentucky

Gary S. Huvard

Huvard Research and Consulting, Chesterfield, Virginia

I. INTRODUCTION

There are a number of separation process that can be used to remove monomers, solvents, and gases from polymers and polymer products. Much information is available for the devolatilization of polymer melts and solutions. However, there are many cases for which it is necessary to remove liquids and gases from polymers in the solid state. Examples of polymer systems devolatilized from the solid phase include semicrystalline barrier copolymers that crystallize during polymerization (Saran™ copolymers are commercial examples*), amorphous polymers sparingly soluble in their monomers (suspension poly(vinyl chloride)), latex homopolymers and copolymers, supercritical fluid precipitations, nonsolvent precipitations for fractionating or purifying polymers, the removal of water or solvents from microcapsule coatings, the removal of spinning solvents from fibers, and the removal of monomers and diluent solvents from many Ziegler–Natta polymerized polyolefins, for example, polypropylene produced by the continuous slurry process.

As discussed in Chapter 3, it is preferable to remove volatiles at and above the glass transition temperature, T_g, or above the melting temperature, T_m, for crystalline polymers. Both the thermodynamic driving force and the diffusion coefficient are much higher above these thermal transitions, leading to very high rates for mass transfer so long as large vapor–polymer melt surface areas can be maintained.

In the case of devolatilization of solid polymers, the surface area of the solid and its characteristic dimensions are already established, and these are likely to be constant during solvent removal if the conditions during treatment (temperature, pressure, and composition) maintain the polymer in its solid state. Also, the diffusion coefficient of the gas or liquid often is constant, or nearly so, over the concentration range of the devolatilization. This occurs when the amount of solute in the polymer phase is low, so that there is little concentration dependence of the diffusion coefficient. The equilibrium driving force is set by the activity of the solvent on the exterior surface of the polymer. Therefore, knowledge of the vapor or liquid-phase conditions in the devolatilizer permits calculation of the devolatilization rate if diffusion through the polymer is rate-limiting.

Knowledge of mass transfer coefficients and phase equilibria alone is usually insufficient for the design and operation of separation processes. A complete analysis will include (1) the physical mechanism(s) for the separation, (2) phase equilibrium data and models, (3) material and energy balances for the process equipment, (4) mass transfer or efficiency coefficients, (5)

---

* Saran™—trademark of the Dow Chemical Company.

other physical property data, and (6) hydrodynamic and other factors affecting the equipment's approach to equilibrium conditions, or its ability to achieve high mass transfer rates. The objective of this chapter is to demonstrate how to integrate all the foregoing elements of the analysis into a sound separation design. The theoretical development in this chapter will emphasize items 2 and 4, phase equilibria and mass transfer. Material and energy balance methods for conventional unit operations are available in the literature. The other elements of design will be discussed as specific examples in the development of the case studies.

II. SORPTION EQUILIBRIA BELOW AND NEAR T_g

This section describes simple methods for measuring the equilibrium solubility coefficient of gases and solvents in solid polymers, two types of sorption in polymers below their glass transition temperature, and a "universal" isotherm that can be used to model the gas solubility coefficient over wide ranges of solvent activity.

A. Measurement of Equilibrium Solubility Coefficient

Of the numerous methods devised for measuring the solubility of gases and vapors in polymers, gravimetric and barometric techniques are most commonly used (Felder and Huvard, 1980). Gravimetric systems, which directly measure the mass of solute sorbed into or desorbed by a polymer, include quartz spring balances (so-called McBain balances) and modern electrobalances like those produced by Cahn Instruments (New Hope, Pennsylvania). Turnkey electrobalance systems designed specifically for polymer sorption and diffusion measurements are now available from Abbess Instruments, (Ashland, Massachusetts) and from VTI Corporation (Hialeah, Florida).

Barometric systems, like that shown schematically in Fig. 1, are also popular since they are easily designed for high-vacuum or high-pressure experimentation. A polymer sample of a known weight, w, and volume, v_p, is placed in a sorption cell of volume V held in a constant temperature bath. The sample, initially vacuum evacuated, is exposed to a known initial pressure of the penetrant vapor ($p_{initial}$), and the pressure decrease due to sorption in the polymer is noted. When the pressure decline ceases, the sample is considered to be in sorption equilibrium with the remaining vapor at the final pressure (p_{final}), and the solubility of the vapor in the polymer at the temperature and final pressure is calculated. High-pressure experiments normally require accounting for gas-phase nonideality. At these conditions, compressibility factors may be included in the following

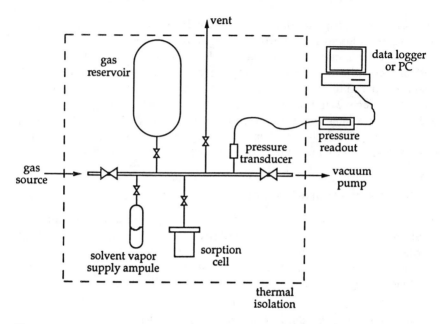

Fig. 1 Barometric sorption system. This basic design is suitable for subatmospheric or high pressure with proper choice of components and materials of construction.

calculations. The amount of gas sorbed into the polymer when the gas phase is ideal is calculated as follows:

$$M_{initial} = \frac{P_{initial}(V - v_p)M_w^{gas}}{R_g T}$$

$$M_{final} = \frac{P_{final}(V - v_p)M_w^{gas}}{R_g T} \tag{1}$$

$$S_i = \frac{M_{initial} - M_{final}}{w}; \qquad \frac{g\,gas}{g\,polymer}$$

where M_j is the mass of gas sorbed at time j, P_j is the pressure reading at time j, M_w^{gas} is the molecular weight of the gas, R_g is the ideal gas constant, and T is the temperature. Often, more penetrant vapor is introduced to a new "initial" pressure and the solubility determined at the (necessarily) higher "final" pressure after a second sorption period. Continuing in this manner allows the determination of the sorption isotherm up to the pressure limit of the experimental system. Methods for estimating the experimental time needed to achieve equilibrium are discussed in the next section.

This experiment gives a direct measurement of the gas solubility coefficient, S_i, if equilibrium has been reached between the gas and polymer phases. Methods for estimating the experimental time needed to achieve equilibrium are discussed in the next section. The solubility coefficient is reported in gravimetric (g gas/g polymer, mg gas/g polymer), weight percent, or volumetric (cm^3 gas (STP)/cm^3 polymer) units. Solvent concentration in the gas phase is reported as pressure (atm, bar, kPa, psia), concentration (mol/L, $kmol/m^3$), or activity (P_i/P_i^0 for many gases at low pressure).

B. Solubility Coefficient of Gases at Low Activities in Rubbery Polymers

Henry's law gives a linear relationship between the partial pressure of the solvent and the polymer phase solubility. The form of Henry's law employed in this chapter uses solvent partial pressure as the independent variable:

$$S_i = kP_i \tag{2}$$

where P_i is the pressure, S_i is the solubility, and k is the Henry's law solubility coefficient (g penetrant/g polymer-atm). Equation (2) defines a Henry's law coefficient that is the inverse of coefficients used in other chapters in this book, but it is useful since the solubility is directly calculated. Care must be taken when using literature values to convert the unit system of the data source to the unit system of the application.

Gas solubility coefficients may increase or decrease with increasing temperature, depending on whether the sorption is endothermic or exothermic, respectively. The heat of solution (ΔH_s) of the gas in the polymer depends on the heat of condensation (ΔH_c) and the heat of mixing (ΔH_m):

$$\Delta H_s = \Delta H_c + \Delta H_m \tag{3}$$

The van't Hoff relationship is most often used to correlate the Henry's law coefficient with temperature:

$$k = k_0 \exp\left(-\frac{\Delta H_s}{R_g T}\right) \tag{4}$$

where k_0 is the preexponential constant. Gases well above their critical temperature at the experimental conditions have small heats of condensation, so that the heat of solution is governed by the heat of mixing, which is typically small and positive (0.5–3.0 kcal/mol). Condensable gases and vapors have negative heats of condensation and, usually, negative heats of solution. In these cases, k decreases as temperature increases. In general, gases that are easier to condense are more soluble in a given polymer. For elastomers and rubbery polymers, the solubility coefficient can be correlated to the enthalpy of sorption (Fig. 2). More complex behavior is observed for glassy polymers for the reasons discussed in Section D.

An analysis of Henry's law with the solvent activity (see Eq. 13, Chapter 2) leads to an approximation useful for correlating solubility coefficient data over reasonable ranges of temperatures. We can write Eq. (2) as

$$S_i = kP_i^0 a_i \tag{5}$$

where P_i^0 is the saturated vapor pressure of i, and a_i is the solvent activity. We substitute the van't Hoff form for the Henry's law coefficient, and the Clausius–Clapeyron equation[†] for the saturated vapor pressure of solute i to get

$$S_i = k_0 \exp\left(-\frac{\Delta H_s}{R_s T}\right) \cdot p_0 \exp\left(-\frac{\Delta H_v}{R_g T}\right) \cdot a_i \tag{6}$$

Combining terms, we get

$$S_i = k_0 p_0 \exp\left(-\frac{\Delta H_m}{R_g T}\right) \cdot a_i$$

$$\approx k_0 p_0 a_i \tag{7}$$

since the enthalpy of condensation is equivalent to the negative of the enthalpy of vaporization and the enthalpy of mixing is small for most solvent–polymer pairs. For many solutes, the solubility coefficient exhibits little temperature dependence when plotted versus activity. This plot can be used to extrapolate solubility coefficient data taken at one temperature to other temperatures. It can also be used to test whether the heat of mixing of the system is small.

C. Solubility Coefficient of Gases and Vapors at High Activities in Rubbery Polymers

At high solvent activities, more vapor will condense into the polymer, causing it to expand and swell. Under these conditions, the solubility isotherm of the solvent is nonlinear. A number of models used to describe such equilibria have recently been reviewed by Danner and High (1993), but the simplest

---

† The form of the Clausius–Clapeyron equation used is obtained assuming ΔH_v is constant. This a good approximation for most solvents over a modest range of temperatures.

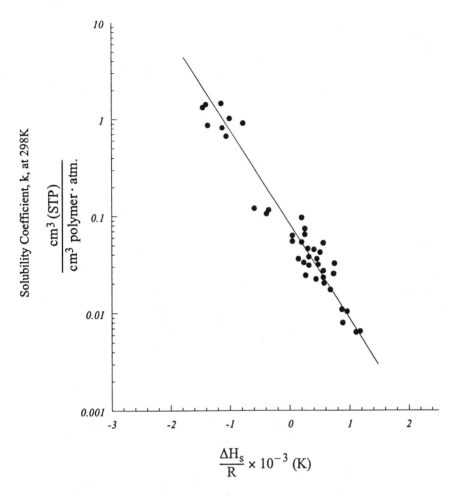

Fig. 2 Correlation of the Henry's law solubility coefficient with the enthalpy of sorption for elastomers. (From van Krevelen, 1990.)

one is Flory–Huggins theory. We will use it here to demonstrate how to model solubility coefficients at high activities, but similar techniques can be used with alternative models. The Flory–Huggins theory is described in Section IV of Chapter 2 (see Eqs. 32 and 33):

$$\ln a_i = \ln \phi_1 + \phi_2\left(1 - \frac{1}{x}\right) + \chi\phi_2^2 \tag{8}$$

where ϕ_1 is the volume fraction of the solvent, ϕ_2 is the volume fraction of the polymer, x is the ratio of the molar volume of the polymer to the solvent (often approximated by the degree of polymerization), and χ is the interaction parameter. Equation (8) can be related to Henry's law by considering the solvent activity as the solvent volume fraction goes to zero:

$$\ln a_i = \ln \phi_1 + 1\left(1 - \frac{1}{\infty}\right) + \chi \cdot 1^2$$

$$= \ln \phi_1 + (1 + \chi) \tag{9}$$

The interaction parameter, χ, can vary considerably with solvent volume fraction and system temperature. The analysis in this chapter is restricted to systems in which the temperature and volume fraction variation of χ is modest, and a constant value gives engineering accuracy ($\pm 15\%$) for phase equilibria calculations. Such accuracy is often lost when the temperature range is more than $40°C$, or when the system is highly polar. As the volume fraction of polymer approaches 1, the interaction parameter approaches a constant value at a given temperature and a single value describes phase equilibria well down to very low solvent volume fractions. However, a number of polymer–solvent pairs are well described by a single χ value over the entire activity range of 0 to 1, for example, poly(vinyl chloride)–vinyl chloride.

1. Simple Methods for Estimating χ

The main advantage of the Flory–Huggins equation is that it can be used to approximate the phase equilibrium to within engineering accuracy ($\pm 15\%$), providing rapid answers until the information needed for improved phase equilibria models can be obtained. The interaction parameter can be estimated from one phase equilibria experiment and then used to predict the entire sorption isotherm. In one typical experiment, a solid polymer sample is soaked in the solvent liquid ($a_1 = 1$) and then is weighed periodically until constant weight has been attained. The weight of solvent added to the polymer phase is converted to solvent volume fraction by using the solvent and polymer densities. Equation (8) is used to find χ.

A phase equilibrium measurement at any activity can generally give a very good estimate of the entire sorption isotherm. Henry's law coefficients are particularly convenient since they are often reported in the literature. We can take the following approximation for the solubility coefficient:

$$S_i \approx \frac{\phi_1}{1 - \phi_1} \frac{\rho_1}{\rho_2} \tag{10}$$

Using Eqs. (5), (9), and (10), we can show that

$$\chi = \ln\left(\frac{k\rho_1}{P_i^0 \rho_2}\right) - 1 \tag{11}$$

and

$$S_i = \left\{\frac{\rho_1}{\rho_2} \frac{\exp[-(\chi + 1)]}{P_i^0}\right\} P_i = kP_i \tag{12}$$

The foregoing equations make it simple to interchange between the Henry's law and Flory–Huggins models. The advantage of the latter is that it can be used over the entire activity range. Figure 3 shows the concentration of carbon dioxide in silicone rubber over a large pressure range. At low pressures, the carbon dioxide concentration increases linearly with partial pressure. At pressures above 200 psia, the relationship is nonlinear, and the concentration increases rapidly as the polymer swells.

Gas solubility data can be plotted versus activity as well. For vapors at temperatures below the critical temperature, the activity is very nearly equal to the relative pressure and may be computed by dividing the vapor partial pressure by the saturated vapor pressure of the solute at the system temperature. Above the critical temperature of the solute, activity should be computed as the fugacity at the system temperature and pressure divided by a reference state fugacity. To maintain consistency below and above T_c, Berens and Huvard (1989) used fugacities along the vapor pressure curve for reference state fugacities. This reference fugacity curve can then be extrapolated to define the reference state values above T_c, where the vapor pressure curve no longer exists (Sandler, 1977).

D. Dual-Mode Sorption

Solvent sorption in glassy polymers is best modeled by a more complicated equation. Figure 4 shows the solubility coefficient of vinyl chloride in poly(vinyl chloride) at a range of temperatures and relative pressures (activities). Vinyl chloride is only partially soluble in its homopolymer, and is typical of a number of polar monomers. At high relative pressures, data over a 30°C temperature range collapse to the same curve (calculated as the Flory–Huggins model). At low relative pressures, the Henry's law coefficient would change for each temperature, with the deviation between the data and the Flory–Huggins curve increasing as temperature decreases.

This phenomenon has been interpreted as two different types of solvent sorption in glassy polymers, that is, dual-mode sorption. One type is nonspecific absorption of the solvent into the polymer. This type can be described by the Flory–Huggins model, or by other models applicable to

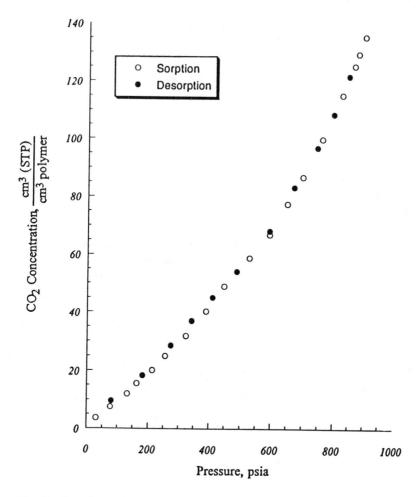

Fig. 3   Sorption and desorption of carbon dioxide in silicone rubber at 35°C. (From Fleming and Koros, 1986.)

the rubbery state. The second type of sorption is thought to be solvent sorbing into the free volume available in the glassy state, created by the "freezing in" of chain motions as the polymer sample is cooled below the glass transition temperature. This site-specific sorption should be larger in magnitude for samples farther below the glass transition temperature *of the mixture* and should essentially disappear at temperatures above T_g. Figure 5 shows plots of the dual-mode sorption model for the sorption data of

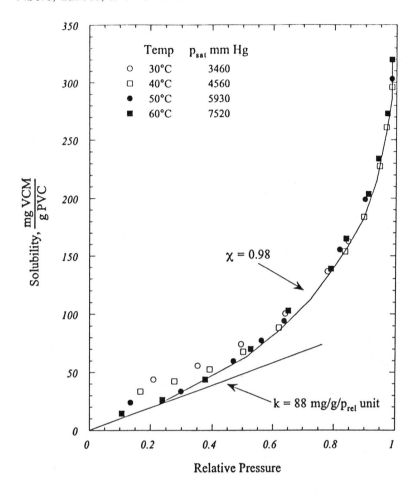

Fig. 4 Vinyl chlorde solubility in poly(vinyl chloride). (From Berens, 1975.)

Fig. 4. The site-specific sorption decreases as temperature increases, and disappears above T_g of the polymer ($\approx 80°C$). The total sorption of the solvent in the polymer phase is

$$S_i = C_D + C_H \tag{13}$$

where C_D is the nonspecific sorption and C_H is the specific sorption contribution (assumed to be independent of each other).

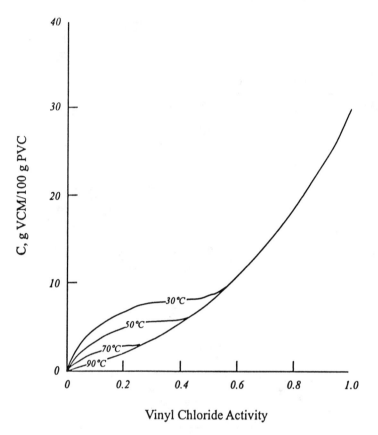

Fig. 5 Sorption of vinyl chloride in poly(vinyl chloride). (From Berens, 1975.)

1. Nonspecific Sorption Models

Any of the nonspecific sorption models used for gases and liquids in rubbery polymers can be used for C_D in Eq. (13). For example, Eqs. (2) or (12) would be appropriate. Henry's law, Eq. (2), would only be applicable at low activities, while the Flory–Huggins equation, Eq. (12), would be applicable over the entire relative pressure range.

2. Site-Specific Sorption Models

As a polymer melt cools, its chains rearrange to achieve an equilibrium between their molecular energy and the amount of space they occupy: the polymer density increases. At the glass transition temperature, long-range motions of the chains are greatly reduced and the state of the polymer is that of a nonequilibrium glass. The glassy polymer occupies a volume larger than that achievable if the material remained flexible while it was being cooled. The excess space exists within the glass as a distribution of tiny holes, or microvoids. While the exact nature of the microvoids is not agreed upon, solute molecules can occupy this space in addition to the volume they occupy as a normal solution.

Microvoids are made as the polymer is quenched below the glass transition temperature, and their number density in the glass depends on the time–temperature history of the sample. The two populations of sorbed solute molecules appear to be in dynamic equilibrium with each other. The microvoids can be saturated with solute molecules as the relative pressure increases. While there are a number of isotherms that might be used to describe the phase equilibrium with the microvoids, the Langmuir isotherm is the most popular and fits most data sets well. Equation (13) can be rewritten as

$$S_i = C_D + C_H = kP_i + \frac{C'_H b P_i}{1 + b P_i} \tag{14}$$

where the first factor on the right-hand side is the Henry's law description for nonspecific sorption and the second factor on the right-hand side is the Langmuir isotherm for site-specific sorption. Figure 6 shows a typical sorption isotherm, along with the contribution of each factor to the material response.

At high relative pressures, the Langmuir factor simplifies to C'_H. This parameter is a saturation constant and represents the capacity of the microvoids for the solute at the system temperature. Since it should be related to the excess volume in glassy polymers, its value can be estimated from the polymer free volume and the molar volume of the solute. The unrelaxed volume should be proportional to the free volume, defined as

$$\frac{V_g - V_{liq}}{V_g} = (\alpha_{liq} - \alpha_g)(T_g - T) \tag{15}$$

where V_g and V_{liq} are the volumes of the glassy and liquid polymer, and α_{liq} and α_g are the thermal expansion coefficients of the rubbery and glassy polymer, respectively, and T_g is the glass transition temperature. The difference between these thermal expansion coefficients is similar for many polymers, and C'_H has been found to be proportional to $T_g - T$. As shown in Fig. 7, the Langmuir capacity goes to zero for any polymer having a T_g below the experimental temperature of 35°C.

The parameter b is an affinity constant. It describes the ratio of the forward and reverse rate constants for solute adsorption into the sites. It can be correlated with the solubility parameter for various polymers (Fig. 8).

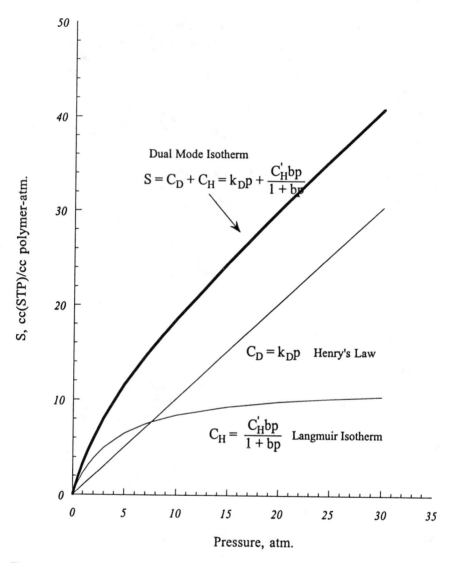

Fig. 6 The dual-mode isotherm is typical of the sorption of low-activity gases and vapors in glassy polymers.

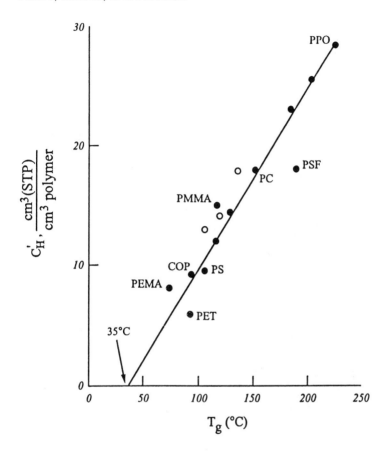

Fig. 7 Correlation of observed Langmuir sorption capacities for CO_2 of various polymers at 35°C with their glass transition temperatures. (From Toi et al., 1982.)

At low relative pressure, Eq. (14) simplifies to an apparent Henry's law isotherm as follows:

$$k_{apparent} = k + C'_H b \tag{16}$$

The Henry's law constant can be correlated with the solubility parameter as well, since it is related to the condensation of the gas (Fig. 9). All of the dual-mode sorption parameters, k, b, and C'_H, scale with temperature via the van't Hoff relationship. However, the "enthalpy of sorption" calculated from the temperature dependence of the Langmuir saturation constant is not really an enthalpy change but merely reflects the dependence of C'_H on temperature due to changes in the amount of frozen-in excess volume with temperature

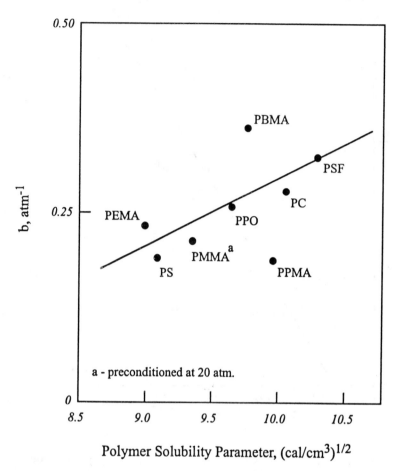

Fig. 8 Correlation of the dual-mode affinity constant for sorption of CO_2 at 35°C with polymer solubility parameter. (Previously unpublished, courtesy of W. J. Koros, 1995.)

(Koros et al., 1979). Nearly every polymer–solvent pair exhibits dual-mode sorption below T_g. One exception is poly(n-butyl methacrylate), possibly because its thermal expansion coefficients are nearly identical for the rubbery and glassy states.

E. Universal Sorption Isotherm

As we have discussed, nonspecific sorption can be described well by Henry's law at low relative pressures, but different models are needed when the

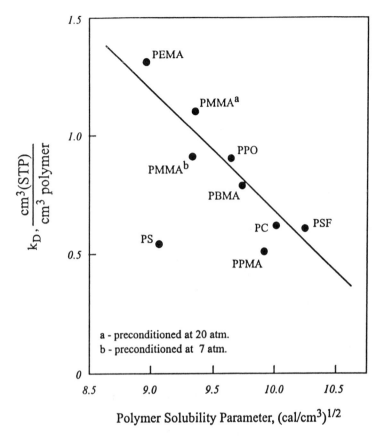

Fig. 9 Correlation of the dual-mode Henry's law constant for sorption of CO_2 at 35°C with polymer solubility parameter. (Previously unpublished, courtesy of W. J. Koros, 1995.)

activity approaches 1. As more solute is introduced into a polymer phase, the glass transition temperature of the mixture decreases, and in many cases, the polymer becomes rubbery at the experimental temperature. Therefore, Eq. (15) is restricted to low relative activities. This restriction can be removed by substituting a different model for the nonspecific sorption term. We will use the Flory–Huggins equation here; however, any other phase equilibrium model can be chosen if it represents the data and simplifies to a Henry's law type performance at low relative pressures. Equation (15) becomes

$$S_i = C_{F-H} + C_H = f'(\chi) + \frac{C'_H bP_i}{1 + bP_i} \tag{17}$$

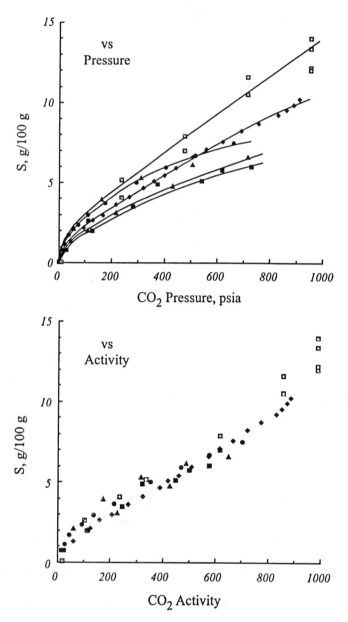

Fig. 10 Solubility of CO_2 in polycarbonate as functions of pressure and activity.
□ 25°C, Berens and Huvard (1989); ◆ 35°C, Fleming and Koros (1986); ● 35°C,
Kamiya et al. (1986); ▲ 45°C, Kamiya et al. (1986); ■ 55°C, Kamiya et al. (1986).

where $f'(\chi)$ is the volume fraction of solute in the polymer calculated by the Flory–Huggins equation and converted to a weight basis. This equation is easy to convert to an activity basis:

$$S_i = C_{F-H} + C_H = f'(\chi) + \frac{C_H' b a_i}{1/P_i^0 + b a_i} \tag{18}$$

This form suggests that families of isotherms be plotted as shown in Figs. 5 and 10. Figure 5 shows the solubility of vinyl chloride monomer in poly(vinyl chloride), and Fig. 10 shows the sorption of carbon dioxide in polycarbonate. The two portions of Fig. 10 show how the isotherms from several different temperatures can often be collapsed onto a single curve at high activities.

There are several other factors that influence the sorption of solutes in glassy polymers. Many polymer products are not pure homopolymer and contain a variety of additives. For example, most pure rubbers and poly-olefins have very low equilibrium solubilities of water. However, commercial compound formulations may contain 5–10 wt % of polar additives, which will enhance the uptake of water into the mixture. Additives in polymer products affect solute sorption, and it is prudent to study the problem with the actual compound to be used.

Crystalline regions in the polymer are generally not accessible to solutes. Sorption equilibria for crystalline materials usually are based on the volume fraction of the amorphous phase. Differential scanning calorimetry can be used to determine this fraction.

Since the glassy state is a nonequilibrium one, the sample history affects the amount of free volume available. Free volume is affected by temperature, mechanical forces including shear, and the sorption of other gases and vapors. Multiple solutes can compete for the same microvoids. This often results in lowering the sorption of the less condensable gas in the Langmuir sites.

III. DIFFUSION BELOW OR NEAR T_g

Chapter 3 describes the diffusion of solvents in polymers above and below the glass transition temperature. The best theoretical descriptions of the process include mutual binary diffusion coefficients and the free volume of the polymer–solvent system. Free-volume theory can be extended to problems below the glass transition temperature. A careful review of Eqs. (13) and (14) in Chapter 3, and a comparison with Eq. (15) of this section, will show the similarity between these approaches. Notice that volumetric contraction of Eq. (13), Chapter 3, includes the difference between the thermal expansion coefficients and the difference between the system temperature and the glass transition temperature. Free-volume theory would be used in preference to dual-mode sorption theory if a continuous function were needed to describe a polymer system as it changed from a melt to a glass.

In general, the diffusion coefficient, D, of solutes in glassy polymers can be described in much simpler ways than suggested by free-volume theory. These descriptions are adequate for polymer phases that are dilute in solute so that there is insignificant effect of the solute concentration on D. In this section, we summarize experimental methods for measuring both diffusion and solubility coefficients, present simple models for D, describe non-Fickian diffusion regions, and demonstrate how to estimate diffusivities and fluxes.

A. Experimental Methods

1. Permeation Methods

a. Permeation into a Closed Container. Permeation into a closed container is recommended for measuring low rates of solute uptake. It measures the cumulative amount of solute sorbed. and long-term experiments are used. It is suggested for easily condensable vapors, but it is difficult to apply for gas mixtures. Permeation methods require that the solubility coefficients be calculated indirectly from transient data.

b. Permeation into a Flowing Stream. This method is suitable for polymers that will sorb solute at moderate to high rates. It can be used for membrane materials that tend to tear or rip when subjected to mechanical pressure, since equal total pressure can be maintained on both sides of the membrane. Gas mixtures can be used, since the analysis (gas chromatography or mass spectroscopy) of the stream will determine the unsteady-state sorption of each component. Table 1 shows several permeation experimental methods and the references reporting their use.

2. Sorption Methods

Sorption methods have the advantage of giving solubilities directly from equilibrium data, and they often have lower experimental error than permeation methods. They are the method of choice when the equilibrium solubility change for the experiment is expected to be low, when the polymer phase swells with the addition of solute, or when the diffusivity of the solute is not constant during the experiment. In the case of very low diffusion rates, microspheres or microfibers can be used as the solid phase to reduce the experimental time. High differential pressures across membranes can lead to breakage and distention, while diffusion into solids usually does not encounter this problem. Cracking and crazing can be observed directly. Table 2 gives several sorption apparatus methods and associated references. The

Table 1   Permeation Experimental Methods

Apparatus	Description	Reference
Daynes–Barrer time-lag	Classical system for measuring the permeation of gases across a flat membrane at subatmospheric pressure	Daynes, 1920; Barrier, 1939
High-pressure permeation	High-pressure, high-vacuum system for gas permeation	Huvard et al., 1980
Linde permeation cell	ASTM volumetric method for permeation rates (D1434-75)	Stern et al., 1963
Dow permeation cell	ASTM D1434-58	
Continuous-flow permeation	Permeation through a polymer tube; compares the dynamic flux with the steady-state flux	Rodes et al., 1973

Table 2   Sorption Apparatus

Apparatus	Description	Reference
Volumetric sorption	Differential gas pressure is measured as a gas sorbs onto a solid sample	Rosen, 1959
Barometric sorption	Low-pressure system; measures sorption by PVT measurements in known volume	
McBain quartz spring balance	Polymer solid is placed in sample pan, which is suspended on a quartz spring in a sorption chamber; cathetometer measures the displacement of the spring as the sample gains weight	McBain and Bakr, 1926; Wossnessensky and Dubinkow, 1936
Electrobalance	Microbalance replaces the quartz spring	
Oscillatory concentration	Sample is exposed to an oscillatory concentration profile	Ju (1991)

details for the measurement of solubility and diffusion coefficients can be found in several reviews, for example, Felder and Huvard (1980).

B. Simple Models

Most diffusion coefficients for solutes at low concentrations in polymers above or below T_g can be modeled by a simple equation (in contrast to Eq. (1) in Chapter 3):

$$D = D_0 \exp \frac{-E_d}{R_g T} \qquad (19)$$

where D_0 is the preexponential or frequency factor, E_d is the activation energy for diffusion, R_g is the ideal gas constant, and T is the absolute temperature. Values of D_0 and E_d can be correlated with simple gas properties for rubbery and glassy polymers. Values of E_d range from 2 kcal/mol for the smallest gases to 45 kcal/mol or higher for large organic molecules in polymer melts. For example, E_d is about 43 kcal/mol for dilute normal pentane in polystyrene above T_g. Values for D_0 range from 10^{-4} to 10^6 cm^2/sec (10^{-8} to 10^2 m^2/sec).

Activation energies for gases in glassy polymers cover a similar range but often are lower that the values above T_g by about 20%. An Arrhenius plot of Eq. (19) will usually show a break in the curve in the vicinity of the glass transition temperature.

Estimating diffusion coefficients of simple gases and organic solvents in rubbery polymers is possible today due to the large amount of data in the literature (using free-volume theory, for example). However, estimating diffusion coefficients of these solutes in glassy polymers is an inexact art and should be done with caution. Mostly, correlations of experimental data must be used. For example, van Krevelen (1990) has developed a useful correlation of E_d for gases with T_g using molecular diameter as a scaling parameter. The correlation is shown in Fig. 11. Van Krevelen (1990) gives similar correlations for organic solvents in rubbery polymers.

1. Effect of T_g and Penetrant Size on D

The rigidity of polymer chains below T_g has a remarkable effect on the diffusion coefficients. Figure 12 shows these effects for a rubbery polymer, natural rubber with a T_g of $-70°C$, and a glassy polymer, poly(vinyl chloride) with a T_g of about 80°C, at the experimental conditions of 25–30°C. For each polymer, there is a nonlinear relationship between van der Waals volume and the diffusion coefficient. For all solutes, the diffusion coefficient in the rubbery polymer is higher than that in the glassy polymer. The diffusion coefficients of gases in natural rubber (and other rubbery polymers)

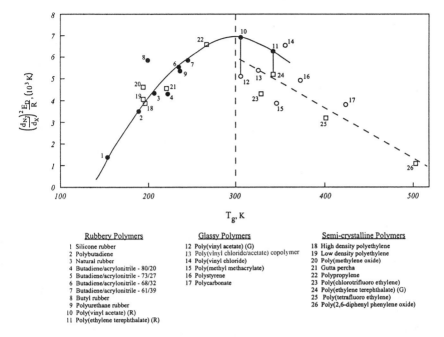

Fig. 11 Correlation between the square of the molecular diameter (relative to N_2) times the activation energy for diffusion and the glass transition temperature for gases diffusing in polymers. (From van Krevelen, 1990.)

fall in the range of 10^{-5}–10^{-7} cm^2/sec, while organic solutes have diffusion coefficients in the range of 10^{-6}–10^{-9} cm^2/sec. A relatively small increase in molecular size of the solute can have a large inverse effect on its diffusion coefficient in glassy polymers. Correlations such as Fig. 12 can be used to estimate the diffusion coefficients of other solutes. However, there are different choices for the molecular diameter data, and these give slightly different results (Berens and Hopfenberg, 1982).

2. Effect of Molecular Weight on D

The molecular weight of rubbery polymers has a modest effect on penetrant diffusion coefficients, with the values decreasing as molecular weight increases. Another factor for rubbery polymers is the chain length between cross links. As the cross link density increases, the diffusivity is reduced. Molecular weight of glassy polymers has a larger effect on penetrant diffusion coefficients since increasing the polymer molecular weight increases T_g and the difference between the experimental temperature and T_g for a specific solvent.

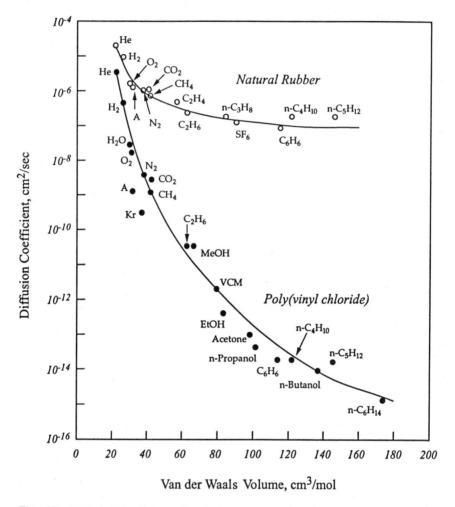

Fig. 12 Diffusion coefficient of various penetrants at 25–30°C in natural rubber and in poly(vinyl chloride) (PVC). Gases in natural rubber: van Amerongen (1964); organic vapors in natural rubber: Michaels and Bixler (1961); gases in PVC: Tikhomirov et al. (1968); organic vapors in PVC: Berens and Hopfenberg (1982).

C. Non-Fickian Diffusion Regions

Diffusion coefficients are nearly constant at the low concentrations typically encountered in devolatilization processes. These are activity–temperature regions possibly very dilute in solute, so that its removal does not change the surrounding polymer's volume or conformation very much. In contrast, diffusivities change markedly with removal of solvent from a swollen polymer. Volume contraction occurs with devolatilization of highly swollen polymer samples, and the diffusion coefficient normally drops dramatically as the volume declines.

A range of behaviors occurs as the temperature and solute activity vary. At high solvent activities and temperatures, the diffusivity may be nearly constant if the polymer is rubbery at the test conditions. At low activities and temperatures, the diffusivity may be nearly constant if the polymer is not changed much by the removal of the solvent. Near the glass transition temperature, any of several types of nonideal anomalies may be observed, and the diffusion coefficient is not constant. These nonideal behaviors are thought to be the result of polymer chain relaxations that occur over a similar time scale as that needed for diffusion. In the extreme, sorption may be completely controlled by relaxation or rearrangement of the polymer chains, and "Case II," or moving-front, transport will be observed. These types of behaviors have been thoroughly discussed by several different authors; references to many original works on these effects may be obtained from the review by Felder and Huvard (1980).

Nonideal, or "anomalous," effects are usually identified by performing sorption and desorption experiments with the same sample. If the sorption and desorption unsteady-state diffusion curves are the same, the diffusion coefficient is constant in the experimental temperature and concentration ranges. If the sorption and desorption curves do not match, then the system is nonideal, and analyses more complex than those following will be required. In most devolatilization problems, anomalous effects are avoided either intentionally or naturally due to low volatile concentrations and the analyses that follow will suffice for process analysis and design.

D. Estimation of Diffusivities and Fluxes

This section provides the background for calculating unsteady-state fluxes of solvents from polymer solids when the diffusion coefficients are constant. A material balance for diffusion in one direction from various geometries gives the following partial differential equations:

Flat Sheet

$$\frac{\partial C(z,t)}{\partial t} = D\,\frac{\partial^2 C(z,t)}{\partial z^2} \tag{20a}$$

Cylinder (Fiber)

$$\frac{\partial C(r,t)}{\partial t} = D\left(\frac{\partial^2 C(r,t)}{\partial r^2} + \frac{1}{r}\,\frac{\partial C(r,t)}{\partial r}\right) \tag{20b}$$

Sphere

$$\frac{\partial C(r, t)}{\partial t} = D\left(\frac{\partial^2 C(r, t)}{\partial r^2} + \frac{2}{r}\frac{\partial C(r, t)}{\partial r}\right) \tag{20c}$$

Diffusion coefficients are often determined experimentally by using solids of known geometry (slabs, fibers, or spheres) with uniform initial concentration profiles and then creating a step change in the solute content of the fluid phase. The change in the solute content in the polymer phase can be matched with the integrated solutions to Eqs. (20a), (20b), or (20c) to estimate the diffusion coefficient. Crank (1975) and Carslaw and Yeager (1959) provide a number of solutions to these equations for a variety of initial and boundary conditions.

The solutions to Eqs. (20a), (20b), and (20c) for sorption experiments are shown in Table 3. They are reported in terms of the ratio of the mass of solute in the polymer at time t, M_t, to the mass of solute in the polymer at infinite time, M_∞. The summations converge after just a few terms for long times. The same equations hold for desorption with M_t taken as the amount of solute to leave a polymer sample initially having M_∞ amount of solute uniformly sorbed.

There are many ways to estimate diffusivities using limiting forms of the solutions to the foregoing equations. Tables 4–7 show a variety of approximations that can be used to estimate diffusion coefficients from data of weight loss or uptake versus time. Half-time formulas (Table 4) are applied by determining the time at which the weight gain (loss) of the sample is half of its ultimate (initial) value, and using a value of the Fourier number for diffusion to find D. At small times, the weight gain versus time curve varies with the square root of time and M_t/M_∞ is proportional to $(Dt)^{1/2}$ (Table 5). At long times, the series solutions in Table 3 converge rapidly and only

Table 3   Solutions to One-Dimensional Sorption in Various Geometries

Sheet	$\dfrac{M_t}{M_\infty} = 1 - \dfrac{8}{\pi^2}\displaystyle\sum_{n=0}^{\infty}\dfrac{1}{(2n+1)^2}\exp\left(\dfrac{-D(2n+1)^2\pi^2 t}{l^2}\right)$
Cylinder (radial)	$\dfrac{M_t}{M_\infty} = 1 - \displaystyle\sum_{n=1}^{\infty}\dfrac{4}{(R\alpha_n)^2}\exp(-D\alpha_n^2 t); \qquad J_0(R\alpha_n) = 0$
Sphere	$\dfrac{M_t}{M_\infty} = 1 - \dfrac{6}{\pi^2}\displaystyle\sum_{n=1}^{\infty}\dfrac{1}{n^2}\exp\left(\dfrac{-Dn^2\pi^2 t}{R^2}\right)$

Table 4 Half-Time Formulas
($t = t_{1/2}$ when $M_t/M_\infty = 0.5$)

Sheet	$\dfrac{Dt_{1/2}}{L^2} = 0.04919$
Cylinder (radial)	$\dfrac{Dt_{1/2}}{R^2} = 0.06306$
Sphere	$\dfrac{Dt_{1/2}}{R^2} = 0.03055$

Table 5 Initial Slope Formulas

Sheet	$\dfrac{M_t}{M_\infty} \approx \dfrac{4}{L}\left(\dfrac{D}{\pi}\right)^{1/2} t^{1/2}$
Cylinder (radial)	$\dfrac{M_t}{M_\infty} \approx \dfrac{4}{R}\left(\dfrac{D}{\pi}\right)^{1/2} t^{1/2}$
Sphere	$\dfrac{M_t}{M_\infty} \approx \dfrac{6}{R}\left(\dfrac{D}{\pi}\right)^{1/2} t^{1/2}$
Arbitrary geometry	$\dfrac{M_t}{M_\infty} \approx 2\,\dfrac{\text{solid surface area}}{\text{solid volume}}\left(\dfrac{D}{\pi}\right)^{1/2} t^{1/2}$

Table 6 Long-Time Formulas

Sheet	$\ln\left(1 - \dfrac{M_t}{M_\infty}\right) \approx \ln\left(\dfrac{8}{\pi^2}\right) - \dfrac{D\pi^2 t}{L^2}$
Sphere	$\ln\left(1 - \dfrac{M_t}{M_\infty}\right) \approx \ln\left(\dfrac{6}{\pi^2}\right) - \dfrac{D\pi^2 t}{R^2}$

Table 7   Zeroth Moment Formulas

$$\tau_s \equiv \int_0^\infty \left(1 - \frac{M_t}{M_\infty}\right) dt$$

Sheet $\qquad\qquad\qquad D = \dfrac{L^2}{12\tau_s}$

Cylinder (radial) $\qquad\quad D = \dfrac{R^2}{8\tau_s}$

Sphere $\qquad\qquad\qquad D = \dfrac{R^2}{15\tau_s}$

Source: Felder (1978).

the first term of the summation is needed (Table 6). Felder (1978) has defined the zeroth moment of sorption/desorption curves as the integral $1 - M_t/M_\infty$ with time (Table 7). The value of the zeroth moment is inversely proportional to the diffusion coefficient for each geometry. The advantage of this method is that the entire sorption/desorption curve is considered in the calculation. However, a complete sorption or desorption to equilibrium is needed to perform the integration.

The preceding methods can be applied when transport through a sample is essentially unidirectional. In many practical systems, diffusion takes place along two or three axes. For example, polymer pellets resembling short cylinders or cubes are dried in air, and the fluxes along each major axis of the solid have similar magnitudes. When the concentration profile is initially uniform and the surface concentration is constant at all times, the solutions for multidimensional diffusion obtained by Newman (1931) are useful. These are solved in graphical form (Fig. 13) with the fractional amount remaining, $E_i = 1 - M_t/M_\infty$, plotted versus the Fourier number for diffusion. The fraction remaining due to diffusion along each axis is computed, and then the total fraction remaining in the solid is the product of all fractions. Table 8 summarizes the computations for various solid geometries. Figure 13 can be used to solve a wide variety of diffusion problems, including one in Section IV.A on polymer leaching.

There are solutions for many other boundary conditions and initial conditions in Crank (1975), and in Carslaw and Yeager (1959). Concentration-dependent diffusion can be approximated by applying the equations over

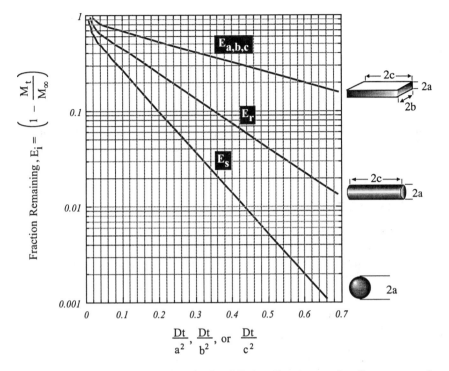

Fig. 13 Fraction remaining versus the diffusion Fourier number for transport in plates (films), cylinders (fibers), and spheres (latex particles). (From Newman, 1931.)

Table 8 Fraction Remaining Computations for Various Geometries ($E_{a,b,c}$ for plate axes, E_r for cylinder radius, E_s for sphere radius)

Solid Geometry	Fraction Remaining
Slabs with two pairs of sealed ends	$E = E_a$
Cylinders (fibers) with sealed ends	$E = E_r$
Spherical particles	$E = E_s$
Rectangular bar with one pair of sealed ends	$E = E_a E_b$
Rectangular parallelpiped	$E = E_a E_b E_c$
Short cylinder	$E = E_a E_r$

When one end of a pair is sealed, the correct dimension to be used is $2a$. This moves the symmetry boundary condition to the sealed face.

small concentration intervals. The diffusion coefficient estimated by this method is an average over the concentration range. Several important practical problems do not have constant solute concentrations in the fluid external to the solid. For example, in fiber dyeing or in latex stripping, the contacting solution may be a finite bath and its concentration will vary during the sorption/desorption process. The references just mentioned also have solutions for these situations. Finally, problems with irregular geometry, nonuniform initial concentration profiles, and variable boundary conditions can be approximated numerically using finite element or finite differences techniques.

IV. DEVOLATILIZATION CASE STUDIES

The equilibrium and diffusion calculations developed in the previous sections have a wide variety of applications. Four case studies are presented in this section to illustrate how to set up solutions for typical devolatilization problems with fibers, latices, and particles.

A. Leaching of Particulate Polymers

Polymers made in latex, dispersion, or suspension processes often contain solvents, initiator fragments, or other agents that need to be removed prior to use. Leaching is one method used to separate soluble material from a mixture containing an insoluble solid.

   A variety of process technologies are available. Solutes can be leached from solids by solvents allowed to percolate through stationary beds of solid particulates. Beds arranged in series permit high removal of the solute while conserving solvent. There are several types of moving bed leachers, for example, Bollman, Hildebrandt, and Kennedy, which move the solid mechanically through the liquid system. Other equipment disperses the solids in a solvent phase and then uses settling or filtration to separate the phases. The solute is recovered from the solvent by a variety of techniques, including distillation, crystallization, ion exchange/adsorption, or membrane separation.

   A number of characteristics are desirable in a leaching solvent. Table 9 gives a partial list. As with other separation processes, solvent choices and optimization of leaching conditions are best done in the context of a complete process design.

1. Mechanism for Separation

   A sketch of the leaching process for polymeric solids is shown in Fig. 14. The solute in the polymer phase partitions into the solvent at the surface

Table 9   Characteristics of Good Leaching Solvents

High solute saturation limit	Solute selectivity
No solvent–solute reactions at the leaching temperature, pressure, and concentration	Low viscosity
Low vapor pressure	Low toxicity
Nonflamable	Low density
Low surface tension	Easy to recover and recycle
Low cost	Little or no swelling of the solid with high transfer rates

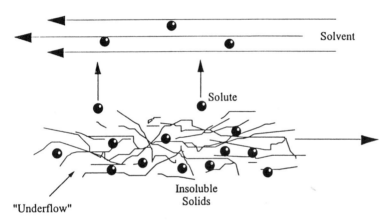

Fig. 14   Leaching of solutes from polymeric solids.

of the polymer and is conveyed away by the bulk liquid. When using conventional design balances, the streams leaving each stage of a multistage process are assumed to be in equilibrium. Most separation processes are designed by using efficiency factors to describe the approach of each stage to equilibrium. Accordingly, designs based on equilibrium partitioning are corrected by dividing the theoretical stage count by an efficiency. Geankoplis (1993) and McCabe et al. (1993) discuss conventional leaching design methods.

     There are two main mechanisms for the removal of the solute from the solid. If the solute is on the surface of porous polymer particles, then the partition coefficient of the solute between the liquid associated with the polymer phase (the "underflow") and the free solvent phase is 1 (when the concentration in each phase is calculated on a solid-free basis). This case is analogous to the leaching of soybean oil from macerated soybeans, in which the cell structure has been broken so that all the oil can dissolve freely in the solvent associated with the plant matter. An example of a polymer leaching problem that follows this mechanism is the removal of surfactant from a partially coagulated latex. The coagulum has an open pore structure that communicates to its interior, and the surfactant is concentrated on the surface of the particles. In these cases, the leaching solvent does not dissolve significantly in the polymer, and it merely acts as a carrier for the solute (the surfactant).

A second leaching mechanism involves the partial swelling of the polymer solid by the solvent and the partitioning of the solute from the polymer phase to the solvent phase. This case is analogous to the leaching of caffeine from coffee beans by partial swelling agents such as water, methylene chloride, or supercritical carbon dioxide. Examples of polymer leaching problems that follow this mechanism include the removal of catalyst from a particulate polymer using water, or the removal of a polymerization solvent from particulate polymer by a second "nonsolvent" or sparingly soluble solvent. In the case of solvent removal, the leaching solvent partially swells the polymer phase, replacing some of the solvent to be removed.

Swelling to a near-T_g state is critical to the process result when glassy polymers are being leached because it greatly increases the diffusion coefficient of the solute in the polymer. Swelling solvents increase the free volume compared with the polymer phase free volume at low solute levels with a concomitant increase in the diffusion coefficients (see Fig. 2 of Chapter 3 for diffusion slightly above T_g). Without this agent, the process would simply be the unsteady-state diffusion of solute from a glassy polymer: potentially a very slow process depending on temperature and geometry. At very low free volume levels (near that of the pure polymer), the remaining solute could be kinetically "trapped" by the polymer and there could be a solute-lean "skin" on the polymer surface, while the center of the particle would be high in solute. The operating temperature for such a leaching system should not be too high above the effective glass transition temperature of the polymer–solvent mixture, to reduce or prevent tacky particles from agglomerating. This mechanism adds extra constraints to the properties of good leaching solvents listed in Table 9.

2. Material and Energy Balances

The material and energy balances for staged leaching processes can be applied to polymer systems (Geankoplis, 1993). Figure 15 shows a sketch of a countercurrent multistage leaching process for removing low levels of

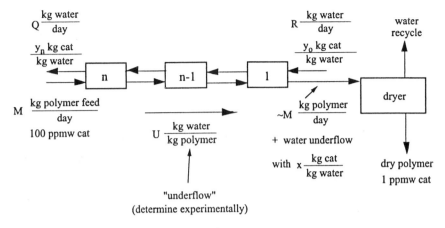

Fig. 15 Process flow sketch for solute leaching from polymeric solids.

catalyst from a polymer solid using water. Counter-current processes are usually the most efficient with respect to minimizing solvent volume, although single-stage and co-current designs may work in some cases. With either leaching mechanism, the polymer phase can be treated as an inert solid not soluble in the solvent, and material balances are done by considering the liquid associated with each phase. Energy balances usually are ignored when the solute is dilute in the polymer phase, or when there is little or no enthalpy change with solution for the phase mixtures.

Conventional graphical techniques can be used to solve these problems; however, stage-to-stage calculations are easy to carry out with computer spreadsheet programs or with other software available for leaching calculations. Models for the phase equilibria may be used as inputs for commercial process simulators as well.

3. Phase Equilibrium Models and Data

Leaching equilibrium can be characterized by experimental partition coefficients or by more complex models such as the Flory–Huggins equation expansion for ternary systems. The first kind of leaching mechanism, removal of materials from the particulate surface, can be characterized by partition coefficients of 1 (when the concentrations are given in terms of kg solute/kg solution in each phase). This simplifies the stage-wise calculations.

Leaching equilibrium models may be much more complex for the partial swelling mechanism. The initial volume fraction of solute in the polymer, the solvent volume fraction in the polymer at equilibrium partial swelling, and the change in the polymer phase volume in every stage are

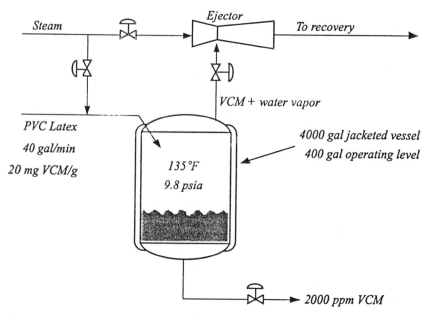

Fig. 16 Sketch of a continuously stirred tank reactor (CSTR) devolatilization system for a PVC latex.

important. Fresh polymer entering the process on stage n of Fig. 15 will sorb some solute. Partition coefficients may be adequate for defining the equilibrium if solute levels are low, the solvent swelling of the polymer is low, and the addition of the solute to the solvent does not change the mixture's solvency properties very much. Partition coefficients can usually be measured by placing known volumes of solvent and solids in a suitable stirred vessel and then measuring the solute concentration in the liquid phase as a function of time. The concentration in the liquid approaches a constant value when equilibrium between the phases is reached. A material balance on the system gives the partition coefficient. When solute levels are low, the partition coefficients may be nearly constant, and phase equilibrium computations are easy.

When the second leaching mechanism applies, the polymer swells in the solvent, and more complex phase equilibria models are needed. A typical starting point would be the Flory–Huggins equation expanded to three components: polymer, solvent, and solute. This equation describes the liquid–liquid demixing equilibrium for the solute partitioning between a swollen polymer phase and the solvent phase. Interaction parameters are needed for each component, but the equations should model any changes in the solvency of the mixture as compositions vary throughout the equipment.

It usually is desirable to keep the temperature of the system below the glass transition temperature of the swollen polymer phase. When $T < T_g$, the polymer is in a glassy state, its surface does not become tacky, and particle agglomeration rates are greatly reduced. If the solvent is a poor one for the polymer and mixing is optimal, it is possible to increase the leaching temperature above T_g for a short period of time. This permits high devolatilization rates but may induce agglomeration, loss of porosity, and the loss of small particles.

4. Mass Transfer Models and Data

Conventional leaching models generally assume equilibrium between the streams leaving each stage. The mass transfer between the solid and the fluid phase must be rapid compared with the particle residence time on each stage for this to occur. Also, for the first mechanism, the flux of solute from the interior of the agglomerate to the particle surface must be high. For the second mechanism, the flux of solute from the polymer phase interior to its surface must also be high. Example 1 illustrates this concept.

5. Physical Properties

As mentioned in Section 3, the polymer phase should remain below its glass transition temperature. This places a constraint on the choices of solvent and temperature, but it ensures that the solid does not agglomerate during the leaching process. The thermal transition needs to be measured in the presence of the solvent and the solute, since both will be present on any given equilibrium stage. This is easier said than done, however, and the process conditions are often established by trial-and-error experimentation.

6. Example 1. Effect of Particle Size on the Approach to Equilibrium

a. Problem Statement. Residual monomer is to be leached from a porous polymer resin in a multistage process using a nonsolvent leaching agent. The polymer resin has been made by a suspension process, and it has primary particles sizes averaging 10 μm in diameter and an agglomerate size of 200 μm. The liquid phase on the exterior of the polymer particles is agitated continuously to suspend the particles and to mix the liquid phase. Suppose that the rate-limiting step of the process would be the diffusion of the monomer from the resin particles. The polymer is a glass at the leaching temperature, but the nonsolvent is thought to swell it slightly. If the leaching process will have an average residence time of one minute per stage, what will be the typical approach to equilibrium on each stage?

b. Solution. This problem is an unsteady-state diffusion problem with a step change in the concentration of the solute at the surface of the porous particle in a given stage. As a first approximation, the solute will be assumed to have a uniform concentration throughout the solid. As shown in Fig. 12, there is a wide range of possible diffusion coefficient values for the solute. If this were a rubber, we could use 1×10^{-8} cm²/sec as an initial estimate for the solute's diffusivity. This value is typical for moderately sized solutes diffusing in polymers above their glass transition temperature. The glassy polymer in Fig. 12 is 50°C below its glass transition temperature. It is likely that we would raise the temperature of the leaching system so that it was just below the glass transition temperature. Under these conditions, we might be able to get diffusivities on the order of 10^{-9} cm²/sec.

Figure 14 for spheres can be used to estimate the fraction remaining inside the particulate. The Fourier number for diffusion is

$$\frac{Dt}{R^2} = \frac{1 \times 10^{-9} \dfrac{\text{cm}^2}{\text{sec}} \, 60 \text{ sec}}{(5 \times 10^{-4} \text{ cm})^2} = 0.24$$

From the long-time formula in Table 6, the fractional amount of monomer remaining (relative to the step change) is 0.057, and so we may reasonably assume that each stage should be close to equilibrium.

Changing the particle diameter changes the Fourier number by the square of the ratio of the new to the old diameter. For example, if the particle were a factor of two larger, the Fourier number would be smaller by a factor of four, and the fractional amount of solute remaining would be 40%.

We would consider several other factors in our analysis and design of this system. A variation in primary particle sizes will affect the leaching process and its approach to equilibrium. The polymer morphology plays an important role in its ability to be treated. It is likely that the system temperature would be set as near as practical to the glass transition temperature without causing polymer agglomeration. Some polymers, for example, suspension poly(vinyl chloride) (PVC), are stripped at temperatures above T_g for short periods of time. Crystallinity in a polymer allows greater freedom to exceed T_g since the crystallites will provide some mechanical strength and lessen the tendency for particles to agglomerate.

B. Latex Stripping

The removal of monomer from latex and suspension resins by causing it to boil into the gas phase is a common operation in the production of these materials. There are a variety of ways in which this can be done, ranging from using a distillation column to remove monomer from suspension resins to using the polymerization reactor for devolatilizing latex. In this section we describe the steam stripping of a poly(vinyl chloride) latex (PVC) in a continuously stirred tank reactor (CSTR). There can be several rate-limiting steps to this process, including diffusion from the solid and removal of monomer from the liquid phase. Both of these possibilities can be treated by conventional chemical engineering techniques.

1. Mechanism for Separation

Figure 16 shows a CSTR system for stripping monomer from a PVC latex using steam. There are several potential rate-limiting steps: the diffusion of monomer from the polymer, the mass transfer of monomer from the polymer surface to the bulk liquid, and the mass transfer of monomer from the bulk liquid to the gas phase. For most latices, the diffusion of monomer from the polymer will not be the rate-limiting step because the particles are so small. This can be checked for any particular case by calculating the Fourier number for diffusion and using Fig. 13. As long as the liquid phase is well mixed, mass transfer from the particles to the bulk phase should be rapid. It often occurs that the rate-limiting step is moving the solute from the liquid phase to the gas phase. Alternatives to steam stripping include the use of air or nitrogen gas. These choices have several potential disadvantages including formation of explosive vapor mixtures, introduction of a polymer oxidant (the oxygen in the air), more difficult condensation of the overhead vapor stream, and lack of adequate energy input for vaporization.

2. Material and Energy Balances

Conventional equipment can be used for polymer stripping. Suspension resins with densities higher than 1 can be stripped using distillation columns. Latices usually require gentler processing since they tend to agglomerate, and their high surfactant content leads to foaming. Figure 16 shows one solution to these problems: the use of steam to preheat the latex followed by gentle agitation in the stirred tank to enhance liquid–gas mass transfer. Energy balances must be included in these analyses since the solute and the steam (a nonsolvent) are being vaporized.

3. Phase Equilibrium Models and Data

Phase equilibrium predictions are needed for the solute partitioning to the gas phase as well as for the solute partitioning between the polymer and the liquid phase. For typical low-pressure stripping conditions, the gas phase is essentially ideal. In general, the liquid–vapor equilibria would be described by activity coefficients. However, under stripping conditions, the liquid-phase composition of the solute would be low, and the equilibrium might be characterized by an infinite dilution activity coefficient or a Henry's law coefficient.

4. *Mass Transfer Models and Data*

For most latices, it is unlikely that the diffusion rate of monomer from the polymer into the liquid phase is rate-limiting. More often the mass transfer within the liquid phase or to the gas phase is rate-limiting (Englund, 1981). The generation of many bubbles in the liquid phase causes foaming. Stable foams usually lead to liquid-to-gas phase mass transfer limitations. However, some latex foam bubbles form and burst rather quickly. As with extruder devolatilization, rapidly forming and bursting bubbles may enhance the mass transfer rate to the vapor phase. In either case, the rate of transfer is enhanced by using an agitator in a tank with a low fill level so that the gas–liquid surface area is increased. Overly aggressive agitation usually leads to excessive foaming, particularly under vacuum conditions.

5. *Physical Properties*

In addition to the normal physical property data needed for energy and material balances, there are special constraints on most stripping problems. Information is needed on the time the latex can spend at stripping conditions without coagulating, and on the level of vacuum that induces foaming.

In the following example, we examine a model for the loss of monomer from PVC latex particles assuming that the concentration of monomer in the water phase is zero. This model will establish whether or not the diffusion of monomer from the solid into the liquid phase is the rate-limiting step.

6. *Example 2.   Latex Devolatilization in a CSTR*

a.   Problem statement.   Vinyl chloride monomer (VCM) is being removed from a 30 wt % solids PVC latex in a continuously stirred tank. The latex contains 20 mg VCM/g latex, and it cannot be heated above 74°C without excessive coagulation. Stripping at 68°C should give enough of a safety margin. Excessive foaming of the latex occurs if it is directly sparged with a gas or if the vacuum is too high. This stripper is operated by injecting steam directly into the feed stream to the CSTR (see Fig. 16). The VCM is to be reduced an order of magnitude, to 2000 ppm in this process, but the unit is not operating to this performance. Determine whether the stripping process is rate-limited by slow diffusion from the latex particles.

b.   Solution.   Physical property data for this problem are given in Table 10, and Fig. 17 is a histogram of the particle size distribution of the latex. Since the latex particles are essentially spherical, we can use the one-dimensional sorption equation for spheres given in Table 3 for each particle size and then sum the residual monomer concentration over all sizes. The material balance for the CSTR shows that the residence time is

Table 10   Physical Property Data for Latex Devolatilization Problem

Parameter	Value	Source
Henry's law coefficient for VCM in PVC	$S = \dfrac{88P_i}{P_i^0} \left(\dfrac{\text{mg}}{\text{g}}\right);$ $\chi = 0.98$	Berens, 1975
Henry's law coefficient for VCM in water, $30 < T < 50°C$	$S = \dfrac{8.6P_i}{P_i^0} \left(\dfrac{\text{mg}}{\text{g}}\right)$	Berens, 1975
Density of PVC	$\rho_{\text{PVC}} \approx 1.4 \left(\dfrac{\text{g}}{\text{cm}^3}\right)$	
VCM vapor pressure	$P_{\text{VCM}}^0 = 1.952 \times 10^7 \cdot \exp\left(\dfrac{-2618}{T}\right)$ (mm Hg)	Berens, 1975
VCM diffusivity in PVC	$D = 3.7 \exp\left(\dfrac{-17{,}000 \text{ cal/mol}}{R_g T}\right)\left(\dfrac{\text{cm}^2}{\text{sec}}\right)$	Berens, 1977

10 min. It is convenient to use an equation for the fraction of solute remaining summed over all the particle sizes. The equation is (Berens and Huvard, 1981)

$$\frac{M_t}{M_\infty} = \sum_{i=1}^{I} w_i \left[1 - \frac{6}{\pi^2} \sum_{n=1}^{\infty} \frac{1}{n^2} \exp\left(\frac{-(n\pi)^2 Dt}{R_i^2} \right) \right]$$

where the weight fraction distribution must be normalized. Figure 18 shows the desorption curve for the latex particles calculated for a temperature of 57°C. Even with the high fraction of large particles, more than two orders of magnitude ($>99\%$) of the VCM can be removed in about 5 min when the concentration of VCM in the liquid phase is zero. Since the residence time in the vessel is 10 min, it is clear that the process is not diffusion-limited based on the PVC particles sizes.

The VCM in the water should be close to equilibrium with that in the exiting vapor stream if the stripper is operating at maximum efficiency. A complete analysis of the operation of the CSTR indicates the rate-limiting step in the PVC latex process is the transfer of monomer from the liquid phase to the vapor phase. Maximum mass transfer to the vapor is achieved by driving the vapor phase partial pressure of VCM down. High steam sparging rates move the process in this direction.

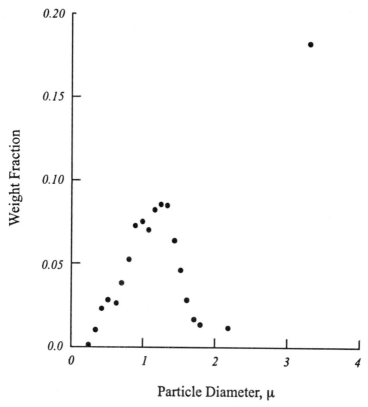

Fig. 17   Example PVC latex particle size distribution.

C. Fiber Drying

There are several devolatilization problems involving solvent removal from spun fibers. The general problem has a number of fibers being rapidly dried in air. The air stream velocity is controlled and, in some cases, is directly transverse to the fiber direction. The temperature of the fibers must be below the melting point to prevent sintering and agglomeration.

1. Mechanism of Separation

In these problems, the rate-limiting steps can include diffusion of the solvent from the cylindrical fibers or convective mass transfer of the solvent from the fiber surface into the bulk gas phase. It is rarely the case that so much solvent would be removed from the

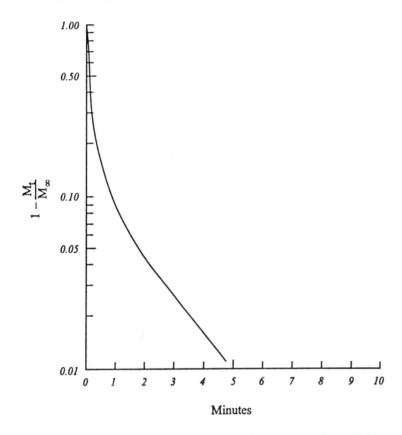

Fig. 18 Calculated desorption of vinyl chloride monomer from PVC latex particles (size distribution of Fig. 17) at 57°C. Curve assumes zero external monomer concentration and zero mass transfer resistance.

fibers that heat transfer to evaporate the solvent would be rate-limiting.

2. Material and Energy Balances

Energy balances are needed around the drying system to help determine the mechanism(s) for vaporization. Fibers containing only several percent solvent probably will provide the energy for vaporization adiabatically, while fibers containing larger amounts may need heat to be transferred from the gas phase. In the former case, the energy balances may not be needed for a satisfactory solution.

3. *Phase Equilibrium Models and Data*

The Flory–Huggins model is adequate to describe the polymer phase equilibrium for many fiber drying problems. Only the solid and gas phases are present, and the partial pressure of the solvent in the exiting gas gives a direct measure of its activity.

4. *Mass Transfer Models and Data*

The mass transfer of solvent within the fibers can be modeled using the equation in Table 3 for cylinders. A mass transfer coefficient is included to describe the flux of solvent from the fiber surface to the bulk gas phase. This coefficient can be a function of the relative velocity of the fiber with respect to the gas, and of the local density of fibers.

5. *Physical Properties*

The fiber surface should not become tacky at any time during the drying process. Therefore, the fiber surface must be kept below either its T_g or T_m. Models that predict solid–liquid transitions for the system can be very valuable for determining workable processing conditions. When air is used, organic vapor concentrations should be kept below the lower explosion limit. Because of this, gas-phase solute concentrations are low (less than about 3%) and solute recovery by condensation is usually not economical. Instead, carbon adsorption or related abatement technology is usually needed.

6. *Example 3. Counter-Current Stripping of Solvent from a Fiber Sheet*

a. Problem Statement. Flash-spun polyethylene sheet is produced by dissolving polyethylene in a solvent at high temperature and pressure, and spinning a web of polymer fibers through a set of tiny orifices. Webs of fibers are cast down into wide sheets, which are passed through consolidation rollers and then into a special chamber where residual solvent is removed. The stripping chamber is shown in Fig. 19. The gas phase is assumed to be plug flow. A boundary layer of stagnant air can exist at the surface of the fiber sheet, and it would reduce the mass transfer rate of solvent leaving the fibers. We wish to determine how much solvent remains in the fibers when they exit the stripper.

b. Solution. We will write the transport equations for this problem and then use appropriate dimensionless numbers to determine which transport mechanism should be rate-limiting. The material balance on the solvent in the fibers can be converted from a time derivative to a space derivative describing the position of the fibers in the stripper:

$$V \frac{\partial C(r, t)}{\partial x} = D \left[\frac{\partial^2 C(r, t)}{\partial r^2} + \frac{1}{r} \frac{\partial C(r, t)}{\partial r} \right]$$

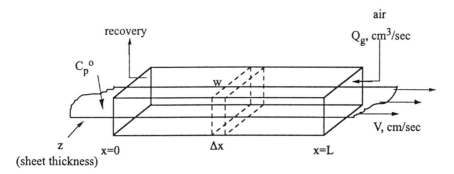

Fig. 19 Sketch of a counter-current stripper for removal of solvent from a fiber sheet. Air may be used with nonflammable solvents; otherwise, nitrogen or carbon dioxide should be used.

where V is the sheet velocity, $C(r, t)$ is the concentration of the solvent in the fibers, r is the radial distance into an individual fiber, x is the distance down the stripper (or sheet), and D is the diffusivity of the solvent in the fiber. The solvent concentration is initially uniform across the fibers as they enter the stripper. The flux is zero through the centerline of the fibers, and the flux at the fiber surface is given by the product of a mass transfer coefficient and the difference between the fiber surface and the vapor concentrations:

$$-D\left(\frac{\partial C(r, t)}{\partial r}\right)_{r=R} = k_m\left(\frac{C_s}{k} - C_{air}\right)$$

where k_m is a mass transfer coefficient for the sheet, k is a Henry's law coefficient for the partition of the solvent from the fiber surface, and C_s and C_{air} are the solvent concentrations at the fiber surface and bulk air, respectively. The concentration at the fiber surface is divided by the Henry's law coefficient to obtain the correct thermodynamic driving force for transfer to the bulk phase. The foregoing equation and boundary conditions describe the flux of solvent from the fibers.

A material balance on the gas phase gives the following equation:

$$Q_g \frac{dC_{air}}{dx} - 2\pi r\left(D \frac{\partial C(r, t)}{\partial r}\right)_{R, x} \cdot N = 0$$

where N is the number of fibers in a control volume and Q_g is the flow rate

Table 11 Data for Example 3

Parameter	Value
D, diffusivity of solvent in the fiber	4×10^{-9} cm^2/sec
L, length of the stripper	500 cm
R, fiber radius	1×10^{-4} cm
V, sheet velocity	115 cm/sec (75 yd/min)
Q_g, air volumetric flow rate	85,000 cm^3/sec
A_s, fiber specific surface area	250,000 cm^2/g (25 m^2/g)
B, sheet bulk density	0.3524 g/cm^3
z, sheet thickness	0.0203 cm (8 mil)
w, sheet width	71.1 cm (28 in.)
k, Henry's law constant	3.5
k_m, mass transfer coefficient	~ 10 cm/sec

of air. We can convert this equation to one that uses properties of the sheet:

$$Q_g \frac{dC_{air}}{dx} - D A_s B z w \left(\frac{\partial C(r, t)}{\partial r} \right)_{R, x} = 0$$

where A_s is the fiber specific surface area, B is the sheet bulk density, z is the sheet thickness, and w is the sheet width. The data for solving this problem are given in Table 11.

The mass transfer coefficient can be estimated as follows. The diffusivity of a gas is on the order of 0.1 cm^2/sec. Gas-phase boundary layers are on the order of 0.01 cm thick (Cussler, 1984), so that the mass transfer coefficient is on the order of 10 cm/sec (the ratio of the diffusivity divided by the film thickness). This estimate will let us evaluate dimensionless numbers in order to determine the rate-limiting step.

Characteristic time of the process. The residence time of the sheet in the stripper is simply L/V. The characteristic time for diffusion from the fiber is given by the Fourier number for diffusion, Dt/R^2. The ratio of these two numbers should tell us whether the process is long or short relative to the diffusion time:

$$\gamma = \frac{\tau_{sheet}}{\tau_{diffusion}} = \frac{L/V}{R^2/D} = \frac{LD}{R^2 V}$$

$$= \frac{500 \cdot 4 \cdot 10^{-9}}{(1 \cdot 10^{-4})^2 \cdot 115} \approx 1.8$$

Thus, the sheet will be in the stripper much longer than the characteristic time for diffusion, and so we should be able to remove a significant amount of solvent.

Sherwood number. The Sherwood number compares the mass transfer resistance of the boundary layer with that of the fiber:

$$Sh = \frac{k_m}{D/R} = \frac{10 \text{ cm/sec}}{4 \cdot 10^{-9-9}/1 \cdot 10^{-4}}$$

$$= 2.5 \cdot 10^5$$

The Sherwood number is quite large. Accordingly, diffusion from the fibers will be the rate-limiting resistance. Also, the unsteady-state diffusion equation for solvent loss from the fiber will suffice to model the process; mass transfer resistance at the fiber–gas boundary can be neglected. This can be done using finite difference methods, finite element methods, or by the Method of Lines (Sincovec and Madsen, 1975), which transforms the partial differential equation to a set of ordinary differential equations. A variety of computational tools is available to solve systems of ordinary differential equations, and so the latter choice is convenient; it was used to obtain the average fiber concentrations shown in Fig. 20.

D. Particulate Drying

Particulate drying is encountered in a number of polymer recovery systems. There are two general classes of application: the removal of water from porous solid particles originally conveyed in a water slurry (see Example 1) or the removal of sorbed water from hygroscopic polymer pellets. Water removal from such pellets may be a necessary part of the manufacturing process, or, as with many step growth polymers, it may be necessary before the pellets are used after shipping and storage. Each of these processes is discussed in this section, and an example problem is given on removing moisture from stored resins.

1. Problems Caused by Moisture in Polymers

A number of melt processing operations are susceptible to the formation of bubbles or "blisters" caused by liquids retained in the polymer. Extruded profiles with small cross-sectional areas are particularly sensitive to this problem. However, a related problem can occur in injection molding, and this sometimes leads to excessive shrinkage of parts on cooling. When the pressure in the melt drops rapidly as the polymer passes through and out of the die, sorbed water can vaporize and form defects on or just under the solid surface.

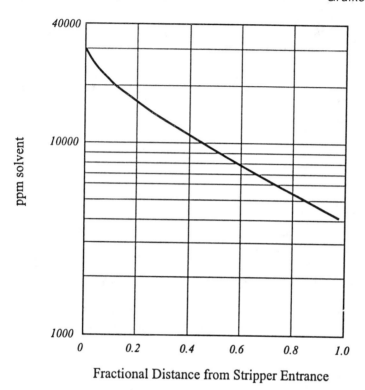

Fig. 20   Average solvent content of the fibers as a function of fractional distance from the stripper entrance.

        Residual water also participates in depolymerization reactions with step polymers and can lead to molecular weight reduction during melt processing. The problem is most acute with nylons and polyesters. The degree of depolymerization depends on the moisture content, temperature and pressure history, and the phase equilibrium of the water–polymer pair. Both problems are corrected in practice by drying polymer pellets to very low water content prior to melt processing.

2.   Minimizing Reversible Reactions in the Melt

        When step polymers are melt processed, the temperatures usually approach those used in second-stage polymerization reactions where high-molecular-weight polymers are formed. If water is present, hydrolysis can occur, and the molecular weight of the polymer can decrease rapidly. The amide group of nylons is especially susceptible to acid- and base-catalyzed hydrolysis. Polycarbonates are fairly stable to hydrolysis as solids at moderate temperatures but are easily attacked at melt processing temperatures ($>300°C$) by even trace amounts of water (<100 ppm).

Depolymerization by water may be autocatalytic since polar end groups of the chains can act as catalysts. Poly(ether urethanes) are susceptible to hydrolysis, but only the urethane linkage hydrolyzes and severe conditions are needed. Poly(ester urethanes) are much more sensitive to water since both the urethane and ester groups can be hydrolyzed, and acidic products can catalyze further reaction.

Moisture levels resulting in depolymerization are usually determined experimentally. On the other hand, several kinetic models for the commercial step polymers are available in the literature and often include coefficients for the reverse reaction of polymer with water, for example, Wajge et al. (1994). These can be used to calculate the expected depolymerization at processing conditions for various levels of water.

3. Mechanism of Separation

The removal of liquid and bound water from porous solids is a traditional problem in unit operation textbooks (Geankoplis, 1993; McCabe et al., 1993). There are several drying mechanisms that can be rate-limiting, as shown in Fig. 21.

During the constant rate period, drying occurs as liquid vaporizes at the liquid–gas interface. The heat duty of the dryer is large at this point since almost all the energy needed for liquid vaporization is supplied by the drying gas (usually air). The surface area to unit volume of most commercial dryers is modest, and the process is essentially adiabatic. Moisture movement is rapid enough to replenish the liquid surface, and drying is controlled by heat transfer. The temperature of the solid approaches the wet bulb temperature of the gas. When the heat is supplied by the gas, a dynamic equilibrium is established between the rate of heat transfer and the rate of heat loss through evaporation. The change in weight of the solid is modeled as

$$\frac{dW}{dt} = \frac{hA\,\Delta T}{\Delta H_v} = k_m\,\Delta P \tag{21}$$

where the left-hand side is the drying rate, g liquid/hour, h is the heat transfer coefficient, cal/m^2-hour-°C, k_m is the mass transfer coefficient, g/hour-atm, ΔT is the difference between the dry bulb temperature of the gas and its surface temperature, ΔH_v is the heat of vaporization, cal/g, and ΔP is the difference between the liquid vapor pressure at the surface temperature and

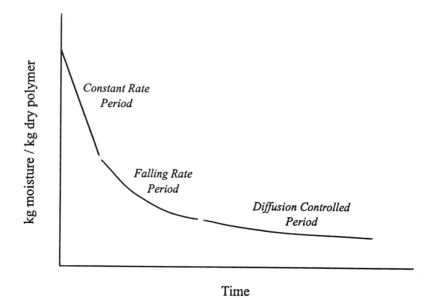

Fig. 21 Unsteady-state drying of a polymer solid containing surface water. (After Wentz and Thygeson, 1979.)

the liquid partial pressure in the gas phase, atm. When heat is supplied only by convection, the surface temperature is the wet bulb temperature. When radiation also supplies heat, the heat transfer coefficient contains the sum of the convective and radiative heat transfer coefficients, and the surface temperature may be higher than the wet bulb temperature. Finally, when heat is also supplied by conduction to the solid (as in vacuum drying), the surface temperature may approach the boiling point of the liquid at the system temperature. Appropriate heat transfer correlations are available in Perry's Handbook.

When the surface moisture has been depleted, the rate-limiting step is the movement of water from the water interface within the pore to the particle surface. Moisture movement can occur by diffusion, capillary transport, or a combination of the two. Perry's Handbook has a discussion of this phenomenon.

At lower water contents, moisture vaporizes in the pores and diffuses to the surface. With polymers, moisture may also need to be removed from the solid itself. In general, particulate drying is quite complex, and dryers usually are designed using experimental drying curves rather than detailed mechanistic models.

4. Material and Energy Balances

Material and energy balance models are available for a variety of dryer configurations (Wentz and Thygeson, 1979). Quite often, designs are based on experimental drying curves to inherently account for rate-limiting behaviors.

5. Phase Equilibrium Models and Data

Bubbles will form in a polymer melt if a volatile component in the polymer exerts a vapor pressure above that of the system. Thus, the vapor pressure of residual water becomes a problem as the melt pressure approaches atmospheric. Consider the profile extrusion of a nylon saturated with water. The typical process conditions might be 225°C and 100 atm in the extruder. At this temperature, water exerts a vapor pressure of about 25 atm. Therefore, even if the sorbed water in the polymer would phase separate at extruder conditions, it would not form bubbles because the system pressure is above its equilibrium vapor pressure.

However, the polymer melt experiences a drop in pressure as it moves through the profile die, and it exits to atmospheric pressure. If the equilibrium vapor pressure of water is above the system pressure at any time, then bubbles can form. This can be predicted by applying a phase equilibria model to the problem, as shown in Example 4. The Flory–Huggins equation is convenient since the values of the interaction parameter can be estimated from water sorption data, and it is easy to scale with temperature.

Equilibrium water uptake by polymers varies widely. Table 12 shows some values for polymers equilibrated at 70°F with air at 65% relative humidity. Polar polymers tend to absorb significant amounts of water, while nonpolar polymers sorb low amounts.

6. Mass Transfer Models and Data

The calculation of the Sherwood number can be used to evaluate the relative resistance of convective to diffusive mass transfer, as demonstrated in the previous problem. For problems in which there is no surface water, the process will likely be controlled by diffusion through the polymer phase. The composition in the gas phase may vary through the equipment, and numerical solutions to the partial differential equations may be needed for problems requiring precise engineering.

Once the target water level has been determined, the drying time can be estimated by applying Newman's method for irregularly shaped solids. Most resin pellets can be approximated as spheres, short cylinders, or boxlike solids. At conditions typical of most counter-current pellet dryers, water activity in the vapor phase at the surface of the pellets is low and can be taken as constant during the drying process. If the pellets move in plug

Table 12 Equilibrium Water Content of Polymer
Exposed to Air at 70°F and 65% Relative Humidity

Polymer	Water absorbed (wt %)
Rayon	11–13
Spandex	1.1–1.3
Polyester	0.4–0.8
Polytetrafluoroethylene	<0.005
Nylon 6	2.8–5.0
Nylon 6,6	4.2–4.5
Aramids: Kevlar	4.5–7.0
Nomex	6.5
Polyethylene	0.005
Polypropylene	0.01
Wool	11–17
Polyurethanes	0.1–0.4

flow through this equipment, then the boundary and initial conditions are such that the solutions plotted in Fig. 13 apply. This figure, coupled with a phase equilibria model such as the Flory–Huggins equation, can be used to estimate the amount of water remaining in the pellets as a function of drying conditions.

7. Physical Properties

The physical properties of the polymer, air, and water needed to solve drying problems include the polymer geometry and morphology (porosity, pore structure, and sieve size distribution or shape), the thermal transitions of the polymer, the saturated vapor pressure of water at the drying conditions, and coefficients of a thermodynamic model such as Flory–Huggins. In addition, the conditions of the drying air (flow rate, relative humidity, and temperature) must be known.

*8. Example 4. Setting a Polymer Drying Target: Estimating
the Water Activity that Will Prevent Bubble Formation*

a. Problem. Nylon 6 containing 3 wt % water will be used in an profile extrusion process at 250°C. Estimate the maximum amount of water that can be tolerated in the nylon pellets in order to prevent bubble formation.

b. Solution. We need to apply a phase equilibria model to estimate the vapor pressure of water in the melt. The Flory–Huggins equation is convenient since it is easily applied to find the value of the interaction parameter from the information in Table 12. The relative humidity can be converted directly to water activity (0.65). Since the density of Nylon 6 is about 1.1 g/cm^3, the volume fraction of water in the polymer is 0.033. Substituting these values into the Flory–Huggins equation gives

$$\ln(0.65) = \ln(0.033) + (1 - 0.033) + \chi(1 - 0.033)^2$$

$$\chi = \frac{\ln(0.65) - \ln(0.033) - (1 - 0.033)}{(1 - 0.033)^2}$$

$$= 2.15$$

This value of the interaction parameter is typical of solvents that are partially soluble in polymers. The actual value of the interaction parameter in the melt will be a function of temperature and pressure. One model for the interaction parameter (Grulke, 1989) is

$$\chi = 0.34 + \frac{V_1}{R_g T}(\delta_1 - \delta_2)^2$$

Applying this equation to our problem gives $\chi = 1.56$ at the extrusion temperature. The saturated vapor pressure of water at 250°C is 40 atm. When the part exits the die, the external pressure will be 1 atm, but the temperature still will be high. If the equilibrium vapor pressure of water in the polymer at these conditions is 1 atm or higher, then bubbles of water vapor could form. We use the Flory–Huggins equation to solve for the volume fraction of water in the polymer at which this could happen:

$$\ln\left(\frac{1}{40}\right) = \ln(\phi_1) + (1 - \phi_1) + \chi(1 - \phi_1)^2$$

The solution is 0.0009 wt fraction water (900 ppm) if χ is 2.15 and 0.0018 (1800 ppm) if χ is 1.56. The water content should be reduced by at least an order of magnitude to prevent bubble formation.

NOMENCLATURE

a_i	activity of component i, equal to P_i/P_i^0 for ideal gas phases
A_s	fiber specific surface area
B	sheet bulk density
b	Langmuir isotherm coefficient, specific sorption model
$C(x, t)$ or $C(r, t)$	solute concentration
C_D	nonspecific sorption concentration, Henry's law model
C_H	specific sorption concentration

C'_H Langmuir isotherm coefficient, specific sorption model
C_{F-H} nonspecific sorption concentration, Flory–Huggins model
D diffusion coefficient
D_0 frequency factor, diffusion coefficient
E_d activation energy, diffusion
E_i fractional amount remaining (Fig. 13)
h heat transfer coefficient
k Henry's law coefficient (Eq. 2), g solvent/g polymer-atm
k_m mass transfer coefficient from sheet to gas phase
k_0 preexponential constant for k
M_{state} mass of gas sorbed at indicated state or time
M_w^{gas} molecular weight of the gas
N number of fibers
P_i partial pressure of i in the gas phase, atm
P_i^0 saturated vapor pressure of i, atm
P_{state} total pressure of the gas at indicated state or time
p_0 preexponential factor for vapor pressure
Q_g gas flow rate
R sphere or fiber radius
r radial distance
R_g ideal gas constant
S_i solubility (g penetrant/g polymer)
Sh   Sherwood number
T temperature, K or °C
t time
T_g glass transition temperature
T_m melting temperature
V system volume (Eqn. 1)
V_g volume of glassy polymer
V_{liq} volume of liquid (molten) polymer
v_p polymer volume (Eqn. 1)
W sample weight
w sheet width
x ratio of the polymer to solvent molar volume
z film thickness, one-dimensional diffusion problem

Greek Letters

α_g thermal expansion coefficient, glassy polymer
α_{liq} thermal expansion coefficient, liquid polymer
γ characteristic process time
δ_i solubility parameter of i; 1 = solvent, 2 = polymer

ΔH_c enthalpy (heat) of condensation
ΔH_m enthalpy (heat) of mixing
ΔH_s enthalpy (heat) of solution
ΔH_v enthalpy (heat) of vaporization
ΔP vapor pressure difference
χ interaction parameter
ϕ_i volume fraction of component i; 1 = solvent, 2 = polymer
ϕ_p volume of the polymer (see Fig. 1)
ρ_i density of component i; 1 = solvent, 2 = polymer
τ_i residence time of process i

REFERENCES

Barrer, R. M. (1939). *Trans. Faraday Soc., 35*: 628.

Berens, A. R. (1975). *Angewandte Makromolekulare Chemie, 27*: 97.

Berens, A. R. (1977). *Polymer, 18*: 697.

Berens, A. R., and Hopfenberg, H. B. (1982). *J. Membrane Sci., 10*: 283.

Berens, A. R., and Huvard, G. S. (1981). *J. Dispersion Sci. Technol., 2*: 359.

Berens, A. R., and Huvard, G. S. (1989). *Supercritical Fluid Science and Technology*, ACS Symposium Series 406 (K. P. Johnston and J. M. L. Penninger, eds.), American Chemical Society, Washington, D.C.

Carslaw, H. S., and Yaeger, J. C. (1959). *Conduction of Heat in Solids*, 2nd ed., Clarendon Press, Oxford, U.K.

Crank, J. (1975). *The Mathematics of Diffusion*, 2nd ed., Oxford University Press (Clarendon), London, U.K.

Cussler, E. L. (1984). *Diffusion: Mass Transfer in Fluid Systems*, Cambridge University Press, New York.

Danner, R. P., and High, M. S. (1993). *Handbook of Polymer Solution Thermodynamics*, Design Institute for Physical Property Data, American Institute of Chemical Engineers, New York.

Daynes, H. A. (1920). *Proc. Royal Soc. London, Ser. A, 97*: 286.

Englund, S. M. (1981). *Chem. Eng. Progr.*, August: 55.

Felder, R. M. (1978). *J. Membrane Sci., 3*: 15.

Felder, R. M., and Huvard, G. S. (1980). *Methods of Experimental Physics*, Vol. 16 (R. A. Fava, ed.), Academic Press, New York.

Fleming, G. K., and Koros, W. J. (1986). *Macromolecules, 19*(8): 2285.

Geankoplis, C. J. (1993). *Transport Processes and Unit Operations*, 3rd ed., Prentice Hall, Englewood Cliffs, New Jersey.

Huvard, G. S., Stannett, V. T., Koros, W. J., and Hopfenberg, H. B. (1980). *J. Membrane Sci., 6*: 185.

Ju, S. T., "Oscillatory and transient sorption studies of diffusion in poly(vinyl acetate)," Ph.D. Dissertation, Pennsylvania State University, May 1991.

Kamiya, Y., Hirose, T., Mizoguchi, K., and Naito, Y. (1986). *J. Polym. Sci., Polym. Phys. Ed., 24*: 1525.

Koros, W. J. (1995). Unpublished correlations. Included with permission by private communication.

Koros, W. J., Paul, D. R., and Huvard, G. S. (1979). *Polymer, 20*: 956.

McBain, J. W., and Bakr, A. M. (1926). *J. Am. Chem. Soc., 48*: 690.

McCabe, W. L., Smith, J. C., and Harriott, P. (1993). *Unit Operations of Chemical Engineering*, 5th ed., McGraw-Hill, New York.

Michaels, A. S., and Bixler, H. J. (1961). *J. Poly. Sci., 50*: 413.

Newman, A. B. (1931). *Trans. AIChE, 27*: 203, 310.

Rodes, C. E., Felder, R. M., and Ferrell, J. K. (1973). *Environ. Sci. Technol., 7*: 545.

Rosen, B. (1959). *J. Poly. Sci., 35*: 225.

Sandler, S. I. (1977). *Chemical and Engineering Thermodynamics*, Wiley, New York.

Sincovec, R., and Madsen, N. (1975). *ACM Trans. Mathematical Software, 1*(3): 232.

Stern, S. A., Gareis, P. J., Sinclair, T. F., and Mohr, P. H. (1963). *J. Appl. Poly. Sci., 7*: 2035.

Tikhomirov, B. P., Hopfenberg, H. B., Stannett, V. T., and William, J. L. (1968). *Makromol. Chem., 118*: 177.

Toi, K., Morei, G., and Paul, D. R. (1982). *J. Appl. Polym. Sci., 27*: 2997.

van Amerongen, G. J. (1964). *Rubber Chem. Tech., 37*: 1065.

van Krevelen, D. W. (1990). *Properties of Polymers*, 3rd ed., Elsevier Scientific, New York.

Wajge, R., Rao, S., and Gupta, S. (1994). *Polymer, 35*(17): 3722.

Wentz, T. H., and Thygeson, J. R. (1979). *Handbook of Separation Techniques for Chemical Engineers*, Section 4.10 (P. A. Schweitzer, ed.), McGraw-Hill, New York.

Wossnessensky, S., and Dubinkow, L. M. (1936). *Kolloid-Z., 74*: 183.

18

The Future of Solvents in the Manufacture of Polymers

E. Bruce Nauman and Timothy J. Cavanaugh

Rensselaer Polytechnic Institute, Troy, New York

This chapter addresses a simple question: Will solvents continue to be used in the manufacture of plastics? To some extreme environmentalists, plastics themselves are bad, let alone in combination with solvents. Joe Thornton, a spokesman for Greenpeace, has said "Next, we'll go after plastic [*sic*] and solvents" (Amato, 1993).

A campaign against plastics and solvents is unlikely to succeed. The world will not willingly return to the low population densities and primitive distribution systems that would be necessary to eliminate plastics from

packaging and other high-volume applications. Plastics will remain, but it is less clear that solvents will continue to be used in their manufacture.
Three concerns about the use of solvents are

1. The effect of emissions on the general environment and on occupational health
2. The health effects of residual solvents in food contact applications
3. The costs of energy and of solvents

The first two concerns are matters of public policy. Many regulations have been promulgated by government agencies such as EPA (Environmental Protection Agency), OSHA (Occupational Safety and Health Administration), and FDA (Food and Drug Administration). The final concern, at least at the moment, is one of economics rather than of government regulations.

Together, these three concerns provide large incentives to eliminate solvents, and the success of gas-phase processes for the polyolefins may suggest that this elimination is possible and perhaps even imminent. Thus, it may seem that devolatilization of polymers is a technology that will soon be unneeded. The actual situation is more complex. We predict that solvents will continue to be widely used in the manufacture of plastics for the foreseeable future, at least until 2050.

The arguments behind this prediction can be summarized as follows:
1. Monomers are small, energetic molecules that must be confined for reasons of environmental and occupational safety. Essentially zero emissions will be required throughout the chemical industry. The cost of confinement is largely independent of the monomer concentration and of the possible presence of a solvent. In essence, monomer confinement will bear the cost of solvent confinement.

2. Packaging plastics made by solvent processes have been accepted as safe for many years. This acceptance occurred when there were much higher levels of residual monomers and solvents than is current practice. Although one can always postulate a scare of the Alar variety (Seltzer, 1989), it seems reasonable to suppose that any substantial health risk would have been uncovered. Insubstantial health risks are routinely publicized but are soon contradicted or forgotten in the face of overwhelming utility and need.

3. The theoretical minimum energy needed to separate polymers from small molecules is extremely low. Practical devolatilization schemes require much more energy that the thermodynamic minimum, but the total cost remains a small fraction of the selling price of the finished polymers. Historically, solvent losses were a major cost; but these become quite small for modern processes with near-zero emissions.

The following sections treat these points in greater detail. Some of the concepts have been discussed elsewhere (Cavanaugh and Nauman, 1995).

I. ENVIRONMENTAL AND OCCUPATIONAL REGULATIONS

The first federal regulations of water pollution were passed in 1899 and have been greatly tightened in the succeeding decades. These laws had major impact on industries such as pulp and paper, leather tanning, and metal plating. The petrochemical industry has been less affected. Aside from the occasionally disastrous spills, water pollution currently originating from the manufacture of chemicals and plastics is no longer considered a serious issue of health and environmental quality. The continuing cost of water pollution avoidance and abatement has been absorbed into the cost of the products; and by and large, these products remain competitive worldwide.

Air pollution poses a more difficult problem for the petrochemical industry. A distinction has been made between occupational exposure affecting workers and atmospheric exposure affecting the general population. The creation of OSHA in 1971 marked the beginning of a long-term effort to limit exposures in the workplace. Permissible exposure limits (PELs) have been specified by OSHA to ensure safety against the detrimental effects of exposure over a working lifetime. The PELs for some compounds of interest in the plastics industry are shown in Table 1.

The PELs are limits for eight-hour, time-weighted average concentrations. A similar set of limits has been developed by the American Conference of Governmental Industrial Hygienists (ACGIH). Known as threshold limit values (TLVs), these concentrations are intended to represent limits below which there are no adverse health effects. The PELs and TLVs are sometimes different due to details in the regulatory process. Still other limits may apply to instantaneous exposure.

The imposition of PELs has required substantial changes in coatings technology. Web-coating applications as used for the manufacture of photographic film, sandpaper, and water-repellent fabrics are very difficult to operate in a closed environment. Materials injurious at extremely low levels, for example, toluene diisocynate, have been effectively precluded from coating applications unless the workers are supplied with external air. As a practical matter, most such applications now use water-borne or powder coatings or less volatile reactants (e.g., pre-polymers or higher-molecular-weight diisocynates).

Polymer manufacturing operations have been less affected. Such operations are much easier to confine, and there are few substitutes for basic building blocks such as benzene and ethylene. Benzene has a very low PEL

Table 1 Exposure Limits for Common Monomers and Solvents

Compound	PEL in ppm[a] (OSHA)	TLV in ppm[b] (ACGIH)
Acetone	1000	750
Acrylonitrile	2	2
Benzene	1	10
Betachlorprene	25	25
1,3-Butadiene	1000	10
Chlorobenzene	75	75
Cresol (all isomers)	5	5
n-Heptane	500	50
Isopropanol	400	400
Methanol	200	200
Methylene chloride	500	50
Methyl methacrylate	100	100
Phenol	5	5
Styrene	100	50
Tetrahydrofurane	200	200
Toluene	200	100
2,4-Toluene diisocynate	0.02[c]	0.005
Vinyl chloride	1	5
Xylene (all isomers)	100	100

[a] Federal Code of Regulations, Title 29, part 1910.1000, July 1, 1994.
[b] *Handbook of Toxic and Hazardous Chemicals*, Noyles Publications, Park Ridge, New Jersey, 1991.
[c] Ceiling value.

but continues to grow as a raw material for monomers such as styrene and phthalic anhydride. Its use as a solvent has declined since alternative solvents can usually be found. However, compliance with the 1 ppm limit is possible at reasonable cost. When there is no substitute, benzene can and will be used.

Although federal regulations on general air pollution date from 1955, the major legislation affecting the chemical industry includes the Clean Air Act of 1970 and the Clean Air Act Amendments of 1990. These regulations are administered by the EPA rather than OSHA and are aimed at decreasing undesired chemical exposures in the general population. The EPA's emphasis is on reducing total emissions rather than on reducing local concentrations. When concentration limits are imposed, for example, at property lines, the limits are typically an order of magnitude or more below the corresponding

PELs. Exposure of the general population is involuntary and unrenumerated and should thus meet even higher standards than workplace exposure.

Ozone is another concern in protecting the general population. Organic compounds in the atmosphere promote the formation of ozone. Ozone may be desirable in the upper atmosphere, but it is a pollutant when breathed. The Clean Air Act seeks to reduce emissions of all volatile organic compounds (VOCs). The definition of a VOC is any organic compound with a vapor pressure greater than 0.1 mm Hg at 20°C. The EPA has compiled a list of 318 industrially important VOCs, which includes the monomers of essentially all commercial polymers and essentially all the solvents used in their manufacture.

Many monomers and solvents pose direct health risks (rather than indirect risks through the formation of ozone). The EPA has published a list of 189 hazardous air pollutants (HAPs), which are subject to special regulations. Most monomers are included on this list. Yet to come is a list of extremely hazardous air pollutants (EHAPs). This list will undoubtedly include benzene and vinyl chloride since these compounds are already subject to special regulations known as NESHAPs (National Emission Standards for Hazardous Air Pollutants).

Regulations aimed at limiting workplace exposure and those aimed at reducing emission are philosophically different. They are administered by different government agencies, and the formal details of compliance are largely independent. However, there is technical overlap. Elimination of fugative emissions helps satisfy both sets of regulations. In particular, leak detection and repair programs are required by some EPA regulations. They are not required by OSHA, but their implementation can have an impact on workplace exposure.

Leak detection and repair programs require periodic monitoring and timely repair of equipment in a particular service. Equipment includes flanges, valves, pumps, and compressors. The current regulatory philosophy is that equipment need not be intrinsically nonleaking (e.g., diaphragm valves and pumps) but, if properly maintained, will leak at acceptably low levels.

Emissions from point sources such as vacuum system vents are also becoming increasingly regulated. New sources must satisfy particularly rigorous standards. Future plants will have essentially zero emissions of air pollutants or must purchase emission rights from current emitters.

The costs of plants with near-zero emissions must be borne by the products they produce. These costs can be high, but in most cases they appear to be acceptable. Benzene and vinyl chloride continue to be used as raw materials for commodity polymers. Domestic manufacture of plastics such as polystyrene, ABS, and poly(vinyl chloride) (PVC) has not been eliminated and, indeed, continues to grow faster than the economy as a whole. Domestic production of all plastics has increased an average of 5% per year for the last decade (Reisch, 1995). Although the Clean Air Act Amendments of 1990 impose significant new economic burdens on the plastics industry, continued growth of the plastics industry appears likely.

Essentially all commercial polymers are made from small, energetic molecules that must be confined for workplace and environmental safety. The confinement techniques required for the monomers are sufficient to adequately confine any solvent used in the process. A perfect example of this statement is the use of ethylbenzene in continuous polystyrene processes. Conversion of styrene is incomplete, and so styrene is recycled. From the viewpoint of regulatory compliance, it makes little or no difference if some ethylbenzene is also recycled.

Gas-phase polymerizations are an exception to the preceding argument. Solvent-based processes have a liquid recycle stream that is missing in the gas-phase process. This adds a compliance cost as well as other costs, and these can be significant when compared with the total monomer-to-polymer conversion cost. For polyethylene, the solvent-based processes typically give differentiated products with higher selling prices. Profitability is similar and perhaps even superior for the solvent-based technology. New catalysts may change this situation, but the final outcome is unclear. Metallocene catalysts, the exciting new development in vinyl polymerizations, have so far been commercialized only in solvent-based systems.

We suggest that, in the reaction portion of a polymerization process, the cost of confining solvent plus monomer will be comparable with that for confining the monomer alone. The presence of a solvent will add some equipment in gas and light liquid service that is needed to recover and recycle the solvent. A point source, typically the vacuum system vent, will also be added. The vent exhaust can be handled is several ways. Feeding it to the burner of a boiler or hot oil heater is generally preferred. Additional equipment requiring leak detection and repair will add cost but will probably add no regulatory complications to the plant.

II. FOOD CONTACT APPLICATIONS

The FDA classifies plastics used in packaging and other food contact applications as an indirect food additive. The general category of indirect food additives includes adhesives and components of coatings, paper and paperboard components, and polymers. Approximately 60 polymers have FDA approval for food contact. See Code of Federal Regulations, 21, part 177 for a comprehensive list. A typical statement has the form "[the specific polymer] may be safely used as articles or components of articles intended for use in contact with food in accordance with the following prescribed condition." Although there is no universal format, the prescribed conditions usually include statements of specification or identity (defining the polymer), extractive or similar limitations (defining the results of certain analytical procedures), and limitations or conditions of use (preventing certain application, for example, to prohibit use with beverages containing more than a specified volume percent ethanol).

It is the extractive limitations that would limit residual solvent concentrations. However, none of the specifications limits them directly. Direct limits are sometimes imposed on residual monomer concentrations. More frequently, limits are prescribed for extractables, which typically include residual solvents, residual monomers, and some oligomers.

As a practical matter, the FDA limitations are easily satisfied by production materials. Modern plastics typically have residual monomer concentrations (or residual volatile concentrations or total extractables) much less than the FDA limits. As one example, rubber-modified polystyrene has an FDA limit of 0.5% residual monomer. (It is 1% for unmodified polystyrene.) Standard production grades of polystyrene contain 700–1300 ppm of total volatiles (mainly ethylbenzene and styrene), and special grades are in the range of 300–500 ppm. Industry has exceeded FDA requirements in order to achieve higher heat distortion temperatures and to avoid taste or odor problems with sensitive foodstuffs such as chocolate. Polypropylene is another example where modern processes substantially exceed FDA requirements. The limit on xylene extractables at 25°C is 9.8%. This limit corresponds to early Ziegler–Natta catalysis, which gave large amounts of atatic, extractable polypropylene. Xylene extractables with current catalysts are typically less than 1%.

The major packaging plastics have been items of commerce for 30 or more years. The compounds that can be extracted from plastics by food are low in absolute amounts and in toxicity. Existing FDA regulations make no distinction between polymers produced by solvent-free or solvent-based processes. Modern devolatilization technology gives such low residual solvent levels that toxicity is unlikely to emerge as a concern, since these concerns have not arisen during previous exposures to much higher concentrations. A new polymer or the use of a new solvent for an existing polymer would require FDA approval before the polymer could be used in contact with food. This requires a petition to the FDA containing information on the nature and amounts of chemicals likely to migrate to food. Data from extractions with water, ethanol, and *n*-heptane are typically required. Obtaining FDA approval is a lengthy process, but the historic success rate has been high.

III. ENERGY AND SOLVENT COSTS

The minimum energy needed to separate a polymer from a solvent is $-\Delta G_{mix}$, where ΔG_{mix} is the free energy of mixing. Flory–Huggins theory gives

$$\frac{\Delta G_{mix}}{RT} = \phi_1 \ln \phi_1 + (1 - \phi_1)\frac{\ln(1 - \phi_1)}{n} + \chi\phi_1(1 - \phi_1) \qquad (1)$$

where ϕ_1 is the volume fraction of the solvent, n is the ratio of polymer to solvent molar volumes, and χ is the Flory–Huggins interaction parameter. The minimum value for χ is 0.35 for a good but nonassociating solvent (Blanks and Prausnitz, 1964). This gives a maximum for $-\Delta G_{mix}$ on the order of 5 cal/g. Thus the thermodynamic minimum energy requirement represents an insignificant cost.

Practical separation schemes, such as devolatilization, require much more energy. The solvents must be vaporized. Part of the latent heat of vaporization, ΔH_v, can be recovered through vapor recompression and other heat recovery techniques used in distillation. However, the more elaborate of these schemes will have high capital and operating costs. Table 2 gives ΔH_v for some common solvents at their normal boiling points. We will use ΔH_v as a rough measure of the energy needed for devolatilization.

Table 3 shows energy costs for solvent vaporization assuming energy is available from natural gas or fuel oil at a cost of $3 per million BTU ($2.84 per million kJ). The energy costs are relatively modest for nonpolar solvents. Water is, of course, special.

Table 2 Latent Heats of Vaporization

Compound	Normal boiling point (°C)	ΔH_v (cal/g)	ΔH_v (J/g)
n-Pentane	36	85	356
n-Hexane	69	80	335
m-Xylene	139	82	343
o-Xylene	144	83	347
Propylene	−48	105	439
Methyl ethyl ketone	78	106	444
Ethanol	78	204	854
Water	100	539	2255
Hexamethyldisiloxane	100	46	192

Table 3 Energy Costs[a] for Devolatilization (in cents per pound of polymer)

Weight percent polymer in solution	Heat of vaporization of solvent (cal/g)		
	46	82	539
2.5	0.97	1.72	11.35
5	0.47	0.84	5.53
10	0.22	0.40	2.62
25	0.07	0.13	0.87
50	0.02	0.04	0.29
75	0.01	0.01	0.10
90	0.00	0.00	0.03

[a] Assumes a unit energy cost of $3 per 10^6 BTU ($2.84 per million kJ).

The costs of Table 3 do not reflect the full costs of devolatilization, particularly at low solvent concentrations. However, they do show that the vaporization energy is a modest contributor to total plant costs and suggest that the incentive for reducing solvent concentrations is small unless devolatilization can be avoided completely. Nearly complete avoidance of devolatilization is possible with the gas-phase processes for polyolefins (although unreacted monomers must still be captured and recycled). Even here, however, solution and solvent–slurry processes operating at about 15 wt % polymer remain competitive.

The cost of lost solvent was once a major consideration. Solvent losses used to be 1–3% of the solvent recirculation rate. Modern controls have reduced losses to 1% of the polymer production rate. Thus a plant producing 100,000,000 kg per year of polymer consumes about 1,000,000 kg per year of solvent. The great majority of this loss is through the vent system and will be burned rather than be emitted to the atmosphere. The simple hydrocarbon solvents' currently cost 12–25 cents per pound (26–55 cents per kilogram). Their consumption would thus contribute 0.12–0.25 cents per pound (0.26–0.50 cents per kilogram) of finished polymer.

Together, energy and solvent costs total something less than 1 cent per pound (2 cents per kilogram) and contribute perhaps 2% of the selling price of commodity polymers when made by a solvent route.

IV. LOW SOLIDS FLASH DEVOLATILIZATION

The benefits of a solution process are typically achieved only at fairly low polymer concentrations, say 15 wt % or less. The solutions have moderate viscosities, and the solvent has enough thermal mass to mitigate the reaction exotherm. Numerous solution and slurry processes operate with 5–15% polymer. Commercial examples include homogeneous Ziegler–Natta and metallocene polymerizations and most anionic polymerizations.

A plastics recycling system using selective dissolution (Nauman and Lynch, 1993), a process for forming polymer-in-polymer microdispersions (Nauman et al., 1988; Lynch and Nauman, 1994), and the blending of ultrahigh-molecular-weight polymers (Ueda et al., 1986) are future possibilities for solution processing.

The commercial examples confirm the conclusion of the previous section: the energy requirements for separation are not excessive even for

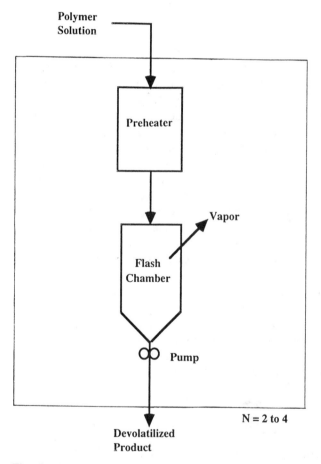

Fig. 1 Staged flash devolatilization sequence.

fairly dilute solutions. This section describes technologies for removing solvent from solutions containing as little as 5 wt % polymer (Nauman, 1989). Most of the technologies described elsewhere in this book are primarily directed toward high solids devolatilization, where the starting material typically contains 70% or less solvent.

In a small pilot plant, it is perfectly feasible to go from 5% polymer solution to solid polymer in a single flash. Staged flashes are used for larger-scale operations. See Fig. 1. The primary consideration in staging is to maintain the polymer solution as a pumpable liquid after each flash. Four stages of flash devolatilization (or two stages plus a rear-vented, devolatilizing extruder) are typically adequate for entering solutions containing 5–15 wt % polymer.

Table 4 gives approximate calculations for polystyrene in a hydrocarbon solvent. The input to the third stage (75% polymer at 173°C) is identical to the reactor effluent in a typical bulk, continuous process for polystyrene. Thus stages 3 and 4 of the sequence represent conventional technology, the details of which have been discussed in other chapters. A significant feature of stages 3 and 4 is that boiling can occur in the

Table 4 A Staged Devolatilization Sequence for Polystyrene

Stage 1		
	Inlet polymer concentration	5%
	Preflash temperature	270°C
	Outlet polymer wt %	35%
	Postflash temperature	114°C
Stage 2		
	Inlet polymer concentration	35%
	Preflash temperature	270°C
	Outlet polymer wt %	75%
	Postflash temperature	173°C
Stage 3		
	Inlet polymer concentration	75%
	Preflash temperature	270°C
	Outlet polymer wt %	97%
	Postflash temperature	229°C
Stage 4		
	Inlet polymer concentration	97%
	Preflash temperature	250°C
	Outlet polymer wt %	99.95%
	Postflash temperature	245°C

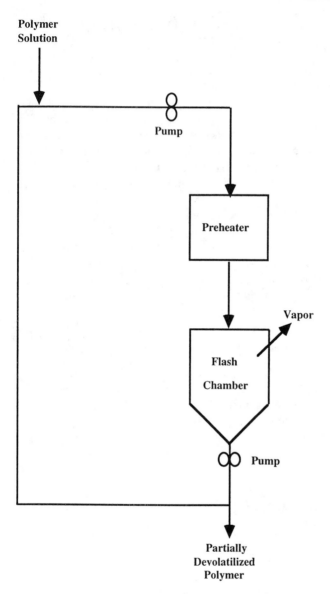

Fig. 2   Recycle flash devolatilization.

shell-and-tube preheaters. It must be suppressed at low solids—particularly in stage 1—to avoid fouling. Some of the tubes plug with devolatilized polymer that is too viscous to be displaced by the entering fluid. An actively controlled, back pressure value is used to maintain sufficient pressure to prevent boiling within the preheater.

Figure 2 shows an alternative method to stabilize the first-stage flash. The reaction engineer will recognize that this approach is analogous to using a stirred tank reactor in place of a tubular reactor. A high recycle rate minimizes the viscosity difference across the preheater so that no back pressure valve is needed.

V. CONCLUSIONS

Technological and environmental trends that eliminate solvents from end-use applications of polymers such as coatings may actually favor the use of solvents for polymerization processes. Rigorous environmental standards force very tight plants with low emissions. This confinement is necessary for essentially all polymerizations, and the addition of a solvent adds reasonably modest incremental costs that can be justified by differentiated products with higher selling prices. The inherent energy cost of devolatilization appears acceptable for commodity plastics such as high-density polyethylene provided the polymer concentration is kept above about 10 wt %. Still lower concentrations could be acceptable for engineering and specialty polymers. The cost of consumed solvent is low for normal hydrocarbon solvents. There is no discernible trend to eliminate solvents in the manufacture of polymers for food contact. Indeed, mechanisms exist for certifying new plastics— whether or not made with solvents—for food contact applications. Historically, petitions to the FDA for new approvals have been time consuming but generally successful.

Solvents will remain part of polymer manufacturing processes for the foreseeable future. The raw materials for almost all plastics are oil and natural gas, and solvents will remain acceptable unless and until the raw materials change. Plastics can be made from biologically derived chemicals such as lactic acid, but the energy cost is high. It is increasingly recognized that the greenness of a process must be judged by what enters or leaves, not by what is confined inside. Solvents can be confined and thus will remain acceptable.

NOMENCLATURE

ABS   acrylonitrile–butadiene styrene
ACGIH   American Conference of Government Industrial Hygienists

BTU   British thermal unit
EHAP   extremely hazardous air pollutant
EPA   Environmental Protection Agency
FDA   Food and Drug Administration
HAP   hazardous air pollutant
NESHAP National Emission Standards for Hazardous Air Pollutant
OSHA   Occupational Safety and Health Administration
PEL   permissible exposure limit
PVC   poly(vinyl chloride)
TDI   2,4-toluene diisocynate
TLV   threshold limit value
VOC   volatile organic compound

Variables

n ratio of polymer to solvent molar volumes
R gas constant
T temperature
ΔG_{mix} Gibbs free energy of mixing
ΔH_v heat of vaporization
ϕ_1 volume fraction of solvent
χ Flory–Huggins interaction parameter

REFERENCES

Amato, I. (1993). The crusade against chlorine, *Science, 261*: 152.
Blanks, R. F., and Prausnitz, J. M. (1964). Thermodynamics of polymer solubility in polar and nonpolar systems, *Ind. Eng. Chem. Fund., 3*(1): 1.
Cavanaugh, T. J., and Nauman, E. B. (1995). The future of solvents in the polymer industry, *Trends Polym. Sci., 3*(2): 48.
Lynch, J., and Nauman, E. B. (1994). The effect of interfacial adhesion on the Izod impact strength of isotactic polypropylene and ethylene–propylene copolymer blends, *Polym. Mater.: Sci. Eng., 71*: 609.
Nauman, E. B. (1989). Flash devolatilization, *Encyclopedia of Polymer Science and Engineering*, supplement volume, Wiley, New York. p. 317.
Nauman, E. B., and Lynch, J. C. (1993). U.S. Patent 5,198,471.
Nauman, E. B., Ariyapadi, M. V., Balsara, N. P., Grocela, T. A., Furno, J. S., Lui, S. H., and Mallikarjun, R. (1988). Compositional quenching: A process for forming polymer-in-polymer microdispersions and cocontinuous network, *Chem. Eng. Commun., 66*: 29.
Reisch, M. S. (1995). Plastics, *Chem. Eng. News, 73*(21): 30.
Seltzer, R. (1989). Alar sales in U.S. for food use halted, *Chem. Eng. News, 67*(24): 6.
Ueda, H., Karasz, F. E., and Farris, R. J. (1986). Characterization of mixtures of linear polyethylenes of ultrahigh and moderate molecular weights, *Polym. Eng. Sci., 26*(21): 1483.

19

The Analysis of Volatiles in Polymers

Thomas R. Crompton*

Shell Research Ltd., Carrington, Cheshire, England

---

* Retired.

I. INTRODUCTION

Polymers usually contain low concentrations of volatile constituents arising from their method of manufacture. The major types of such substances include unreacted monomers, nonpolymerizable components of the original charge stock, residual polymerization solvents, and water. The concentrations of these substances usually range from a few tens to several hundred parts per million. Frequently, complex mixtures are present. Thus, the nonpolymerizable fractions of styrene monomer (usually 1–2%) consist of several dozen aromatic hydrocarbons such as ethylbenzene. The polymerization solvent used in the manufacture of high-density polyethylene and polypropylene by the low-pressure catalyzed route is usually a crude petroleum distillation cut with a complex composition.

One must determine the concentration of these substances for many reasons, two being the effects they have on the mechanical properties of polymers and the risk of tainting in the case of foodstuff- or beverage-packaging grades of polymers.

Polymers often contain substances of medium volatility such as residual monomers, residual polymerization solvents, and expanding agents. In addition, when polymers are heated they may release volatiles as a result of the thermal degradation of either the polymers themselves or their additives or catalyst residues. These volatiles can bear on such properties as processability, the tendency to form voids, and in the case of foodstuff-packaging grades, the possible tendency to impart taste or odor to the packed commodity.

One way of identifying nonpolymeric constituents of polymers is to extract the polymer with a low-boiling-point solvent, remove the solvent from the extract by evaporation or distrillation, and anayze the residue. This procedure is, of course, inapplicable to the analysis of extracted polymer constituents volatile enough to be lost during the solvent removal stage. Alternatively, an extract or a solution of the polymer may be examined directly for volatile constituents by gas chromatography, in which case losses of volatiles are less likely to occur. In such a procedure, however, the large excess of solvent used for extraction or solution might interfere with the interpretation of the chromatogram, obscuring some of the peaks of interest. Trace impurities in the solvent may also interfere with the chromatogram. Of course none of these procedures is suitable for studying the nature of volatile breakdown products that are produced only upon heating a polymer. For the identification or determination of residual solvents in polymers it is mandatory, therefore, to use solventless methods of analysis that is, there must be no risk of confusing solvents in which the sample is dissolved for analysis with residual solvents in the sample. Most methods for the determination of residual solvents are based on the technique of heating the solid polymer and examining the head space over the polymer by various analytical techniques, as discussed here.

II. GAS CHROMATOGRAPHY

Due to their volatility and complex composition, methods based on gas chromatography have emerged as the most suitable way of analyzing for these parameters. Basically, three different approaches have evolved in the application of gas chromatography:

1. Solution of the polymer in a solvent and injection into the gas chromatograph (Section II.A.1)
2. Heating the dry polymer and sweeping the volatiles released into a gas chromatograph using the carrier gas (Section II.A.2)
3. Head-space analysis, that is, heating the polymer in a closed system, then withdrawing the head space with a syringe for direct injection into the gas chromatograph (Section VII)

A. Volatiles

1. Monomers

a. Determination of monomers in a solvent solution. Crompton et al. (1965) and Crompton and Myers (1968a, b) have applied gas chromatography to the determination of polystyrene in styrene and a wide range of other aromatic volatiles in amounts down to the 10 ppm level. In this method a weighed portion of the sample is dissolved in propylene oxide containing a known concentration of pure *n*-undecane as an internal standard. After allowing any insolubles to settle, an approximately measured volume of the

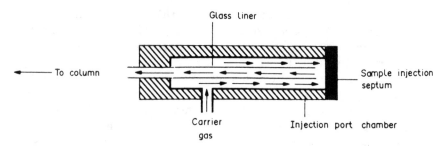

Fig. 1 Injection port glass linear fitted to F. & M. Model 1609 gas chromatograph. The glass linear measures 60 mm × 40 mm o.d. × 2 mm i.d. and is very loosely packed with glass fiber.

solution is injected into the chromatographic column, which contains 10% Carbowax 15–20 M supported on 60–70 BS Celite. Helium is used as carrier gas, and a hydrogen flame ionization detector is employed. Figure 1 shows a device that is connected to the injection port of the gas chromatograph to prevent the deposition of polymeric material in the injection port with consequent blockages. When a solution of polystyrene is injected into the liner, polymer is retained by the glass fiber, and volatile components are swept onto the chromatographic column by the carrier gas.

Figure 2 illustrates, by means of a synthetic mixture, the various aromatics that can be resolved, and Fig. 3 illustrates a chromatogram obtained with a polystyrene sample, indicating the presence of benzene, toluene, ethylbenzene, xylene, cumene, propyl benzenes, ethyl toluenes, butyl benzenes, styrene, and α-methyl styrene.

2. Volatiles Other than Monomers

Expandable grades of polystyrene are manufactured by steeping volatile polystyrene granules in a low-boiling hydrocarbon solvent until the polymer becomes saturated with the solvent. Commonly used are *n*- and isopentane; in addition, isobutane, isohexane, neohexane, isooctane, and cyclopentane have been used. When the granules are put in a mold and are treated with steam, they expand to many times their original volume, and the expanded granules coalesce. This process is used for the manufacture of insulating polystyrene board and insulated hot beverage vending vessels. One must determine the concentration of residual volatiles in the original and the expanded polymer; methods for carrying out this analysis are discussed here.

In a typical method for *n*- and isopentane, the polystyrene (2 g) is dissolved in 25 ml propylene oxide containing 1% internal standard. A portion of the solution is chromatographed, and the amounts of the pentanes

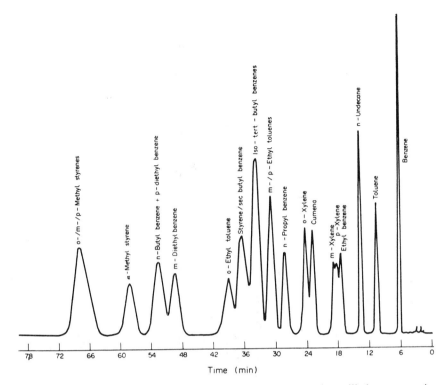

Fig. 2 Gas chromatogram of a synthetic blend of hydrocarbons likely to occur in polystyrene on a Carbowax 15–20 M column at 80°C.

are determined from previously calculated calibration factors, using the 2:2 dimethyl butane or cyclohexane as the internal standard. To calibrate, weigh in turn into a 100-ml volumetric flask 0.40 ml isopentane, 0.50 ml n-pentane, and 1 ml internal standard. Seal the flask with a serum cap during each weighing. Dilute the mixture to 100 ml with propylene oxide, seal with a serum cap and mix thoroughly. Chromatograph 10 µl and measure either the peak heights or the integrated areas of the isopentane, n-pentane, and internal standard.

Then

$$\% \text{ w/w isopentane} = \frac{25 W_1 W_4 P_3 P_4}{W_3 W_5 P_1 P_6}$$

$$\% \text{ w/w } n\text{-pentane} = \frac{25 W_2 W_4 P_3 P_5}{W_3 W_5 P_2 P_6}$$

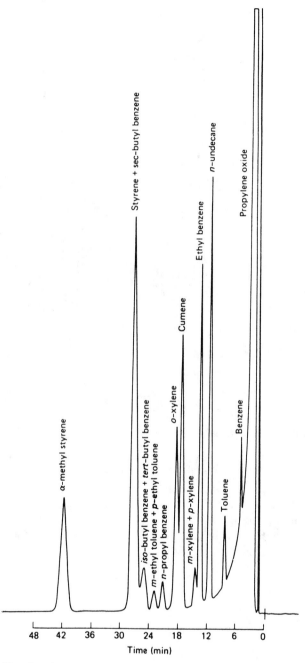

Fig. 3 Gas chromatogram of blend of aromatic hydrocarbons in propylene oxide.

where

W_1, W_2 = respective weights of iso- and n-pentane in calibration blend

W_3, W_4 = respective weights of internal standard in calibration blend and sample solution

W_5 = weight of sample

P_1, P_2, P_3 = respectively, peak heights or area of isopentane, n-pentane, and internal standard in calibration blend.

P_4, P_5, P_6 = respectively, peak height or area of isopentane, n-pentane, and internal standard in sample solution.

Figure 4 shows a gas chromatogram obtained in the analysis of an

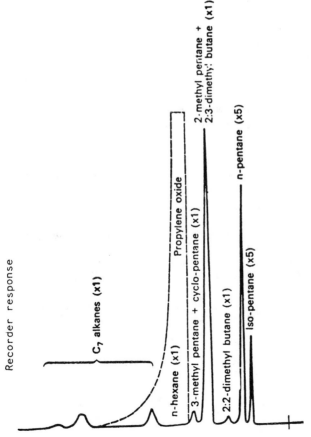

Fig. 4   Gas chromatogram of pentanes in polystyrene.

expandable polystyrene sample containing n- and isopentane. In this determination, cyclohexane was used as the internal standard.

Wide-bore fused silica column capillary gas chromatography has been used by workers at Perkin-Elmer (1981) to determine impurities in vinyl chloride monomer (VCM) used in the manufacture of poly(vinyl chloride) (PVC). The analysis was performed on a Perkin-Elmer Model 8500 gas chromatograph using a 25 m × 0.53 mm fused silica column coated with 5 µm immobilized methyl silicone phase. The column being run at $-10°C$ for 4 min programmed at 10°C/min to 10°C for 1 min and programmed at 30°C/min to 170°C for 2 min. A flame ionization detector was employed. The three principal impurities present in the monomer and, consequently, expected to occur in the PVC were methyl chloride, monovinyl acetylene, and ethyl chloride.

Gas chromatography coupled to mass spectrometry has been used by Baba (1977) and Gilbert et al. (1975) to determine VCM in PVC and by Tengler and Von Falkai (1972) to determine diols, adipic acid, and cyclo-monodiol adipate in aliphatic polyesters.

Iedrezejczak and Gaind (1993) developed a method for the determination of toluene diisocyanate in polyurethanes. To a methanol extract of the foam was added deuterated toluene 2,4-diyl dimethyl carbamate internal standard. The extract was derivatized with pentafluoropropionic anhydride at 70°C and analyzed by capillary gas chromatography using selected ion mode. Down to 10 µg/kg of the isocyanate could be determined by this procedure.

Other methods for the gas chromatographic determination of volatiles in polymers are summarized in Table 1.

B. Instrumentation

The basic requirements required of a high-performance gas chromatograph are as follows:

1. Sample is introduced to the column in an ideal state, that is, uncontaminated by septum bleed or previous sample components, without modification due to distillation effects in the needle and quantitatively, that is, without hold-up or adsorption prior to the column.
2. The instrument parameters that influence the chromatographic separation are precisely controlled.
3. Sample components do not escape detection; that is, highly sensitive, reproducible detection, and subsequent data processing are essential.

There are two types of separation columns used in gas chromatography, capillary column and packed column. Packed columns are still used extensively, especially in routine analysis. They are essential when sample components have high partition coefficients or high concentrations. Capillary columns provide a high number of theoretical plates, hence a very high resolution, but they cannot be used in all applications because there are not many types of chemically bonded capillary columns.

Recent advances in capillary column technology presume stringent performance levels for the other components of a gas chromatograph, as column performance is only as good as that of the rest of the system. Crucial in capillary column gas chromatography is that a high repeatability of retention times be ensured even under adverse ambient conditions.

Another important factor for reliable capillary column gas chromatography is the sample injection method. Various types of sample injection port are available. The split/splitless sample injection port unit series is designed so that the glass insert is easily replaced and the septum is continuously purged during operation. This type of sample injection unit is quite effective for the analysis of samples having high-boiling-point compounds as the major components.

1. Shimadzu Gas Chromatographs

This is a typical high-performance gas chromatograph version (see Table 2 for further details). The inner chamber of the oven has curved walls for smooth circulation of air; the radiant heat from the sample injection port units and the detector oven is completely isolated. These factors combine to provide demonstrably uniform temperature distribution. (The temperature variance in a column coiled in a diameter of 20 cm is less than ± 0.75 K at a column temperature of 250°C.)

When the column temperature is set near ambient temperature, external air is brought into the oven via a computer-controlled flap, providing rigid temperatue control stability. (The lowest controllable column temperature is 24°C when the ambient temperature is 18°C and the injection port temperature is 250°C. The temperature fluctuation is less than ± 0.1 K even when the column temperature is set at 50°C.)

This instrument features five detectors (Table 2). In the flame ionization detector, the high-speed electrometer, which ensures a very low noise level, is best suited to trace analysis and fast analysis using a capillary column.

Samples are never decomposed in the jet, which is made of quartz.

Carrier gas, hydrogen, air and makeup gas are separately flow-controlled. Flow rates are read from the pressure flow rate curves.

In the satellite system, one or more satellite gas chromatographs (GC-14 series) are controlled by a core gas chromatograph (e.g., GC 16A series). Since the control is made externally, the satellite gas chromatographs

Table 1 Summary of Methods

Monomer	Polymer	Method	References
Styrene, ethyl benzene	Polystyrene	Solution in o-dichloro-benzene or methylene dichloride, GLC	Carsen et al. (1962)
Styrene, α-methyl styrene benzene, toluene, ethyl benzene, xylene, cumene, propyl benzenes, ethyl toluenes, butyl benzenes	Polystyrene	Solution in propylene oxide, GLC	Markelov and Semenko (1973) De Forero et al. (1971)
Styrene	Polystyrene	Solution in THF, GLC	Nowak and Klemett (1962) Shanks (1975a) Shanks (1975b)
Styrene	Polystyrene	Solution in DMF, GLC	
Styrene, acrylonitrile butadiene	Polystyrene	Solution in DMP, GLC	Klesper and Hartman (1978)
Styrene	Polystyrene	Solution in benzene, GLC	Klesper and Hartman (1978)
Styrene ethyl benzene	Polystyrene	Solution, GLC	Rohrachneider (1971)
Styrene	Polystyrene	Solution in o-dichloro-benzene or methylene dichloride, GLC	Pfab and Noffz (1963)
Styrene	Polystyrene	Solution in DMF, GLC	Shapras and Claver (1964)
Styrene, o-methyl styrene	Polystyrene	Solution in propylene oxide, GLC	Crompton et al. (1965)

Benzene, toluene, ethyl benzene, xylene, cumene, propyl benzenes, ethyl toluenes, butyl benzenes, styrene	Polystyrene	Solution in THF, GLC	Crompton and Myers (1968b)
Alkanes	Polystyrene	Heating sample, GLC	Schwoetzer (1972) Podzeeva et al. (1971) Crompton and Myers (1968a)
Volatiles	Poly α-methyl styrene	GLC	Shanks (1975b) Streichen (1976)
Volatiles	Styrene–butadiene	GLC	Shapras and Claver (1964) Shanks (1975b)
Ethyl acrylate styrene	Polyacrylates	Distilled in presence of toluene, GLC	Shapras and Claver (1962) Tweet and Miller (1963)
Methyl methacrylate	Polymethacrylates	GLC	Cobler and Samsel (1962)
Ethylacrylate styrene	Polyacrylates	Distilled in presence of toluene solution	Shapras and Claver (1962)
Methyl acrylate, ethyl acrylate	Polyacrylates	Solution in methylene chloride, GLC	Shapras and Claver (1962)
2-Ethylhexyl acrylate, butyl acrylate, ethyl acrylate, vinyl propionate	Polyacrylates	Solution in solvent, GLC	Wilkinson et al. (1964)
2-Ethylhexyl acrylate, vinyl acetate	Polyacrylates	Solution in propyl acetate cyclohexanol, GLC	Nowak and Klemett (1962)
Butyl acrylate, methyl acrylate, ethyl acrylate	Polyacrylates	Solution in isopropanol, GLC	Adcock (1962)
Methyl methacrylate, ethyl acrylate, styrene	Polyacrylates	Solution in solvent, GLC	Ragelis and Gajan (1962) Shapras and Claver (1962)

Table 1 *Continued*

Monomer	Polymer	Method	References
Vinyl acetate, 2-ethylhexyl acrylate	Polyacrylates	Solution in propyl acetate, GLC	Ragelis and Gajan (1962)
Ethyl acrylate, styrene, vinyl acetate	Styrene–ethyl acrylate	Distill in presence of toluene, GLC	Tweet and Miller (1963)
Methyl acrylate, ethyl acrylate, 2-ethylhexyl acrylate, butyl acrylate, ethyl acrylate, vinyl propionate	Mixed polyacrylates	Solution, GLC	Schwoetzer (1972) Streicken (1976)
2-Ethylhexyl acrylate, vinyl acetate, butyl acrylate, methyl acrylate, methacrylic acid	Mixed polyacrylates	Solution in propyl acetate cyclohexanol, GLC; solution in isopropanol, GLC	Shiryaev and Kozmenkhova (1972) Mel'inkova et al. (1977)
Methyl methacrylate, ethyl acrylate, styrene	Styrene acrylate and styrene methacrylate	Solution, GLC	Rosenthal et al. (1963)
Vinyl acetate, 2-ethylhexyl acrylate	Vinyl acrylate-2-ethyl hexyl acrylate	Solution in propyl acetate, GLC	Schwoetzer (1972)
Benzene and toluene	Rubber adhesives	Gas chromatography	Takashima et al. (1960)

Styrene	Latex	Gas chromatography	Brodsky (1961) Neldon et al. (1961)
Styrene, Dialkyl phthalate	Polyester resin moldings	Gas chromatography	Holtmann and Souren (1977) Yamaoka and Matsui (1978)
Styrene acrylonitrile butadiene	Acrylonitrile–butadiene–styrene and styrene acrylonitrile	Solution in dimethyl-formamide, GLC	Oprea and Pogorevic (1974) Doak (1948)
Ethylene glycol, dimethyl terephthalate, dimethyl isophthalate, dimethyl adipate	Isophthalic acid, terephthalic acid, ethylene glycol	Gas chromatography	Oprea and Pogorevic (1974)
Vinyl chloride	Polycarbonates	GLC	Hartley et al. (1968)
Volatiles	Polyethylene	GLC	Jeffs (1969) Crompton and Myers (1968a)
Volatiles	Polypropylene	GLC	Maltese et al. (1969)
Volatiles	Rubber	GLC	Takashima et al. (1960) Haslam et al. (1962) Haslam and Jeffs (1957) Haslam and Jeffs (1958)
Volatiles	Polyether-coated film	GLC	Welsch et al. (1976)

Table 2   Commercial Gas Chromatographs

Manufacturer	Model	Packed column	Capillary column	Detectors	Sample injection port system
Shimadzu	GG-14A	Yes	Yes	FID ECD FTD FPD TCD (all supplied)	1. Split–splitters 2. Glass insert for single column 3. Glass insert to dual column 4. Cool on column system unit 5. Moving needle system 6. Rapidly ascending temperature vaporizer
Shimadzu	GC15A	Yes	Yes	FID ECD FTD FPD TCD	1. Split–spliters 2. Direct sample injection (capillary column) 3. Standard sample injector (packed column) 4. Moving precolumn system (capillary columns) 5. On column (capillary columns)
Shimadzu	GC 16A	Yes	Yes	FID FCD FTD FPD TCD (all supplied)	Split–splitless
Shimadzu	GC 8A	Yes	Yes	FID FCD FPD TCD Single-detector instruments (detector chosen on purchase)	1. Point for packed columns 2. Point for capillary columns 3. Split–splitless
Varian	3400GC	Yes	Yes	Up to 4 types	Temp. prog. SPI system
Varian	3600	Yes	Yes	FID ECD TCD TSD EPD Photoionization Hall EC	
Perkin-Elmer	8410	Yes	Yes	Single-detector instrument (detector chosen on purchase) FID ECD FTD FPD TCD	1. Flash vaporization 2. Split–splitless injector 3. Manual or automatic gas sampling values 4. Manual or automatic liquid sampling valves

Keyboard control	Link to computer	Visual display	Printer	Core instrument amonable to tap automation	Temperature programming/ isothermal	Cryogenic unit (subambient chromatography)
Yes	Yes	No	No	No	Yes/Yes	No
No	Yes	Yes	Yes	No	Yes/Yes	No
Yes	Yes	Yes	Yes	No	Yes/Yes	No
Yes	Optional	No	Optional	No	Yes temp. programming GC 8APT (TCD detector) GC 8APF (FID detector) GV 8APFD (FID detector) isothermal: GC 8AIT (TCD detector) GC 8AIF (FID detector) GC 8AIE (ECD detector)	No
Yes	Yes	No	Yes	No	Yes/Yes	No
Yes	Yes	No	Yes	No	Yes/Yes	Yes
No	No	No	No	No	Yes/Yes	Yes, down to $-80°C$

Table 2 *Continued*

Manufacturer	Model	Packed column	Capillary column	Detectors	Sample injection port system
Perkin-Elmer	8420	No	Yes	Single-detector instrument (chosen on purchase) FID ECD FTD FPD TCD	1. Programmable temperature vaporizer 2. Split–splitless injector 3. Direct on column injector
Perkin-Elmer	8400 and 8500	Yes	Yes	Dual-detector instrument (detectors chosen from following) FID ECD FTD FPD TCD	Can be fitted with any combination of above injection systems
Perkin-Elmer	8700	Yes	Yes	FID ECD EPD TCD Hall E. C. photoionization dual-detector instrument (Detectors chosen from above list)	1. Flash vaporization 2. Split-splitless 3. Programmable temperature vaporizer 4. Gas sampling valve 5. Liquid sampling valve
Nordion	Micromat HRGC 412	No	Yes	Dual simultaneous detector combinations from the following: FID ECD FTD Photoionization Hall E.C.	1. Split–splitless 2. On-column injector
Siemens	SiChromat 1–4 (single oven) SiChromat 2–8 (dual oven) for multidimensional GC)	Yes	Yes	FID ECD FTD FPD TCD Helium detector	1. Liquid–liquid packed columns 2. Split–splitless 3. Temperature programmable 4. On-column 5. Liquid injector valve on-line 6. Gas injection valve 7. Rotary as injection

Keyboard control	Link to computer	Visual display	Printer	Core instrument amonable to tap automation	Temperature programming/ isothermal	Cryogenic unit (subambient chromatography)
No	No	No	No	No	Yes/Yes	Yes, down to $-80°C$
Yes	Yes	Yes	Yes (GP100 printer plotter)	Yes	Yes/Yes	Yes, down to $-80°C$
Yes	Yes	Yes	Yes	—	Yes/Yes	Yes
Yes	Yes	Yes	Yes	—	Yes/Yes	No
No	No	Yes	Yes	—	Yes/Yes	No

are not required to have control functions (the keyboard unit is not necessary).

When a GC 16A series gas chromatograph is used as the core, various laboratory-automation-oriented attachments such as a bar-code reader and a magnetic-card reader become compatible: a labour-saving system can be built, with the best operational parameters automatically set. Each satellite gas chromatograph (GC 14A series) operates as an independent instrument when a keyboard unit is connected.

The internal computer (IC) card operated gas chromatography system consists of a GC-14A series gas chromatograph and a C-R5A Chromatopac data processor. All of the chromatographic and data processing parameters are automatically set simply by inserting the particular IC card. This system is particularly convenient when one GC system is used for the routine analysis of several different types of samples.

One of the popular trends in laboratory automation is to arrange for a personal computer to control the gas chromatograph and to receive data from the GC to be processed as desired. Bilateral communication is made via the RS-232C interface built in a GC 14A series gas chromatograph. A system can be built to meet requirements.

A multidimensional gas chromatography system (multistage column system) is effective for analysis of difficult samples and can be built up by connecting several column ovens, that is, tandem GC systems, each of which has independent control functions such as for temperature programming.

The Shimadzu GC 15A and GC 16A systems are designed not only as independent high-performance gas chromatographs but also as core instruments (see above) for multi–gas chromatography systems or computerized laboratory automation systems.

Other details of these instruments are given in Table 2. The Shimadzu GC 8A range of instruments do not have a range of built-in detectors but are ordered either as temperature-programmed instruments with thermal conductivity detector (TCD), flame ionization defector (FID), or flame photometric detector (FPD) detectors or as isothermal instruments with, TCD, FID, or electron capture (ECD) detectors (Table 2).

Perkin-Elmer gas chromatographs supply a range of instruments including the basic models 8410 for packed and capillary work and the 8420 for dedicated capillary work, both supplied on purchase with one of the six different types of detection (Table 2). The 8400 and 8500 models are more sophisticated capillary column instruments capable of dual-detection operation with the additional features of keyboard operation; screen graphics method storage, host computer links, data handling, and compatibility with laboratory automation systems. Perkin-Elmer supplies a range of accessories for these instruments including an autosampler (AS-8300), an infrared spectrometer interface, an automatic head-space accessory (HS101 and H5-6), an autoinjector device (AI-I), a catalytic reactor and a pyroprobe (CDS 190), and an automatic thermal desorption system (ATD-50) (both useful for examination of sediments).

The Perkin-Elmer 8700, in addition to the features of the models 8400 and 8500, has the ability to perform multidimensional gas chromatography. Other applications of the 8700 system include fore-flushing and back-flushing of the pre-column, either separately or in combination with heart cutting, all carried out with complete automation by the standard instrument software.

There are many other suppliers of gas chromatography equipment, some of which are listed in Table 2.

III. ULTRAVIOLET SPECTROSCOPY

A. Volatiles

1. Monomers

Ultraviolet (UV) spectroscopy suffers from several disadvantages in the determination of monomers in polymers, particularly in the case of styrene monomer in polystyrene and its copolymers. In addition to lack of sensitivity, which limits the lower detection limit to about 200 ppm styrene in polymer under the most favorable circumstances, UV spectroscopic methods are subject to interference by some of the types of antioxidants included in polystyrene formulations. Such interference can be overcome only by lengthy pretreatment of the sample to remove antioxidants. In addition to residual styrene monomer, polystyrene may also contain traces of other aromatic hydrocarbons such as benzene, toluene, xylenes, ethyl benzene, and cumene, which originate either as impurities in the styrene monomer employed to manufacture the polystyrene or from use in small quantities as dilution solvents at some stage of the manufacturing process. Ultraviolet spectroscopic methods for determining styrene cannot differentiate among the various volatile substances present in polystyrene.

For some types of polymer additives, interference effects can be overcome by the use of a baseline correction techniques. Thus polystyrene contains various nonpolymer additives (e.g., lubricants) that result in widely different and unknown background absorptions at the wavelength maximum at which styrene monomer is evaluated (292 nm). The influence of the background absorptions on the evaluation of the optical density due to styrene monomer is overcome by the use of an appropriate baseline technique, claimed to make the method virtually independent of absorptions due to polymer additives. In this technique, a straight line is drawn on the recorded spectrum across the absorption peak at 292 nm so that the baseline

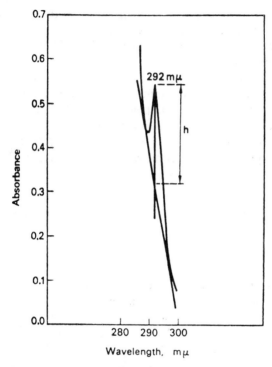

Fig. 5 Typical ultraviolet absorption curve of polystyrene containing styrene monomer.

is tangential to the absorption curve at a point close to the absorption minima occurring at 288 nm and 295–300 nm (Fig. 5). A vertical line is drawn from the tip of the styrene absorption peak at 292 nm to intersect the baseline, and the height of this line is a measure of the optical density due to the true styrene monomer content of the test solution.

This baseline correction technique can obviously be applied to the determination of styrene monomer in polystyrene only if any other UV-absorbing constituents in the polymer extract (e.g., lubricant, antioxidants) absorb linearly in the wavelength range 288–300 nm. If the polymer extract contains polymer constituents other than styrene with nonlinear absorptions in this region, then incorrect styrene monomer contents will be obtained. An obvious technique for removing such nonvolatile UV-absorbing compounds is by distillation of the extract followed by UV spectroscopic analysis of the distillate for styrene monomer as discussed below.

In the distillation technique (Crompton et al., 1965) the polystyrene is dissolved in chloroform or ethylene dichloride (20 ml) in a stoppered flask and the solution is poured into an excess of methyl alcohol (110 ml) to reprecipitate dissolved polymer. The polymer is filtered off and washed with methanol (120 ml), and the combined filtrate and washings gently distilled to provide 200 ml of distillate containing styrene monomer and any other distillable component of the original polystyrene sample. Nonvolatile polymer components (viz. stabilizers, lubricants, and low-molecular-weight polymer) remain in the distillation residue. The optical density of the distillate is measured at 292 nm or by the baseline method against the distillate obtained in a polymer-free blank distillation. Calibration is performed by applying the distillation procedure to solutions of known weights of pure styrene monomer in the appropriate quantities of methyl alcohol and the chlorinated solvent. Tables 3 and 4 show results obtained for styrene monomer determinations carried out on samples of polystyrene by the direct UV method, and by the distillation modification of this method. The distillation method gives results consistently higher than those obtained by direct spectroscopy, indicating that additives present in the polystyrene are interfering in the latter method of analysis.

Similar effects to these were observed with polystyrene containing other UV-absorbing additives. Thus, the influence of a mixture of 0.4% w/w *tris*-(nonylated phenyl phosphite) (Polygard) and 0.2% w/w 2,6-di-*tert*-butyl-*p*-cresol (Ionol CP) on the determination of styrene monomer is shown in Table 5.

Table 3 Comparison of Direct Ultraviolet and Distillation/Ultraviolet Methods for the Determination of Styrene Monomer

| | | Styrene monomer (% w/w) | |
| | | Polystyrene sample | |
Method	Solvent	No. 1	No. 2
Direct UV method	Chloroform	<0.05	0.13
	Carbon tetrachloride	<0.05	0.13, 0.14, 0.16, 0.18
	Ethyl acetate	<0.05	0.14, 0.12, 0.14, 0.14, 0.15
Distillation/UV	Ethylene dichloride/methanol	0.16, 0.18, 0.16, 0.18, 0.20	0.27, 0.29, 0.26, 0.29, 0.29

Table 4 Influence of Phenolic Antioxidant[a] on the Determination of Styrene Monomer by Direct Ultraviolet and by Distillation/Ultraviolet Methods

	Styrene found (% w/w)			
	Direct UV method[b]		Distillation/UV method	
Styrene added to polystyrene (% w/w)	No phenolic antioxidant[a] addition	0.5% phenolic antioxidant[a] added on polymer	No phenolic antioxidant[a] addition	0.5% phenolic antioxidant[a] addition on polymer
0.11	0.11, 0.11	0.04, 0.08	0.12	0.12
0.22	0.22	0.11	—	—
0.27	0.26	0.18	0.29	0.26
0.41	0.41, 0.40	0.30, 0.27	0.40	0.40

[a] Wingstay T.
[b] Chloroform used as a sample solvent.

Table 5 Influence of Polygard/Ionol CP Mixture on Determination of Styrene Monomer by Direct Ultraviolet and by Distillation/Ultraviolet Methods

	Styrene found (% w/w)			
	Direct UV method[a]		Distillation/UV method[a]	
Styrene added to polystyrene (% w/w)	No additive	0.4% w/w Polygard and 0.2% w/w Ionol CP added on polymer	No additive	0.4% w/w Polygard and 0.2% w/w Ionol CP added on polymer
0.10	0.10, 0.10	—[b]	0.11	0.11, 0.09
0.26	0.25, 0.26	0.16, 0.16	0.26	0.27, 0.25
0.41	0.42, 0.41	0.29, 0.29	0.41	0.40, 0.42

[a] Chloroform used as a sample solvent.
[b] Not measurable due to strong interference by additives.

Table 6 Visible–Ultraviolet–Near Infrared Spectrophotomers

Spectral region	Range (nm)	Manufacturer	Model	Single or double beam	Cost range
UV/visible	—	Philips	PU 8620 (optional PU 8620 scanner)	Single	Low
Visible	325–900	Celcil Instruments	CE 2343 Optical Flowcell	Single	Low
Visible	280–900	Celcil Instruments	CE 2393 (grating, digital)	Single	High
Visible	280–900	Celcil Instruments	CE 2303 (grating, nondigital)	Single	Low
Visible	280–900	Celcil Instruments	CE 2373 (grating, linear)	Single	High
UV/visible	190–900	Celcil Instruments	CE 2292 (digital)	Single	High
UV/visible	190–900	Celcil Instruments	CE 2202 (nondigital)	Single	Low
UV/visible	190–900	Celcil Instruments	CE 2272 (linear)	Single	High
UV/visible	200–750	Celcil Instruments	CE 594 (microcomputer controlled)	Double	High
UV/visible	190–800	Celcil Instruments	CE 6000 (with CE 6606 graphic plotter option)	Double	High
UV/visible	190–800	Celcil Instruments	5000 series (computerized and data station)	Double	High
UV/visible	—	Philips	PU 8800		High
UV/visible	—	Kontron	Unikon 860 (computerized with screen)	Double	High
UV/visible	—	Kontron	Unikon 930 (computerized with screen)	Double	High
UV/visible	190–1100	Perkin-Elmer	Lambda 2 (microcomputer electronics screen)	Double	High
UV/visible	190–750 or 190–900	Perkin-Elmer	Lambda 3 (microcomputer electronics)	Double	Low to High

Table 6 *Continued*

Spectral region	Range (nm)	Manufacturer	Model	Single or double beam	Cost range
UV/visible	190–900	Perkin-Elmer	Lambda 5 and Lambda 7 (computerized with screen)	Double Double	High
UV/visible	185–900 and 400–3200	Perkin-Elmer	Lambda 9 (computerized with screen)	UV/visible/ near infrared	High
UV/visible	190–900	Perkin-Elmer	Lambda Array 3840 (computerized with screen)	Photodiode	High
UV/Visible near IR	175–900	Varian	Cary 4E Cary 5E (computerized with screen)	Double	High
High-performance UV/visible	190–900	Varian	Cary 1E Cary 3E (computerized with screen)	Double	High

Direct UV spectroscopic methods are reliable only in the case of polystyrene samples not containing UV-absorbing antioxidants or any other type of strongly UV-absorbing additive. The more lengthy distillation/ UV spectroscopic method, however, will give correct results in the presence of such additives unless the additive is both sufficiently volatile to distill and absorbs in the same region of the spectrum as styrene monomer.

Urbanski (1977) has also used UV spectroscopy to determine styrene monomer in chloroform extracts of polystyrene.

2. Volatiles Other than Monomers

Ultraviolet spectroscopy has also been used by Miller et al. (1961) to determine propylene oxide and epichlorohydrin in polyesters.

B. Instrumentation

Suppliers of visible spectrophotometers are reviewed in Table 6.

For many applications, a simple basic UV/visible single-beam instrument such as the Phillips PU 8620 or the Cecil Instrument CE 2343,

CE 2303, or CE 2202 will suffice. If better instrument stability is required, then a double-beam instrument is preferable and can be purchased at little extra cost.

Moving upmarket, spectrophotometers are available with computer interfaces that enable reaction kinetics studies to be carried out (e.g., the Cecil CE 2202, CE 2272, CE 2292, CE 2303, CE 2373, and CE 2393 single-beam instruments and the Cecil CE 594 double-beam instruments). These instruments also have an autosampler facility capable of handling up to 40 samples. Even more sophisticated is the Cecil 6000 double-beam instrument equipped with a real-time graphics plotter. In this instrument, sets of parameters, that is, methods, can be stored and curve fitting carried out. It also enables multicomponent analysis to be carried out. Analyses of mixtures of up to nine different materials may be carried out.

IV. POLAROGRAPHIC METHODS

Polarography can be used to determine electroreducible substances such as monomers and organic peroxides in solvent extracts of polymers as discussed below. The basic process of electron transfer at an electrode is a fundamental electrochemical principle and for this reason polarography can be used over a wide range of applications.

A. Volatiles

1. Monomers

a. Styrene and acrylonitrile. Residual amounts of styrene and acrylonitrile monomers usually remain in manufactured batches of styrene–acrylonitrile copolymers. As these copolymers have potential use in the food-packaging field, one must ensure that the content of both of these monomers in the finished copolymer is below a stipulated level.

In a polarographic procedure described by Claver and Murphey (1959) and Crompton and Buckley (1965) for determining acrylonitrile (down to 2 ppm) and styrene (down to 20 ppm) monomers in styrene–acrylonitrile copolymer, the sample is dissolved in 0.2 M tetramethyl-ammonium iodide in dimethylformamide base electrolyte and polarographed at start potentials of -1.7 and -2.0 V respectively for the two monomers (Fig. 6). Excellent results are obtained by this procedure.

Table 7 shows the results obtained in determinations of acrylonitrile monomer in some copolymers by the polarographic procedure. Comparison of these results with those obtained by the dodecyl mercaptan procedure show that acrylonitrile contents some 30% higher are obtained by the latter method. The cause of the high results obtained by the dodecyl mercaptan

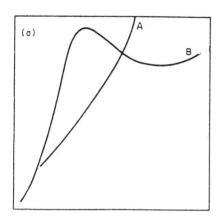
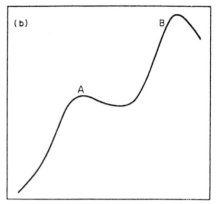

Fig. 6 (a) Cathode-ray polarogram of a synthetic solution of acrylonitrile in tetrabutyl ammonium iodide–dimethylformamide base electrolyte. Curve A: base electrolyte blank solution; curve B: 4 ppm of acrylonitrile in base electrolyte. Start potential -1.7 V. (b) Cathode-ray polarogram of a synthetic solution of 11.2 ppm of styrene and 9.3 ppm of acrylonitrile monomer in tetrabutyl ammonium iodide–dimethylformamide base electrolyte. Wave A: acrylonitrile wave; wave B: styrene wave. Start potential -1.7 V.

method was not discovered. However, it could be due to the presence of interfering substances (e.g., polymerization additives such as residual traces of organic peroxide or of low polymers of acrylonitrile) in the copolymers, which might also react with dodecyl mercaptan and lead to high results in this method of analysis.

The polarographic method for determining acrylonitrile was also checked on some styrene–acrylonitrile copolymers containing less than 50 ppm of the monomer. Determination of acrylonitrile at these levels in copolymer is necessary when they are being considered for food-packaging applications. Acrylonitrile was determined in styrene–acrylonitrile copolymers that had been subjected to a treatment designed to reduce their monomer content, namely reprecipitation with methanol from chloroform solutions with subsequent removal of residual solvent from the reprecipitated polymer by drying under vacuum. Polymers that initially contained several hundred parts per million of acrylonitrile were shown by the polarographic method to contain about 25 ppm of monomer after a single reprecipitation step, which reduced to about 5 ppm of monomer after a further reprecipitation from chloroform.

The acrylonitrile and styrene monomer contents of several copolymers were determined by the polarographic procedure. The styrene monomer

Table 7 Determination of Acrylonitrile:
Comparison of Polarographic and Chemical
Methods

Acrylonitrile content, % w/w, determined by	
Polarograph	Titration with dodecyl mercaptan
0.06, 0.07	0.09, 0.09
0.08	0.12
0.11	0.15
0.11	0.19
0.12	0.15
0.12	0.17
0.14	0.18
0.21	0.31, 0.29

content of these samples was also determined by a procedure involving evaluation by UV spectroscopy at 292 nm of solutions of the samples in carbon tetrachloride. The results obtained (see Table 8) show the styrene contents determined by the two procedures are in good agreement.

    b. Acrylamide and Acrylic acid monomers. Betso and McLean (1976) have described a differential pulse polarographic method for carrying out the determination of these monomers. A measurement of the acrylamide electrochemical reduction peak current is used to quantitate the acrylamide concentration. The differential pulse polarographic technique also yields a well-defined acrylamide reduction peak at ca. -2.0 V versus the standard calomel electrode (SCA), suitable for qualitatively detecting the presence of acrylamide. The procedure involves an extraction of the acrylamide monomer

Table 8 Determination of Acrylonitrile and Styrene
Monomers in Copolymers

Acrylonitrile determined by polarography % w/w	Styrene, % w/w, determined by	
	Polarography	Spectroscopy
0.015, 0.014	0.20	0.21
0.015, 0.015	0.15	0.13, 0.15
0.035, 0.036	1.60	1.53

from the polyacrylamide, a treatment of the extraction solution on mixed resin to remove interfering cationic and anionic species, and polarographic reduction in an 80/20 (v:v) methanol/water solvent with *tert-n*-butylammonium hydroxide as the supporting electrolyte. Acrylic acid is polarographically distinguishable from acrylamide in a neutral medium. Ethyl acrylate is an interference in the analysis. Acrylonitrile is removed from interfering substances by treatment on mixed resin. The detection limit of acrylamide monomer by this technique is less than 1 ppm.

The recovery of acrylamide from the polymer totally depends on the extraction efficiency and the physical chemical interaction (e.g., absorption) between the monomer and the polymer. In evaluating the extraction procedure for an unknown polymer, the polymer extraction solution must be spiked with a known quantity of acrylamide monomer, and its recovery through extraction and resin treatment must be quantitated. Recovery of spiked acrylamide from polyacrylamides was greater then 90% with the extraction and resin treatment reported herein. The overall precision of the analysis was $\pm 5\%$.

The acrylamide reduction peak is well defined and well resolved from the background: no difficulty is encountered in either the detection or measurement of this peak. The differential pulse polarographic acrylamide reduction current is directly proportional to concentration, as shown in Table 9. The polarographic detection limit for acrylamide in a clean system is less then 1 pg acrylamide/ml. Even at this low concentration, the acrylamide reduction peak is well defined and resolved from the background.

The reduction of the acrylamide monomer occurs at ca. -2.0 V versus SCE. In this region of the polarogram, reduction of alkali cations occurs. The mixed resin treatment is used to remove polarographically interfering ionic species, such as sodium and potassium cations, which are major interferences. This is shown clearly in Fig. 7a, which shows a polarogram of a methanolic solution containing sodium ions and acrylamide in a weight ratio of 4:1. After a mixed resin treatment for 20 min all the interfering sodium species are removed; as seen in Fig. 7b, no acrylamide loss is detectable. Acrylic acid monomer does not interfere in the determination of acrylamide. Figure 7c shows a polarogram of acrylic acid and acrylamide in an 80/20 methanol/water solution with tetra-*n*-butyl ammonium chloride as the supporting electrolyte. In this medium, the reduction of the associated acid occurs at ca. -1.7 V versus SCE, 0.3 V more positive than the acrylamide reduction. The reduction peak of acrylic acid is easily resolved from acrylamide. The reduction current of this peak is directly proportional to the associated acid concentration. Acrolein, acetone, vinyl benzyl alcohol, and vinyl benzyl chloride also do not interfere. Acrylonitrile, nonionic species, nitrilotrispropionamide, and some esters of acrylic acid do interfere.

Table 9 Differential Pulse Polarographic Response to
Acrylamide Monomer

Acrylamide conc.[a], ppm	Peak current[b], μA	Peak potential, V vs. SCE	Peak/conc., μA/ppm
0	0.00	—	—
1	0.35	−2.01	0.35
2	0.68	−2.01	0.34
3	1.07	−2.01	0.36
4	1.37	−2.02	0.34
5	1.67	−2.02	0.33
11	3.61	−2.02	0.33
15	5.51	−2.02	0.37
21	7.44	−2.02	0.35
31	11.30	−2.03	0.36
41	15.00	−2.03	0.37
51	19.20	−2.03	0.38
96	37.3	−2.04	0.39

[a] 9.5 ml 80/20 methanol/water + 0.5 ml 1 N tetra-n-butyl-ammonium hydroxide.
[b] Peak current is calculated as current above baseline.

Other applications of polarography to the determination of monomers in polymers are summarized in Table 10.

B. Instrumentation

Three basic techniques of polarography are of interest and the basic principles of these are outlined.

1. Universal: Differential Pulse (DP)

In this technique a voltage pulse is superimposed on the voltage ramp during the last 40 msec of controlled drop growth with the standard dropping mercury electrode; the drop surface is then constant. The pulse amplitude can be preselected. The current is measured by integration over a 20 msec period immediately before the start of the pulse and again for 20 msec as the pulse nears completion. The difference between the two current integrals $(l_2 - l_1)$ is recorded, and this gives a peak-shaped curve. If the pulse amplitude is increased, the peak current value is raised but the peak is broadened at the same time.

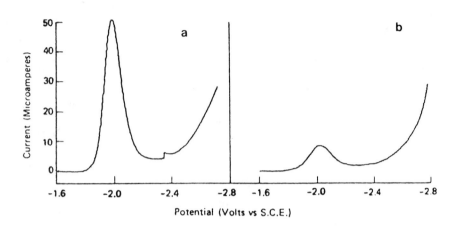

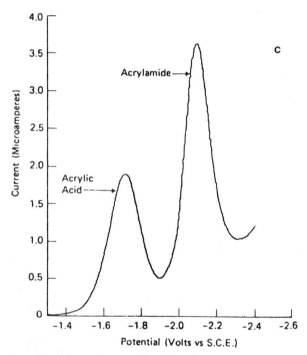

Fig. 7 Removal of sodium ion by resin treatment: (a) 1 mH Na$^+$ + 10 ppm acrylamide; (b) after 20 min on mixed resin; (c) differential pulse polarogram of a mixture of 10 ppm acrylic acid + 10 ppm acrylamide in methanolic solvent with tetra-*n*-butyl ammonium chloride as the supporting electrolyte.

Table 10 Polarographic Methods for the Determination of Monomers

Monomer	Polymer	Method	References
α-Methylstyrene	Polystyrene	Solvent solution polarography	Novak (1972) Podzeeva et al. (1971)
Styrene	Polystyrene	Solvent solution polarography	Novak (1972) Podzeeva et al. (1971) Crompton et al. (1965)
Methyl methacrylate	Polymethacrylates	Polarography	Katelin et al. (1977) Crompton and Myers (1968a)
Acrylonitrile	Styrene–acrylonitrile	Polarography	Schwoetzer (1972)
Styrene–acrylonitrile	Styrene–acrylonitrile	Solution in dimethyl formamide, polarography	Claver and Murphy (1959) Mayo et al. (1948) Shapras and Claver (1964)
Styrene	Polyester laminates, styrene–butadiene acrylics	Polarography	Novak and Seidl (1978)

2. *Classical: Direct Current Tast (DCT)*

In this direct current method, integration is performed over the last 20 msec of the controlled drop growth (Tast procedure); during this time, the drop surface is constant in the case of the dropping mercury electrode. The resulting polarogram is step-shaped. Compared with classical DC polarography according to Heyrovsky, that is, with the free-dropping mercury electrode, the DCT method offers great advantages: considerably shorter analysis times, no distrubance due to current oscillations, simpler evaluation, and larger diffusion-controlled limiting current.

3. *Rapid: Square Wave (SQW)*

Five square-wave oscillations of frequency around 125 Hz are super-imposed on the voltage ramp during the last 40 msec of controlled drop growth; with the dropping mercury electrode, the drop surface is then constant. The oscillation amplitude can be preselected. Measurements are performed in the second, third, and fourth square-wave oscillation; the current is integrated over 2 msec at the end of the first and the end of the second half of each oscillation. The three differences of the six integrals $(l_1 - l_2)$, $(l_3 - l_4)$, $(l_5 - l_6)$ are averaged arithmetically and recorded as one current value. The resulting polarogram is peak shaped.

Metrohm is a leading supplier of polarographic equipment. They supply three main pieces of equipment: the Metrohm 646 VA processor, the 647 VA stand (for single determinations), and the 675 VA sample changer for a series of determinations). Some features of the 646 VA processor are

Optimized data acquisition and data processing
High-grade electronics for a better signal-to-noise ratio
Automatic curve evaluation as well as automated standard addition for greater accuracy and smaller standard deviation
Large, nonvolatile methods memory for the library of fully developed analytical procedures
Connection of the 675 VA sample changer for greater sample throughput
Connection of an electronic balance
Simple, perfectly clear operation principle via guidance in the dialogue mode, yet at the same time high application flexibility thanks to the visual display and alphanumeric keyboard
Complete and convenient result recording with built-in thermal recorder/printer

The 675 VA sample changer is controlled by the 646 VA processor, on which the user enters the few control commands necessary. The 646 VA processor also controls the 677 drive unit and the 683 pumps. With these auxiliary units, the instrument combination becomes a polarographic analysis station that can be used to carry out on-line measurements.

The 646 VA processor is conceived as a central, compact component for automated polarographic and voltametric systems. Thus, two independent 647 VA stands or a 675 VA sample changer can be added. Up to four multidosimats of the 665 type for automated standard additions and/or addition of auxiliary solutions can be connected to each of these wet-chemical workstations. Connection of an electronic balance for direct transfer of data is also possible.

Program-controlled automatic switching and mixing of these three electrode configurations during a single analysis occur via software commands. The complete electrode is pneumatically controlled. A hermetically sealed mercury reservoir of only a few milliliters suffices for approximately 200,000 drops. The mercury drops are small and stable; consequently, there is a good signal-to-noise ratio. Mercury comes into contact only with the purest inert gas and plastic free of metal traces. Filling is seldom required and very simple to carry out. The system uses glass capillaries, which can be exchanged simply and rapidly.

Up to 30 complete analytical methods (including all detailed information and instructions) can be filed in a nonvolatile memory and called up.

Table 11 Polarographic Differential Pulse Direct Current Square-Wave and Anodic Scanning

Supplier	Type	Model no.	Detection limits
Metrohm	Differential Pulse Direct current Square wave	646 VA processor 647 VA stand 675 sample changer 665 Dosimat (motor-driven piston burettes for standard additions)	0.05 μm L^{-1} quoted 2–10 μg L^{-1} for nitriloacetic acid
	Direct current normal pulse differential pulse 1st harmonic a.c. 2nd harmonic a.c. Kalousek	506 Polarecord	
Chemtronics Ltd.	Direct current sampled differential pulse On-line voltammetric analyzer for metals in effluents and field work	DC 626 Polarecord PDV 2000	~0.1 μg L^{-1}
RDT Analytical Ltd.	Differential pulse anodic stripping on-line voltammetric analyzer for metals in effluents and field work On-line voltammetric analyzer for continuous measurement of metals in effluents and water	ECP 100 plus ECP 104 programs ECP 140 PDV 200 OVA 2000	— —
EDT Analytical Ltd.	Cyclic voltametry, differential pulse voltametry, linear scam voltametry, square-wave voltametry, single- and double-step chronompotentiometry, and chronocoulometry	Cipress Model CYSY-1B (basic system) CY57-1H (high-sensitivity system)	

Various suppliers of polarographs are summarized in Table 11.
Metrohm, in addition to the 646 VA processor, which can carry out differential pulse direct current and square-wave measurements, also supplies two other instruments capable of carrying out different kinds of measurements:

The SO6 Polarecord

Direct current
Normal pulse
Differential pulse
1st harmonic alternating current
2nd harmonic aalternating current
Kalousek

The 626 Polarecord

Direct current sampled direct current
Differential pulse

The latter is a basic instrument. It does not have sensitivity of the 646 VA.

V. HIGH-PERFORMANCE LIQUID CHROMATOGRAPHY

A. Volatiles

1. Monomers

High-performance liquid chromatography (HPLC) has found a limited number of applications in the determination of monomers including acrylic acid monomer in polyacrylates (Brown, 1979) and acrylamide (Husser et al., 1977). It has been used for the determination of oligomers in polyethylene terephthalate (Van der Maeden et al., 1978; Zaborsky, 1977) and epoxy resins (Van der Maeden et al. 1978).

Brown has described an HPLC method for the determination of acrylic acid monomer in polyacrylates in which a known mass of polymer (500 mg) is added to 50 ml of a methanol: distilled water mixture (1 + 1) and allowed to stand overnight to complete the extraction of the monomer from the polymer. Various mixtures of methanol and distilled water (range 1 + 99 to 99 + 1, 50 ml) are added to the polymer for 1, 2, 8, and 24 hours.

An aliquot (12 µl) of the sample is injected into the liquid chromatograph's Rheodyne valve and chromatographed under the following conditions:

column Whatman PXS 10/25 µm PAC (250 × 4 mm inner diameter [i.d.]); mobile phase, 0.01% v/v orthophosphoric acid in distilled water; flow rate, 4 ml/min: pressure, 138 kg cm^{-2}; detector wavelength, 195 nm; chart speed, 0.5 cm/min; and absorbance scale, 0.02. According to the sensitivity required, duplicate aliquots of the sample (1–100 µl) are injected. The concentration of acrylic acid is found by comparison with a previously prepared calibration graph of total absorbance versus original acrylic acid concentration.

High-performance liquid chromatography using the reverse-phase mode has been used by Skelly and Husser (1978) and Husser et al. (1977) to determine acrylamide monomer and related compounds, including methacrylonitrile in polyacrylamide. Water-soluble compounds such as acrylamide and methacrylamide have sufficient lipophilic character to be retained and separated on HPLC reverse-phase columns using water as the eluent. By employing a low-wavelength UV detector, these compounds can be measured with high sensitivity. The relative precision of the 95% confidence level for acrylamide is ±7.5%.

In this procedure the polymer is extracted for 4 hours with 80 + 20 methanol/water and the extract injected on a Partisil^{-10} OD5^{-2} 4 × 250 mm reverse-phase column. These extracts can also be examined by ion exclusion liquid chromatography under the following conditions:

Column: Partial 10 PAC (250 × 4.6 mm), octadecyl loading = 5, acrylonitrile retention for 5 min
Mobile phase: 15% methanol, 85% methylene chloride
Flow rate: 1 ml/min
Pressure: 35 kg cm^{-2}
Detector: 240 nm

The acrylamide response is linear from 1 to 500 ppm in solution using a 20-µl injection. This equates to 0.02–10 µg acrylamide injected. Area response obtained from a computing integrator is also found to be linear. Sensitivity of acrylamide detection is about 0.1 ppm in solution based on a 20-µl injection.

The retention times for acrylamide and related compounds are given in Table 12. No known impurities are observed at the retention time of acrylamide.

High-performance liquid chromatography has been used by Ernes and Hashumaker (1983) to determine traces of methylene *bis* (aniline) in polyurethanes.

2. Oligomers

Oligomers are very low molecular weight polymers. Thus, polystyrene usually contains low concentrations of monomer, dimer, trimer,

Table 12 HPLC Retention Times[a]
for Acrylamide and Related
Compounds

Compound	Minutes
Acrylic acid	1.4
β-Hydroxypropanamide	2.1
Acetamide	3.0
Acrylamide	5.4
Propanamide	7.3
Acrylonitrile	11.8
Methacylamide	18.0
Butanamide	20.8
Methacrylonitrile	46.0

[a] Partisil-10 ODS-2, water, 2.0 ml/min, 208 nm, 0.04 aufs.

and tetramer of the general formula $(C_6C_5—CH{=}CH_2)_n$, where $n = 1$ to 4. Various techniques have been used for the determination of oligomers, including gel permeation chromatography (Shiono, 1979; Zaborsky, 1977) (polyesters), thin-layer chromatography (Gankina et al., 1976) (poystyrene and poly α-methyl styrene), liquid chromatography (Ludwig and Bailie, 1984; Mourey and Smith, 1984; Mourey, 1984), and gas chromatography (Utterback et al., 1984). High-performance liquid chromatography has been used for the determination of oligomers in polyethylne terathalate and epoxy resins (Van der Maeden et al., 1978; Zaborsky, 1977).

Boehm and Martire (1989) have confirmed the feasibility of HPLC fractionation for the determination of flexible oligomers and homopolymers. Sample concentration has a significant effect on the retention behavior of high-molecular-weight homopolymers.

B. Instrumentation

Modern HPLC has been developed to a very high level of performance by the introduction of selective stationary phases of small particle sizes, resulting in efficient columns with large plate numbers per liter.

There are several types of chromatographic columns used in HPLC. The most commonly used chromatographic mode in HPLC is reversed-phase chromatography, used for the analysis of a wide range of neutral and polar organic compounds. Most common reversed-phase chromatography is performed using bonded silica-based columns, thus inherently limiting the operating pH range to 2.0–7.5. The wide pH range (0–14) of some columns (e.g., Dionex Ion Pac NSI and NS 1–5 μm columns) removes this limitation; consequently, they are ideally suited for ion-pairing and ion-suprression reversed-phase chromatography: the two techniques that have helped extend reverse-phase chromatography to ionizable compounds.

Typically, reversed-phase ion-pairing chromatography is carried out using the same stationary phase as reversed-phase chromatography. A hydrophobic ion of opposite charge to the solute of interest is added to the mobile phase. Samples determined by reversed-phase ion-packing chromatography are ionic and thus are capable of forming an ion pair with the added counterion. This form of reversed-phase chromatography can be used for anion and cation separations and for the separation of surfactants and other ionic types of organic molecules.

Ion suppression is a technique used to suppress the ionization of compounds (such as carboxylic acids) so that they will be retained exclusively by the reversed-phase retention mechanism and chromatographed as the neutral species.

1. Elution Systems

Four basic types of elution system are used in HPLC. This is illustrated below by the systems offered by LKB, Sweden.

a. The isocratic system. The isocratic system consists of a solvent delivery for isocratic reversed-phase and gel filtration chromatography. It provides an economic first step into HPLC techniques. The system is built around a high-performance, dual-piston, pulse-free pump providing precision flow from 0.01 to 5 ml/min.

Any of the following detectors can be used with this system:

Fixed-wavelength UV detector (LKB Unicord 2510)
Variable UV visible (190–600 nm)
Wavelength monitor (LKB 2151)
Rapid diode array spectral detector (LKB 2140) (discussed later)
Refractive index detector (LKB 2142)
Electrochemical detector (LKB 2143)
Wavescan EG software (LKB 2146)

b. Basic gradient system. This is a simple upgrade of the isocratic system with the facility for gradient elution techniques and greater functionality. The basic system provides for manual operating gradient techniques such as reversed-phase, ion-exchange, and hydrophobic interaction chromatography. Any of the detectors listed for the isocratic system can be used.

c. Advanced gradient system. For optimum functionality in automated systems designed primarily for reversed-phase chromatography and other gradient techniques, the LKB advanced-gradient system is recommended. Key features include the following:

A configuration that provides the highest possible reproducibility of results

A two-pump system for highly precise and accurate gradient formation for separation of complex samples

Full system control and advanced method development provided from a liquid chromatography controller

Precise and accurate flows ranging from 0.01 to 5 ml/min

This system is ideal for automatic method development and gradient optimization.

  d. The inert system. By a combination of the use of inert materials (glass, titanium, and inert polymers) this system offers totally inert fluidics. Primary features of the system include the following:

The ability to perform isocratic or gradient elution by manual means

Full system control from a liquid chromatography controller

Precise and accurate flows from 0.01 to 5 ml/min

This is the method of choice when corrosive buffers, for example, those containing chloride or aggressive solvents, are used.

2. Chromatographic Detectors

  Details concerning the types of detectors used in HPLC are given in Table 13. The most commonly used detectors are those based on spectrophotometry in the region 185–400 nm, visible UV spectroscopy in the region 185–900 nm, post-column derivativization with fluorescence detection (see next section), conductivity, and those based on multiple-wavelength UV detectors using a diode array system detector (see Section 4). Other types of detectors available are those based on electrochemical principles, refractive index, differential viscosity, and mass detection.

3. Post-Column Derivatization—Fluorescence Detectors

  Modern column liquid chromatography has been developed to a very high level by the introduction of selective stationary phases of small particle sizes, resulting in efficient columns with large plate numbers per meter. The development of HPLC equipment has been built on the achievements in column technology, but the weakest part is still the detection system. Fluorescence and UV/visible detectors offer tremendous possibilities, but because of their specificity it is possible to detect components only at very low concentrations with a specific chromophore or fluorophore. The lack of a sensitive all-purpose detector in liquid chromatography, like the flame ionization detector in gas chromatography, is still disadvantageous for the detection of important groups of compounds, such as amino acids. Consequently, chemical methods are increasingly used to enhance selectivity and sensitivity of detection. On-line post-column derivatization started with the classic work of Spackmann et al. (1958) and has recently found increasing interest and use (Frei and Lawrence, 1981a, b; Krull, 1986; Englehardt, 1979; Englehardt and Newe, 1982; Englehardt and Lilling, 1985; Englehardt et al., 1985; Uehlein and Swab, 1982).

With on-line post-column detection, the complexity of the chromatographic equipment increases. An additional pump is required for the pulseless and constant delivery of the reagent.

4. Diode Array Detectors

With the aid of a high-resolution UV diode array detector, the eluting components in a chromatogram can be characterized on the basis of their UV spectra. The detector features high spectral resolution (comparable with that of a high-performance UV spectrophotometer) and high spectral sensitivity. The high spectral sensitivity permits the identification of spectra near the detection limit, that is, within the submilliabsorbance range.

Several manufacturers (Varian, Perkin-Elmer, LKB, and Hewlett Packard; see Table 13) have developed diode array systems. In the polychromator incorporated in the Perkin-Elmer LC 480 diode array system, the light beam is dispersed within the range 190–430 nm onto a diode array consisting of 240 light-sensitive elements. This effects a digital resolution of 1 nm, which thus satisfies the spectral resolution determined by the entrance slit.

The stand-alone design of the detector used in the Perkin-Elmer LC-480 instrument includes all the functions chromatographers need: four-channel detection, analog or digital on-line output of chromatograms, arithmetic combination of two channels each, for example, for peak purity confirmation or enhanced selectivity. In addition to the ability of the system to run wavelength programs, the built-in gradient compensation software is an important feature; a baseline drift caused by a solvent gradient can be stored internally for all four channels. As samples are run, the baseline will be corrected on-line and thus can be fed directly into a data handling device.

The Perkin-Elmer model LC-235, designed for routine applications, combines the advantages of a variable UV detector with those of diode array technology. Conversely, the LC 480 AutoScan diode array detector optimizes the sensitivity of self-scanning diode arrays to meet HPLC requirements and facilitate LC–UV coupling by achieving high spectral resolution.

Table 13 Detectors Used in HPLC

Type of detector		Supplier	Detection part no.	HPLC instrument part no.
Spectrophotometric (variable wavelength)	190–390 nm	Perkin-Elmer	LC-90	—
	195–350 nm	Kontron	735 LC	Series 400
	195–350 nm	Shimadzu	SPD-7A	LC-7A
	195–350 nm	Shimadzu	SPD-6A	LC-8A
	195–350 nm	Shimadzu	SPD-6A	LC-6A
	206–405 nm (fixed wavelength choice of 7 wavelengths between 206 and 405 nm)	LKB	2510 Uvicord SD	—
Variable wavelength UV-visible	190–370 nm	Cecil Instruments	Model 1937	Chrom-A-Scope
	190–400 nm	Cecil Instruments	CE 1220	Series 1000
	190–600 nm	Varian	2550	2500
	190–600 nm	LKB	2151	Uvicord SD
	190–700 nm	Kontron	432	Series 400
	190–800 nm	Kontron	430	Series 400
	185–900 nm	Kontron	720 LC	Series 400
	200–570 nm	Kontron	740 LC	Series 400
	190–800 nm	Dionex	VDM II	Series 400
	190–750 nm	Isco	V4	Series 400
	214–660 nm (18 preset wavelengths)	Isco	UAS and 228	Microbo system
	195–700 nm	Shimadzu	SPD 7A	LC-7A
	195–700 nm	Shimadzu	SPD-6AV	LC-8A

Detector	Spectral range	Manufacturer	Model	Model
	195–700 nm	Shimadzu	SPD-6AV	LC-6A
	193–350 nm	Shimadzu	SPD6A/SPDM6A	LC-9A
	196–700 nm	Shimadzu	SPDEA V	LC-9A
	190–900 nm	Shimadzu	—	LC-10A
	190–900 nm	Varian	—	star 9060
	190–900 nm	Pharmacic	—	
	190–900 nm	ICI	—	LC/1210/1205
	190–600 nm	Hewlett Packard	Programmably variable wavelength detector	9050 series
	380–600 nm	Cecil Instruments	CE 1200	Series 1000
	190–800 nm	Applied Chromatography Systems	750/16 and 5750/11	—
Conductivity	—	Dionex	CDM 11	4500 i
	—	Roth Scientific	—	Chrom-A-scope
Electrochemical detector		Dionex	PAD-11	4500 i
		LKB	2143	Wave-scan EG
		Roth Scientific	—	Chrom-A-scope
		Cecil Instruments	CE 1500	—
		PSA Inc.	5100A	—
		Applied Chromatography Systems	650/350/06	—
Refractive index detector		LKB	2142	Wavescan E.G.
		Roth Scientific	—	Chrom-A-Scope
		Cecil Instruments	CE 1400	Series 1000

Table 13 *Continued*

Type of detector		Supplier	Detection part no.	HPLC instrument part no.
Differential viscosity mass detection (evaporative)		Roth Scientific Applied Chromatography Systems	—	Chrom-A-Scope
		Varian	750/14	—
Diode array		Varian	9060	2000L and 5001 5500 series
		Perkin-Elmer	LC135, LC 235, and LC 480	—
		LKB	2140	—
		Hewlett-Packard	Multiple-wavelength detector	1050 series
Fluorescence	220–900 nm	Shimadzu	RF-551	LC-9A
	—	Shimadzu	FLD-ʌA	LC-9A
	220–630 nm	Shimadzu	RF 335	LC-9A

5. *Electrochemical Detectors*

These are available from several suppliers (Table 13). ESA supplies the model PS 100A coulochem multielectrode electrochemical detector. Organics, anions, and cations can be detected by electrochemical means.

VI. MISCELLANEOUS METHODS

Guthrie and McKinney (1977) have described a thin-layer chromatography–fluorimetric method for the determination of 2,4- and 2,6-diaminotoluene in methanol extracts of flexible polyurethane foams. The precision of this method at the 20 ppm level is $\pm 30\%$. Diaminotoluenes can be detected in amounts down to 1 ppm. The thin-layer plates are developed with 120:33:20:7 v/v chloroform: ethyl acetate: ethanol: glacial acetic and after drying are sprayed with 0.015% fluram in acetone. The separated aminotoluene spots can be visualized under a UV lamp and are evaluated within 1 hour of development using an Aminco scanner with excitation and emission wavelengths of 390 and 500 nm, respectively.

Other methods for the determination of monomers in polymers are tabulated in Table 14.

Table 14   Miscellaneous Methods for the Determination of Monomers

Monomer	Polymer	Method	References
Styrene ethyl acrylate	Styrene ethyl acrylate copolymer	Radioactive tracer methods	Acosta and Sastre (1975) Tweet and Miller (1963)
Styrene methacrylic acid	Styrene methacrylic acid copolymer	Mercuimetric	Kreshkov et al. (1972, 1973)
Styrene acrylonitrile	Styrene acrylonitrile copolymer	Bromometric titration	Roy (1977)

VII. HEAD-SPACE ANALYSIS

A. Solution Head-Space Gas Chromatographic Methods

In this procedure a solution of the polymer in a suitable solvent is placed in a closed container and allowed to equilibrate at at controlled temperature so that volatile monomers or other impurities dissolved in the polymer solution partition between the solution and the gas phase. Subsequent analysis of the gas phase enables the concentration of monomers to be calculated. In a variant of this method, the solid polymer is allowed to equilibrate with the head-space gas.

Although the solid head-space method, discussed later, provides about 10-fold more sensitivity than the solution head-space method (assuming a 10% sample solution), the solid method may be applied only to sample systems where equilibration with the head space is rapid and complete. For example, residual styrene monomer in polystyrene does not reach equilibrium with the head space after 20 hours (Rohrachneider, 1971) and thus may not be determined by the solid head-space method. Furthermore, even if equilibration between the solid and head space is obtained, the partition coefficient must also be determined for the component of interest in each type of sample matrix.

The solution head-space approach is applicable to a much wider range of samples than the solid approach. When working with sample solutions, head-space equilibrium is more readily attained and the calibration procedure is simplified. The sensitivity of the solution method depends on the vapor pressure of the constituent to be analyzed and its solubility in the solvent phase. Vinyl chloride, butadiene, and acrylonitrile are readily promoted from polymer solutions into the head space by heating to 90°C. The head space/solution partioning for these constituents is not appreciably affected by changes in the solvent phase (viz. addition of water) since the more volatile materials favor the head space at 90°C. Less volatile monomers such as styrene (boiling point = 145°C) and 2-ethylhexyl acrylate (boiling point = 214°C) may not be determined using head-space techniques with the same sensitivities realized for the more volatile monomers. By altering the composition of the solvent phase to decrease the monomer solubility, the equilibrium monomer concentration in the head space can be increased. This results in a dramatic increase in the detection sensitivity for styrene and 2-ethylhexyl acrylate. Based on these principles, a procedure will be described for the gas chromatographic analysis of residual vinyl chloride, butadiene, acrylonitrile, styrene, and 2-ethylhexyl acrylate in polymers by head-space analysis.

The more volatile monomers vinyl chloride, butadiene, and acrylonitrile can be determined by dissolution of the polymer in *N,N'*-dimethylacetamide in closed vials and analysis of the equilibrated head space above the polymer solution. By this method Streichen (1976) showed it was possible to determine vinyl chloride and butadiene at the 0.05 ppm level, and acrylonitrile down to 0.5 ppm. The injection of water into polymer solutions containing styrene and 2-ethylhexyl acrylate monomers prior to head-space analysis greatly enhanced the detection capability for these monomers, making it possible to determine styrene down to 1 ppm and 2-ethylhexyl acrylate at 5 ppm. Incorporation of polymer into the calibration standards compensates for the effect that the polymer matrix has on the equilibrium partitioning of the monomer between the solution and head space. The relative precision and error in the determination of these monomers near the quantitation limit was found to be less than 7%.

In the Streichen (1976) method, weighed portions of the polymer were dissolved in septum-sealed vials containing measured aliquots of *N,N'*-dimethylacetamide (DMA). The vials were heated to 90°C to aid dissolution of the polymer. When solution was complete, the vials were cooled to room temperature. The solutions were swirled to mix, and an aliquot of distilled water was forcibly injected into each polymer solution to decrease the solubility of monomers. The vials were shaken briefly to assure complete mixing of the water with the organic phase, and to prevent the precipitated polymer from forming a film on top of the solution. The vials were equilibrated at 90°C for 60 min prior to head-space sampling and analysis by flame ionization–gas chromatography. Standards comprising monomer-free polymer and known additions of standard monomer solutions were run in parallel. Figure 8 shows a typical head-space calibration curve.

Greater sensitivities and shorter analysis times were obtained using the head-space analysis methods than were possible by the direct injection of polymer solutions into a gas chromatograph (Table 15).

The response obtained for a given monomer using the head-space techniques depended on several factors: the relative volatility of the monomer, the concentration of polymer in solution, and the solubility of the monomer in the solvent phase.

The greatest detection sensitivities using solution head-space analysis were obtained for monomers having relatively low boiling points. These values were determined experimentally using polymer solutions with known amounts of monomer added.

The equilibrium head-space concentration for 2-ethylhexyl acrylate at 90°C was not sufficient to allow the determination of residual 2-ethylhexyl acrylate in the polymer, even when present at the 1000 ppm level. The sensitivity achieved by the head-space method can be improved by decreasing the solubility of the 2-ethylhexyl acrylate monomer in the solution of polymer

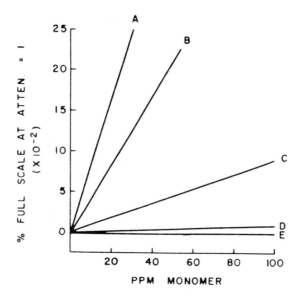

Fig. 8 Head-space calibration curves for monomers in DMA–polymer solutions: (A) butadiene, (B) vinyl chloride, (C) acrylonitrile, (D) styrene, (E) 2-ethylhexyl acrylate.

Table 15 Comparison of Quantitation Limited[a] for Residual Monomers Using Conventional and Head-Space GC Methods

Monomer	Boiling point	Direct solution injection[b]	Solution head space	Modified solution head space
Vinyl chloride	−13°C	1–2 ppm	0.05 ppm	—[c]
Butadiene	−4°C	5 ppm	0.05 ppm	—[c]
Acrylonitrile	76°C	10 ppm	0.5 ppm	—[c]
Styrene	145°C	10 ppm	20 ppm	1 ppm
2-Ethylhexyl acrylate	214°C	200 ppm	1000 ppm	5 ppm

[a] The quantitation limit is defined as the monomer concentration necessary to produce a peak at least three times the baseline noise of 3% of full scale.
[b] Injection of a 10% polymer solution into a gas chromatograph.
[c] A two- to threefold-fold increase in monomer peak height resulted from the injection of water into the polymer solution. A baseline disturbance due to elution of water negated any real improvement in detection limit for these monomers.

in N,N'-dimethylacetamide through the introduction of a second solvent. Water is the most effective solvent for this purpose. A greater than 200-fold increase in the 2-ethylhexylacrylate equilibrium head-space concentration resulted when water is injected into its polymer solutions.

Variations in the polymer weight had a much greater effect on the monomer response for 2-ethylhexylacrylate and styrene than was observed for the other monomers. This greater dependence on polymer weight is apparently related to the precipitation of the polymer that occurs upon the injection of water. This may be attributed to either coprecipitation of the monomer with the polymer or entrapment of water in the polymer precipitate. Regardless of the mechanism, the precision was not affected.

The time required for the monomer in solution to equilibrate with the head space was less then 60 min in all cases. When using the solution method (no water added), equilibrium was reached in less than 30 min.

B. Solid Polymer—Head-Space Analysis

This technique has been applied to the determination of styrene monomer in polystyrene and its copolymers. The polymer (2 g) is placed in a 250-ml screw-top glass jar with a Teflon seal and left in an oven for 4–6 hours at 110°C prior to withdrawal of a portion of the head space for gas chromatographic analysis using a flame ionization detector. The chromatogram in Fig. 9 shows the volatiles from a sample of the terpolymer polystyrene–α-methylstyrene–acrylonitrile.

A sample of monomer-free polystyrene was spiked with styrene monomer at concentrations of 0.5, 1.0, 1.5, and 2.0 ppm and 2 ppm of an internal standard of o-xylene was added. The ratio of the peak height of styrene to o-xylene was used to form a calibration curve (Fig. 10). This curve has some scatter, but it allows adequate accuracy for this concentration range.

1. Water (and Volatiles)

Jeffs (1969) has described a rather complicated piece of equipment for the gas chiromatographic determination of water and other volatiles in vinyl, acrylic, and polyolefin powder polymers. The instrument is shown diagrammatically, in Fig. 11 and consists essentially of a sample tube, forming an external loop, connected to a gas chromatograph. This loop can be isolated and the sample heated to the required temperature. After an initial heating period, the volatile constituents liberated from the sample are "flushed" onto the chromatographic column by a flow of carrier gas through, or over, the sample, and the required components are separated and determined quantitatively. A pneumatic switch valve located in the chromatographic oven to prevent the condensation of volatile constituents within the valve and a split

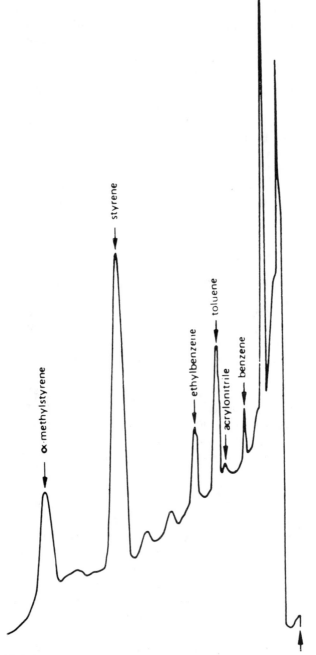

Fig. 9 Chromatogram of head-space vapor.

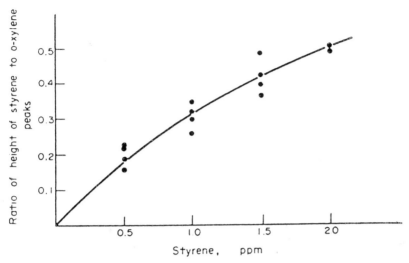

Fig. 10 Internal standard calibration for determination of styrene monomer.

heater mounted on a horizontal travel in a plane at right angles to the sample tube are essential parts of the apparatus. The instrument is semiautomatic. The carrier gas flows through the copper sulfate in tube D, which imparts a constant amount of water (about 3 ppm w/v) to the helium. The "wet" carrier gas prevents the gas-flow lines from "drying out." Dry pipework tends to adsorb moisture, which can then be desorbed, thus leading to spurious results. The determination of water in a sample is unaffected as the "wet" carrier gas flows continuously through both the reference and analysis cells of the katharometer.

The pneumatic sample valve, F, operates as a switch valve, directing the carrier-gas flow either around the internal loop, G, or the external loop, H. The pilot valve N_1 operates sample valve F. The sample split heater, J, consists of a cylindrical aluminium block 15.2 cm long with a 9-mm hold through the center. The block is split axially and the two halves hinged. Each half of the block contains two cartridge-heater elements, each 14 cm long, 0.95 cm outer diameter (o.d.), and one half contains a thermocouple pocket to accommodate the thermocouple, P. The cartridge heaters are supplied by a semiconductor energy controller contained in R, which is controlled by a galvanometer, two-position temperature controller, Q. The heater is mounted on a horizontal travel in a plane at right angles to the sample tube. A jig for mounting the sample tubes is also part of the heater assembly. The 0.95 cm, coupling K_1 is brazed to its mounting bracket, which is rigidly

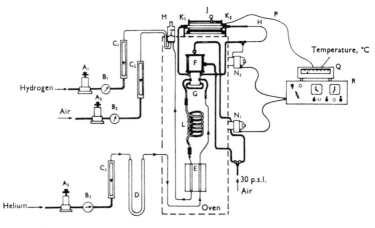

A₁, A₂ and A₃ = Edwards VPC I pressure controller

B₁, B₂ and B₃ = Pressure gauges 0 to 30 p.s.i.

C₁, C₂ and C₃ = Rotameter-type flow gauges

D   = Clear plasticised PVC tubing (5 foot long × ¼ inch bore), packed with copper sulphate crystals, $CuSO_4.5H_2O$, >44 mesh

E   = Katharometer

F   = Pneumatic sample valve (Pye Cat. No. 12900), fitted with P 9904 change-over block

G   = Internal loop

H   = External loop made in part of 18-gauge stainless-steel capillary tubing

J   = Sample split heater

K₁ and K₂ = Straight reducing couplings, captive seal type, for ⅜ to ¼ inch o.d. tubing (Drallim, Cat. No. L/50/D/B)

L   = Chromatographic column

M   = Flame-ionisation detector

N₁ and N₂ = Electrically actuated 3-port pilot valves (Martinair, Type 557C/IZ)

P   = Nickel - chromium/nickel - aluminium thermocouple embedded in the split heater

Q   = Electronic temperature controller, proportional type

R   = Sample-valve time-delay unit, containing three synchronous timers and a semi-conductor, proportional energy controller for the sample heater

Fig. 11   Details of general-purpose instrument for gas chromatographic determination of water and volatiles in polymers.

attached to the heater base. This coupling is accurately centered, with the central hole through the aluminium block. The coupling K_2 is brazed to the flexible carrier-gas inlet tube and rests loosely in the second mounting bracket. The 0.95 cm couplings are supplied with neoprene or butyl rubber captive seals.

The sample-heater assembly is placed on top of a Griffin and George oven, so that the L-bend of 0.31 cm o.d. S/S tubing disappears almost immediately into an opening on the top of the oven. In practice, both the inlet tube (15.2 cm long, 0.63 cm o.d., and 0.31 cm i.d. copper) and the exit tube (0.31 cm o.d.) are wrapped with heating tape and lagging to maintain the temperature of the whole assembly at about 100°C. The inlet tube is wrapped to a length of 10.1–12.7 cm and the L-bend of the exit tube is wrapped to a point 7.6 cm inside the oven. Both tubes are wrapped up to, and including, the 0.63 cm thread of the Drallim coupling, leaving only the center nut and the 0.31 cm coupling nut exposed so that the sample tubes can be readily changed. N_2 acts as a pressure release valve to the external loop H.

The apparatus shown in Fig. 11 is fitted with katharometer and flame-ionization detectors. Although only one detector is necessary for any one specific method, for example, a katharometer for the determination of water in polymer powder, it is invaluable to have both available (with separate recorders) to establish the conditions, for instance, in the above case to ensure that no organic components are being eluted at the same time as water and thus contributing to the peak measurement. Figure 12 shows chromatograms obtained simultaneously from the katharometer and the flame-ionization detectors on a partially dried poly(vinyl chloride) (PVC) powder.

Figures 13 and 14 show some typical results obtained when this method is applied to vinyl polymers and polyolefins. The peaks due to water and monomers are clearly visible.

Jeffs (1969) recommends that before carrying out any quantitative work

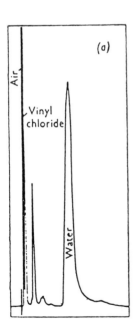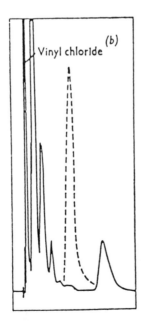

Fig. 12 Gas chromatograms obtained simultaneously with (a) katharometer and (b) flame-ionization detectors on a partially dried PVC powder. In (b) the dotted line represents the portion of the water peak as transposed from (a).

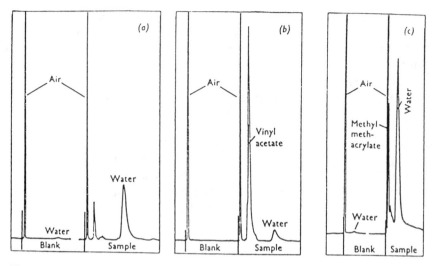

Fig. 13 Typical gas chromatograms obtained with method for (a) polyvinyl chloride (PVC) powder, (b) PVC–polyvinyl acetate (PVA) copolymer power, and (c) acrylic molding powder.

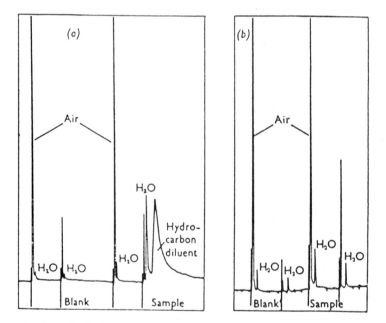

Fig. 14 Typical gas chromatograms obtained with method for (a) polypropylene powder and (b) high-pressure polyethylene.

on the volatile constituents obtained from a polymer powder, a preliminary gas-chromatographic investigation should be carried out of their complexity. For example, Fig. 12(b) (flame-ionization detector trace) shows nine components other than air, water, and the original monomer. These components are chlorinated hydrocarbons such as 1,1- and 1,2-dichloroethane and *cis*- and *trans*-dichloroethylene that are present as impurities in the original monomer.

Although these impurities are present only in ppm amounts in the original monomer, they can be readily detected in the polymer. If water is to be determined, then a suitable column has to be chosen so that the water is eluted free from organic constituents. At this stage the recording of simultaneous chromatograms with the two different detectors is invaluable.

Other head-space analysis methods for the determination of monomers in polymers are summarized in Table 16.

2. Volatiles Other than Monomers

Solid polymer head-space analysis: A simple and inexpensive apparatus has been described (Crompton et al., 1965; Crompton and Myers, 1968a) for liberating both existing volatiles in polymers and those produced by thermal degradation from polymers by heating at temperatures up to 300°C, in the absence of solvents, prior to their examination by gas

Table 16 Head-Space–Gas Chromatographic (GLC) Methods for the Determination of Monomers

Polymer	Monomers	References
Polystyrene	Styrene	Shanks (1975a)
		Rohrachneider (1971)
		Shapras and Claver (1962)
		Tweet and Miller (1963)
		Wilkinson et al (1964)
		Nowak and Klemett (1962)
		Adcock (1962)
PVC	Vinyl chloride	Berens et al. (1974)
		Berens (1974)
Styrene–butadiene	Styrene–butadiene	Shapras and Claver (1964)
		Shanks (1975b)
Styrene–butadiene α-methyl styrene–acrylonitrile	Styrene–butadiene acrylonitrile	Shanks (1975b)

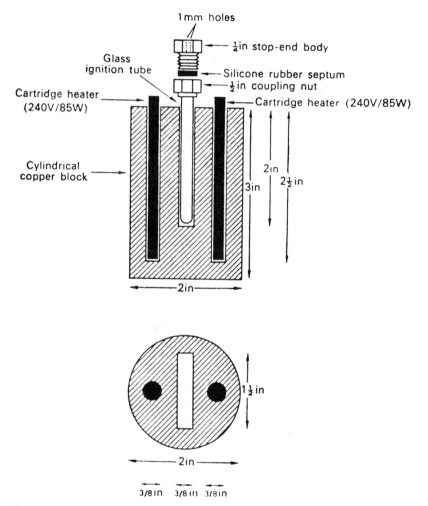

Fig. 15 Apparatus for liberating volatiles from polymers.

chromatography. The technique avoids the disadvantages resulting from the use of extraction or solution procedures.

The apparatus illustrated in Fig. 15 consists of a glass ignition tube, supported as shown in a Wade 0.63 cm diameter brass coupling nut, covered with a silicone rubber spectrum and sealed with a Wade 0.63 cm brass stop-end body. The stop-end body has two 1-mm diameter holes drilled through the cap. The whole unit is placed in a slot in a cylindrical copper block (7.6 cm long × 5.1 cm dia.), which is heated by two (240 V, 85 W) cartridge heaters and controlled at temperatures up to 300°C from a variable transformer. The temperature is measured with a thermocouple capable of accurately measuring temperatures in the 100–300°C range with a maximum error of ±5%. The thermocouple is inserted in the slot adjacent to the ignition tube; it has been shown that under these conditions the thermocouple records the true temperature of the contents of the tube. The provision of a slot in the copper block enables more than one ignition tube to be heated simultaneously if required.

A sample of the polymer (0.25–0.50 g) is placed in an ignition tube and sealed with Wade fittings and a septum, as described. If necessary the tube is then purged with a suitable gas by inserting two hypodermic needles through the septum via the holes in the cap of the stop-end body and passing the gas into the tube through one hole and allowing it to vent through the other. After purging, the two needles are removed simultaneously and the tube is then heated in the copper block for 15 min at required temperature. A sample (1–2 ml) of the head-space gas is withdrawn from the ignition tube into a Hamilton gastight hypodermic syringe via the septum, and injected into a gas chromatograph. It is advisable to fill the syringe with the gas used initially in the ignition tube and to inject this into the tube before withdrawing the sample. This facilitates sampling by preventing the creation of a partial vacuum in the ignition tube or the syringe or both. It also minimizes any undesirable entry of air into the ignition tube.

With the apparatus, a polymer may be heated under any desired gas; while this may frequently be the carrier gas used with the gas chromatograph, it is also possible to carry out studies in oxidizing or reducing atmospheres. A polymer may also be heated to any temperature, and samples of the head-space gas may be withdrawn at intermediate temperatures and times to determine under what conditions any particular volatile is liberated.

By using gas chromatography detectors of suitable sensitivities and selectivities, one can examine polymers for the presence or formation of volatiles at both the percentage and the ppm levels. For example, traces of organic halogen compounds lend themselves to analysis with an electron capture detector. Thermal conductivity cells of the hot-wire or thermistor type are suitable for the detection of inorganic volatiles, and a helium ionization detector could be used for analyzing trace amounts of permanent gases.

Figure 16 shows some results obtained by applying this technique to a sample of solid polyethylene at 125°C and 200°C. Evidently low-molecular-weight paraffin solvents are present.

Food and drink containers extruded or molded from polyethylene sometimes possess unpleasant odors that are likely to taint the packaged

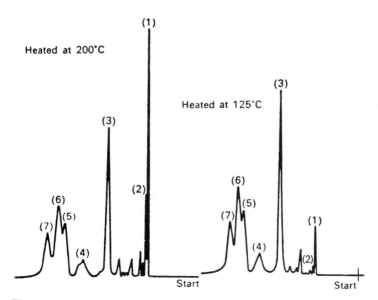

Fig. 16 Gas chromatogram of volatiles liberated from polyethylene heated at different temperatures for 15 min in air. Chromatographed on 61 m × 0.16 cm i.d. dibutyl phthalate coated copper column at 30°C and 100 ml/min helium flow, with flame-ionization detector.

product and are unacceptable to the consumer. In one such case it was found that, by heating a sample of an odor-producing polyethylene for 15 min at 200°C under helium, the chromatogram of the liberated volatiles contained certain peaks absent from the corresponding chromatogram from a polyethylene that produced nonodorous food containers. The temperature of 200°C was chosen to simulate extrusion temperature. The two chromatograms are shown in Fig. 17, from which it may be seen that components A, B, D, and I are present in the odorous sample but are absent from the nonodorous sample. These substances were always associated with the odorous polyethylene.

The quantitative aspects of this technique itself were investigated by analyzing the volatiles liberated by heating polystyrene at 200°C and then reanalyzing the polymer as a solution in propylene oxide. Several samples of polystyrene were examined in this way, and in each case component peak areas were normalized and compared. The results (Table 17) show that the values obtained by the head-space technique differ by up to 20% from those obtained by the solution procedure, which are known to be correct.

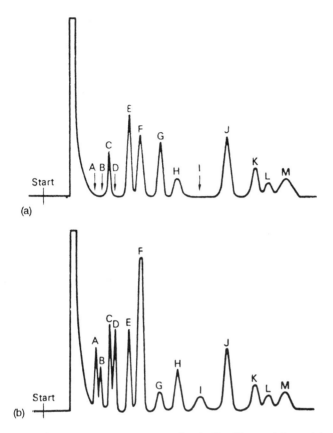

Fig. 17 Gas chromatograms of volatiles liberated from (a) nonodorous and (b) odorous polyethylenes at 200°C for 15 min in helium. Chromatographed on 61 m × 0.16 cm i.d. dibutyl phthalate coated copper column at 30°C and 100 ml/min helium flow, with flame-ionization detector.

Figure 18 shows a gas chromatogram obtained when this technique is applied to a sample of polystyrene heated to 200°C. A wide range of aromatic volatiles is present.

Solid polymer head-space analysis—alternative procedure: The polymer is heated and the head-space atmosphere analyzed by gas chromatography as described by Davies and Bishop (1971) and Schmidt (1961, 1962). The film or sheet, together with internal standard, is placed in a 250-ml sealed container and heated at 100°C for 90 min.

The following results were obtained for the determination of toluene and ethyl acetate on adjacent pieces of a polyethylene adhesive laminated

Table 17 Comparison of Normalized Peak Areas Calculated from Chromatograms of Polystyrene Solutions and Polystyrene Volatiles Liberated by Heating at 200°C

	Normalized peak area							
	Analysis of polymer as a solution in polypropylene oxide				Analysis of volatiles liberated by heating polymer at 200°C			
Component	Manufacturer				Manufacturer			
	A	B	C	D	A	B	C	D
Ethyl benzene	16.4	17.4	27.2	27.5	20.2	23.4	35.4	34.4
n-Propyl benzene	5.5	5.3	3.5	3.2	3.2	2.5	1.9	2.3
Cumene	17.7	18.1	10.4	10.8	15.4	12.7	10.2	10.7
m-/*p*-Xylenes	4.1	2.5	2.7	2.0	1.1	1.3	0.6	0.7
m-/*p*-Ethyl toluenes	1.7	1.7	2.0	1.5	0.8	0.7	0.5	0.6
Styrene	54.6	55.0	54.2	55.0	59.3	59.4	51.4	51.3

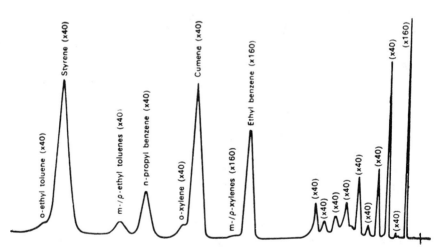

Fig. 18 Gas chromatograms of volatiles liberated from different polystyrenes at 200°C for 15 min in helium. Chromatographed on 15 ft × 3/16 in. 10% Carbowax 15–20 m on 60–72 Celite at 90°C and 100 ml/min helium flow, with flame-ionization detector.

to polypropylene double-coated with Saran. The solvents originate from the adhesive.

									Mean	S.D.
Toluene (mg/m^2)	127	135	119	127	140	122	130	119	127	7
Ethyl acetate (mg/m^2)	136	145	124	141	136	133	138	128	135	6

Results as follows were obtained for the determination of toluene on adjacent pieces of printed film (polypropylene with a single Saran coating). The toluene originated from the printing ink.

Toluene (mg/m^2)   11.9, 11.4, 12.3, 11.9, 12.5, 11.9, 12.3
Mean: 12.0 mg/m^2;   standard deviation: 0.34 mg/m^2

This method has been used to determine many different solvents in several different substrates. The solvents include ethanol, ethyl acetate, ethyl methyl ketone, 2-ethoxyethanol, propan-l-ol, and toluene. The substrates include polythene, polypropylene, and cellophane, which occur individually, coated with Saran or combined in laminates.

Figure 19 shows typical chromatograms obtained from a film and a calibration jar.

A variant on this method that has been applied to polystyrene is depicted in Fig. 20. The gas chromatograph carrier gas is fed to the instrument via a gas sampling valve, coupled as shown. The sampling valve is attached to a glass tube, also as shown, and the latter is heated by means of a removable hot copper block.

The heater block is lowered and the sample valve turned to the "bypass" position. Then 5–50 mg of the polymer is weighed into a rimless ignition tube, which connects to the apparatus. The sample valve is turned to the "analysis" position for 1 min to remove air; then the sample valve is turned to the "bypass" position and the heater block (at 240°C) is raised to surround the ignition tube, which is then heated for 5 min. The sample valve (Fig. 20) is then turned to the "analysis" position to allow the chromatogram to develop.

Determine (F) for each aromatic hydrocarbon as follows:

$$F = \frac{\text{wt of hydrocarbon}}{\text{wt of } n\text{-undecane}} \times \frac{\text{pk. ht. of } n\text{-undecane}}{\text{pk. ht. of hydrocarbon}}$$

From the analysis chromatogram, calculate the concentration of each aromatic hydrocarbon as follows:

$$\% \text{ wt/wt hydrocarbon} = \frac{\text{pk. ht. of hydrocarbon}}{\text{pk. ht. of } n\text{-undecane}} \times \frac{10\,FP}{\text{wt of sample}}$$

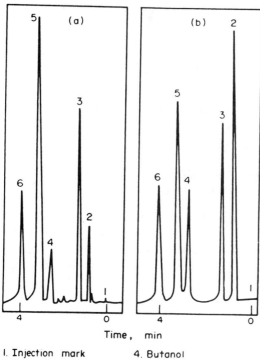

Time, min

1. Injection mark
2. Ethanol
3. Ethyl methyl ketone
4. Butanol
5. Toluene
6. Butyl acetate (internal standard)

Fig. 19 (a) Chromatogram of solvents from a sample of printed polythene–polypropylene laminate. Chromatograph. Pye 104 with flame-ionization detector, column, glass 152 cm × 0.63 cm packed with 9% silicone oil and 3% UCON HB2000 on Chromosorb W; oven, 80°C; attenuation × 20,000; and chart speed, 1 cm/min. (b) Calibration jar prepared for sample shown in (a).

where $P = \%$ wt/vol concentration of *n*-undecane in the polymer solvent.

Column	Copper tube (460 cm × 0.48 cm i.d.) packed with 10% wt/wt Carbowax 15–20 M on 60–62 BS mesh acid-washed Celite.
Gas flows	Helium, 30 psig, rotameter = 10.0 (100 ml/min); Hydrogen, 13 psig, rotameter = 10.0 (75 ml/min); Air, 7 psig, rotameter = 10.0 (650 ml/min).
Temperatures	Injection 155°C, Column 80°C, Detector 125°C, Flame 200°C.
Recorder	Honeywell-Brown, 1 mV full-scale deflection, 1 sec response, 10 in./hour chart speed.

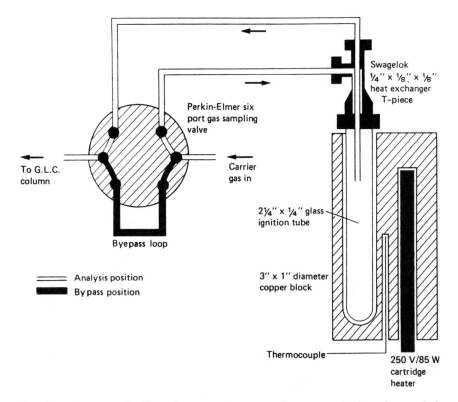

Fig. 20 Apparatus for liberating expanding agents from expandable and expanded polystyrene.

Hagman and Jacobsson (1989) have developed a theoretical model for the quantitative determination of volatile compounds in solid polymers by dynamic head-space sampling. They correlated desorption parameters (temperature and flow rate) with the component–polymer properties (diffusion and distribution coefficients). Using this treatment they quantitatively predicted concentrations and recovery of hexane, tridecane, and butylated hydroxy toluene antioxidant from polypropylene. Di Pasquale et al. (1978) have described head-space analysis techniques for the determination of dichloroethane in polycarbonate and styrene and *o*-xylene in polystyrene.

VIII. PURGE AND TRAP METHODS

This is an alternative to head-space analysis for the identification and determination of volatile organic compounds in polymers. The sample is swept with an inert gas for a fixed period of time. Volatile compounds from the sample are collected on a solid sorbent trap—usually activated carbon. The trap is then rapidly heated, and the compounds collected and transferred as a plug under a reversed flow of inert gas to an external gas chromatograph. Chromatographic techniques are then used to quantify and identify sample components.

A. Volatiles

Roper (1970) has discussed the problem of determining very low concentrations of volatiles in polymers. Methods for the determination of such volatiles frequently include application of heat to the sample and the sweeping action of an inert gas to separate the volatile components from the polymer. The volatiles are then analyzed by gas chromatography. When there is a low concentration of volatile material, it is advantageous to concentrate it to improve the shapes of the chromatographic peaks.

To achieve this Roper employed a trap tube, shown in Fig. 21a. The capillary portion of the tube is packed with a gas chromatographic column packing consisting of 20–30% of a suitable liquid phase on a granular diatomaceous type support. The tube is fitted with a suitable hypodermic needle so that it can be connected to the gas chromatograph through the injection port. The apparatus is arranged as in Fig. 21b. The gas chromatograph is equipped with a temperature program and any suitable column to separate the various volatile components.

The polymer sample to be analyzed is weighed into the trap tube, the sample size being chosen to give suitably sized peaks for measurement. The tube is connected to an inert-gas stream by means of butyl rubber tubing and heated to the proper temperature while the gas stream sweeps the volatile material into the packed section of the tube, which is usually cooled with dry ice. The temperature, sweeping rate, and sweeping time should be determined experimentally and will vary with the type of polymer being investigated. Melting the polymer is often necessary to rid it of volatiles. A microcombustion furnace is a suitable heater for the trap tube.

After the volatiles have been adsorbed in its packed section, the trap tube is disconnected from the butyl rubber tube and moved to the chromatograph. With valve A open and valve B closed, the tube is connected to the butyl rubber tube with the heater positioned away from the capillary end. Then, with valve A closed too, the hypodermic needle is inserted in the injection port and valve B is opened. The heater, already at the proper temperature, is then moved to the capillary portion of the tube, where the heat and carrier-gas flow sweep the volatiles from the packing through the hypodermic needle and into the gas chromatograph. In some cases it is

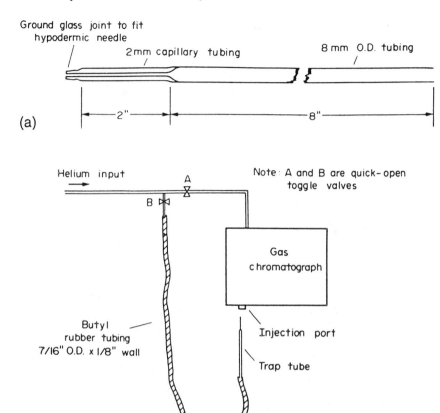

Fig. 21 Head-space method for volatiles in polymers: (a) borosilicate glass trap tube; (b) arrangement of apparatus.

necessary to program the oven temperature of the chromatograph to get a suitable chromatogram.

It is advantageous to use an electronic integrator to determine the response in counts per microgram for the various components to be measured. With this information the percentage of each volatile component of the sample can be calculated from the number of integrator counts in each peak.

B. Instrumentation

OIC Analytical Instrument supplies the 4460 A purge and trap concentrator. This is a microprocessor-based instrument with capillary column capability.

It is supplied with an autosampler capable of handling 76 sample vials. Two automatic rinses of sample lines and vessel purge are carried out between sample analyses to minimize carryover.

Tekmar is another supplier of purge and trap analysis equipment. Their LSC 200 purge and trap concentrator features glass-lined stainless steel tubing, menu-driven programming with four-method storage, and a cryo-focusing accessory.

Cryofocusing is a technique in which only a short section of the column or a pre-column is cooled. In its simplest form a section of the column near the inlet is immersed in a flask of coolant during desorb. After desorb, the coolant is removed and the column allowed to return to the over temperature.

CDS Analytical also supplies a microprocessor-controlled purge and trap concentrator, which features cryogenic trapping, cryogenic refocusing, and thermal desorption to a trap or directly to a gas chromatograph.

IX. DETERMINATION OF WATER

Conventional weight loss methods for determining water in polymers have several disadvantages, not the least of which is that any other volatile constituents of the polymer such as solvents, dissolved gases, volatile substances produced by decomposition of the polymer, or additives therein are included in the determination.

Specific methods for the determination of water in polymers can be based on the principle of heating a weighed amount of polymer in a boat in a tube furnace through which flows a gentle purge of nitrogen (Maltese et al., 1969). The wet nitrogen passes into an automated Karl Fischer titration unit, which estimates the water as it is released from the polymer. In addition to being absolutely specific for water, this method can provide information on the effects of temperature and time on the release of water from the polymers.

The head-space analysis technique described by Jeffs (1969) is applicable to the determination of water and other volatiles in polymers.

Hoffman (1976) has described a procedure for the determination of water in polyesters. In this method, water is allowed to react quickly with a mixture of hexamethyldisilazane and trimethylchlorosilane (2:1) in the presence of pyridine to form hexamethylidisiloxane:

$$2(Me - Si_2)_2NH + 3H_2O + 2Me_3SiCl$$

$$= 3Me_3Si - O - SiMe_3 + 2NH_4Cl$$

The hexamethyldisiloxane is separated from other silanized active hydrogen compounds and determined by gas chromatography using an internal standard. The sensitivity and specificity is good. The limit of

Table 18 Head-Space Samplers

Supplier	Head-space analyzer model no.	Capillary/packed	Compatible gas chromatograph model no.	Automated	Microprocessor control	Multiple head-space analysis option	Pressure option	Sample carousel	Isothermal/temp. programming at 9°C	Thermostatic sample temperature
Perkin-Elmer	HS-101	Capillary	8000 Series	Yes	Yes	Yes	Yes	100 sample	Yes/Yes	35–150
Perkin-Elmer	HS-100	Capillary	Sigma 2000	Yes	Yes	Yes	Yes	600 sample	Yes/Yes	35–150
Perkin-Elmer	HS-6	Packed or capillary	Sigma 2000 Models 3920, 900, 900, F-22 8-30	No	No	Yes	Yes	No	—	40–196
Perkin-Elmer	HS-6B	Packed or capillary	8000 Series	Yes	Yes	Yes	Yes	No	Yes/Yes	190
Perkin-Elmer	HS-40 (automated)	Packed or capillary	—	Yes	Yes	Yes	Yes	Yes	Yes/Yes	150
Sievers	Headspace	Packed or capillary	Sichromat 1-4 Sichromat 2-8	No	No	—	—	No	Yes/Yes	
Shimadzu	HSS-2A	Packed or capillary	GC-9A	Yes	Yes	No	No	40 sample	Yes/Yes	40–150
Shimadzu	HSS-2B (automated)	Packed or capillary	GC-14A	Yes	Yes	Yes	No	100 sample	Yes/Yes	150
Shimadzu	HSS-3A	Packed or capillary	GC-16A GC-15A	Yes	Yes	Yes	No	40 sample	Yes/Yes	150
Shimadzu	HSS-2B	Packed or capillary	GC-14A							
01 Analytical	4460A	Packed or capillary								
Hewlett Packard	19395A	Packed or capillary	HP7890A							

detection is 50 ppm, and the relative standard deviation at the 200 ppm level is 13%.

Squirrel (1981) has also reported on the phenomenon that with some polymers water, in addition to the initial free water content of the polymer, is produced during heating. He observed this in the case of polyethylene terphthalate and, for this reason, prefers to determine water by Karl Fischer titration.

Dry methanol is placed in the sample flask, excess of dilute Karl Fischer reagent (1 ml = 1 mg of water) is added, and the contents are boiled under reflux for 20 min to dry the apparatus. Slightly wet methanol is then added to the flask, and the contents are immediately titrated to a visual endpoint by means of the Karl Fischer reagent. The flask and contents are then refluxed for 6 hours, and any water picked up during this time is again titrated with the Karl Fischer reagent. This establishes that the equipment is giving a low blank value, which should be less then the equivalent of 2 ml of Karl Fischer reagent.

For the analysis of the sample, the apparatus is again predried and the sample introduced via the 70-ml top cup. The cup is filled to a 50-g mark with polymer granules, and then the latter are allowed to drop into the flask by releasing the clip on the wide-bore rubber tubing. In this way the exposure of the sample to the atmosphere is minimal. The 6-hour reflux period is then repeated, and the water content of the sample is calculated from the difference between the titer for sample and blank experiments.

Other methods for the determination of water in polymers are reviewed in Table 19.

Table 19 Determination of Water in Polymers

Polymer	Procedure	References
PVC	Devolatilization–gas chromatography	Jeffs (1969)
Polypropylene	Heating under nitrogen–Karl Fischer titration	Maltese et al. (1969)
Anionic resins	Oven drying–Karl Fischer titration	Armieau and Costain (1972)
Polyamide and polyethane diol terephtalate	Removal of water in vacuum or a stream of nitrogen, then Karl Fischer titration	Armieau and Costain (1972) Maltese et al. (1969)
Several polymers		Harrington and Keister (1975) Rice and Trowell (1967)

REFERENCES

Acosta, J. L., and Sastre. R. (1975). *Rev.*, *29*(224): 212.

Adcock, L. H. (1962). *Patra*, *3*: 5.

Armieau, V., and Costain, D. (1972). *Mater. Plast.* (*Bucharest*), *9*: 606.

Baba, T. (1977). *Shokuhin Eisegaku Aasshi.*, *18*(6): 500.

Berens, A. R. (1974). "The Solubility of Vinyl Chloride in Polyvinylchloride," paper delivered to the 168th ACS Meeting, Atlantic City, New Jersey, September.

Berens, A. R., Brider, L. B., Tomanek, C. M., and Whitney, J. M. (1974). "Analysis of Vinyl Chloride in PVC Powders by Head-Space Chromatography," manuscript circulated to members of the Vinyl Chloride Safety Association, November 14, BP Goodrich tire Center, Brecksville, Ohio.

Betso, S. R., and McLean, J. D. (1976). *Analyt. Chem.*, *48*: 766.

Boehm, R. E., and Martire, D. E. (1989). *Analyt. Chem.*, *61*: 471.

Boettner, E. A., Ball, G., and Weiss, B. (1969). *J. Appl. Polym. Sci.*, *13*: 377.

Brodsky, J. (1961). *Kunststoffe*, *51*: 20.

Brown, L. (1979). *Analyst* (*London*), *104*: 1165.

Carsen, B. B., Haintzelman, W. J., Moe, H., and Rousseau, C. R. (1962), *J. Organic Chem.*, *27*: 1636.

Claver, G. C., and Murphy, M. E. (1959). *Analyt. Chem.*, *31*: 1682.

Cobler, J. G., and Samsel, D. D. (1962). *SPE Trans.*, April.

Crompton, T. R., and Buckley, D. (1965). *Analyst* (*London*), *90*: 76.

Crompton, T. R., and Myers, L. W. (1968a). *Plastics and Polymers*, *205* (June).

Crompton, T. R., and Myers, L. W. (1968b). *European Polym. J.*, *4*: 355.

Crompton, T. R., Myers, L. W., and Blair, D. (1965). *British Plastics*, December.

Davies, J. T., and Bishop, J. R. (1971). *Analyst* (*London*), *96*: 55.

De Forero, I. B., De Rascouvsky, E. G., De Ruiz, M., and Del Carmen, S. (1971). *Plasticos*, *19*: 122.

Di Pasquale, C., Di Iorio, G., and Capaccioli, T. (1978). *J. Chromatogr.*, *152*: 538.

Doak, K. W. (1948). *J. Amer. Chem. Soc.*, *70*: 1525.

Englehardt, H. (1979). In *High Performance Liquid Chromatography*, Springer, Berlin.

Englehardt, H., and Lillig, B. (1985). *J. High Resolution Chromatogr. Column Chromatogr.*, *8*: 531.

Englehardt, H., and Newe U. D. (1952). *Chromatographia*, *15*, 403.

Englehardt, H., Klinkner, R., and Lillig, B. (1985). In *Kopplungver fahren in der HPLC-GIT*, Verlag Darmstadt.

Ernes, D. A., and Hashumaker, (1983). *Analyt. Chem.*, *55*: 408.

Frei, R. W., and Lawrence, J. F. (1981a). In *Chemical Derivatization in Analytical Chemistry*, vol. 1, Plenum Press, New York.

Frei, R. W., and Lawrence, J. F. (1981b). In *Chemical Derivatization in Analytical Chemistry*, vol. 2, Plenum Press, New York.

Gamble, L. W., and Jones, W. H. (1955). *Anal. Chem.*, *27*: 1456.

Gankina, E. S., Val'chikhina, M. D., and Belen'kii, G. (1976). *Vysokomol. Soedin. Ser. A*, *18*(5): 1170.

Gilbert, S. G., Giacin, J. R., Morano, J. R., and Rosen, J. D. (1975). *Package Dev. Syst.*, *5*: 20.

Guthrie, J. L., and McKinney, R. W. (1977). *Analyt. Chem.*, *49*: 1676.

Hagman, A., and Jacobsson, S. (1989). *Analyt. Chem.*, *61*: 102.

Harrington, R. C., and Keister, D. D. (1975). Paper presented at the Plastics Paper Conference of the Technical Association of the Pulp and Paper Manufacturers Association, New York.

Hartley, A. J., Lueng, Y. K., McMahon, J., Booth, C., and Shepherd, I. W. (1968). *Polymer*, *18*: 336.

Haslam, J., and Jeffs, A. R. (1957). *J. Appl. Chem.*, *7*: 24.

Haslam, J., and Jeffs, A. R. (1958). *Analyst* (*London*), *83*: 455.

Haslam, J., Jeffs, A. R., and Willis, H. A. (1962). *J. Oil Col. Chem. Assoc.*, *45*: 325.

Hileman, F. D., Vornees, K. J., Wojeik, L. H., Birky, M. M., Ryan, P. W., and Einhorn, I. N. (1975). *J. Polvm. Sci.*, *13*: 571.

Hoffman, E. R. (1976). *Analyt. Chem.*, *48*: 445.

Holtmann, R., and Souren, J. R. (1977). *Kunststoffe*, *67*: 776.

Husser, E. R., Stehl, R. H., and Price D. R. (1977). *Analyt. Chem.*, *49*: 154.

Iedrezejczak, K., and Gaind, U.S. (1993). *Analyst* (*London*), *118*: 149.

Jeffs, A. K. (1969). *Analyst* (*London*), *94*: 249.

Joseph, K. T., and Browner, R. F. (1980). *Analyt. Chem.*, *52*: 1083.

Katelin, A. I., Komleva, V. N., and Mal'kova, L. N. (1977). *Metody. Anal. Kontrolya, Proizvod. Khim. Prom.—St.*, *11*: 62.

Klesper, E., and Hartmann, W. (1978). *European Polym. J.*, *14*: 77.

Kreshkov, A. P., Balyatinskaya, L. N., and Chesnokova, S. M. (1972). *Tr. Mosk. Khim Teknol. Inst.*, *70*: 146.

Kreshkov, A. P., Balyatinnskaya, L. N., and Chesnokova, S. M. (1973). *Zh. Anal. Khim.*, *28*(8): 1571.

Krull, L. S., ed. (1986). *Reaction Detection Gas Chromatography*, Marcel Dekker New York.

Liao, J. C., and Browner, R. F. (1978). *Analyt. Chem.*, *50*: 1683.

Ludwig, J. F., and Bailie, A. G. (1984). *Analyt. Chem.*, *56*: 2081.

Maltese, P., Chementini, L., and Panizza, S. (1969). *Mater. Plaste. Elastomerie*, *35*: 1669.

Markelov, M. A., and Semenko, E. (1973). *Plast. Massy.*, (8): 65.

Mayo, F. R., Lewis, F. M., and Walling C. (1948). *J. Amer. Chem. Soc.*, *70*: 1529.

Mel'nikova, S. L., Tishehenko, U. T., and Sazonenko, U. V. (1977). *Lakokras Mater. Ikh. Primen.*, *4*: 56.

Michal, J., Mitera, J., and Tardon, S. (1976). *Fire & Materials*, *1*: 160.

Miller, D. F., Samuel, E. P., and Cobler, J. G. (1961). *Analyt. Chem.*, *33*: 677.

Mourey, T. H. (1984). *Analyt. Chem.*, *56*: 1777.

Mourey, T. H., and Smith, G. A. (1984). *Analyt. Chem.*, *56*: 1773.

Neldon, F. M., Eggertsen, F. T., and Holst, J. J. (1961). *Analyt. Chem.*, *33*: 1150.

Novak, V. (1972). *J. Chem. Prumsyl.*, *22*: 298.

Novak, V., and Siedl, J. (1978). *J. Chem. Prumsyl.*, *28*: 186.

Nowak, P., and Klemett, O. (1962). *Kunststoffe*, *52*: 604.

Oprea, N., and Pogorevic, A. (1974). *Rev. Chim. Bucharest,* 25: 244.

Perkin-Elmer (1981). *Gas Chromatography,* No. 3.

Pfab, W., and Noffz, D. (1963). *Z. Anal. Chem.,* 37: 195.

Podzeeva, R. M., Lukhovitskii, U. I., and Forpov, V. L. (1971). *Zavod. Lab.,* 37: 168.

Ragelis, E. P., and Gajan, R. J. F. (1962). *Assoc. Offic. Agric. Chem.,* 45: 918.

Rice, D. D., and Trowell, J. M. (1967). *Analyt. Chem.,* 39: 157.

Rohrachneider, L. (1971). *Z. Anal. Chem.,* 255: 345.

Roper, J. N. (1970). *Analyt. Chem.,* 42: 688.

Rosenthal, R. W., Schwartzman, L. H., Greco, N. P., and Proper, P. J. (1963). *J. Org. Chem.,* 28: 2835.

Roy, S. S. (1977). *Analyst (London),* 102: 302.

Schmalz, E. O. (1969). *Anal. Abstracts,* 17: 2449.

Schmalz, E. O. (1970). *Faserforsch Tex. Tech.,* 21: 209.

Schmidt, W. (1961). *Beckman Report, Beckman Associates,* 4: 6.

Schmidt, W. (1962). *Beckmann Report, Beckman Associates,* 3: 13.

Schwoetzer, G. (1972). *Z. Anal. Chem.,* 260: 10.

Shanks, R. A. (1975a). *Pye Unicam Newsletter.*

Shanks, R. A. (1975b). *Scan.,* 6: 20.

Shapras, P., and Claver, G. C. (1962). *Analyt. Chem.,* 34: 433.

Shapras, P., and Claver, G. C. (1964). *Analyt. Chem.,* 36: 2282.

Shiono, S. J. (1979). *Polymer Sci., A-1,* 17: 4120.

Shiryaev, B. V., and Kozmenkhova, E. B. (1972). *Zavod. Lab.,* 38: 1303.

Skelly, N. F., and Husser, E. R. (1978). *Analyt. Chem.,* 50: 1959.

Spackman, D. H., Stein, W. H., and Moore, S. (1958). *Analyt. Chem.* 30: 1190.

Squirrel, D. C. M. (1981). *Analyst (London),* 106: 1042.

Streichen, R. J. (1976). *Analyt. Chem.,* 48: 1398.

Takashima, S., Okada, F., and Nimeji, O., (1960). *Kogyo Daigaku Kenkyu Hokoku,* 12: 34.

Tengler, H., and Von Falkai, B. (1972). *Kunststoffe,* 62: 759.

Tweet, O., and Miller, W. K. (1963). *Analyt. Chem.,* 35: 852.

Uehlein, M., and Schwab, E. (1982). *Chromatographia,* 15: 140.

Urbanski, J. (1977). *Anal. Chem. (Warsaw),* 22: 749.

Utterback, D. F., Millington, D. S., and Gold, A. (1984). *Analyt. Chem.,* 56: 470.

Van der Maeden, F. P. B., Biemond, M. E. F., and Janssen, P. C. G. M. (1978). *J. Chromatog.,* 149: 539.

Welsch, T., Engewald, W., and Kowash, E. (1976). *Plaste. u Kaut.,* 23: 584.

Wilkinson, L. R., Norman, C. W., and Brettner, N. P. (1964). *Analyt. Chem.,* 36: 1759.

Wylie, P. L., Perkins, P. D., and Kanahan, G. E. (1986). "Identification of Volatile Residues in Polymers by Headspace. Gas Chromatography–Mass Spectrometry," Hewlett Packard Gas Chromatography Application Brief, January.

Yamaoka, A., and Matsui, T. (1977). *Himeji Kogyo Daigaku Kenkyu Hokoku,* 30: 101.

Yamaoka, A., and Matsui, T. (1978). *Chem. Abstracts,* 89: 7000w.

Zaborsky, L. M. (1977). *Anal. Chem.,* 49: 1166.

Appendix A

The Vapor Pressure of Some Common Solvents and Monomers

This appendix provides data for determining the vapor pressure of some common solvents and monomers as a function of temperature. The following table contains the coefficients $A-E$ to be used in the correlation

$$P_1^0 = \exp\left[A + \frac{B}{T} + C \ln T + DT^E \right]$$

for calculating the vapor pressure in pascals as a function of temperature in degrees Kelvin.

Also given are the minimum and maximum temperatures in degrees Kelvin (T_{min} and T_{max}) at which the correlation is valid for each substance, along with the calculated vapor pressures at these temperatures. These values are provided as a means by which the reader may check for errors in any computation scheme using the correlation.

Following the tables are plots of vapor pressure versus temperature for the range 0°C and above. These are provided as a quicker but less accurate method to determine and compare vapor pressures. The upper curve

The data in these tables are reproduced by permission of Hemisphere Publishing Corporation from *Physical and Thermodynamic Properties of Pure Chemicals* by T. E. Daubert and R. P. Danner, Hemisphere Publishing Corporation, Philadelphia (1989). The assistance of T. E. Daubert in providing updated data is gratefully acknowledged.

in each plot indicates the vapor pressure in millimeters of mercury. The lower curve in each plot indicates the vapor pressure in kPa.

RELATED CONVERSION FACTORS

Temperature

degrees Celsius to Kelvin:	$T[\mathrm{K}] = T[^\circ\mathrm{C}] + 273.16$
degrees Fahrenheit to Kelvin:	$T[\mathrm{K}] = (T[^\circ\mathrm{F}] + 459.67)/1.8$
degrees Fahrenheit to Celsius:	$T[^\circ\mathrm{C}] = (T[^\circ\mathrm{F}] - 32)/1.8$
degrees Rankine to Kelvin:	$T[\mathrm{K}] = T[^\circ\mathrm{R}]/1.8$
degrees Rankine to Celsius:	$T[^\circ\mathrm{C}] = T[^\circ\mathrm{R}]/1.8 - 273.16$

Pressure

Pascals to psi:	$P[\mathrm{psi}] = P[\mathrm{Pa}] \times 1.4504 \cdot 10^{-4}$
Pascals to mm Hg:	$P[\mathrm{mm\ Hg}] = P[\mathrm{Pa}] \times 7.5006 \cdot 10^{-3}$
mm Hg to psi:	$P[\mathrm{psi}] = P[\mathrm{mm\ Hg}] \times 1.9337 \cdot 10^{-2}$

Table 1 Vapor Pressure Coefficients of Common Solvents and Monomers

	A	B	C	D	E	T_{min} [K]	P_1^0 @ T_{min} [Pa]	T_{max} [K]	P_1^0 @ T_{max} [Pa]
ACETIC ACID	5.0691E+01	-6.1779E+03	-3.9199E+00	7.1932E-18	6.0000E+00	289.81	1.2810E+03	592.71	5.6792E+06
ACETONE	6.9006E+01	-5.5996E+03	-7.0985E+00	6.2237E-06	2.0000E+00	178.45	2.7851E+00	508.20	4.7091E+06
ACRYLONITRILE	8.7604E+01	-6.3927E+03	-1.0101E+01	1.0891E-05	2.0000E+00	189.63	3.6829E+00	535.00	4.4800E+06
BENZALDEHYDE	1.1628E+02	-9.3312E+03	-1.4639E+01	1.1932E-02	1.0000E+00	216.02	4.8388E-02	695.00	4.6400E+06
BENZENE	8.3918E+01	-6.5177E+03	-9.3453E+00	7.1182E-06	2.0000E+00	278.68	4.7620E+03	562.16	4.8819E+06
BENZYL ALCOHOL	9.9357E+01	-1.1032E+04	-1.0499E+01	2.5779E-18	6.0000E+00	257.85	1.7885E-01	720.15	4.5010E+06
1,3-BUTADIENE	7.3522E+01	-4.5643E+03	-8.1958E+00	1.1580E-05	2.0000E+00	164.25	6.9110E+01	425.17	4.3041E+06
n-BUTYL ACETATE	7.1340E+01	-7.2858E+03	-6.9459E+00	9.9895E-18	6.0000E+00	199.65	1.4347E-01	579.15	3.1097E+06
n-BUTYL ACRYLATE	8.1598E+01	-7.6129E+03	-8.7765E+00	5.9550E-06	2.0000E+00	208.55	2.1980E-01	598.00	2.9098E+06
n-BUTYL ALCOHOL	9.3173E+01	-9.1859E+03	-9.7464E+00	4.7796E-18	6.0000E+00	184.51	5.7220E-04	563.05	4.3392E+06
CARBON TETRACHLORIDE	7.8441E+01	-6.1281E+03	-8.5766E+00	6.8465E-06	2.0000E+00	250.33	1.1225E+03	556.35	4.5436E+06
CHLOROFORM	1.4643E+02	-7.7923E+03	-2.0614E+01	2.4578E-02	1.0000E+00	207.15	5.2512E+01	536.40	5.5543E+06
m-CRESOL	9.5403E+01	-1.0581E+04	-1.0004E+01	4.3032E-18	6.0000E+00	285.39	5.8624E+00	705.85	4.5221E+06
CUMENE	1.4362E+02	-9.6877E+03	-1.9305E+01	1.7703E-02	1.0000E+00	177.14	3.8034E-04	631.00	3.1837E+06
CYCLOHEXANE	1.1651E+02	-7.1033E+03	-1.5490E+01	1.6959E-02	1.0000E+00	279.69	5.3802E+03	553.58	4.0958E+06
CYCLOHEXANONE	9.5118E+01	-8.3004E+03	-1.0796E+01	6.5037E-06	2.0000E+00	242.00	6.9667E+00	653.00	4.0126E+06
n-DECANE	1.2125E+02	-1.0115E+04	-1.4543E+01	8.2302E-06	2.0000E+00	243.51	1.3279E+00	618.45	2.1297E+06

	A	B	C	D	E	T_{min} [K]	P_1^0 @ T_{min} [Pa]	T_{max} [K]	P_1^0 @ T_{max} [Pa]
DI-n-BUTYLAMINE	6.0838E+01	-7.0928E+03	-5.4295E+00	9.3799E-18	6.0000E+00	211.15	1.6296E-01	602.30	2.5630E+06
DIETHYLAMINE	4.9634E+01	-4.9537E+03	-3.9783E+00	1.0421E-17	6.0000E+00	223.35	3.7941E+02	496.60	3.6808E+06
DIETHYL ETHER	1.3690E+02	-6.9543E+03	-1.9254E+01	2.4508E-02	1.0000E+00	156.85	3.9545E-01	466.70	3.6412E+06
DIMETHYLAMINE	7.1738E+01	-5.3020E+03	-7.3324E+00	6.4200E-17	6.0000E+00	180.96	7.5575E+01	437.20	5.2583E+06
DIMETHYLFORMAMIDE	8.2762E+01	-7.9555E+03	-8.8038E+00	4.2431E-06	2.0000E+00	212.72	1.9532E-01	649.60	4.3653E+06
DIMETHYL SULFOXIDE	5.6273E+01	-7.6206E+03	-4.6279E+00	4.3819E-07	2.0000E+00	291.67	5.0229E+01	729.00	5.6477E+06
1,4-DIOXANE	4.4494E+01	-5.4067E+03	-3.1287E+00	2.8913E-18	6.0000E+00	284.95	2.5325E+03	587.00	5.1577E+06
ETHANOL	7.4475E+01	-7.1643E+03	-7.3270E+00	3.1340E-06	2.0000E+00	159.05	4.8459E-04	513.92	6.1171E+06
ETHYL ACETATE	6.6824E+01	-6.2276E+03	-6.4100E+00	1.7914E-17	6.0000E+00	189.60	1.4318E+00	523.30	3.8502E+06
ETHYL ACRYLATE	5.4005E+01	-5.9639E+03	-4.4734E+00	5.0315E-18	6.0000E+00	201.95	2.0728E+00	553.00	3.6606E+06
ETHYLBENZENE	8.8090E+01	-7.6883E+03	-9.7708E+00	5.8844E-06	2.0000E+00	178.15	4.0140E-03	617.20	3.5968E+06
FORMIC ACID	5.0323E+01	-5.3782E+03	-4.2030E+00	3.4697E-06	2.0000E+00	281.45	2.4024E+03	588.00	5.8074E+06
n-HEPTANE	1.5462E+02	-8.7931E+03	-2.1684E+01	2.3916E-02	1.0000E+00	182.56	1.2372E-01	540.26	2.7534E+06
n-HEXANE	1.6547E+02	-8.3533E+03	-2.3927E+01	2.9496E-02	1.0000E+00	177.84	8.0387E-01	507.43	3.0338E+06
METHANOL	8.1768E+01	-6.8760E+03	-8.7078E+00	7.1926E-06	2.0000E+00	175.47	1.1147E-01	512.64	8.1402E+06
METHYL ACRYLATE	1.0769E+02	-7.0272E+03	-1.3916E+01	1.5185E-02	1.0000E+00	196.32	4.0696E+00	536.00	4.2767E+06
METHYL ETHYL KETONE	7.2698E+01	-6.1436E+03	-7.5779E+00	5.6476E-06	2.0000E+00	186.48	1.3904E+00	535.50	4.1201E+06

	A	B	C	D	E	T_{min} [K]	P_1^0 @ T_{min} [Pa]	T_{max} [K]	P_1^0 @ T_{max} [Pa]
METHYL METHACRYLATE	1.0736E+02	-8.0853E+03	-1.2720E+01	8.3307E-06	2.0000E+00	224.95	1.9086E+01	566.00	3.6744E+06
n-NONANE	6.3634E+01	-7.2006E+03	-5.8097E+00	4.8890E-18	6.0000E+00	219.66	6.2345E-01	595.65	2.2847E+06
n-OCTANE	1.0047E+02	-8.1004E+03	-1.1663E+01	7.6624E-06	2.0000E+00	216.38	1.9747E+00	568.83	2.4797E+06
n-PENTANE	8.3143E+01	-5.6201E+03	-9.4863E+00	1.0001E-05	2.0000E+00	143.42	5.2631E-02	469.65	3.3443E+06
PHENOL	9.5444E+01	-1.0113E+04	-1.0090E+01	6.7603E-18	6.0000E+00	314.06	1.8798E+02	694.25	6.0585E+06
n-PROPANOL	8.8134E+01	-8.4986E+03	-9.0766E+00	8.3303E-18	6.0000E+00	146.95	3.0828E-07	536.78	5.1214E+06
STYRENE	1.0593E+02	-8.6859E+03	-1.2420E+01	7.5583E-06	2.0000E+00	242.54	1.0613E+01	636.00	3.8234E+06
TETRAHYDROFURAN	5.4898E+01	-5.3054E+03	-4.7627E+00	1.4291E-17	6.0000E+00	164.65	1.9554E-01	540.15	5.2026E+06
TETRALIN	1.3665E+02	-1.0600E+04	-1.7811E+01	1.4391E-02	1.0000E+00	237.38	1.3424E-01	720.15	3.6410E+06
TOLUENE	8.0877E+01	-6.9024E+03	-8.7761E+00	5.8034E-06	2.0000E+00	178.18	4.2348E-02	591.80	4.1012E+06
VINYL ACETATE	5.7406E+01	-5.7028E+03	-5.0307E+00	1.1042E-17	6.0000E+00	180.35	7.0586E-01	519.13	3.9298E+06
VINYL CHLORIDE	9.1432E+01	-5.1417E+03	-1.0981E+01	1.4318E-05	2.0000E+00	119.36	1.9178E-02	432.00	5.7495E+06
VINYLIDENE CHLORIDE	7.2641E+01	-5.4481E+03	-7.5697E+00	7.0922E-17	6.0000E+00	150.59	2.2419E-01	482.00	5.1889E+06
VINYL PROPIONATE	6.5239E+01	-6.4649E+03	-6.1255E+00	1.2302E-17	6.0000E+00	192.05	5.3294E-01	546.00	3.6776E+06
WATER	7.3649E+01	-7.2582E+03	-7.3037E+00	4.1653E-06	2.0000E+00	273.16	6.1056E+02	647.13	2.1940E+07
p-XYLENE	8.5475E+01	-7.5958E+03	-9.3780E+00	5.6875E-06	2.0000E+00	286.41	5.8144E+02	616.23	3.4984E+06

acetic acid

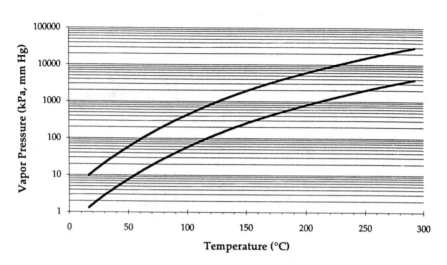

acetone

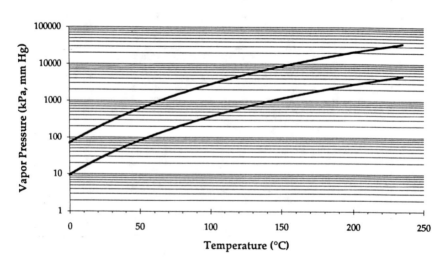

acrylonitrile

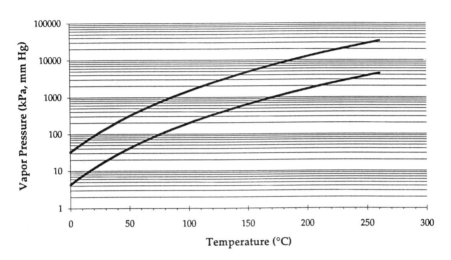

benzaldehyde

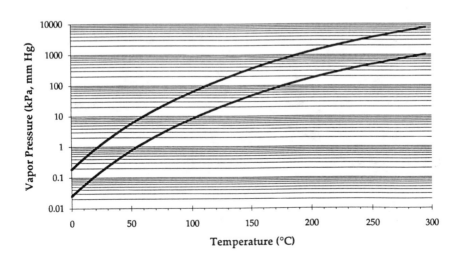

benzene

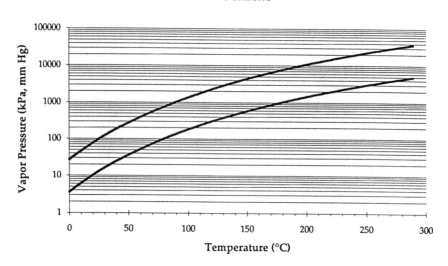

benzyl alcohol

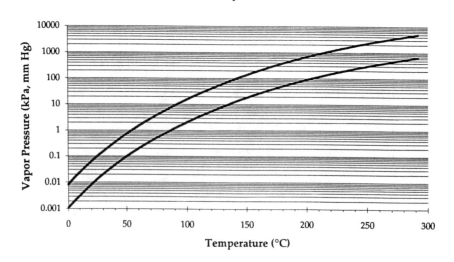

1,3-butadiene

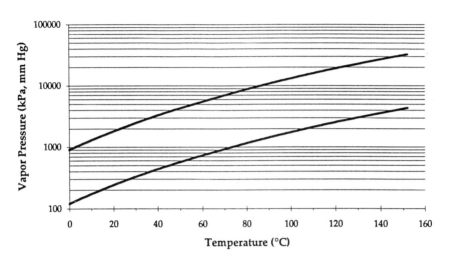

n-butyl acetate

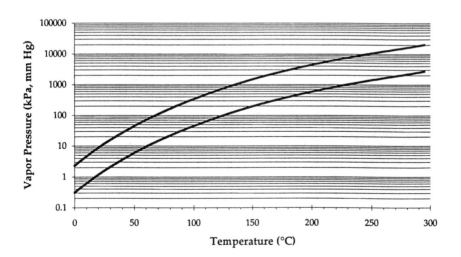

n-butyl acrylate

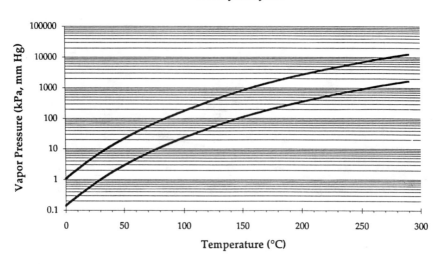

n-butyl alcohol

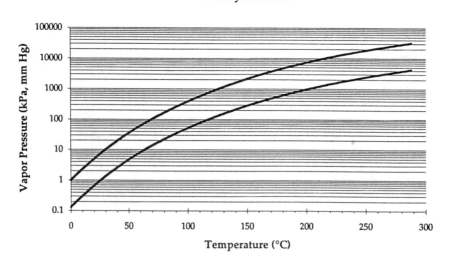

carbon tetrachloride

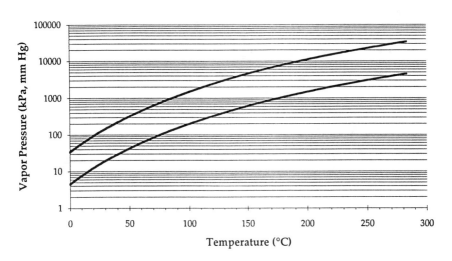

chloroform

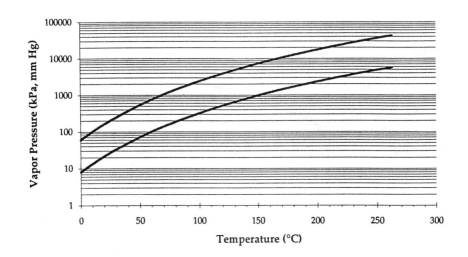

m-cresol

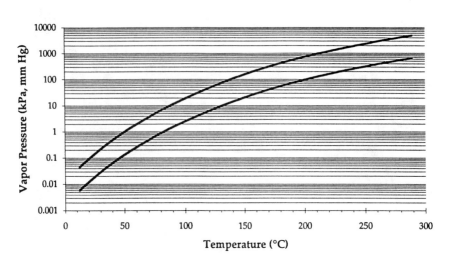

cumene

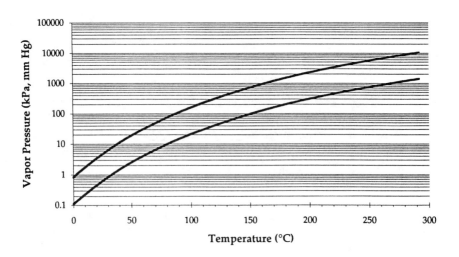

cyclohexane

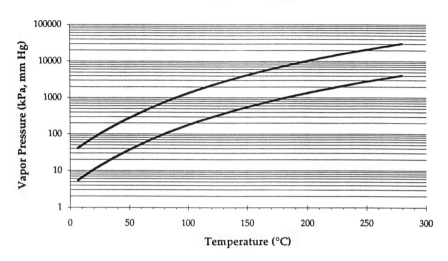

cyclohexanone

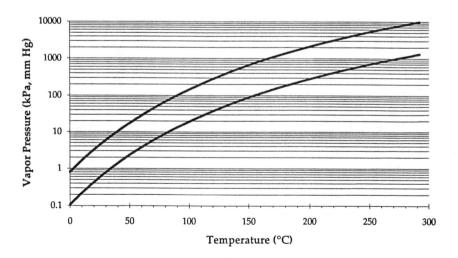

n-decane

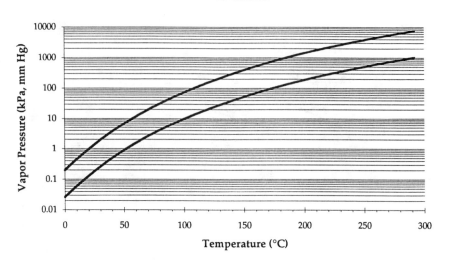

di-n-butylamine

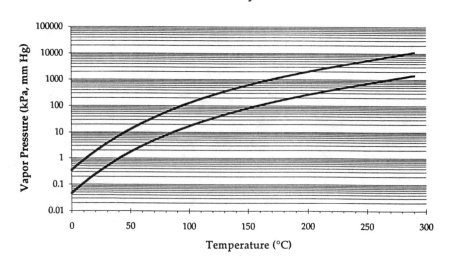

diethylamine

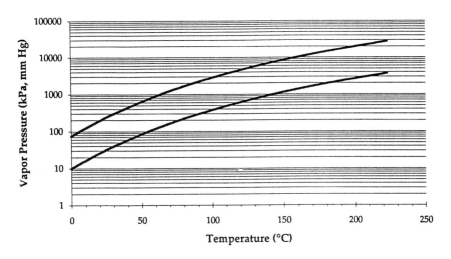

diethyl ether

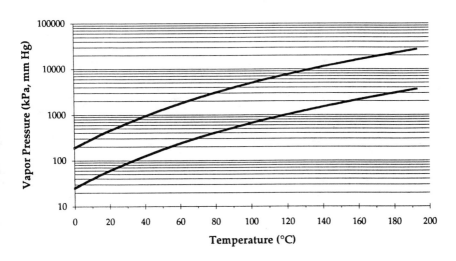

dimethylamine

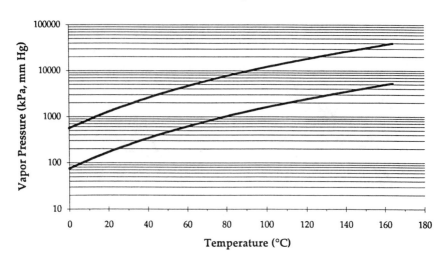

dimethylformamide

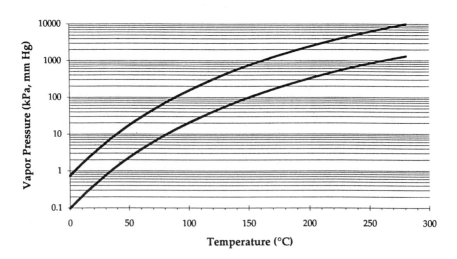

dimethyl sulfoxide

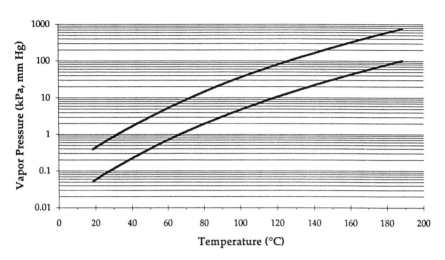

1,4-dioxane

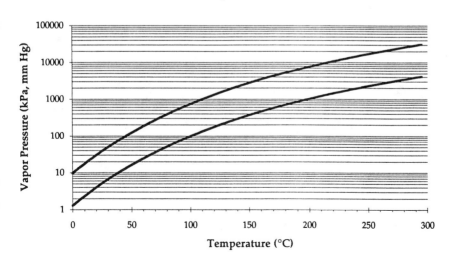

ethanol

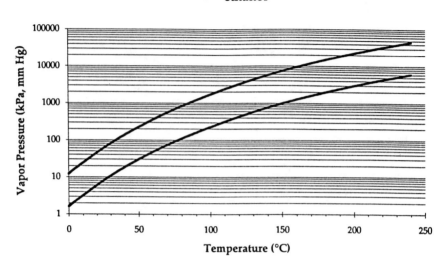

ethyl acetate

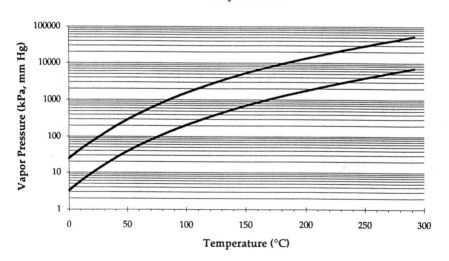

ethyl acrylate

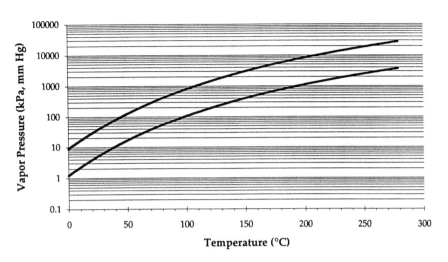

ethylbenzene

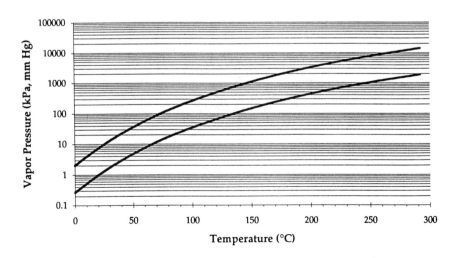

formic acid

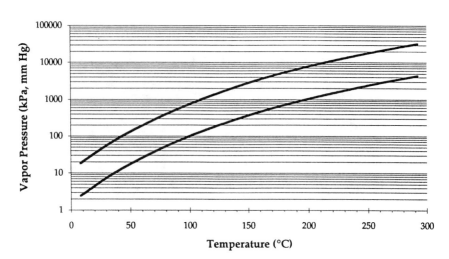

n-heptane

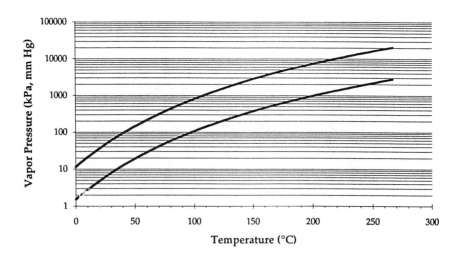

n-hexane

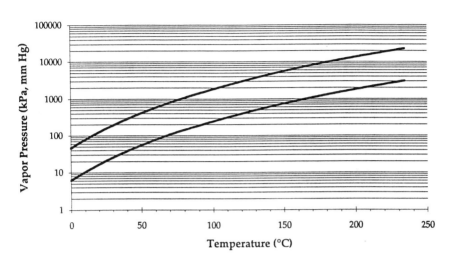

methanol

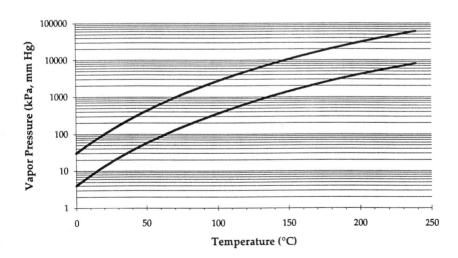

methyl acrylate

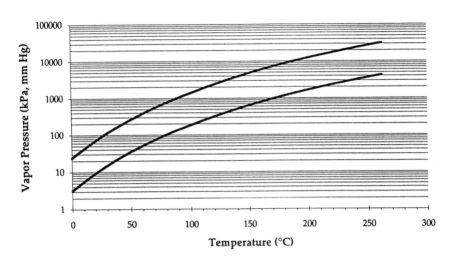

methyl ethyl ketone

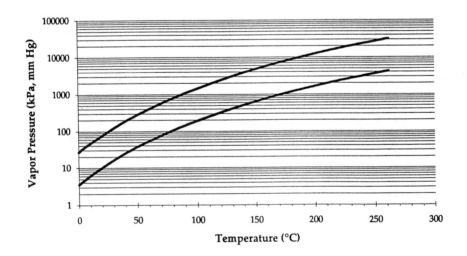

methyl methacrylate

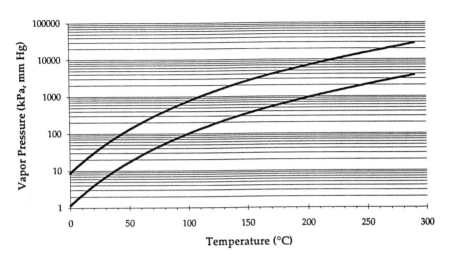

n-nonane

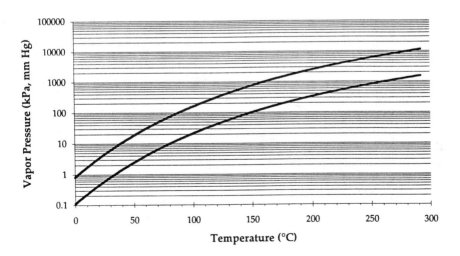

n-octane

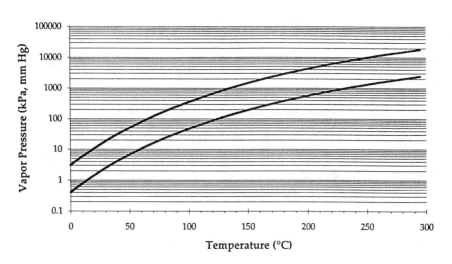

n-pentane

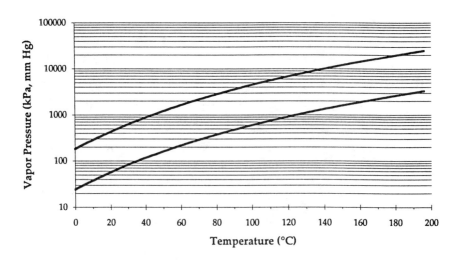

phenol

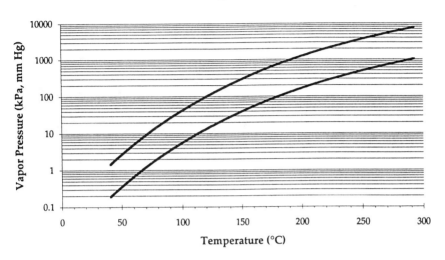

n-propanol

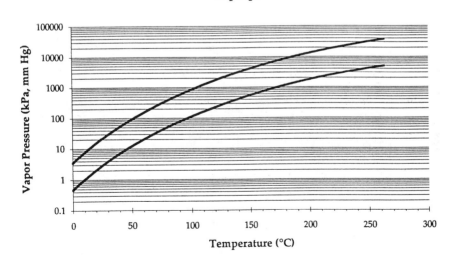

styrene

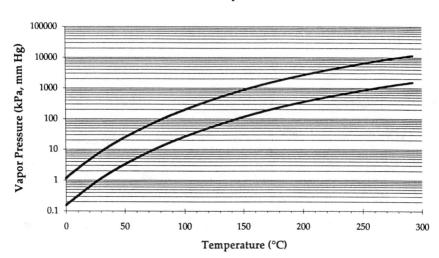

tetrahydrofuran

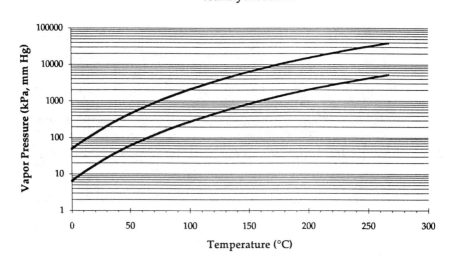

tetralin

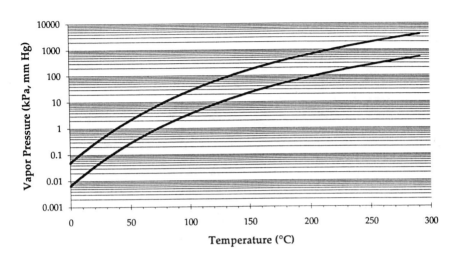

toluene

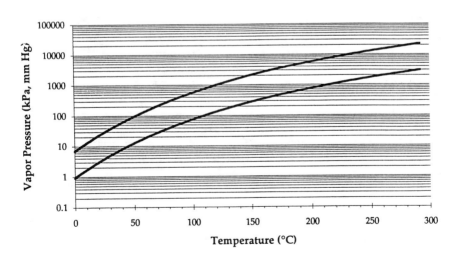

vinyl acetate

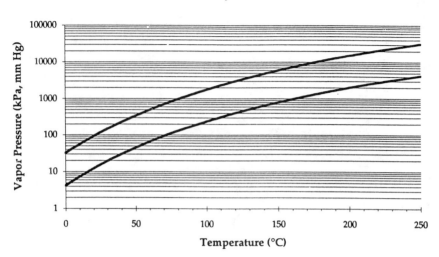

vinyl chloride

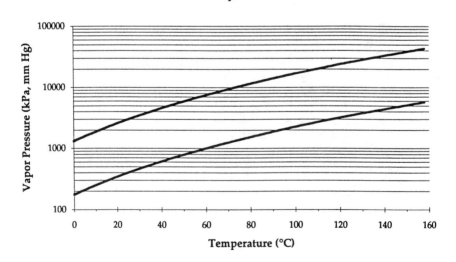

vinylidene chloride

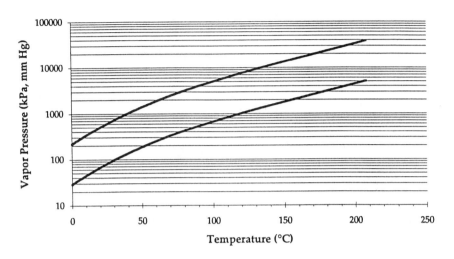

vinyl propionate

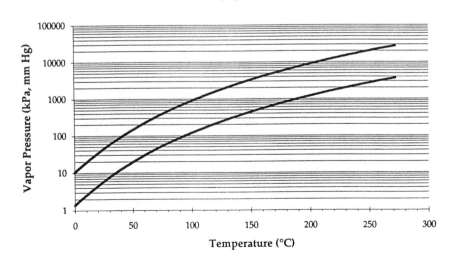

water

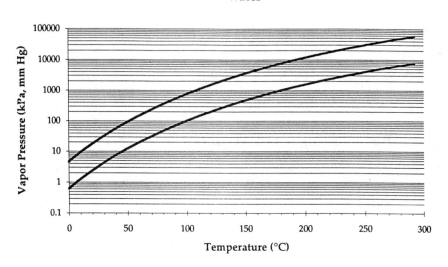

p-xylene

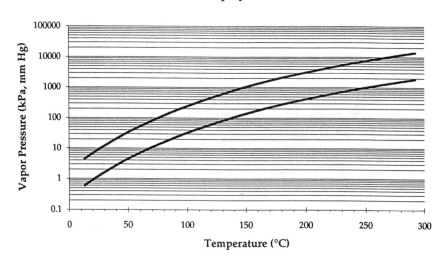

Appendix B

Polymer–Solvent Interaction Parameters

The following tables contain values of the polymer–solvent interaction parameter χ (also known as the Flory–Huggins parameter) for several systems. The data are tabulated as a function of concentration at various temperatures.

In Table 1 the method used to determine a given value of χ is indicated by the following abbreviations:

LS	Light scattering
OS	Osmosis
CM	Critical miscibility
VP	Vapor pressure methods
IGC	Inverse gas chromatography (gas–liquid chromatography)

When inverse gas chromatography was the only method applied at a certain temperature or temperature interval, the data are given in Table 2.

Table 1 Polymer–Solvent Interaction Parameters χ

Solvent	$T(^\circ C)$	Volume Fraction of Polymer						Method	Refs.
		0	0.2	0.4	0.6	0.8	1		
CELLULOSE ACETATE (D.S. = 2.5)[a]									
Acetone	30		0.30	0.51				VP	1
Dioxane	30		0.31	0.51				VP	1
Methyl acetate	30		0.43	0.59				VP	1
Pyridine	30		0.07	0.09				VP	1
CELLULOSE ACETATE (D.S. = 3.0)[a]									
Chloroform	30		0.36	0.45	0.51			VP	2
Methylene chloride	25		0.38	0.45	0.49			VP	2
CELLULOSE NITRATE (D.S. = 2.6)[b]									
Acetone	20		0.14	0.06	−0.37	−1.24		VP	3
Acetonitrile	20			0.59	0.42	0.12	−0.1	VP	3
Cyclopentanone	20		0.42	0.07	−0.71	−2.4		VP	3
Dioxane	20			1.2	−0.25	−1.7		VP	3
Ethyl acetate	20		0.04	−0.43	−1.35			VP	4
Ethyl formate	20		−0.08	−0.14	−0.42	−3.2		VP	4
Isopentyl acetate	20		−0.89	−1.8	−3.3			VP	4
Isopropyl ketone	20		0.62	−0.08	−1.7			VP	3
Methyl t-butyl ketone	20		0.16	−1.5	−2.8	−3.7		VP	3
Methyl iso-propyl ketone	20		−0.5	−0.52	−1.6			VP	3
Nitromethane	20		0.66	0.64	0.60	0.45		VP	3
n-Propyl acetate	20		−0.38	−0.83	−2.0	−4.1		VP	4

Solvent	Temp	χ	χ	χ	χ	χ	Method	Ref
POLY(p-CHLOROSTYRENE)								
Toluene	22	0.53	0.58	0.55			VP	5
POLY(ETHYLENE) (LOW DENSITY)								
n-Heptane	108.9	0.29	0.31	0.34	0.2		VP	6
POLY(ETHYLENE OXIDE)								
Benzene	50.3	0.18	0.14	0.1			VP	7
Benzene	70	0.19	0.14	0.12	0.09		VP	7
POLY(DIMETHYL SILOXANE)								
Benzene	20		0.64	0.71	0.77	0.85	VP	8
Benzene	25		0.63	0.66	0.73	0.79	VP	9, 10
Benzene	25			0.76	0.82	0.81	VP, IGC	11, 12
Cyclohexane	20	0.56	0.48	0.50	0.41	0.47	OS	13
Cyclohexane	25	0.44	0.43	0.43	0.42	0.42	OS, VP	9, 10
Cyclohexane	30	0.46					VP	14
Diisobutyl ketone	35	0.49					OS	15
Ethylbenzene	23.5		0.54	0.63	0.70	0.77	VP	8
Ethylbenzene	25		0.45				IGC	10
Ethyl n-butyl ketone	35	0.52					OS	15
n-Heptane	20			0.45	0.45	0.48	VP	8
n-Heptane	25			0.51			IGC	10
Hexamethyl disiloxane	23.2		0.30	0.28	0.26	0.25	VP	8
Hexamethyl disiloxane	25					0.28	IGC	10
Methyl ethylketone	20	0.57	0.57				OS	15
Methyl ethylketone	30			0.64	0.79		VP	16
Methyl ethylketone	35	0.57					OS	15
Methyl ethylketone	50	0.56					OS	15

Table 1 (continued)

Solvent	$T(°C)$	\multicolumn{6}{c}{Volume Fraction of Polymer}	Method	Refs.
		0	0.2	0.4	0.6	0.8	1		
\multicolumn{10}{c}{POLY(DIMETHYL SILOXANE) *cont'd*}
Methyl iso-butyl ketone	20		0.54					OS	15
Octamethyltrisiloxane	23.2			0.22	0.19	0.17	0.14	VP	8
n-Octane	20		0.42	0.50	0.49	0.50	0.51	OS, VP	13, 8
	25						0.52	IGC	10
	35		0.42				0.54	OS, IGC	17, 12
n-Pentane	20			0.43	0.41	0.40	0.40	VP	8
	25						0.45	IGC	10
Toluene	20	0.45	0.50	0.59	0.64	0.72	0.82	OS, VP	8
	25						0.75	IGC	10
p-Xylene	25				0.58	0.67	0.78	VP	8
	25						0.80	IGC	12
	25						0.75	IGC	10
\multicolumn{10}{c}{POLY(ISOBUTYLENE)}
Benzene	10			0.67	0.78	0.92	1.06	OS, VP	18
	25	0.50	0.57	0.66	0.75	0.88	0.91	OS, VP	19, 18
	25						0.87	IGC	20
	25						0.86	IGC	21
	25							IGC	22
	26.9				0.73	0.82	1.07	VP	23
	39.6				0.70	0.80		VP	8
Cyclohexane	25	0.44		0.42	0.42	0.41	0.43	OS, VP, IGC	24, 25, 20

Solvent	T (°C)	χ	χ	χ	χ	χ	χ	Method	Ref.
n-Octane	25	0.48		0.48	0.62	0.63	0.55	IGC	20
n-Pentane	25		0.44	0.57	0.62	0.68	0.54	IGC	21
	25			0.57				IGC	22
	25						0.68	OS	26
	25						0.75	VP	27, 28
	25						0.70	VP	29
	25				0.62	0.62	0.73	IGC	20
	25			0.62	0.63	0.63	0.72	IGC	21
	35			0.58			0.62	IGC	22
	35		0.53	0.57	0.62	0.63		VP	28
	40		0.53	0.56	0.63	0.63		VP	27
	45							VP	27
	55							VP	27
	25						0.28	IGC	10
POLY(ISOPRENE) (NATURAL RUBBER)									
Acetone	25	0.40		0.41	0.42	1.27	1.8	VP	30
Benzene	10					0.43	0.41	VP	31
	25		0.41	0.41	0.41	0.41	0.46	OS, VP	31, 32
	25				0.43	0.43	0.43	VP, IGC	33, 34
	25			0.41	0.41	0.42	0.44	VP	31
	40				0.41	0.41		VP, IGC	31, 34
Ethyl acetate	25			0.69	0.82	0.96	1.24	VP	35
	25			0.68	0.81	0.96	1.23	VP	30
	50			0.68	0.77	0.91	1.0	VP	35
	50				0.79	0.92	1.0	VP	30
Methyl ethyl ketone	25				0.86	1.05	1.43	VP	30
	45				0.83	0.99	1.2	VP	30

Table 1 (continued)

Solvent	T(°C)	Volume Fraction of Polymer						Method	Refs.
		0	0.2	0.4	0.6	0.8	1		
		POLY(STYRENE)							
Acetone	25				0.81	0.94	1.1	VP	36
	50				0.80	0.92		VP	36
Benzene	25		0.42	0.37	0.32	0.26		VP	37
t-Butyl acetate	10	0.50	0.54	0.64				LS, CM	38, 39
	30	0.50	0.53	0.63				LS, CM	38, 39
	50	0.49	0.53	0.63				LS, CM	38, 39
Chloroform	25		0.52	0.32	0.23	0.17		VP	36
	50		0.45	0.34	0.23	0.14		VP	36
Cyclohexane	30		0.57					CM	40
	34	0.50	0.57	0.64	0.75	0.93		OS, VP	41
	35		0.56					LS	42
	44	0.49	0.57	0.63	0.72	0.93		OS, VP	41
	45		0.55					LS	42
	65		0.53					LS	42

Solvent	Temp (°C)	χ	χ	χ	χ	χ	χ	Method	Ref.
Methylcyclohexane	72	0.49	0.58	0.67	0.69	0.77		OS	43
Methyl ethyl ketone	25			0.63	0.63	0.72		VP	44
	70			0.66				VP	44
Propyl acetate	25			0.66	0.66	0.66		VP	36
	70			0.61	0.60	0.59		VP	36
Toluene	22		0.40	0.42	0.39	0.31		VP	5
	25		0.37	0.42	0.37	0.16		VP	42, 44
	25		0.37	0.30	0.23			VP	37
	65		0.37					LS	42
	80			0.40	0.35			VP	44
POLY(VINYL ACETATE)									
Acetone	30				0.33	0.33	0.31	VP	45
	40				0.34	0.34	0.35	VP	45
Benzene	30			0.45	0.37	0.29	0.30	VP	45, 46
Vinyl acetate	30			0.41	0.29	0.22		VP	46

[a] D.S.: Degree of substitution, i.e., number of acetate groups per glucose residue.

[b] D.S.: Degree of substitution, i.e., number of nitro groups per glucose residue.

Table 2   Polymer–Solvent Interaction Parameters χ Obtained by Inverse Gas Chromatography (IGC)

Polymer	Solvent	Temperature range (°C)	χ	Refs.
Poly(arylate)	Chlorobenzene	140–200	0.54–0.30	47
(Arylef/Solvay)	o-Dichlorobenzene	140–200	0.59–0.27	47
	Diethylene glycol diethyl ether	140–200	1.08–0.56	47
	n-Dodecane	140–200	1.34–1.24	47
	Ethylbenzene	140–200	0.82–0.48	47
	Tetralin	140–200	0.75–0.57	47
	p-Xylene	140–200	0.84–0.39	47
Poly(butene-1)	n-Heptane	115–135	0.38	48
	n-Octane	115–135	0.36	48
	n-Nonane	115–135	0.32	48
	n-Decane	115–135	0.30	48
Poly(ethylene)	cis-Decalin	120–145	0.03	49
(low-density)	trans-Decalin	120–145	0.02–0.00	49
	n-Decane	120–145	0.25–0.26	49
	2,4-Dimethylhexane	120–145	0.34–0.32	49
	2,5-Dimethylhexane	120–145	0.35	49
	3,4-Dimethylhexane	120–145	0.25	49
	n-Dodecane	120–145	0.23–0.24	49
		110–145	0.18	50
	Ethylbenzene	120–145	0.33	49
	3-Methylheptane	120–145	0.31–0.30	49
	Mesitylene	120–145	0.24–0.25	49
	3-Methylhexane	120–145	0.33–0.34	49
	n-Nonane	120–145	0.28	49
	n-Octane	120–145	0.31–0.30	49
	Tetralin	120–145	0.29–0.28	49
	Toluene	120–145	0.34	49
	2,2,4-Trimethylhexane	120–145	0.29–0.28	49
	2,2,4-Trimethylpentane	120–145	0.34	49
	m-Xylene	120–145	0.29	49
	p-Xylene	120–145	0.27–0.28	49
Poly(ethylene)	cis-Decalin	149	0.07	49
(high-density)	trans-Decalin	149	0.05	49
	n-Decane	149	0.31	49
		185	0.12	51
		145–190	0.18	50
	2,4-Dimethylhexane	149	0.38	49
	3,4-Dimethylhexane	149	0.31	49
	n-Dodecane	149	0.28	49

Table 2 (continued)

Polymer	Solvent	Temperature range (°C)	χ	Refs.
	Ethylbenzene	149	0.37	49
	Mesitylene	149	0.28	49
	2-Methylheptane	149	0.39	49
	3-Methylhexane	149	0.40	49
	n-Nonane	149	0.34	49
	n-Octane	149	0.36	49
	Tetralin	149	0.32	49
	Toluene	149	0.39	49
	2,2,4-Trimethylhexane	149	0.35	49
	2,2,4-Trimethylpentane	149	0.40	49
	m-Xylene	149	0.34	49
	p-Xylene	149	0.32	49
Poly(dimethyl siloxane)	Benzene	25–70	0.81–0.75	12
		25–70	0.79–0.74	10
	Cycloheptane	25–70	0.56–0.53	10
	Cyclohexane	25–70	0.47–0.49	10
	Cyclooctane	25–70	0.66–0.61	10
	Cyclopentane	25–70	0.42–0.46	10
	n-Decane	20	0.64	52
	p-Dioxane	25–70	1.32–1.18	10
	Ethylbenzene	25–70	0.83–0.78	12
		25–70	0.77–0.73	10
	n-Heptane	20	0.50	52
		25–70	0.50–0.49	12
		25–70	0.48–0.50	10
	Hexamethyl disiloxane	25–70	0.28–0.34	10
	n-Hexane	20	0.50	52
		25–70	0.45	12
		25–70	0.46–0.47	10
	Mesitylene	25–70	0.95–0.86	10
	2-Methylbutane	25	0.39	12
	2-Methylheptane	25–70	0.52–0.50	12
		25–70	0.49–0.47	10
	2-Methylhexane	25–70	0.46	12
		25–70	0.45–0.46	10
	3-Methylhexane	25–70	0.45–0.43	10
	2-Methylpentane	25–70	0.45–0.43	12
	o-Xylene	25–70	0.44–0.43	10
	2,2,4-Trimethylpentane	25–70	0.45–0.43	10
	n-Octane	20	0.53	52
		25–70	0.56–0.54	12

Table 2   (continued)

Polymer	Solvent	Temperature range (°C)	χ	Refs.
		25–70	0.56–0.52	12
	n-Pentane	20	0.48	52
		25–70	0.41–0.43	12
		25–70	0.45–0.49	10
	Toluene	25–70	0.80–0.75	12
		25–70	0.75–0.71	10
	2,2,4-Trimethylpentane	20	0.43	52
	o-Xylene	25–70	0.86–0.80	10
	m-Xylene	25–70	0.82–0.76	10
	p-Xylene	25–70	0.80–0.77	12
Poly(isobutylene)	Benzene	25–50	0.91–0.79	20
		25–65	0.86–0.61	22
		25	0.87	21
		40–65	0.87–0.77	53
	Cyclohexane	25	0.57	21
		25–50	0.43–0.40	20
		25–65	0.54–0.42	22
		40–65	0.47–0.42	53
	n-Heptane	25	0.57	21
		25–65	0.57–0.47	22
	n-Hexane	25	0.65	21
		25–65	0.63–0.50	22
	n-Nonane	25	0.49	21
	n-Octane	25	0.52	21
		25–65	0.52–0.43	22
	n-Pentane	25	0.73	21
		25–50	0.70–0.60	20
		25–65	0.72–0.57	22
Poly(isoprene)	Benzene	25–55	0.46–0.43	34
	Ethylbenzene	25–55	0.34–0.30	34
	n-Heptane	25–55	0.51–0.49	34
	n-Hexane	25–55	0.54–0.50	34
	2-Methylheptane	25–55	0.50–0.47	34
	2-Methylhexane	25–55	0.52–0.50	34
	2-Methylpentane	25–55	0.56–0.52	34
	n-Octane	25–55	0.49–0.46	34
	n-Pentane	25–55	0.61–0.53	34
	Toluene	25–55	0.36–0.32	34
	2,2,4-Trimethylpentane	25–55	0.49–0.46	34
	p-Xylene	25–55	0.28–0.26	34

Table 2 (continued)

Polymer	Solvent	Temperature range (°C)	χ	Refs.
Poly(methyl	Benzene	90–110	0.51–0.46	54
acrylate)	*n*-Butylbenzene	90–110	1.15–1.05	54
	t-Butylbenzene	90–110	1.03–0.95	54
	n-Butylcyclohexane	90–110	2.34–2.13	54
	Cyclohexane	90–110	1.72–1.51	54
	cis-Decalin	90–110	2.07–1.84	54
	trans-Decalin	90–110	2.13–1.90	54
	n-Decane	88–100	2.68–2.43	54
	n-Dodecane	90–110	3.00–2.75	54
	Ethylbenzene	90–110	0.83–0.75	54
	Naphthalene	100–110	0.49–0.47	54
	n-Octane	90–110	2.38–2.19	54
	n-Tetradcane	90–110	3.36–3.06	54
	Tetralin	90–110	1.04–0.95	54
	3,3,4,4-Tetramethylhexane	90–110	2.17–1.95	54
	Toluene	90–110	0.67–0.62	54
	3,4,5-Trimethylheptane	90–110	2.44–2.20	54
	2,2,5-Trimethylhexane	90–110	2.48–2.21	54
	2,2,4-Trimethylpentane	90–110	2.35–2.07	54
Poly(propylene)	*n*-Hexane	80	0.18	56
Poly(styrene)	Acetic acid	162–229	3.00–2.15	55
	Acetone	162–229	1.30–0.56	55
	Acetonitrile	162–229	2.02–0.93	55
	n-Amyl alcohol	162–229	1.75–0.86	55
	Aniline	162–229	1.11–0.68	55
	Benzaidehyde	162–229	1.22–0.80	55
	Benzene	120	0.33	51
		150–200	0.38	20
		160–180	0.29–0.24	54
		162–229	0.66–0.13	55
	Benzyl alcohol	162–229	1.42–0.65	55
	n-Butyl acetate	162–229	1.01–0.45	55
	t-Butyl acetate	143–183	0.73–0.52	39
	n-Butyl alcohol	162–229	1.47–0.82	55
	i-Butyl alcohol	162–229	1.71–0.81	55
	n-Butylbenzene	183–203	0.38–0.35	54
	n-Butylcyclohexane	160–180	0.77–0.71	54
	Carbon tetrachloride	162–229	0.90–0.26	55
	Chlorobenzene	162–229	0.68–0.28	55
	Chloroform	162–229	0.43– −0.01	55
	Cyclohexane	160–180	0.62–0.53	54

Table 2 (continued)

Polymer	Solvent	Temperature range ($^\circ$C)	χ	Refs.
		162–229	1.11–0.46	55
	Cyclohexanone	162–229	1.08–0.49	55
	cis-Decalin	183–203	0.47–0.42	54
	trans-Decalin	183–203	0.52–0.46	54
	n-Decane	183–203	1.01–0.94	54
	1,2-Dichloroethane	162–229	0.85–0.22	55
	Diethyl ether	162–229	0.78–0.71	55
	Diisopropyl ether	162–229	1.42–0.41	55
	Dioxane	162–229	0.95–0.42	55
	n-Dodecane	183–203	1.09–1.00	54
	Ethyl acetate	162–229	1.14–0.35	55
	Ethyl alcohol	162–229	1.80–0.43	55
	Ethylbenzene	120–185	0.22–0.14	56
	Ethylene glycol	162–229	3.77–2.23	55
	Formamide	162–229	4.11–3.16	55
	n-Heptane	162–229	1.33–0.25	55
	n-Hexadecane	183–203	1.22–1.14	54
	n-Hexane	162–229	1.35– −0.03	55
	Methyl alcohol	162–229	2.19–0.44	55
	Methylene chloride	162–229	0.62– −0.21	55
	Methyl ethyl ketone	162–229	1.16–0.36	55
	Naphthalene	183–203	0.13–0.11	54
	Nitrobenzene	162–229	1.18–0.72	55
	n-Octane	162–229	2.19–0.80	55
	i-Octane	162–229	1.72–0.35	55
	n-Octyl alcohol	162–229	1.14–0.55	55
	n-Pentane	162–229	1.12–0.83	55
	n-Propyl alcohol	162–229	1.71–0.27	55
	i-Propyl alcohol	162–229	1.74– −0.15	55
	Pyridine	162–229	1.02–0.23	55
	n-Tetradecane	183–203	1.14–1.08	54
	Tetrahydrofuran	162–229	0.70– −0.16	55
	Tetralin	183–203	0.20–0.11	54
	3,3,4,4-Tetramethylhexane	160–180	0.90–0.76	54
	Toluene	162–229	0.67–0.04	55
	Trichloroethylene	162–229	0.69–0.12	55
	Water	162–229	4.40–3.10	55
	o-Xylene	162–229	0.72–0.26	55

For additional information on determining values of χ the reader is referred to the source of these tables given in the footnote to page 675 and to Chapter 2 of this book.

CONVERTING CONCENTRATION VARIABLES

The concentrations in Tables 1 and 2 are given in terms of the polymer volume fraction ϕ_2. If the density of the polymer (ρ_2) and the density of the solvent (ρ_1) are known, the volume fraction of the polymer may be expressed in terms of the weight fraction of the polymer (W_2):

$$\phi_2 = \frac{W_2}{\rho_2\left(\dfrac{W_2}{\rho_2} + \dfrac{1 - W_2}{\rho_1}\right)}$$

REFERENCES

1. W. R. Moore and R. Shuttleworth, (1963). *J. Polym. Sci., Part A*1 , 733.
2. W. R. Moore and R. Shuttleworth, (1963). *J. Polym. Sci., Part A1*, 1985.
3. E. C. Baughan, A. L. Jones, and K. Stewart, (1954). *Proc. Roy. Soc. London, Ser. A 225*, 478.
4. A. L. Jones, (1956). *Trans. Faraday Soc. 52*, 1408.
5. R. Corneliussen, S. A. Rice, and H. Yamakawa, (1963). *J. Chem. Phys. 38*, 1768.
6. J. H. van der Waals and J. J. Hermans, (1950). *Rec. Trav. Chem. Pays Bas. 69*, 971.
7. C. Booth and C. J. Devoy, (1971). *Polymer 12*, 309.
8. R. S. Chahal, W.-P. Kao, and D. Patterson, (1973). *J. Chem. Soc., Faraday Trans. 1*(69), 1834.
9. P. J. Flory and H. Shih, (1972). *Macromolecules 5*, 761.
10. R. N. Lichtenthaler, D. D. Liu, and J. M. Prausnitz, (1974). *Ber. Bunsengesell. 78*, 470.
11. M. J. Newing, (1950). *Trans. Faraday Soc. 46*, 613.
12. W. R. Summers, Y. B. Tewari, and H. P. Schreiber, (1972). *Macromolecules 5*, 12.
13. N. Kuwahara, T. Okazawa, and M. Kaneko, (1968). *J. Polym. Sci., Part C 23*, 543.
14. R. W. Brotzman and B. E. Eichinger, (1982). *Macromolecules 15*, 531.
15. T. Shiomi, Z. Izumi, F. Hamada, and A. Nakajima, *Macromolecules* (1980). *13*, 1149.
16. A. Muramoto, (1982). *Polymer 23*, 1311.
17. K. Sugamiya, N. Kuwahara, and M. Kaneko, (1974). *Macromolecules 7*, 66.
18. B. E. Eichinger and P. J. Flory, (1968). *Trans. Faraday Soc. 64*, 2053.
19. P. J. Flory and H. Daoust, (1957). *J. Polym. Sci. 25*, 429.
20. R. D. Newman and J. M. Prausnitz, (1972). *J. Phys. Chem. 76*, 1492.

21. Y.-K. Leung and B. E. Eichinger, (1974). *J. Phys. Chem. 78*, 60.
22. Y.-K. Leung and B. E. Eichinger, (1974). *Macromolecules 7*, 685.
23. R. S. Jessup, (1958). *J. Res. Nat. Bur. Stand. 60*, 47.
24. P. J. Flory, (1943). *J. Am. Chem. Soc. 65*, 372.
25. B. E. Eichinger and P. J. Flory, (1968). *Trans. Faraday Soc. 64*, 2061.
26. P. J. Flory, J. L. Ellenson, and B. E. Eichinger, (1968). *Macromolecules 1*, 279.
27. C. H. Baker, W. B. Brown, G. Gee, J. S. Rowlinson, D. Stubley, and Y. E. Yeadon, (1962). *Polymer 3*, 215.
28. S. Prager, E. Bagley, and F. A. Long, (1953). *J. Am. Chem. Soc. 75*, 2742.
29. B. E. Eichinger and P. J. Flory, (1968). *Trans. Faraday Soc. 64*, 2066.
30. C. Booth, G. Gee, G. Holden, and G. R. Williamson, (1964). *Polymer 5*, 343.
31. B. E. Eichinger and P. J. Flory, (1968). *Trans. Faraday Soc. 64*, 2035.
32. G. Gee, *J. Chem. Soc.* (1947). *280.*
33. G. Gee, J. B. M. Herbert, and R. C. Roberts, (1965). *Polymer 6*, 541.
34. Y. B. Tewari and H. P. Schreiber, (1972). *Maromolecules 5*, 329.
35. C. Booth, G. Gee, and G. R. Williamson, (1957). *J. Polym. Sci. 23*, 3.
36. C. E. H. Bawn and M. A. Wajid, (1956). *Trans. Faraday Soc. 52*, 1658.
37. I. Noda, Y. Higo, N. Ueno, and T. Fujimoto. (1984). *Macromolecules 17*, 1055.
38. B. A. Wolf and H. J. Adam, (1981). *J. Chem. Phys. 75*, 4121.
39. K. Schotsch, B. A. Wolf, H.-E. Jeberien, and J. Klein, (1984). *Makromol. Chem. 185*, 2169.
40. R. Koningsveld and L. A. Kleintjens, (1971). *Macromolecules 4*, 637.
41. W. R. Krigbaum and D. O. Geymer, (1959). *J. Am. Chem. Soc. 81*, 1859.
42. T. G. Scholte, (1970). *Eur. Polym. J 6*, 1063.
43. K. Kamide, K. Sugamiya, T. Kawai, and Y. Miyazaki, (1980). *Polym. J. 12*, 67.
44. C. E. H. Bawn, R. F. J. Freeman, and A. R. Kamaliddin, (1950). *Trans. Faraday Soc. 46*, 677.
45. R. J. Kokes, A. R. DiPietro, and F. A. Long, (1953). *J. Am. Chem. Soc. 75*, 6319.
46. A. Nakajima, H. Yamakawa, and I. Sakurada, (1959). *J. Polym. Sci. 35*, 489.
47. J. I. Eguiazabal, M. J. Fernandez-Berridi, J. J. Iruin, and J. M. Elorza, (1985). *Polym. Bull. 13*, 463.
48. G. Charlet, R. Ducasse, and G. Delmas, (1981). *Polymer 22*, 1190.
49. H. P. Schreiber, Y. B. Tewari, and D. Patterson, (1973). *J. Polym. Sci., Polym. Phys. Ed. 11*, 15.
50. D. Patterson, Y. B. Tewari, H. P. Schreiber, and J. E. Guillet, (1971). *Macromolecules 4*, 356.
51. N. F. Brockmeier and R. W. McCoy, and J. A. Meyer, (1972). *Macromolecules 5*, 130.
52. W. W. Hammers and C. L. Ligny, (1974). *J. Polym. Sci., Polym. Phys. Ed. 12*, 2065.
53. R. N. Lichtenthaler, D. D. Liu, and J. M. Prausnitz, (1974). *Macromolecules 7*, 565.
54. G. DiPaola-Baranyi and J. Guillet, (1978). *Macromolecules 11*, 228.
55. S. Gündüz and S. Dincer, (1980). *Polymer 21*, 1041.
56. N. F. Brockmeier, R. W. McCoy, and J. A. Meyer, (1972). *Macromolecules 5*, 464.

Appendix C

Selected Abstracts of Publications on Polymer Devolatilization

This appendix presents a list of selected abstracts of publications dealing with different aspects of polymer devolatilization that have appeared in the literature over the past decade or so. All the abstracts, with the exception of those marked with an asterisk (*) following the title, have been selected from the CD-ROM version of the Rapra Abstracts Database. This material is © copyright of Rapra Technology Ltd. and is reproduced here with the permission of

Rapra Technology Ltd.
Shawbury, Shrewsbury
Shropshire SY4 4NR UK
Fax: 0939 251118

The abstracts with an asterisk (*) following the title are reproduced from *AIChE J.* by permission of the American Institute of Chemical Engineers.

The abstracts are listed in reverse chronological order.

Devolatilization of Polyimide Fibre Composites: Model and Experimental Verification

Authors	Yoon, I. S.; Yang, Y; Dudukovic, M. P.; Kardos, J. L.
Affiliation	St. Louis, Washington University
Journal Name	*Polym. Composites*

Citation   15, No. 3, June 1994, pp. 184–196

Abstract   A one-dimensional devolatilization model was developed for autoclave processing of graphite fiber-reinforced polyimide. Model calculated temperature profiles in the laminate during polymerization, and the predicted removal rates of volatile species were compared with experimental data. 17 refs.

Devolatilization of PDMS Gums: A Performance Comparison of Co- and Counter-Rotating Twin-Screw Extruders

Author   Powell, K. G.
Affiliation   General Electric Co., Corporate R & D
Doc. Name   Antec '94. Conference Proceedings
Citation   San Francisco, Calif., 1st–5th May 1994, Vol. I, pp. 234–238

Abstract   A study was made of the performance of a co-rotating twin-screw extruder and a nonintermeshing counter-rotating twin-screw extruder in the devolatilization of polydimenthyl siloxane gums. Hydrodynamic differences between the extruders were reflected in their devolatilization performance as a function of operating parameters. The concept of normalized residuals based on Latinen's wiped-film model was a useful method for classifying devolatilization problems and simplifying process optimization. 11 refs.

Foam-Enhanced Devolatilization of Polystyrene Melt in a Vented Extruder*

Authors   Tukachinsky, A.; Tadmor, Z.; Talmon, Y.
Affiliation   Technion-Israel Institute of Technology
Journal Name   *AIChE J.*
Citation   Vol. 40, April 1994, pp. 670–675

Abstract   Vented extruder devolatilization (DV) of PS melt containing 6000 ppm styrene was studied by scanning electron microscopy (SEM) and video photography. Vacuum DV of a polymer is accompanied by foaming, which starts instantaneously upon supersaturation of the stretched melt and is enhanced at higher speeds of the vented extruder screw. As the volatiles are removed from the melt, foaming gradually ceases, starting with the pushing flight of the screw. The experimental installation design allowed us to quench the polymer melt in the DV zone at various stages of the process. Samples taken from four areas of the channel width were investigated by SEM. Bubble nucleation in the melt appears to take place mainly in the border area adjoining the gas phase. In the shear field caused by screw rotation, large bubbles become noticeably elongated. Their surface, as well as the free surface of the melt, is covered with blisters, 1–100 µm in size.

Microblisters are often concentrated in areas subjected to stretching. Calculations of cooling due to volatile evaporation and of heating due to viscous dissipation near a growing bubble show that the process of foam-enhanced DV of a PS/styrene system can be regarded isothermal if the initial volatile concentration does not exceed approximately 1%. 12 refs. (Abstract reproduced from Tukachinsky et al. (1994) by permission of the American Institute of Chemical Engineers. © 1994 AIChE.)

Devolatilization of Polymer Solutions

Authors	Chen, L.; Hu, G. H.; Lindt, J. T.
Affiliation	Pittsburgh University
Journal Name	*Int. Polym. Processing,*
Citation	9, No. 1, March 1994, pp. 26–32
Abstract	A simulator of the operation of a counter-rotating twin-

screw extruder was constructed for studying devolatilization of reactive polymer solutions. The rate of the release of the volatiles was measured by an in-line mass flow meter. The operating pressure and solution temperature were measured using a pressure transducer and a thermocouple. The vapor sample compositions were determined by UV spectrophotometry. The transesterification of EVA with 3-phenyl-1-propanol in 1-nonane solution was chosen as a model reaction. The volumetric mass transfer coefficient was determined at various temperatures, pressures, shaft speeds, and sample sizes. The effect of the reaction on this coefficient was also examined, and it was found that the coefficient of the solvent increased significantly with a decrease in operating pressure and that it also increased appreciably with shaft speed, whereas temperature and initial sample size had little effect. 15 refs.

Experimental Study on Dewatering and Devolatilizing Operations Using Intermeshed Twin Screw Extruders

Authors	Sakai, T; Hashimoto, N.; Kataoka, K.
Affiliation	Japan Steel Works Ltd.
Journal Name	*Int. Polym. Processing*
Citation	8, No. 3, Sept. 1993, pp. 218–223
Abstract	The dewatering and devolatilizing extrusion behavior of

elastomers, such as SBR and NBR, in intermeshing twin-screw extruders is discussed. Various twin-screw extruders with long segmented barrels having several sections with slits and venting ports were used. The successful production of polymer crumbs containing high concentrations of water in an intermeshing co-rotating twin-screw extruder, coagulation extrusion of rubber latex and devolatilization extrusion of PETP without pre-drying are described. 5 refs.

Acceleration of Chemical Reaction in Boiling Polymer Solutions

Authors   Hu, G. H.; Chen, L. Q.; Lindt, J. T.
Affiliation   Pittsburgh University
Doc. Name   Antec '93. Conference Proceedings
Citation   New Orleans, La., 9th–13th May 1993, Vol. I, pp. 37–40
Abstract   The transesterification of EVA with phenyl propanol in boiling inert hydrocarbon solvents was studied in a batch apparatus simulating combined devolatilization and reactive extrusion occurring in a counter-rotating nonintermeshing twin-screw extruder. The results showed that the overall reaction rate of transesterification was substantially increased by devolatilization, and that the enhancement in the overall reaction rate increased with increased rate of devolatilization. The acceleration of the reaction rate continued for a certain time after devolatilization was stopped. 4 refs.

Effect of Stripping Agents for the Devolatilizing of Highly Viscous Polymer Melts

Authors   Mack, M. H.; Pfeiffer, A.
Affiliation   Berstorff Corp.
Doc. Name   Antec '93. Conference Proceedings
Citation   New Orleans, La., 9th–13th May 1993, Vol. II, pp. 1060–1068
Abstract   Results are presented of a study of devolatilization in the extrusion of octene-based linear LDPE resins using water as a stripping agent. The mixing element in the single-screw extruder was optimized for the dispersion of water droplets and for generating a fine foam structure prior to the vacuum vent section. The devolatilization process was shown to follow the principles of mass transfer by bubble growth rather than by drag flow of the melt pool. A fiber optic probe was used to measure the quality of mixing during the water dispersion process, and the foam quality data from these experiments could be directly linked to degassing results for the different resins. The best foam structure, and hence the best degassing results, were obtained with the resins of the highest viscosity. 4 refs.

Devolatilization of Reacting Polymer Solutions ⸱

Authors   Chen, L.; Hu, G. H.; Lindt, J. T.
Affiliation   Pittsburgh University
Doc. Name   Antec '93. Conference Proceedings

Citation   New Orleans, La., 9th–13th May 1993, Vol. II, pp. 1055–1059

Abstract   Mass transfer in reactive polymer solutions during devolatilization was investigated using the transesterification of EVA with phenyl propanol in nonane solution as the reaction model. The study was undertaken using an apparatus simulating a counter-rotating twin-screw extruder and that allowed visualization and in-line measurement of the flow rate of volatiles. 13 refs.

Application of Engineering Science to the Extruder Isolation of Polymers from Solution

Author   Powell, K.

Affiliation   GE Corporate R & D

Doc. Name   Antec '93. Conference Proceedings

Citation   New Orleans, La., 9th–13th May 1993, Vol. II, pp. 1046–1054

Abstract   The extruder isolation of polyphenylene oxide was studied in a counter-rotating, nonintermeshing twin-screw extruder by dividing the complex process into a number of simpler process steps. Following a study of precipitation, additional extruder length was added and deliquefaction was considered. Knowledge of the precipitation step over a broad range of conditions provided input for the subsequent unit operations, including deliquefaction, washing, and devolatilization. 5 refs.

Polymer Devolatilization: State of the Art

Authors   Astarita, G.; Maffettone, P. L.

Affiliation   Naples University

Journal Name   *Makromol. Chem., Macromol. Symp.*

Citation   Vol. 68, April 1993, pp. 1–12

Abstract   The current state of the art on polymer devolatilization is reviewed. After a brief description of existing processes, two aspects are considered, namely the modeling and optimization of actual processes with emphasis on the relative importance of momentum, mass, and heat transfer phenomena, and physicochemical data needed for analysis. (3rd AIM Conference, Italy, 1992). 62 refs.

Removal of Residual Volatiles from Linear Low-Density Polyethylene (LLDPE)

Author   Pfeiffer, A.

Affiliation   Berstorff, H., Maschinenbau GmbH

Doc. Name   Polyolefins VIII. Conference Proceedings

Citation   Houston, Tex., 21st–24th Feb. 1993, pp. 132–173

Abstract   Based on the prior art of devolatilization, a process for removing the residual volatiles from a special LLDPE was developed. A detailed account is given with particular reference to extruders for LDPE production, single-screw extruder designs, devolatilization extruder design, devolatilization using single-screw extruders, extruders with rearward devolatilization systems, extruders with forward devolatilization systems, forward and/or rearward devolatilization, operation of devolatilization extruders, test setup and performance, LDPE devolatilization with a two-stage devolatilization system, LLDPE devolatilization with a one-stage forward devolatilization system, and a fiber optic method for mixing quality analysis. 11 refs.

Ultrasound-Enhanced Devolatilization of Polymer Melt*

Authors   Tukachinsky, A.; Tadmor, Z.; Talmon, Y.
Affiliation   Technion-Israel Institute of Technology
Journal Name   *AIChE J.*
Citation   Vol. 39, February 1993, pp. 359–360
Abstract   An experimental system was constructed in which acoustically treated (or untreated) polystyrene melt was extruded into a vacuum chamber. The 20 mm diameter single-screw extruder used was equipped with a crosshead die incorporating a waveguide attached to an ultrasonic generator and acoustic transformer. A series of experiments was carried out to study the effect of acoustic treatment on the DV process. Acoustically treated samples visibly foamed up even at atmospheric pressure, whereas untreated samples did not. Analysis of the samples by gas chromatography indicated that acoustic treatment significantly reduced the level of residual styrene in the extruded samples and that the effect was increased with increasing vacuum levels. 5 refs. (Abstract reproduced from Tukachinsky et al. (1993) by permission of the American Institute of Chemical Engineers. © 1993 AIChE.)

Stability of Multislit Devolatilization of Polymers*

Authors   Ianniruberto, G.; Maffetone, P. L.; Astarita, G.
Affiliation   Naples University
Journal Name   *AIChE J.*
Citation   Vol. 39, January 1993, pp. 140–148
Abstract   The stability of operation of a realistic multislit devolatilization process is considered. It is shown that the flow rate per slit that can be stably operated may be significantly less than the optimal one for single-slit operation. Stability improves as the concentration of volatiles in the feed solution decreases. The non-Newtonian character of the feed solution at high shear rates is discussed in qualitative terms. 13 refs. (Abstract reproduced

Experimental Study on Dewatering and Devolatilization Operations Using Intermeshed Twin Screw Extruder

Authors   Sakai, T.; Hashimoto, N.; Kataoka, K.
Affiliation   Japan Steel Works Ltd.
Doc. Name   Antec '92. Plastics: Shaping the Future. Volume 1. Conference Proceedings
Citation   Detroit, Mich., 3rd–7th May 1992, pp. 7–14
Abstract   The results are reported of dewatering and devolatilization studies carried out mostly on rubbers using an intermeshed twin-screw-extruder that is changeable to either co- or counter-rotation. Data on the dewatering of SBR, NBR, and ABS, the coagulation extrusion of synthetic rubber latex, and the direct devolatilization twin-screw extrusion of PETP without any pre-drying of the raw materials are presented and discussed. 5 refs.

Devolatilization of Polymer Melts. II. More Machine Geometry effects

Authors   Beisenberger, J. A.; Wang, N.; Dey, S. K.; Lu, Y.
Affiliation   Polymer Processing Institute
Doc. Name   Antec '91. Conference Proceedings
Citation   Montreal, 5th–9th May 1991, pp. 119–121
Abstract   The effects are examined of varying helix angle (screw lead) on separation of volatile component at various screw speeds and throughputs, utilizing only a single-screw extruder, and an attempt is made to interpret the data in terms of defined parameters. 10 refs.

Slit Devolatilization of Polymers*

Authors   Maffetone, P. L.; Astarita, G.; Cori, L.; Carnelli, L.; Balestri, F.
Affiliation   Naples University; Enimont Anic Research Center; Montedipe Research Center
Journal Name   *AIChE J.*
Citation   Vol. 37, May 1991, pp. 724–734
Abstract   A model of the process of polymer devolatilization in a heated slit is presented. Momentum, heat, and mass transfer in the slit are taken into consideration, and only one adjustable parameter is used. The model predicts pressure and temperature profiles, and residual volatile contents at the exit that are in good agreement with those measured in a

single instrumental slit. 26 refs. (Abstract reproduced from Maffetone et al. (1991) by permission of the American Institute of Chemical Engineers. © 1991 AIChE.)

Application of an Enhanced Flash-Tank Devolatilization System to a Degassing Extruder

Author   Craig, T. O.
Affiliation   Fordel Ltd..
Journal Name   *Adv. Polym. Technol.*
Citation   10, No. 4, Winter 1990, pp. 323–325
Abstract   The design and operation of a foam-enhanced flash-tank devolatilization system applied to a conventional degassing extruder are described. The extruder degenerates to a melt feed device, and devolatilization takes place in a static mixer in which foaming is induced. The flash tank provides vapor disengagement space. Improved residual volatile levels and throughput rates are achieved, as well as the elimination of downgraded polymer that is caused by vent port flooding in degassing extruders. 4 refs.

Direct Injection Gas-Liquid Chromatographic Method to Monitor Polymer Melt Devolatilization. Methyl Methacrylate in PMMA

Authors   Edkins, T. J.; Notorgiacome, V. J.; Biesenberger, J. A.
Affiliation   Stevens Institute of Technology
Journal Name   *Polym. Engng. Sci.*
Citation   30, No. 23, Dec. 1990, pp. 1500–1503
Abstract   A new capillary gas chromatography method to monitor a residual monomer in a polymer is described. A short (5 m) cross-linked methyl silicone fused silica capillary column with flame ionization detection gave good separation efficiency and ppm detection limits. Capillary gas–liquid chromatography was superior to packed column analysis (SP 1000) in terms of both analytical utility and ease of use. This technique may prove useful for monitoring other monomer/polymer systems. 13 refs.

Devolatilization of Polymer Melts. Machine Geometry and Scale Factors

Authors   Biesenberger, J. A.; Dey, S. K.; Brizzolara, J.
Affiliation   Stevens Institute of Technology
Journal Name   *Polym. Engng., Sci.*
Citation   30, No. 23, Dec. 1990, pp. 1493–1499
Abstract   Data are presented to compare single with co-rotating twin-screw extruders for devolatilization of methyl methacrylate from PMMA. In both machines, a melt seal was required upstream of the vented section to

contain the applied vacuum. This was achieved with reverse-flighted elements in twin-screw and with a blister (or dam) in the single-screw machine. The material was PMMA contaminated with 0.65 wt. % monomer. Parameters are identified that characterize dynamic similarity for potential use in design and scale-up. These parameters include conventional process variables such as temperature, throughput, screw rev/min and degree of fill. 19 refs.

Polymer Melt Devolatilization Mechanisms*

Authors   Albalak, R. J.; Tadmor, Z.; Talmon, Y.
Affiliation   Technion-Israel Institute of Technology
Journal Name   *AIChE J.*
Citation   Vol. 36, September 1990, pp. 1313–1320
Abstract   Scanning electron microscopy (SEM) was used to study the mechanism of falling-strand devolatilization of polymer melts. Polystyrene and low-density polyethylene were enriched with styrene and hexane, respectively, and were subsequently extruded as thin strands at various conditions. The polymer strands were exposed to superheat conditions for preset periods of time. The strands were then frozen, fractured, and studied by SEM. Devolatilization was found to proceed through a blistering mechanism, both on the lateral surface of the strands and on the surfaces of volatile bubbles formed within the core of the melt. 20 refs. (Abstract reproduced from Albalak et al. (1990) by permission of the American Institute of Chemical Engineers. © 1990 AlChE.)

Twin Screw Extrusion Devolatilization: From Foam to Bubble-Free Mass Transfer

Authors   Foster, R. W.; Lindt, J. T.
Affiliation   Pittsburgh University
Journal Name   *Polym. Engng. Sci.*
Citation   30, No. 11, Mid-June 1990, pp. 621–634
Abstract   The results are reported of a study of the performance of a first forward vent in the counter-rotating, nonintermeshing twin-screw extrusion devolatilization process. The process is shown to undergo a transition from the bubble growth controlled mass transfer mechanism to the relatively bubble-free diffusion controlled problem. This transition is represented by a combination of theoretical modeling used to describe both of the limiting problems. 31 refs.

Devolatilization of Polymer Melts: Machine Geometry Effects

Authors   Biesenberger, J. A.; Dey, S. K.; Brizzolara, J.
Affiliation   Stevens Institute of Technology

Doc. Name   Antec '90. Plastics in the Environment: Yesterday, Today &
Tomorrow. Conference Proceedings
Citation   Dallas, Tex., 7th–11th May 1990, pp. 129–134
Abstract   The aim of this study was to subject a polymer melt, flowing
through the vented sections of the different extruders, to the same dynamic
experience so that the material would not know the difference. The aim was
to identify parameters and develop algorithms to guide the design and
scale-up of devolatilization stages. PMMA was investigated contaminated
with 0.65% residual MMA. 10 refs.

Bubble Free Devolatilization in Counter Rotating Non-Intermeshing Twin Screw Extruder

Authors   Foster, R. W.; Lindt, J. T.
Affiliation   Pittsburgh University
Journal Name   *Polym. Engng. Sci.*
Citation   30, No. 7, Mid-April 1990, pp. 424–430
Abstract   A theoretical analysis of the diffusion-controlled, essentially
bubble-free, mass transfer regime of polymer solution devolatilization
has been performed for a counter-rotating, nonintermeshing twin-screw
extruder. The commonly used penetration diffusion approach is applied
specifically to this geometry. Mass transfer from the recirculating pool
and barrel film in the closed channel and pool only in the open channel is
taken into account. Use is made of the recently determined pumping
characteristics allowing the estimation of the length of the extraction
section. Mass transfer rate measurements have been taken on a 20-mm
extruder using the PS/ethylbenzene model system. Very good agreement
between the mass transfer calculations and experimental results was obtained.
14 refs.

Mixing, Devolatilization, and Reactive Processing in the Farrel Continuous Mixer

Authors   Valsamis, L. N.; Canedo, E. L.
Affiliation   Farrel Corp.
Journal Name   *Int. Polym. Processing*
Citation   4, No. 4, Dec. 1989, pp. 247–254
Abstract   The mixing characteristics of the Farrel Continuous Mixer
are discussed in detail, including recent developments. Flow patterns
inside the machine are established, and mathematical models for mixing
and devolatilization are developed and verified. Reactive processing is
exemplified by data on the visbreaking of polypropylene with organic
peroxides. 24 refs.

Evaluation of the Performance of a Commercial Polystyrene Devolatilizer

Authors   Meister, B. J.; Platt, A. E.
Affiliation   Dow Chemical Co.
Journal Name   *IEC Res.*
Citation   28, No. 11, Nov. 1989, pp. 1659–1664

Abstract   Residual styrene, ethylbenzene, dimers, and trimers were measured as a function of flash tank vacuum and product viscosity during the devolatilization of PS in a commercial flash tank devolatilizer. The results demonstrated that Flory–Huggins multicomponent equilibrium calculations coupled with styrene generation and mass transfer control provided an adequate description of the devolatilization process. Product viscosity was shown to have little effect on the devolatilization. The degree of mass transfer control was found to be significantly related to the molecular weight of the volatile component. 6 refs.

Fundamental Study of Polymer Melt Devolatilization.
IV. Some Theories and Models for Foam Enhanced Devolatilization

Authors   Lee, S. T.; Biesenberger, J. A.
Affiliation   Stevens Institute of Technology
Journal Name   *Polym. Engng. Sci.*
Citation   29, No. 12, June 1989, pp. 782–790

Abstract   Some model experiments on polymer melt devolatilization are reported. Some theories are examined and mathematical models proposed in pursuit of an explanation of these experimental observations. Bubble nucleation, free-volume theories, continuum models for bubble growth in quiescent fluids, and fluid flow in rolling pools without bubbles are included. 17 refs.

Devolatilization and Degassing in Continuous Mixers

Authors   Valsamis, L. N.; Canedo, E. L.
Affiliation   Farrel Corp.
Doc. Name   Antec '89. Conference Proceedings
Citation   New York 1st–4th May 1989, pp. 257–259

Abstract   A diffusion model of devolatilization in the Farrel Continuous Mixer was developed, based on a simplified melt transport model. Expressions for the extraction efficiency were obtained in terms of geometric parameters, operational conditions, and physical properties of the polymer–volatile system. Theoretical predictions were compared with pilot plant experimental data for the system PP/tert-butyl alcohol, and scale-up to production-size machines was considered as an example of application of the model. 10 refs.

Stochastic Simulation of Foaming Devolatilization

Authors   Foster, R. W.; Lindt, J. T.
Affiliation Pittsburgh University
Doc. Name   Antec '89. Conference Proceedings
Citation   New York, 1st–4th May 1989, pp. 150–153
Abstract   An attempt at a stochastic representation of bubble dynamics within the extrusion foaming process is discussed, with reference to nucleation, bubble growth, bubble motion, and bubble rupture. Application of the simulation to the continuous foaming devolatilization process is considered. 22 refs.

Comprehensive Model of Devolatilization in Twin-Screw Extruders

Authors   Lindt, J. T.; Foster, R. W.
Affiliation Pittsburgh University
Doc. Name   Antec '89. Conference Proceedings
Citation   New York, 1st–4th May 1989, pp. 146–149
Abstract   The devolatilization process involving large amounts of volatiles was successfully analyzed, and the overall transport pattern was broken down into four elementary mechanisms that fully described the devolatilization process, i.e., liquid flow in the screw channels, flow past the liquid seals, bubble growth kinetics and mixing-controlled mass transfer, and drag flow–controlled mass transfer. A mathematical model combining the analysis of these four interacting mechanisms was applied to the geometry of the nonintermeshing counter-rotating twin-screw extruder and was shown to provide meaningful predictions of the devolatilization process. 13 refs.

Bubble Growth Controlled Devolatilization in Twin-Screw Extruders

Authors   Foster, R. W.; Lindt, J. T.
Affiliation   Pittsburgh University
Journal Name Polym. Engng. Sci.
Citation   27, No. 3, Mid-Feb. 1989, pp. 178–185
Abstract   An experimental study of polymer solution devolatilization in a counter-rotating twin-screw extruder is described. The work analyzes the experimental results to determine the controlling mechanisms in the separation of volatile components from the polymer. Recent and previously reported data on the PS/ethyl benzene system and the PMMA/MMA system are analyzed during the heavily foaming stages of the process. 14 refs.

Comparative Study of Single and Twin Screw Devolatilizers

Authors   Notorigiacomo, V.; Biesenberger, T.
Affiliation   Stevens Institute of Technology
Doc. Name   ANTEC '88. Proceedings of the 46th Annual Technical Conference
Citation   Atlanta, 18–21 April 1988, pp. 71–75
Abstract   The performance of single- and twin-screw extruders as devolatilizers was compared using PMMA. Parameters that were varied included rpm, temperature, and throughput rate. The degree of fill and residence time in the vented sections were estimated by computation, and the fraction of contaminant removed was measured by gas chromatography. The results indicated that, with proper design of the vented section and at sufficiently high operating temperatures, the single-screw extruder could achieve separations comparable with the co-rotating twin-screw extruder. 6 refs.

Polymer Melt Devolatilization: On Equipment Design Equations

Author   Biesenberger, J. A.
Affiliation   Stevens Institute of Technology
Journal Name   *Adv. Polym. Technol.*
Citation   7, No. 3, Fall 1987, pp. 267–278
Abstract   The development of engineering algorithms for devolatilization equipment design, selection, and operation is described, with emphasis on mass transfer models. 17 refs.

Counter-Rotating Twin-Screw Extruder as a Devolatilizer and as a Continuous Polymer Reactor

Authors   Shah, S.; Wang, S. F; Schott, N.; Grossman, S.
Affiliation   General Motors Corp.; Lowell University.
Doc. Name   ANTEC '87. Plastics—Pioneering the 21st Century. Proceedings of the 45th Annual Technical Conference and Exhibit, held Los Angeles, 4–7 May 1987
Citation   Brookfield Center, Conn., 1987, pp. 122–127.
Abstract   Results are presented of an investigation of the use of the above type of extruder for devolatilization studies on Upjohn's PA 7030 polyamide, for residence time distribution studies and for studies of the reactive extrusion of polyamide with MDI. 2 refs.

Devolatilization of Concentrated Polymeric Solutions in Extensional Flow

Authors   Nangeroni, J. F.; Denson, C. D.
Affiliation   Delaware University

Doc. Name   ANTEC '87. Plastics—Pioneering the 21st Century. Proceed-
             ings of the 45th Annual Technical Conference and Exhibit,
             held Los Angeles, 4–7 May 1987
Citation   Brookfield Center, Conn., 1987, pp. 87–89.

Abstract   Theoretical aspects of the above are discussed and results of
experimental studies, using the tubular film blowing of HDPE to perform
extensional flow on a concentrated polymer solution, are presented. 11 refs.

Fundamental Study of Polymer Melt Devolatilization. IV. Some Theories and Models for Foam-Enhanced Devolatilization

Authors   Lee, S. T.; Biesenberger, J. A.
Affiliation Sealed Air Corp.; Stevens Institute of Technology
Doc. Name   ANTEC '87. Plastics—Pioneering the 21st Century. Proceed-
             ings of the 45th Annual Technical Conference and Exhibit,
             held Los Angeles, 4–7 May 1987
Citation   Brookfield Center, Conn., 1987, pp. 81–86

Abstract   Some theories and mathematical models relating to the devolatilization process are reviewed and applied, particular attention being paid
to bubble formation, bubble growth, diffusion models, and flow models. 15
refs.

Progress in Devolatilization of Polymer Melts in Injection Moulding and Extrusion

Author   Meder, S.
Affiliation Stuttgart University
Doc. Name   ANTEC '87. Plastics—Pioneering the 21st Century. Proceed-
             ings of the 45th Annual Technical Conference and Exhibit,
             held Los Angeles, 4–7 May 1987
Citation   Brookfield Center, Conn., 1987, pp. 77–80.

Abstract   Progress in the above is discussed with particular reference to
the techniques developed at the Institut fuer Kunststofftechnologie of the
University of Stuttgart. 7 refs.

Scanning Electron Microscopy Studies of Polymer Melt Devolatilization*

Authors   Albalak, R. J.; Tadmor, Z.; Talmon, Y.
Affiliation   Technion-Israel Institute of Technology
Journal Name *AIChE J.*
Citation   Vol. 33, May 1987, pp. 808–818

Abstract   Scanning electron microscopy (SEM) was used to study the
mechanism of falling-strand devolatilization of molten polystyrene. Polymer
strands containing 2300 ppm styrene and polymer strands containing 5%

pentane were extruded into a heated vacuum chamber. The strands were abruptly frozen, then fractured under liquid nitrogen and their morphology studied with SEM. A rich variety of morphological features in the core and on the surface of the strands was discovered. The source of these findings and their relevance to devolatilization is discussed. 6 refs. (Abstract reproduced from Albalak et al. (1987) by permission of the American Institute of Chemical Engineers. © 1987 AIChE.)

Fundamental Study of Polymer Melt Devolatilization. III. More Experiments on Foam-Enhanced DV

Authors   Biesenberger, J. A.; Lee, S.-T.
Affiliation   Stevens Institute of Technology
Journal Name   *Polym. Engng. Sci.*
Citation   27, No. 7, Mid-April 1987, pp. 510–517
Abstract   A systematic, experimental study was conducted on the elementary process of devolatilization (DV) of volatile contaminants from polymer melts. Emphasis was placed on foam-enhanced DV from rolling melt pools. PDMS was used to simulate the melt, and methyl chloride was the contaminant. Various physical properties and process parameters were examined. 6 refs.

Diffusion Coefficient Model for Polymer Devolatilization

Authors   Misovich, M. J.; Grulke, E. A.; Blanks, R. F.
Affiliation   Amoco Chemicals Corp.; Michigan University
Journal Name   *Polym. Engng. Sci.*
Citation   27, No. 4, Feb. 1987, pp. 303–312
Abstract   Three models that can be used for diffusion coefficients in devolatilizer design are discussed, i.e., the free volume model developed by Duda, Vrentas et al., a new linear model proposed in this study, and a constant diffusivity model. The linear model is obtained by combining a new correlation for solvent activity coefficients in molten polymers with free-volume theory and linearizing the resulting equation. A model is presented for determining whether the complete model, the linear model, or the constant diffusivity model is appropriate for a given devolatilizer design. 29 refs.

Devolatilization

Author   O'Brien, K. T.
Affiliation   Celanese Engineering Resins Co.
Doc. Name   Developments in Plastics Technology—3

Citation Barking, Elsevier Applied Science Publishers Ltd., 1986, pp. 47–85

Abstract Removal of volatiles (gas or vapor), including moisture, from plastics, mainly molten, was reviewed in sections on an overview of polymer production and conversion (compounding), fundamental mechanisms and theory (equations, practical implications), vacuum systems (pump types, operating performance), equipment (selection criteria, operating windows, flash evaporator, falling-film and -strand evaporators, thin-film vaporizer: basic construction, flow configurations, operating characteristics; single-screw extruder: devolatilization zone, model, point of devolatilization; disk-pack, kneaders, multiscrew extruders, multistage units, injection molding machines) and vent-port design. 30 refs.

Fundamental Study of Polymer Melt Devolatilization. I. Some Experiments on Foam-Enhanced Devolatilization

Authors Biesenberger, J. A.; Shau-Tarng Lee
Affiliation Stevens Institute of Technology
Journal Name *Polym. Engng. Sci.*
Citation 26, No. 14, Mid-Aug. 1986, pp. 982–988

Abstract A study was made of the elementary process of devolatilization and the various parameters affecting it using a specially designed apparatus, which is devoid of the flow complexities present in industrial equipment and simulates the rotating melt pool. Parameters investigated included exposure time, agitation rate surface-to-volume ratio, melt viscosity, vapor pressure of volatile contaminants, applied vacuum level, concentration level of contaminant and addition of inert substances for devolatilization enhancement. The material examined was a polysiloxane in methyl chloride and Freon. 14 refs.

Choosing an Extruder for Melt Devolatilization

Author Mack, M. H.
Affiliation Berstorff Corp.
Journal Name *Plast. Engng.*
Citation 42, No. 7, July 1986, pp. 47–51

Abstract The selection of single-screw or twin-screw extruders for efficient degassing of polymer melts is discussed with reference to the optimization of devolatilization by controlling melt temperature, vacuum pressure, and screw speed. Rear and front degassing methods are described with cross-sectional diagrams of sections of the extruder showing the optimum configurations for devolatilization of resins.

Polymer Devolatilization. II. Model for Foaming Devolatilization

Authors   Chella, R.; Lindt, J. T.
Affiliation   Pittsburgh University
Doc. Name   Antec '86. Plastics—Value Through Technology. Proceedings
                  of the 44th Annual Technical Conference
Citation   Boston, April 28–May 1, 1986, pp. 851–854

Abstract   A model for describing the kinetics of devolatilization from moderately concentrated (less than 50% solvent) polymer solutions is proposed. It accounts for interactions between closely spaced bubbles and cooling resulting from vaporization, and incorporates realistic formulations for thermodynamic and transport properties. It is considered suitable for preliminary scale-up and design of the upstream sections of a devolatilizing extruder and as a basic building block of a model for devolatilization influenced by an external flow field.

Application of Novel Counter-Rotating Intermeshed Twin Extruder for Degassing Operation

Authors   Sakai, T.; Hashimoto, N.
Affiliation   Japan, Steel Works Ltd.
Doc. Name   Antec '86. Plastics—Value Through Technology. Proceedings
                  of the 44th Annual Technical Conference
Citation   Boston, April 28–May 1, 1986, pp. 860–863

Abstract   The results are reported of a study of the efficient devolatilization method using Tex 65, a high-speed counter-rotating intermeshed twin-screw extruder fitted with 65 mm diameter screws and cylinders of L/D of 30–40. Test results are presented for LLDPE containing a medium amount of residual solvent (octene-1 and n-hexane), polychloroprene containing a large amount of residual solvent (tetrachlorocarbon), and the reclaiming of PETP scrap film and trimmed film in an undried condition. 1 ref.

Single Screw Versus Twin Screw Extruder for Polymer Melt Devolatilization

Author   Mack, M.
Affiliation   Berstorff Corp.
Doc. Name   Antec '86. Plastics—Value Through Technology. Proceedings
                  of the 44th Annual Technical Conference
Citation   Boston, April 28–May 1, 1, 1986, pp. 855–859

Abstract   The advantages and limitations of single-screw and corotating, intermeshing twin-screw extruders are evaluated by comparing their design features for selecting the most economic system. Special design features are presented that allow rear degassing and forward degassing on both machine types. Results for PMMA beads and two LDPE copolymers (EVA) having melt flow indexes of 0.5 and 150 are discussed. 3 refs.

Fundamental Study of Polymer Melt Devolatilization. II. Theory for Foam-Enhanced DV

Authors   Biesenberger, J. A.; Shau-Tarng Lee
Affiliation   Stevens Institute of Technology
Doc. Name   Antec '86. Plastics—Value Through Technology. Proceedings of the 44th Annual Technical Conference
Citation   Boston, April 28–May 1, 1986, pp. 846–850
Abstract   A study was made of foam-enhanced devolatilization in a new test apparatus using samples of polysiloxane contaminated with methyl chloride, Freon 12, Freon 13, Freon 22, and Freon 114. The apparatus is briefly described, and a metastable cavity model that describes the formation of bubbles, known to be active loci in foam devolatilization of polymer melts, is proposed.

Mechanism of Foam Devolatilization in Partially Filled Screw Devolatilizers

Authors   Han, H.-P.; Han, C. D.
Affiliation   New York Polytechnic Institute
Journal Name   *Polym. Engng. Sci.*
Citation   26, No. 10, May 1986, pp. 673–681
Abstract   An experimental investigation was conducted into elucidating the mechanism of foam devolatilization. The test fluids were aqueous solutions of polyacrylamide having various concentrations. 16 refs.

Rotating-Drum Devolatilizer

Authors   Kearney, M.; Hold, P.
Affiliation   Emhart Corp.; Farrel Machinery Group
Journal Name   *Plast. Compounding*
Citation   9, No. 2, March/April 1986, pp. 48–56
Abstract   The results of investigations by several researchers on the process of devolatilizing a polymer melt are described, and a design description of a new devolatilizer intended to be part of a single-screw extruder is presented. The use of specially conditioned polymeric materials with added volatile components is detailed in tests on the 6-inch rotating-drum devolatilizer. Comparison tests are discussed using various grades and copolymers of PE on the new and standard devolatilizers. 3 refs.

Solvent Removal from Ethylene–Propylene Elastomers. I. Determination of Diffusion Mechanism

Authors   Matthews, R. J.; Fair, J. R.; Barlow, J. W.; Paul, D. R.; Cozewith, C.
Affiliation   Exxon Chemical Co.; Texas University
Journal Name   *IEC Prod. Res. Dev.*
Citation   25, No. 1, March 1986, pp. 58–64

Abstract   An examination was made of the devolatilization step in a commercial process for producing ethylene–propylene elastomers by solution polymerisation using a Ziegler–Natta catalyst in order to elucidate the mechanism of mass transport and to point out opportunities for increasing commercial process efficiency. Transport mechanisms and rates were determined for the case of hexane solvent being removed by steam. The overall rate of removal was found to be controlled by particle structure, with surface-connected pores playing a major part. An expression for the diffusion coefficient and a model for a crumb particle that allows calculation of diffusion rates are proposed. 21 refs.

Solvent Removal from Ethylene–Propylene Elastomers. II. Modelling of Continuous-Flow Stripping Vessels

Authors   Mathews, F. J.; Fair, J. R.; Barlow, J. W.; Paul, D. R.; Cozewith, C.
Affiliation   Exxon Chemical Co.; Texas University
Journal Name   *IEC Prod. Res. Dev.*
Citation   25, No. 1, March 1986, pp. 65–68

Abstract   A mathematical diffusion model for solvent removal from porous rubber particles in aqueous slurries, proposed in pt. I, ibid., p. 58–64, was applied to the correlation of data from a commercial-scale, continuous-flow stripper for three different EPR grades and found to give a good correlation of the data. A set of transport, energy, and material balance equations, which permits the effect of operating conditions to be estimated and solvent removal to be optimized, is also proposed. 2 refs.

Rotating Drum Devolatilizer

Authors   Kearney, M.; Hold, P.
Affiliation   Farrel Co.
Doc. Name   ANTEC '85; Proceedings of the 43rd Annual Technical Conference
Citation   Washington D.C., April 29–May 2, 1985, pp. 17–22

Abstract   Specifications for designing a melt devolatilizer were outlined.

A design that was generally an integral part of a single-screw extruder was described, and test results for LDPE, LDPE/EMA copolymer, and various grades of LLDPE were given. The operation was analyzed mathematically. Volatile contents were reduced to desired levels with melt temperatures closely controlled. Channel flow-patterns were well-ordered and predictable by simulation. Low vacuum levels were used economically. Melt could not escape through the vent openings at the output rates and operating speeds specified. 3 refs.

Mechanism of Foam Devolatilization in Partially Filled Screw Devolatilizers

Authors   Han, H.-P.; Han, C. D.
Affiliation   New York Polytechnic Institute
Doc. Name   ANTEC '85; Proceedings of the 43rd Annual Technical Conference
Citation   Washington D.C., April 29–May 2, 1985, pp. 8–11
Abstract   The nucleation of gas bubbles, their growth and their motion in polymeric liquid subject to flow fields, simulating the situations in single- and twin-screw devolatilizes, were studied using aqueous solutions of polyacrylamide of various concentrations with controlled amounts of entrapped air, in a Plexiglas apparatus. The importance of stripping agents in improving the efficiency of foam devolatilization was discussed. 9 refs.

Fundamental Study of Polymer Melt Devolatilization

Authors   Biesenberger, J. A.; Lee, S.-T.
Affiliation   Stevens Institute of Technology
Doc. Name   ANTEC '85; Proceedings of the 43rd Annual Technical Conference
Citation   Washington D.C., April 29–May 2, 1985, pp. 2–7
Abstract   Foam-enhanced devolatilization of polymeric fluids under vacuum was considered in developing mathematical models to ease the selection and design of equipment such as vented extruders and thin-film evaporators. Mass transfer rates were measured for PDMS samples of different viscosities contaminated with methyl chloride or, in one case, Freon 12. Samples were analyzed before and after devolatilization for contaminant content by gas chromatography. 14 refs.

Extrusion Isolation of Polymers from Solution

Authors   Nichols, R. J.; Lubiejewski, P. E.
Affiliation   Welding Engineers Inc.

Doc. Name ANTEC '85; Proceedings of the 43rd Annual Technical Conference

Citation Washington D.C., April 29–May 2, 1985, pp. 12–16

Abstract The effect of various operating conditions (feed temperature, barrel temperature, product temperature, rate) on the devolatilization of low-solids styrene–butadiene copolymer solutions (in 70/30 benzene/cyclohexane) of different ratios and concentrations was studied using a counter-rotating, tangential, twin-screw extruder. Theory and rear vent operation were considered. The latter was related to feed temperature, but no explicit mathematical function relating rear vent yield to solution feed pressure was found. 7 refs.

Two Routes to Effective Resin Devolatilization

Author Biesenberger, J. A.

Affiliation Stevens Institute of Technology

Journal Name *Plast. Engng.*

Citation 40, No. 6, June 1984, pp. 43–46

Abstract A review is given of two approaches to devolatilization of contaminants from polymers. In the first, called flash evaporation, the polymer feed stream is molten and contains large quantities of solvent. The application of vacuum causes foaming and evaporation. In the second approach, the feed is solid or, if molten, sufficiently dilute to make the develolatilization rate diffusion controlled.

Basis for the Design of Single-Screw Devolatilizing Extruders

Author Potente, H.

Affiliation Paderborn Universitat

Journal Name *Adv. Polym. Technol.*

Citation 4, No. 1, Spring 1984, pp. 61–67

Abstract Approximate equations are presented that enable devolatilizing extruders to be designed by relatively simple methods utilizing computer simulation. Good agreement was obtained between numerical and experimental results. Isothermal throughput equations for a single-zone screw and a three-zone screw and equations for calculating minimal and maximal pressures are tabulated, and experimental and theoretical throughput values compared. 22 refs.

Devolatilizing in Single Screw Extruders

Authors Scharer, H. R.; Rizzi, M. A.

Affiliation Emhart Corp.; Farrel Co.

Doc. Name   Polyolefins IV: Innovations in Processes, Products, Processing
            and Additives; RETEC
Citation   Westchase, Tex., February 27–28, 1984, Paper 25, pp. 313–324
Abstract   The mechanism of polymer devolatilization is defined, and
factors that control the degree of devolatilization achievable for a given
system are defined and evaluated. Tests conducted on a melt fed single-screw
extruder with several types of linear LDPE are reported. 4 refs.

Foam Devolatilization in a Multichannel Corotating Disk Processor

Authors   Mehta, P. S.; Valsamis, L. N.; Tadmor, Z.
Affiliation   Emhart Corp.; Farrel Machinery Group
Journal Name *Polym. Proc. Engng.*
Citation   2, No. 2/3, 1984, pp. 103–128
Abstract   A study was made of polymer devolatilization in a co-
rotating disk devolatilizer (diskpack) where the polymer melt is coated on
disk surfaces and transported from one chamber to the next while the
chamber is exposed to vacuum. The results obtained are analyzed in terms
of a modified version of the diffusion film-surface renewed model and a
hypothesis based on bubble transport. 16 refs. (75th Annual AIChE Meeting,
Washington, D.C., 1983).

**Length of a Transfer Unit (LTU) for Polymer Devolatilization Processes
in Screw Extruders**

Authors   Collins, G. P.; Denson, C. D.; Astarita, G.
Affiliation   Delaware University; Naples University
Doc. Name   Plastics—Engineering Today for Tomorrow's World; ANTEC
              '83, 41st Annual Technical Conference
Citation   Chicago, Ill., May 2–5, 1983, pp. 138–141
Abstract   A discussion is presented on polymer devolatilization, a process in which a low-molecular-weight volatile component is transferred from
a liquid phase to a gas phase. Particular attention is paid to the method
involving sweeping an inert gas through the extruder, as opposed to
devolatilization under vacuum. It is considered that counter-current gas flow
is preferable to vacuum operation; it is, however, suggested that combining
the advantages of the two processes could offer a possible alternative. 8 refs.

Devolatilization of Polymers: Fundamentals–Equipment–Applications

Author   Biesenberger, J. A. (ed.)
Citation Munich, Hanser C., Verlag, 1983, viii, 204 pages
Abstract This book deals, in detail, with the devolatilization of polymers.

It contains a wide spectrum of pertinent topics, including equipment (flash evaporators, thin-film evaporators, vented single- and twin-screw extruders), process mechanisms and models, thermodynamic and transport properties, and analytical methods. 153 refs.

Single-Screw Devolatilizer for Polymer Melts

Authors   Scharer, H. R.; Rizzi, M. A.; Hold, P.
Affiliation   Emhart Corp.; Farrel Connecticut Div..
Journal Name   *Adv. Polym. Technol.*
Citation   3, No. 2, 1983, pp. 131–136
Abstract   In order to be able to develop analytical methods for predicting performance and scaling up from small-scale test stands, some elementary devolatilizing studies were carried out. A revised model of the actual melt devolatilizing process was evolved. Based on this model, several novel single-screw devolatilizing concepts were proposed, built, and tested. 3 refs.

Index